Principles of Biological Chemistry

Second Edition

David S. Page

Bowdoin College

Willard Grant Press

Boston, Massachusetts

This book is dedicated to the memory of William K. Page, Jr.

Library of Congress Cataloging in Publication Data

Page, David S
  Principles of biological chemistry.

  Includes bibliographical references and index.
  1. Biological chemistry. I. Title. [DNLM:
1. Biochemistry.   QU 4 P132p]
QP514.2.P33 1981   574.19′2   80-39968
ISBN 0-87150-740-4

Cover design by John Servideo. Text design by David Foss in collaboration with the Willard Grant Press production staff. Composed in Monophoto Times Roman and Univers by Composition House Limited. Artwork for this edition drawn by Phil Carver & Friends, Inc. Cover printed by New England Book Components, Inc. Text printed and bound by Halliday Lithograph Corporation.

Preface

The first edition of this book was written in response to the need for a concise text that satisfactorily related the basic concepts of general and organic chemistry to the chemistry of living cells. In this, the second edition, the overall pedagogical objective remains unchanged.

Students majoring in one of the many disciplines collectively referred to as life sciences often are required to take an introductory course in biological chemistry. The diversity of this audience has had an effect on the form of this edition. In addition, two other considerations have influenced the general contents and level.

First, many of these students go on to more specialized courses in their specific fields of study. Therefore it is crucial that they have a good understanding of the central ideas of biochemistry before proceeding to courses where these ideas will be applied and expanded upon. Second, the amount of class time available in an introductory course is limited. An instructor must choose either to present many topics briefly or to discuss fewer points in greater depth.

In the second edition emphasis has been placed on presenting the basic principles of biochemistry in a clear and readable fashion. Material is included that is necessary to the topics being developed; nonessential discussion beyond the scope of the book is avoided. A deliberate effort has been made to illustrate the material introduced using examples from areas where these students will be working, such as the allied health sciences, nutrition, and the plant and animal sciences. The physical presentation of the contents has also been improved. A two-color format has been adopted for clarity and pedagogical effectiveness, and many new figures have been added throughout the text.

As in the previous edition the book is divided into three parts, representing a logical flow of ideas from small molecules to metabolism and finally to protein synthesis. Part I discusses the chemistry of molecules found in living cells, Part II presents the reactions by which cells degrade molecules and extract energy, and Part III shows how cells produce new molecules from basic raw materials. Throughout the text the interrelationships among various metabolic pathways and the factors that control chemical processes in living cells are stressed.

Each chapter begins with a short introductory section that places the topics to be discussed into the overall context of biochemistry. At appropriate points within chapters short optional reviews are provided of topics from general chemistry that students often find difficult. These include acids and bases, thermodynamics and chemical equilibrium, reaction kinetics and oxidation-reduction. In addition, a brief review of some basic organic functional-group reactions is provided in an appendix.

Suggestions for further reading are listed at the end of each chapter. An effort has been made to give a fairly large number of references that are readily available in most libraries and that are at a level consistent with the text as a whole. Many of these references can serve as a basis for outside reading assignments should the instructor wish to pursue a topic in greater detail than is given in the text.

Problems are provided at the end of each chapter, and many new ones have been added for the second edition. In each problem set, the first few problems are designed to permit the student to review the important terms, topics and concepts presented in the chapter. The remaining problems serve to amplify material from the text and to aid in applying topics from the chapter and relating them to other topics previously discussed.

In addition to the review of function-group reactions, the appendix also contains answers to problems selected from those that appear at the ends of chapters and a glossary of important terms which appear throughout the text. Finally, a chart of the central processes of intermediary metabolism is included at the back of the text. It is designed to clarify the interrelationships among the major biochemical pathways. At many points in the text, the student is referred to this chart in order to appreciate how the sequence being discussed fits into the overall scheme of intermediary metabolism.

As an additional aid to the student, a study guide which follows exactly the chapters in the book has been written to accompany this text. For each chapter, there is a review of important things to study, a self-test section, and worked-out solutions to the problems in the text.

Acknowledgments

The second edition of this book owes a great deal to those who used the first and took the trouble to share their experiences with me. Their comments have had a profound effect on the form and content of this edition. The revision manuscript was reviewed during its preparation by the following people whose comments and suggestions also added to its final form: Ivan Kaiser, University of Wyoming; Duane LeTourneau, University of Idaho; Edwin A. Lewis, University of Alabama; Oliver G. Lien, San Jose State University; Scott Mohr, Boston University; Sal Russo, Western Washington State University; Karl Speckhard, Clemson University; and Carl Tipton, Iowa

State University. I would like to express particular thanks for their efforts to Dr. Tina Bailey, California Polytechnic State University, Dr. Delano Young, Boston University, and Dr. Bernard White, Iowa State University.

This text owes its genesis and revision to the patience, perseverance, guidance, enthusiasm and good personship of Mr. Bruce Thrasher of Willard Grant Press, all of which is deeply appreciated. The book's final form is the result of the painstaking and scholarly editorial efforts of Mr. David Foss, also of Willard Grant Press.

Finally, I would like to thank Barbara Page, Matthew Page, and David C. Page for their patience during the many months spent working on this second edition.

David S. Page
Bowdoin College, 1981

Contents

CONTENTS

Molecules of Living Systems

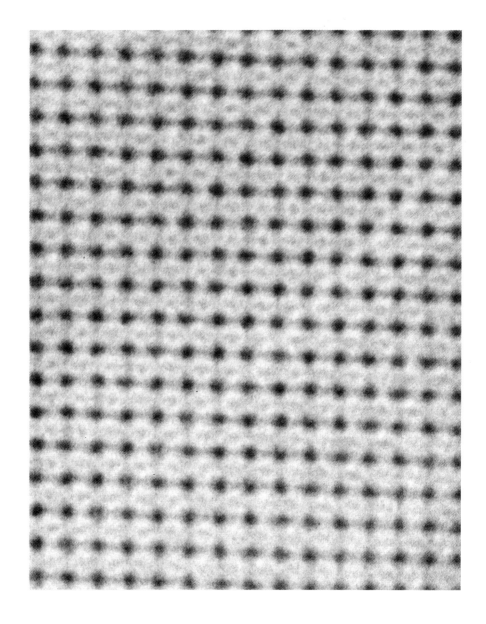

CHAPTER 1

Simple Molecules and Cellular Diversity

This text is about biochemistry, the chemistry of living organisms. As you will discover throughout the book, the basic general chemistry and organic chemistry of living cells are no different from that of inanimate matter. This is an important point because living cells are made up of molecules, which, by themselves, are not alive. Yet somehow, by participating in the complex set of chemical processes that comprises the biochemistry of a cell, these molecules form a living entity.

We will begin our study of biochemistry by considering the basic types of cells and their structures. Then we will examine the atoms and molecules commonly found in biological systems, the energy requirements of these systems, and the properties of one of the crucial components of any cell—water.

1.1

Characteristics of Living Matter

Before we can look closely at what a cell is and what it does, we must first discuss what we mean by the term "living cell." There are certain criteria we can use to differentiate living matter from inanimate matter.

1. **Living matter is highly organized** on both the macroscopic and microscopic levels. The basic organizational unit of all living matter is the cell. Many organisms are unicellular, such as bacteria. Many higher organisms are collections of cells (multicellular) that have aggregated together and differentiated with respect to form and function.

    An individual living cell contains internal structures with different specialized functions. Functional differentiation is also apparent on the molecular level. For example, protein and nucleic acid molecules each play a distinct part in the chemical affairs of the cell. The function of a given substance isolated from a living organism is a fundamental question for the biochemist to determine.

2. **Living matter can extract energy and material** from its environment for purposes of growth and for performing mechanical and chemical work. Scientists can synthesize inanimate systems which will "grow" by the accretion of material from the environment. However, as we will see, the intake of material by living organisms takes place in a specific and controlled manner. This is not the case for inanimate systems. The creation of highly organized internal structures and of complex molecules are not spontaneous processes: work must be done to make such processes occur. The driving force needed to *produce* and *maintain* a living cell from simple molecules arises from the chemical free energy obtained when the cell breaks down food molecules taken from its environment. This constant input of energy is required to create and to maintain order in the cell. When an organism dies, this energy input ceases. Without any further input of energy, the complex structure of the organism decomposes back into the simple molecules from which it was created. Thus, we say that living matter is *thermodynamically unstable*. It is well that this is so, for successive generations of cells can only continue to exist by the constant *recycling* of basic materials.

3. **Living matter is capable of reproducing itself** over and over again in a precise way from generation to generation. This is a key criterion for defining living matter. We could conceivably synthesize nonliving systems that approximate one or both of the preceding criteria. However, we would be hard pressed to design an inanimate system that could *precisely* reproduce itself over many generations. A living cell is able to do this because it contains the genetic information it needs to direct the synthesis of proteins and nucleic acids that define the overall biochemistry of the cell and thus define what the cell is. Genetic information is transmitted from one generation of cells to the next in the form of the molecule **deoxyribonucleic acid (DNA)**. The basic biochemistry of this information transfer and utilization is described in Chapter 13.

This three-part definition of a living system also defines the organization of this text rather well. In Part I we will consider the organization of living matter on both the cellular and molecular levels. Most of this first part discusses the properties of the inanimate molecules that comprise the living organism. Since many of these molecules are unique to living systems, we refer to such molecules as **biomolecules**. In Part II we will describe how chemical energy and materials are extracted from the environment and converted into forms which the cell can use. Part III will discuss the utilization of chemical energy and materials in the cell to synthesize biomolecules and to perpetuate the living organism through successive generations.

1.2

General Classification of Cells

There are thousands upon thousands of different species of living organisms, ranging from simple unicellular microorganisms to higher animals. An important part of biology involves sorting out and interrelating these diverse species. It is a formidable job, but if we look at living organisms at the cellular level, many unifying features become apparent.

All organisms consist of at least one cell. Each cell is a basic living unit capable of satisfying all the criteria for living matter which we have discussed.

We can categorize cells into two broad classes: **procaryotes** and **eucaryotes**. Procaryotic cells are small, single-cell organisms such as bacteria and blue-green algae. Most procaryotic cells contain all the biochemical apparatus required for self-reproduction and for extracting and utilizing energy and materials from the environment. Because the biochemistry of procaryotes is comparatively simple, it has been extensively studied by biochemists and molecular biologists. Much of our current understanding of how living systems reproduce and maintain themselves has been gained from such studies.

Eucaryotes are far more complicated cells than procaryotes. They are larger in overall size than procaryotes and we can observe within the eucaryotic cell many ordered structures which have specialized functions. Eucaryotic cells include all cells of higher animals, plant cells, fungi, most algae, and protozoa. Since eucaryotes are highly specialized with respect to complexity of internal structure, it is reasonable to assume that eucaryotes evolved from the more rudimentary procaryotic cells.

Before we consider the detailed structures of eucaryotes and procaryotes, we must make one further classification of cells, based on how they extract energy from their environment. There are basically two ways in which cells do this:

Autotrophic cells are "self-feeding" in that they can use CO_2 as a carbon source and sunlight as an energy source. Thus, autotrophic cells are largely self-sufficient. All *photosynthetic* cells are autotrophic because they can transform sunlight into the chemical energy contained in such highly reduced organic molecules as carbohydrates. In this capacity, photosynthetic organisms constitute one-half of the carbon cycle shown in Figure 1.1.

Figure 1.1

The Carbon Cycle. Autotrophs and heterotrophs require each other in order to exist. Such "feeding together" is termed **syntrophy**.

Photosynthetic
autotrophic cells:

$$6\,CO_2 + 6\,H_2O \xrightarrow{\;h\nu\;} \underset{\text{(sugars)}}{C_6H_{12}O_6} + 6\,O_2$$

Heterotrophic cells:

$$\underset{\text{(sugars)}}{C_6H_{12}O_6} + 6\,O_2 \longrightarrow 6\,CO_2 + 6\,H_2O$$

Heterotrophic cells "feed on others" because they extract energy from the environment by taking in and breaking down reduced energy-rich organic molecules. Most microorganisms and all animal cells are heterotrophic.

We summarize our broad classification of cells in Figure 1.2.

Autotrophs

Photosynthetic procaryotes include such organisms as photosynthetic bacteria and blue-green algae, which contain photosynthetic pigment molecules in their cell membranes.

Photosynthetic eucaryotes include plant cells and algae that contain one or more internal structures known as *chloroplasts*. Photosynthetic pigments located in the chloroplast make it the site of photosynthesis. In the absence of light, these cells become heterotrophic.

Heterotrophs

Procaryotic heterotrophs include all nonphotosynthetic bacteria. These organisms produce energy by the degradation of reduced molecules absorbed from the environment. Some procaryotes, such as the *rickettsiae* and *mycoplasma*, seem to require for reproduction parts of the biochemical apparatus of the host organisms in which they must exist.

Eucaryotic heterotrophs include cells which produce energy by breaking down ingested "food" molecules. In contrast to procaryotes, much of this breakdown and energy conservation takes place in a specialized subcellular structure called the *mitochondrion*. Eucaryotes can have one or more mitochondria.

Figure 1.2

Classification of types of cells.

1.3

Basic Structure and Function of Cells

We have learned much about the internal organization of cells by direct microscopic observation. The differences between the types of cells we have discussed become readily apparent by close examination of these cells using an electron microscope. Figures 1.3 and 1.4 show electron micrographs of a representative procaryote and eucaryote, respectively. Note the comparative simplicity of the procaryote in Figure 1.3. Figure 1.5 shows an electron micrograph of an autotrophic eucaryotic cell. Note the presence of internal structures that enable the autotroph to transform sunlight into chemical energy that the cell can use. You should carefully study the electron micrographs and the accompanying legends of these three figures.

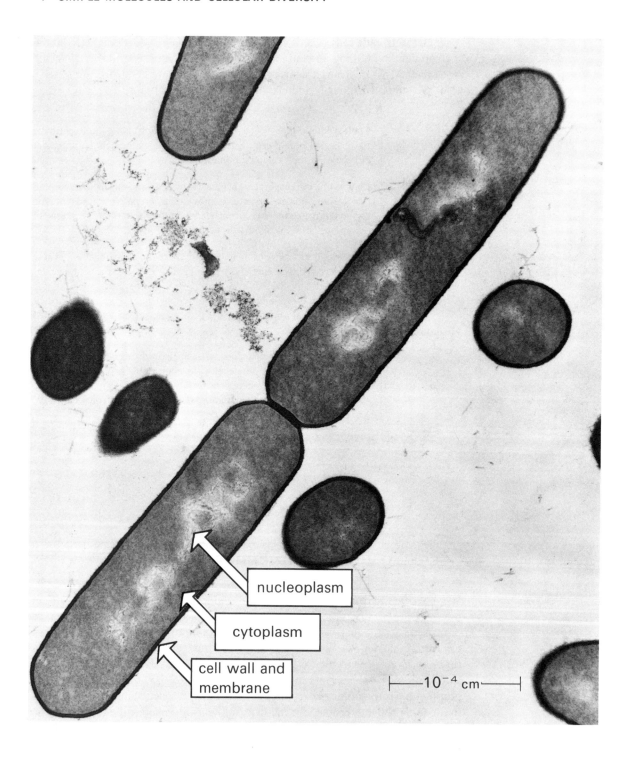

The structures noted in Figure 1.3 are:

Nucleoplasm   The nucleoplasm consists of a single coiled double-helical DNA molecule, the carrier of genetic information in all cells. The DNA in the two joined daughter cells is identical, having been formed by the replication of DNA in the original parent cell which then underwent cell division to produce the two cells shown.

Cytoplasm   The cytoplasm is a structureless solution of proteins, nucleic acids, and other substances involved in the metabolism of the cell. The cytoplasm is the site of protein synthesis (see Chapter 13).

Cell Wall and Membrane   The outer envelope of the bacterial cell is a carbo-hydrate-protein substance possessing great mechanical strength. Since the bacterium has no control over its environment, the cell wall serves an important protective function (see Chapter 4). Within the cell wall is the cell membrane. This acts as a semipermeable barrier where enzymes for respiration, transport, and cell wall synthesis are located (see Chapter 6).

Figure 1.3

Electron micrograph of a procaryotic cell in thin section. The photograph shows two daughter cells of a strain of *Bacillus subtilis*, × 39,000. The round cells are transverse sections of the same organism. The internal structure of bacteria is quite simple compared to animal and plant cells. (Electron micrograph courtesy of R. M. Cole, Ph.D., M.D., Chief, Lab. of Streptococcal Diseases, National Institute of Allergy and Infectious Diseases, Bethesda, Md.)

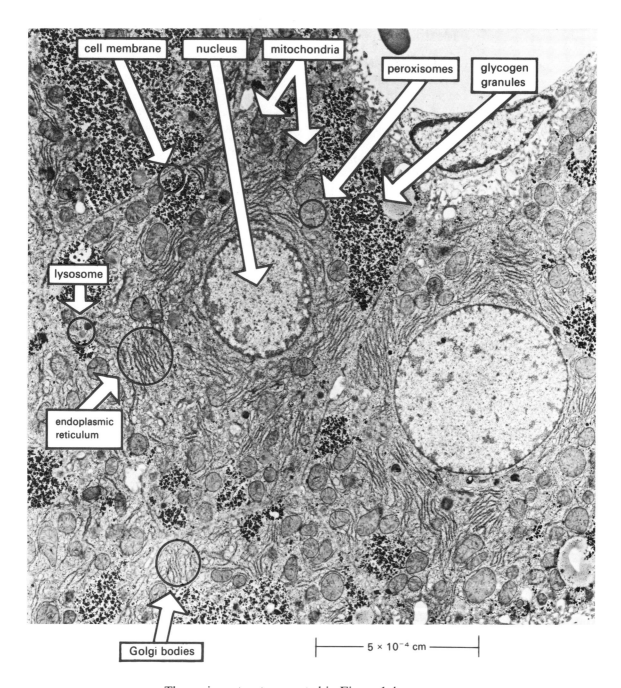

The various structures noted in Figure 1.4 are:

Nucleus   The nucleus of a eucaryotic cell contains the genetic apparatus of the cell. The DNA strands are longer than those in procaryotic DNA, and there is also more than one DNA molecule per eucaryotic cell. The DNA combines with certain proteins to form structures called chromosomes. The dark spots in the nucleus in this figure are sections of chromosomes. The nuclei of human cells ordinarily contain 46 chromosomes, each with a separate store of genetic information (see Chapter 13).

Endoplasmic Reticulum The cytoplasm of a eucaryotic cell is highly structured. Although the endoplasmic reticulum appears to consist of channels through the cytoplasm, it is actually a system of interconnected flattened vesicles. Close examination reveals small dense spots which appear to line the inner surface of the endoplasmic reticulum. These are ribosomes (see Chapter 13), which are the sites of protein synthesis in all cells.

Cell Membrane Since the environment of an animal cell is highly controlled, there is no need for a protective cell wall. The plasma membrane of the eucaryotic cell acts as a permeability barrier, as does the procaryotic cell membrane. Here, however, the eucaryotic cell membrane is more complex, both with respect to the variety of lipids that make up the basic structure of the membrane (see Chapter 6), and also with respect to the membrane-bound enzyme systems it contains. Beyond the cell membrane is a fuzzy cell coating composed of carbohydrate and protein derivatives. This outer layer is responsible for the immunological properties of the cell and for the adhesive properties that enable tissues to be made up of aggregates of cells.

Figure 1.4

Electron micrograph of a eucaryotic cell. The figure shows a human *hepatocyte* (liver cell) in thin section, $\times 17,000$. In contrast with the procaryotic cell shown in Figure 1.3, this cell is both larger and much more complex. (Electron micrograph courtesy of Solon Cole, M.D., Director of Electron-microscopy, Department of Pathology, Hartford Hospital, Hartford, Conn.)

Mitochondria Liver tissue is the site of a great deal of metabolic activity. Therefore a large number of mitochondria exist in a hepatocyte (about 800). The mitochondria are approximately the size of the bacterium shown in Figure 1.3. The mitochondrion is said to be the "powerhouse" of the eucaryotic cell because it is the site of oxidation of carbohydrates, fats, and amino acids to CO_2 and H_2O.

Peroxisomes These small bladder-like cavities, also called *vesicles*, contain various oxidative enzymes such as catalase. The dark spot in the center of the peroxisome noted in the figure is probably crystalline protein material.

Glycogen Granules The darkest spots are granules of the polysaccharide glycogen. This is a particular characteristic of liver cells since the liver is the major site of glycogen synthesis in the body. The glycogen represents stored carbohydrate.

Golgi Bodies The Golgi bodies are flattened membrane vesicles that serve to package substances secreted by the liver cell. Material is released by vacuoles formed at the periphery of the Golgi bodies. A better example of these vesicular structures is shown in Figure 1.5.

Lysosomes Lysosomes are small membrane-surrounded vesicles or packages that contain hydrolyzing enzymes, which can break down material brought into the cell from outside.

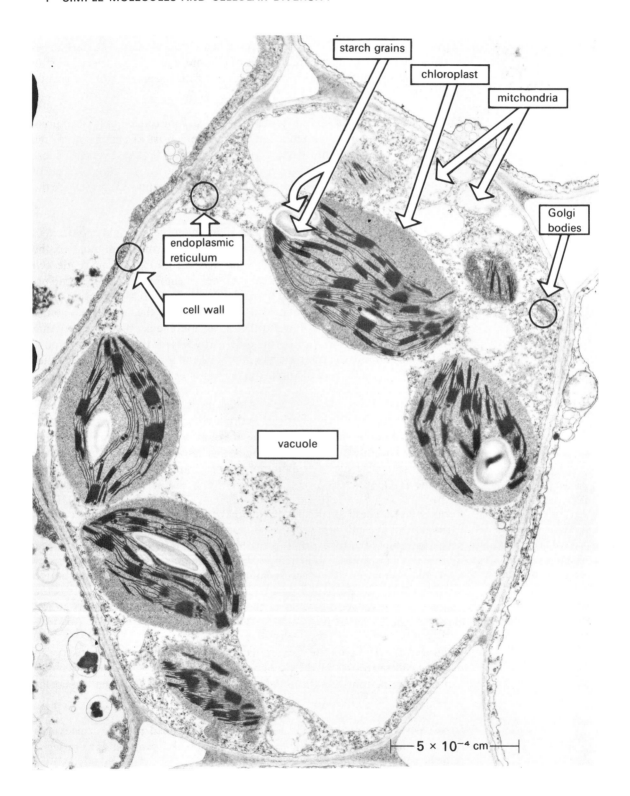

The following structures are shown in Figure 1.5:

Vacuole   As a plant cell matures, a vacuole grows. The vacuole, an internal volume containing dissolved metabolites and waste products, is surrounded by a semipermeable membrane. The cytoplasm and other structures of the plant cell are pushed against the inner cell wall, as shown.

Cell Wall   The plant cell wall must be very strong. The weight of a tree, for example, is supported by the cell walls of living and dead cells in the tree trunk. The thickness and rigidity of the cell wall are evident in the figure. The primary structural material of the plant cell wall is cellulose, bound in a matrix of carbohydrate and protein.

Endoplasmic Reticulum   The endoplasmic reticulum is visible in areas of the cytoplasm near the cell wall. The function of the endoplasmic reticulum is the same here as in the hepatocyte.

Golgi Bodies   A multilayer Golgi complex consisting of flattened vesicles can be seen here.

Chloroplast   The site of photosynthesis is the highly structured subcellular organelle called the chloroplast. The substructure of the chloroplast is readily apparent in the figure. We will consider in detail the relationship of these structures to photosynthesis in Chapter 10. The chemical result of photosynthesis is the conversion of CO_2 and H_2O into carbohydrate. The stored carbohydrate produced as a result of photosynthesis is easily seen in the figure as **starch grains**.

Mitochondria   In the dark, the plant cell obtains energy from the breakdown of carbohydrate molecules in a manner similar to heterotrophic cells. In fact, mitochondrial respiration, or oxidation of "food" molecules, occurs concurrently with photosynthesis in many plants.

Figure 1.5

Electron micrograph of an autotrophic cell in thin section. The figure shows the internal structure of a mesophyll cell from a leaf of timothy (*Phleum pratense*). These particular cells are very active in photosynthesis. All of the internal structures associated with heterotrophic cells can be found in photosynthetic eucaryotes. The nucleus of this particular cell is not apparent in the particular section shown in the figure. However, mitochondria are observable, as are some of the other structures noted in Figure 1.4. (Electron micrograph courtesy of Dr. Myron Ledbetter, Biology Dept., Brookhaven National Laboratory. Prior publication in M. C. Ledbetter and K. Porter, *Introduction to the Fine Structure of Plant Cells*, Springer-Verlag, 1970.)

1.4

Water

Water comprises about 70% of the net weight of a cell and is a crucial component of all cells. It is a unique substance in many ways. A large number of processes in living systems utilize H_2O as a reactant, and it is a ready source of H_3O^+ and of OH^- ions, both through self-dissociation and through reaction with an acid or a base. The physical properties of water are also unusual for a substance of molecular weight 18. Some of these properties are summarized in Table 1.1.

Table 1.1 Properties of Water Pertinent to Living Systems.

Property	Value	Consequence
Heat of vaporization (heat required to vaporize 1 mole of liquid water)	9700 cal/mole	This is very high for liquids, making water an especially good evaporative coolant (i.e., perspiration).
Specific heat (heat required to raise the temperature of 1 gram of H_2O 1 degree centigrade)	1 cal/g · deg	Water is a good heat buffer—the temperature of an aqueous system does not fluctuate rapidly despite rapid temperature changes of the environment.
Boiling point	100°C	This is very high for a substance of molecular weight 18. Water remains a liquid at physiological temperatures.

The bulk structure of water is largely responsible for the values of the physical properties listed in Table 1.1. The structure of a single water molecule is shown below; note that it is a bent molecule.

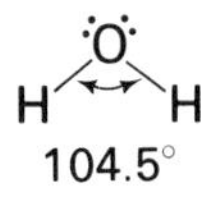

$$H \overset{\ddot{O}}{\underset{104.5°}{\diagdown\ \diagup}} H$$

Due to the electronegativity difference between hydrogen and oxygen, water is a dipolar molecule. Oxygen attracts the electrons in the molecule to a greater extent than hydrogen does, making the oxygen partially negative and the hydrogen partially positive. The total partial negative charge on the oxygen in water is about $\frac{2}{15}$ of the charge of an electron. In the liquid and solid states extensive association of water molecules occurs by dipole–dipole attraction between the positive hydrogens and the negative oxygens. This type of inter-molecular association is called **hydrogen bonding**. Figure 1.6 shows how a maximum of four H_2O molecules can be hydrogen-bonded to a central water molecule.

Figure 1.6

A water molecule shown with the maximum number of four other H_2O molecules hydrogen-bonded to it. The four hydrogen-bonded water molecules lie at the corners of a tetrahedron.

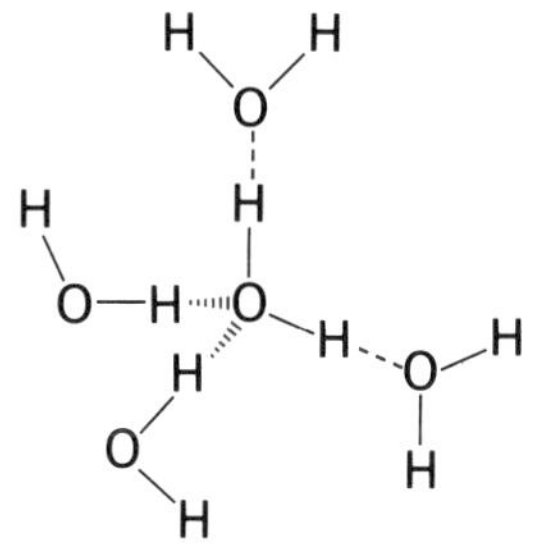

In ice, H_2O molecules are hydrogen-bonded to the maximum number of four others in a regular three-dimensional array. In liquid water about 20–30% of the hydrogen bonds are disrupted, and even though there is a considerable amount of ice-like clusters of hydrogen-bonded water molecules in liquid water, the overall structure is more compact. Therefore, ice is less dense than liquid water: it floats. This is a fortunate fact, for if ice were heavier than water, the bottoms of many bodies of water would remain frozen the year round and aquatic life, as we know it, could not exist.

Hydrogen bonds are crucially important in associating many molecules in living systems. This is particularly true in the case of proteins (Chapter 2) and the nucleic acids DNA and RNA (Chapter 13). Figure 1.7 shows some of the types of hydrogen bond interactions which are common in biochemical systems.

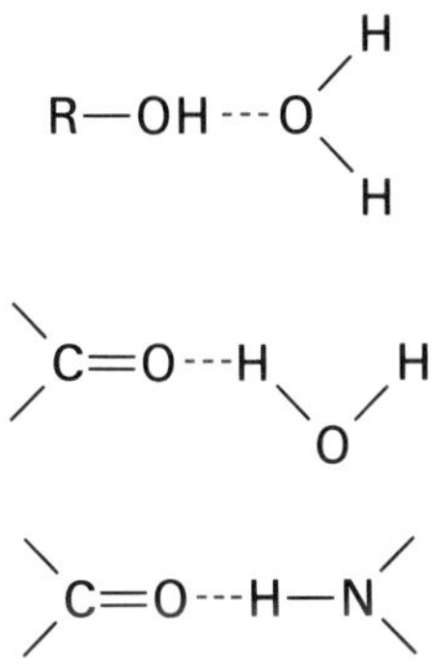

Figure 1.7

Some hydrogen-bonded systems.

Water is a good solvent for ionic and polar solutes and a poor solvent for nonpolar solutes. The inability of water to dissolve nonpolar materials is the primary factor responsible for the structure of membranes, as we will see in Chapter 6.

1.5

Atoms and Molecules in Biological Systems

The term *biomolecule* is used to denote the large and small molecules that are intimately involved in the chemistry of living systems. Most of the atoms found in biomolecules are below atomic number 30. In fact, over 99% of the atomic constituents of biomolecules are *carbon, nitrogen, oxygen,* and *hydrogen.* Of the less abundant components, *phosphorus* and *sulfur* are very important. In Chapter 3 we will see how metal ions play an important role in the function of certain biomolecules. Table 1.2 shows some of the atomic constituents of living cells.

Table 1.2   The Major Atomic Constituents of Cells.

Major elements (listed in approximate order of abundance)			Trace elements (The relative distribution of these depends on the particular organism.)			
O	Ca (ion)	Na(ion)	Mn	Zn	Mo	Se
C	P	Mg (ion)	Fe	B	I	Ni
H	K (ion)	Cl (ion)	Co	Al	Si	
N	S		Cu	V	Cr	

There is a definite order of increasing molecular complexity in all living cells. All biomolecules are produced from *precursors*, which are simple compounds of low molecular weight. Basically, a cell living in an aqueous environment requires among these precursors a *carbon source* and a *nitrogen source* to survive and to reproduce. Autotrophs can use CO_2 as a carbon source, which acts as a precursor for more complex carbon-containing biomolecules. Heterotrophs generally use a sugar as a carbon source. The source of nitrogen can range from N_2 for nitrogen-fixing organisms to NH_3 and other simple organic nitrogen-containing compounds for less self-sufficient heterotrophs.

We can define various levels of size and complexity in the transition from simple molecules to the overall structure of a cell. The increasing order of complexity of substances, shown in Figure 1.8, is general *to all living cells.* At the molecular level, we can see extensive similarities between primitive and highly evolved organisms. For example, proteins are high molecular weight macromolecules found in all living things, and even though the total number of different proteins in all living systems may range as high as 10^{12}, the same 20 amino acids constitute the primary building blocks for all proteins. Also, all organisms require deoxyribonucleic acid (DNA) in order to store, transfer, and transcribe genetic information both within the organism and from one

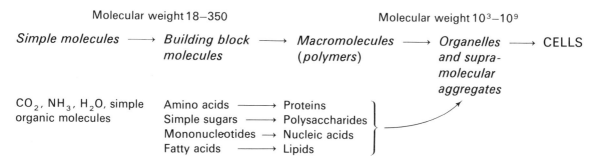

Molecular weight 18–350   Molecular weight $10^3–10^9$

Simple molecules ⟶ *Building block* ⟶ *Macromolecules* ⟶ *Organelles* ⟶ CELLS
                              molecules   *(polymers)*   *and supra-*
                                                                              molecular
                                                                              aggregates

CO_2, NH_3, H_2O, simple   Amino acids ⟶ Proteins ⎫
organic molecules   Simple sugars ⟶ Polysaccharides ⎪
                                Mononucleotides ⟶ Nucleic acids ⎬
                                Fatty acids ⟶ Lipids ⎭

Figure 1.8

Order of complexity
of biomolecules.

generation of the organism to the next. Even though a given DNA molecule is unique to each species of cell, the same four nitrogen-containing purine and pyrimidine bases are found in all DNA molecules.

We can list a relatively small number of biomolecules which are representative of the types of substances common to all living species. The structures of some of these biomolecules are shown in Figure 1.9. What is important to bear in mind is that, considering the diversity and complexity of living systems, there are really *so few* different types of simple molecules. The remainder of the first part of this book discusses these families of biomolecules in detail. It will become apparent that many seemingly specialized biomolecules with more complicated structures are, in fact, derivatives of the simple substances shown in Figure 1.9, (pp. 16–17).

Nature relies heavily on polymeric materials of large molecular weight, ranging from 1000 to over 10^9. As the legend in Figure 1.8 indicates, these large molecules are termed **macromolecules** to connote their large size. Even though macromolecules are often very large and have extremely complex three-dimensional structures, the covalent nature of the repeating units of these large polymeric species can be quite simple. Table 1.3 illustrates this fact and summarizes the relationship between the important macromolecules and their building block molecules.

Table 1.3   Biomolecules as Building Blocks for Macromolecules.

Monomer unit	Polymers
Amino acids	Proteins
Sugars	Polysaccharides
Mononucleotides (consisting of a purine or pyrimidine base, a 5-carbon sugar and a phosphate group)	Nucleic acids
Fatty acids and phospholipids	Cell membranes (noncovalent association of monomer units)

Amino acids

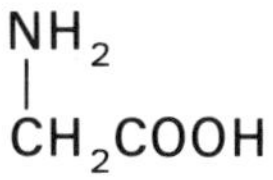

NH_2
|
CH_2COOH

Glycine

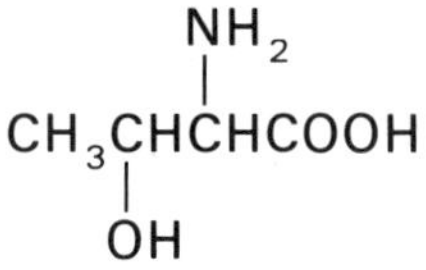

NH_2
|
$CH_3CHCHCOOH$
|
OH

†Threonine

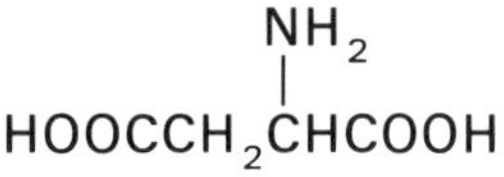

NH_2
|
$HOOCCH_2CHCOOH$

Aspartic acid

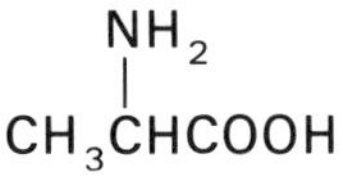

NH_2
|
$CH_3CHCOOH$

Alanine

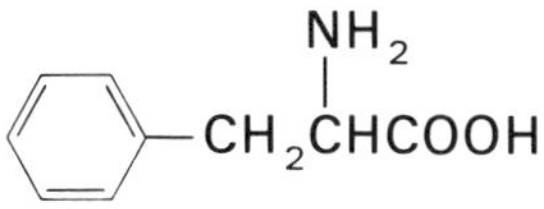

NH_2
|
⟨benzene ring⟩—$CH_2CHCOOH$

✗Phenylalanine

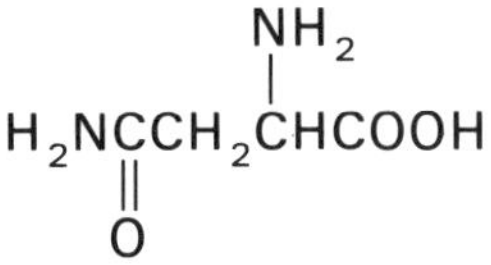

NH_2
|
$H_2NCCH_2CHCOOH$
‖
O

Asparagine

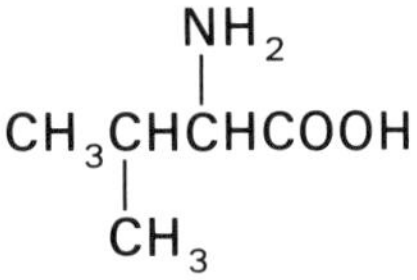

NH_2
|
$CH_3CHCHCOOH$
|
CH_3

✗Valine

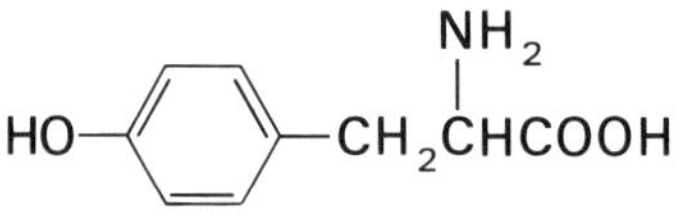

NH_2
|
HO—⟨benzene ring⟩—$CH_2CHCOOH$

Tyrosine

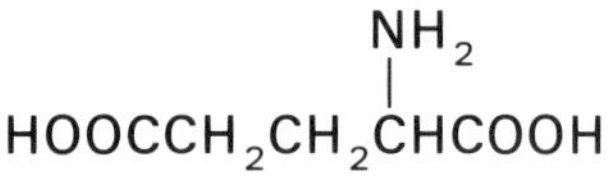

NH_2
|
$HOOCCH_2CH_2CHCOOH$

Glutamic acid

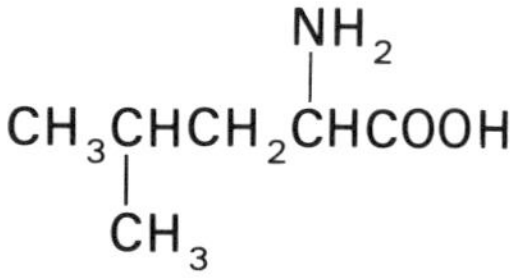

NH_2
|
$CH_3CHCH_2CHCOOH$
|
CH_3

✗Leucine

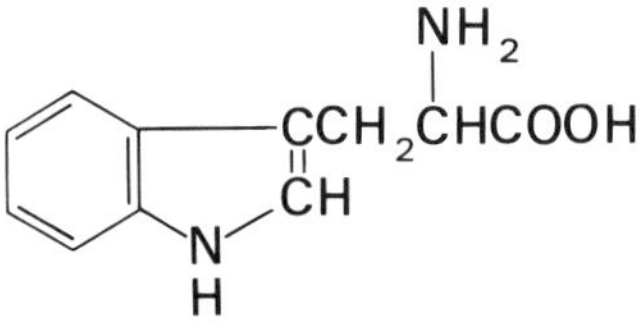

NH_2
|
⟨indole ring⟩$CCH_2CHCOOH$
CH
|
N
|
H

✗Tryptophan

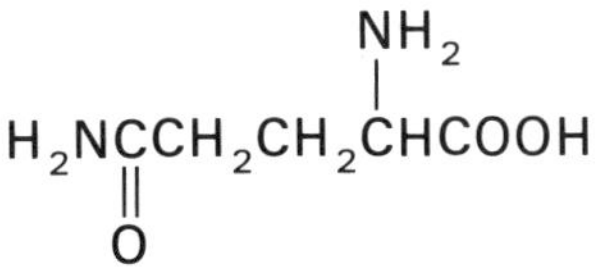

NH_2
|
$H_2NCCH_2CH_2CHCOOH$
‖
O

Glutamine

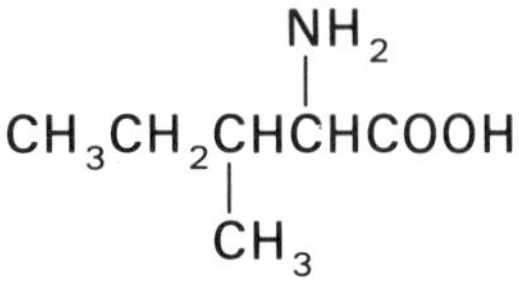

NH_2
|
$CH_3CH_2CHCHCOOH$
|
CH_3

✗Isoleucine

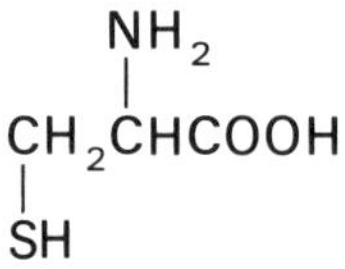

NH_2
|
$CH_2CHCOOH$
|
SH

Cysteine

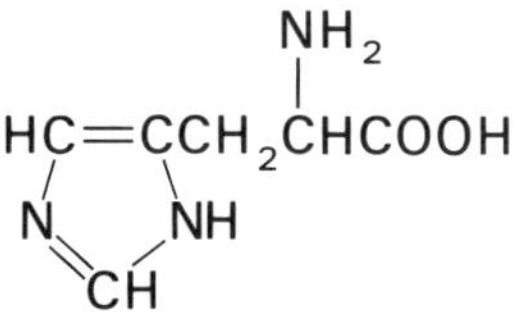

NH_2
|
HC=$CCH_2CHCOOH$
N NH
CH

✗ Histidine

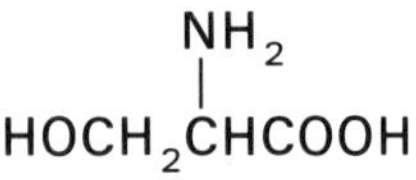

NH_2
|
$HOCH_2CHCOOH$

Serine

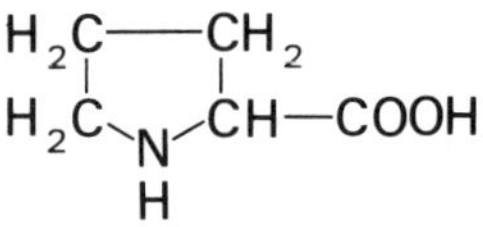

H_2C—CH_2
H_2C CH—$COOH$
N
|
H

Proline

NH_2
|
$H_2NCNHCH_2CH_2CH_2CHCOOH$
‖
NH

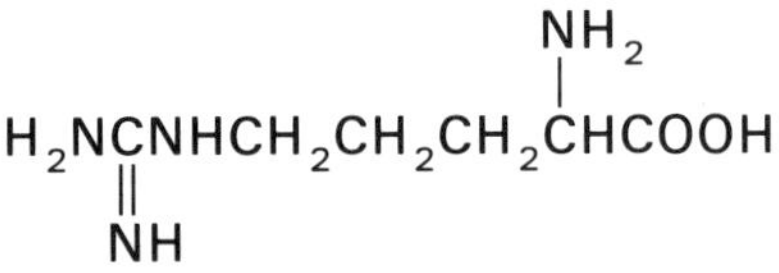

✗ Arginine

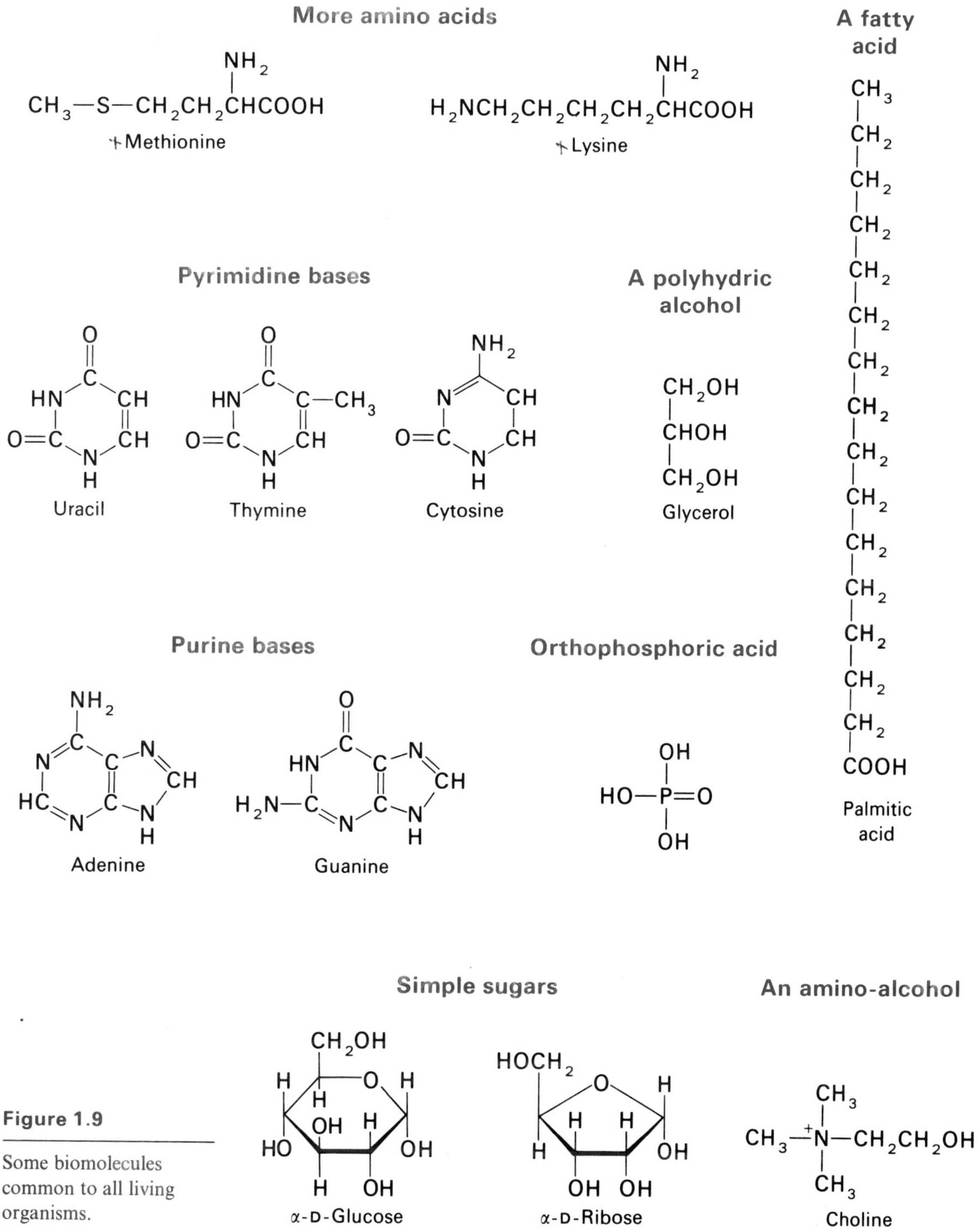

More amino acids

A fatty acid

$$CH_3-S-CH_2CH_2\overset{\overset{\displaystyle NH_2}{|}}{C}HCOOH$$

+Methionine

$$H_2NCH_2CH_2CH_2CH_2\overset{\overset{\displaystyle NH_2}{|}}{C}HCOOH$$

+Lysine

CH₃
|
CH₂
|
CH₂
|
CH₂
|
CH₂
|
CH₂
|
CH₂
|
CH₂
|
CH₂
|
CH₂
|
CH₂
|
CH₂
|
CH₂
|
CH₂
|
CH₂
|
COOH

Palmitic acid

Pyrimidine bases

Uracil

Thymine

Cytosine

A polyhydric alcohol

CH₂OH
|
CHOH
|
CH₂OH

Glycerol

Purine bases

Adenine

Guanine

Orthophosphoric acid

$$HO-\overset{\overset{\displaystyle OH}{|}}{\underset{\underset{\displaystyle OH}{|}}{P}}=O$$

Simple sugars

α-D-Glucose

α-D-Ribose

An amino-alcohol

$$CH_3-\overset{+}{\underset{\underset{\displaystyle CH_3}{|}}{\overset{\overset{\displaystyle CH_3}{|}}{N}}}-CH_2CH_2OH$$

Choline

Figure 1.9

Some biomolecules common to all living organisms.

1.6

Energy Requirements of Living Systems

Living matter is thermodynamically unstable. It is well that this is so, for otherwise the biosphere might have run out of carbon, nitrogen, hydrogen, and oxygen long ago. This constant recycling of atoms and simple molecules is made possible by the fact that dead organisms *spontaneously* decay into simple organic and inorganic molecules, thus releasing them to the biosphere for reuse.

Without a constant input of energy from the environment, living matter cannot maintain its internal order. A cell requires energy in the form of *chemical work* to synthesize a protein molecule or a nucleic acid molecule from monomeric precursors. Both proteins and nucleic acids contain hydrolyzable linkages which are thermodynamically unstable in aqueous solution. However, this should not cause us alarm because the hydrolysis of these macromolecules is kinetically extremely slow in the absence of a catalyst.

A major use of cellular energy and material is the **biosynthesis** of macromolecules such as proteins to replace those which have degraded, or to increase their amount in the cell. The turnover of molecular components in a cell is continuous. Some macromolecules, such as DNA, are reproduced only once during a cell division cycle; others, such as many enzyme proteins, may have lifetimes as short as hours, minutes, or seconds. Such short-lived molecules are said to have a *high rate of turnover*.

Living matter follows the same physical and chemical laws as inanimate matter. In carrying out the chemical processes associated with life, a cell obeys the principles of conservation of matter and energy. It may appear at first that the cell violates the second law of thermodynamics, which says that real physical and chemical processes proceed in such a way as to increase the disorder of the universe (see Chapter 7). A cell, which is highly ordered, seems to contradict this principle. However, *work* must be performed to create this order and overcome the tendency of things toward chaos. For example, in an aqueous environment, the lipid units of a membrane will spontaneously assemble to form a membrane structure (see Chapter 6). This *self-assembly* of subunits to form complex ordered structures in a cell is very common. Here, the attractive forces of binding between subunits serve to overcome the tendency toward randomness and provide a *driving force* for the self-assembly of the subunits.

As we will see in Chapter 7, energy exists in two forms: **work** and **heat**. In a chemical process, such as the breakdown of foodstuffs by a cell, the energy change in each of the steps of the breakdown involves both energy in the form of work, which is utilizable by the cell, and energy in the form of heat, which is released to the environment. Your body, for example, radiates a considerable amount of heat. In fact, we must have mechanisms, such as perspiration, to facilitate this transfer of heat. We cannot utilize heat as a form of energy—sitting on a hot radiator is no substitute for a good meal.

All of the chemical processes that occur in a cell constitute the **metabolism** of that cell. This includes processes associated with the breakdown of food molecules to gain energy and the processes of biosynthesis, in which biomolecules and macromolecules are synthesized from simple precursors. These two types of metabolic processes are called *catabolism* and *anabolism*.

Anabolism is the total of all biosynthetic processes where structural and functional biomolecules and macromolecules are produced from precursors. *Anabolic* (biosynthetic) reaction sequences *require energy* in an overall sense: *order is being created and thus work must be done.*

Catabolism is the total of all processes involving the degradation of complex organic molecules into simpler organic and inorganic molecules. *Catabolic* (degradative) reaction sequences *yield energy*, part of which is conserved in a form that the cell can use for such purposes as biosynthesis, motion, and transport of substances across membranes.

Organic and inorganic molecules involved in the reactions of metabolism are called **metabolites**. Overall, the processes of metabolism, catabolism and anabolism, must run concurrently because each set of processes provides the energy or material required by the other set.

The processes of metabolism are catalyzed and controlled by **enzymes**, which are catalytic proteins. Most individual reactions of metabolism are far too slow in the absence of a catalyst to permit life as we know it to exist. The most remarkable thing about enzymes is their ability to provide tremendous rate enhancements under mild conditions of physiological temperature, pH, and atmospheric pressure. Furthermore, the specific relationship between a given enzyme and the structures of the particular reactant(s) catalyzed permits many related and unrelated chemical processes to occur in the same reaction medium—all within the confines of the cell. This is done without these many processes interfering with each other except in a controlled and organized way. The structure and action of enzymes will be considered in detail in Chapters 2 and 3.

Suggestions for Further Reading

The Living Cell, Readings from "Scientific American." San Francisco: W. H. Freeman and Co., 1965. This collection of reprints from *Scientific American* is an excellent first step beyond the level of the material in this chapter.

Scientific American. This readily available periodical has published many excellent articles dealing with the topics discussed in this chapter. They include:
Affrey, V. G. and Mirsky, A. E. "How Cells Make Molecules." September 1961.
Brachet, Jean. "The Living Cell." September 1961.
Gibor, A. "Acetablularia: A Useful Giant Cell." November 1966.
Lehninger, A. L. "How Cells Transform Energy." September 1961.
Margulis, L. "Symbiosis and Evolution." August 1971.
Morowitz, H. J. and Tourtellotte, M. E. "The Smallest Living Cells." March 1962.

Problems

1. For each of the terms given below, provide a brief definition or explanation.

   Anabolism   Heterotrophic
   Autotrophic   Hydrogen bond
   Catabolism   Macromolecule
   Cell   Metabolism
   Eucaryote   Procaryote

2. Refer to Figures 1.3, 1.4, and 1.5 and answer these questions:

   (a) What are the differences between procaryotes and eucaryotes?
   (b) What are the similarities and differences between the photosynthetic autotrophic cell of Figure 1.5 and the heterotrophic cell shown in Figure 1.4?

3. If macromolecules are thermodynamically unstable, how do these macromolecules exist in a cell?

4. What are the characteristics of living matter that differentiate it from inanimate matter?

CHAPTER 2

Amino Acids and Proteins

An important part of biochemistry involves the study of protein structure and function. Proteins are nitrogen-containing macromolecules with molecular weights ranging from 5,000 to over 1,000,000. Proteins are a major cellular component, comprising about 50% of the dry weight of a cell. Protein functions range from catalytic, in the case of enzymes, to toxic, in the case of bacterial and snake poisons. Table 2.1 summarizes some major types of proteins and their functions.

Table 2.1 Protein Functions.

Function	Type	Examples
Catalytic	Enzymes	Catalase, pepsin
Structural	Structural proteins	Collagen (connective tissue and bone), elastin, keratin (hair, skin)
Motile (mechanical)	Contractile proteins	Actin, myosin (muscle)
Storage (of nutrients)	Storage proteins	Casein (milk), ovalbumin (egg), ferritin (iron storage)
Transport (of nutrients)	Transport proteins	Serum albumin (fatty acids), hemoglobin (oxygen)
Regulatory (of cell metabolism)	Protein hormones Regulatory enzymes	Insulin Phosphofructokinase
Protection (immunity; blood)	Antibodies Clot-forming proteins	Immune globulins Thrombin, fibrinogen
Toxic response	Protein toxins	Snake venom toxins, bacterial toxins (botulism, diphtheria)

The term **protein** originated in the 1830s with the Dutch chemist Mulder, who was one of the first to systematically study the chemistry of proteins. He correctly inferred the central role of proteins in living systems by deriving the name from the Greek *proteios*, "of first rank." Because of the significance of proteins in biochemistry, we will begin our study of biomolecules with these substances. This chapter discusses the importance of proteins in living cells.

2.1

Covalent Linkage of Proteins

In Chapter 1 we saw that many key substances in living systems are polymeric in character. Proteins are polymers consisting of covalently-linked amino acid units, as shown in Figure 2.1. The basic covalent connection is simply an amide linkage, formed by the condensation of the amino group of one amino acid with the carboxylic acid group of another. (These functional groups are discussed in Appendix B.) This amide linkage has been given a special name: **peptide bond**. Some plastics, such as nylon, are also polyamides, similar to proteins. The important difference is that the basic units of nylon or other synthetic polyamides are all identical. The synthetic polymer produced, however, has a range of molecular sizes and molecular weights. In the case of a given protein, all molecules in a pure sample have the same size and weight.

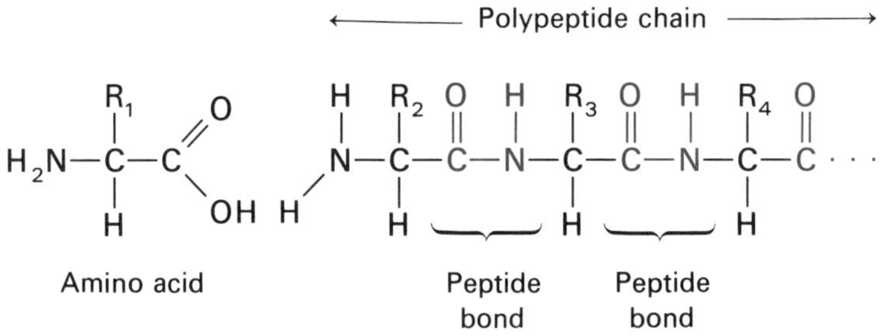

Figure 2.1

Covalent backbone of proteins.

In proteins there are 20 amino acids that predominate as building blocks, irrespective of the source of the protein. For any particular protein, the sequence in which the amino acid units appear in the protein chain is unique. This enables us to distinguish a great many different proteins. For example, if you had an unlimited supply of beads of 20 different colors and wished to string them in varying orders so as to make as many different necklaces as you could, each 100 beads long, you would have to make 20^{100}, or 10^{130} necklaces before you exhausted all the possible arrangements! This would be the number of different proteins possible with a polypeptide chain 100 amino acid units long. We can, however, have proteins with fewer or greater than 100 amino acid units. Since the number of unique proteins in all living organisms has been estimated to be between 10^{10} and 10^{12}, nature has obviously evolved a system that allows for virtually unlimited variability and adaptability. Before discussing proteins in detail, we must first consider some of the physical and chemical properties of amino acids.

2.2

Functional Identity of Amino Acids

The complete hydrolysis of a pure protein results in a mixture of the 20 "common" amino acids plus, perhaps, one or more of the "less common" amino acids. Most amino acids have the same general structure, shown in Figure 2.2 in its un-ionized form. The only major exception is proline and its derivatives, whose structure is shown in Table 2.2. Most of the amino acids in living organisms are α-amino acids; that is, the amino function is on the carbon atom which is next to the carboxylic acid functional group. Because the basic structure of all α-amino acids is the same, a given amino acid establishes its identity by the nature of its side chain group (**R**). Since the covalent backbone of a protein is invariant (see Figure 2.1) and involves the amino acid carboxyl and amino functions, *it is the R group which gives a particular position in the protein chain its physical and chemical properties.*

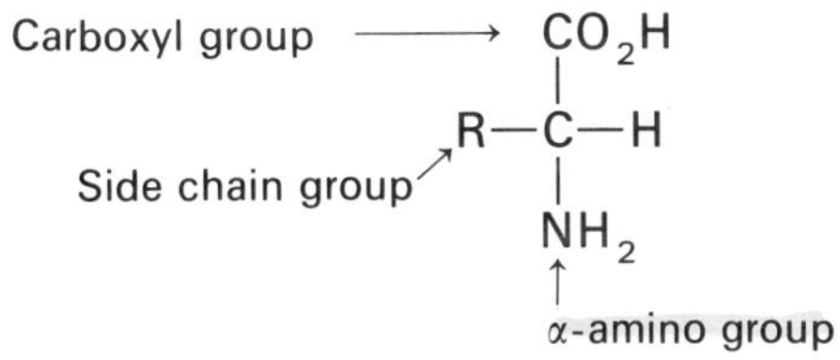

Figure 2.2

General structure of amino acids found in proteins.

$$
\begin{array}{c}
\text{Carboxyl group} \longrightarrow \quad CO_2H \\
\nearrow \quad\quad R{-}C{-}H \\
\text{Side chain group} \quad\quad NH_2 \\
\uparrow \\
\text{α-amino group}
\end{array}
$$

We can classify amino acids conveniently and meaningfully using the physical properties of the side chain group. Because pH 7 approximates physiological pH, we make it a standard reference point for H_3O^+ in biological systems. Using this point of reference, we can classify amino acids which have side chain groups that gain or lose protons with respect to their net charge at pH 7. The four different classes of amino acids and the structures of the individual amino acids are summarized in Table 2.2 (pp. 24–25).

The organization of amino acids into these four groups has fundamental functional significance. For example, the side chains of the *hydrophobic* amino acids generally form the interior of globular protein molecules where the polar solvent, water, is excluded. The *uncharged polar* amino acids have side chains which can enter into hydrogen bonding (see Chapter 1). Polar side chains, with the exception of glycine, can also participate in metal ion bonding through the interaction of the lone pair(s) of electrons of the O, N, or S atom of the side chain with the metal ion. Polar groups are often located on the surface of a protein molecule exposed to the polar aqueous solvent medium. Amino acid side chains that can carry a full positive or negative charge can enter into ionic interactions with groups of opposite charge, or with uncharged polar molecules or groups.

Table 2.2 Classification of the Twenty Amino Acids Commonly Found in Proteins. (Each is shown in its major ionic form at pH 7.)

1. *Hydrophobic side chain groups* (*hydrocarbon-like*)

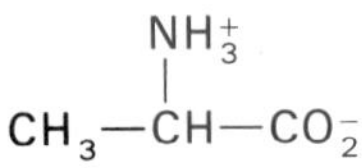

Alanine 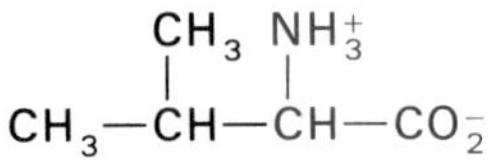 Valine 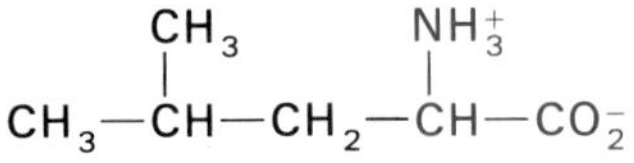 Leucine

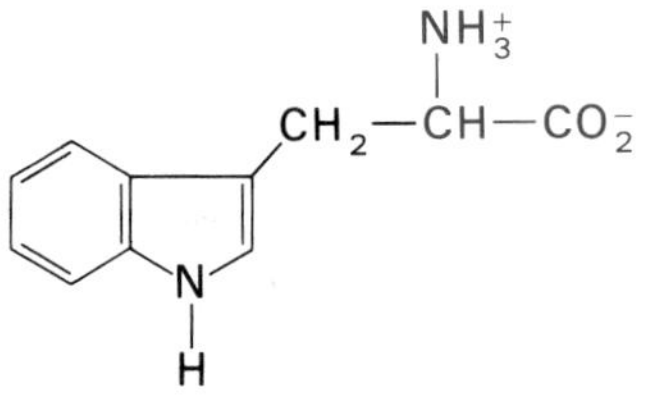

Tryptophan 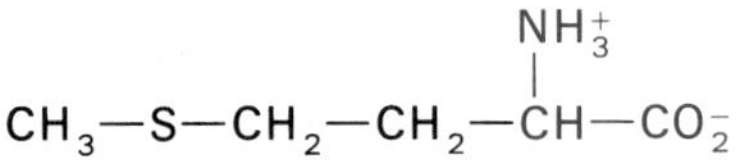 Methionine

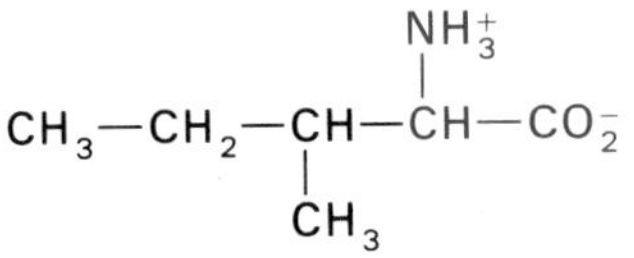

Isoleucine 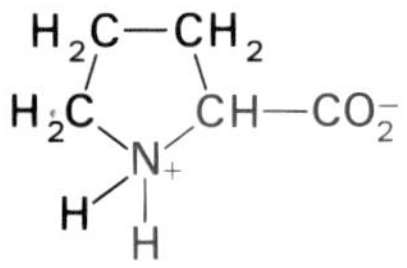 Proline (an *imino* acid) 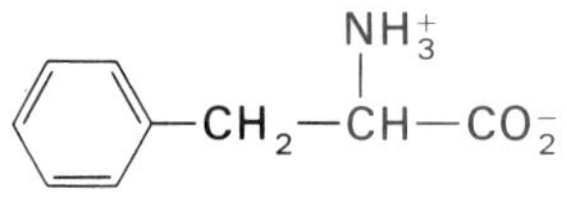 Phenylalanine

2. *Uncharged polar side chain groups* (*no charge on the side chain at pH 7*)

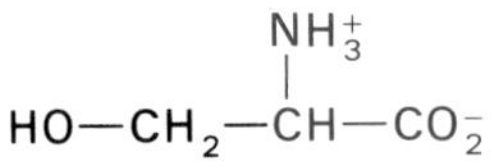

Serine 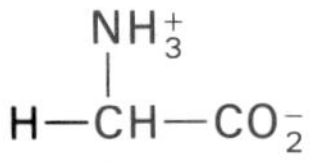 Glycine

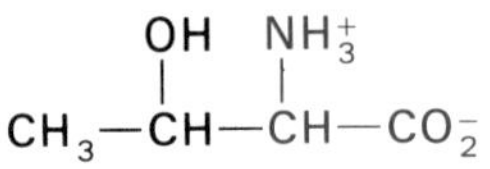

Threonine 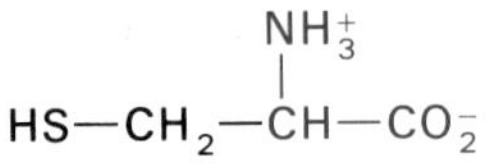 Cysteine

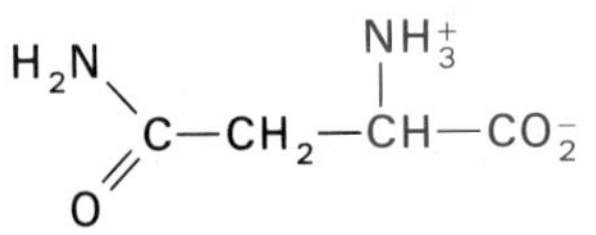

Asparagine 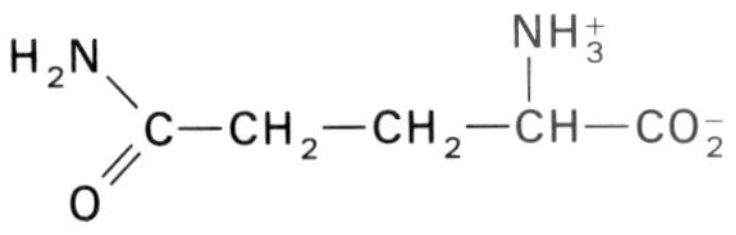 Glutamine

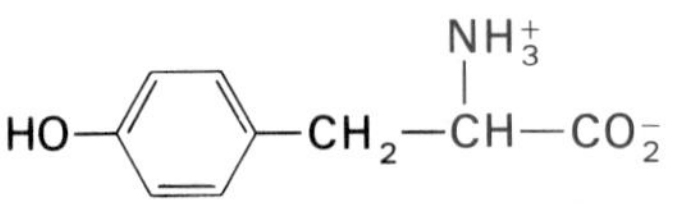

Tyrosine

3. *Negatively charged side chain groups at pH 7*

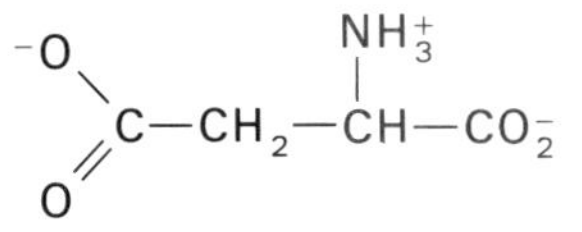

Aspartic acid

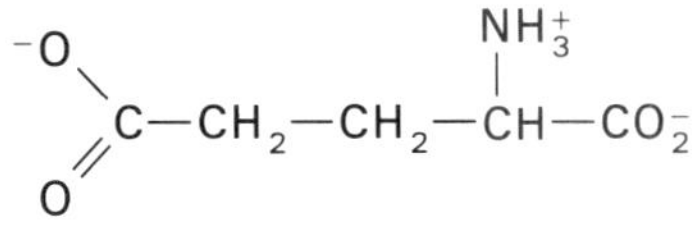

Glutamic acid

4. *Positively charged side chain groups at pH 7*

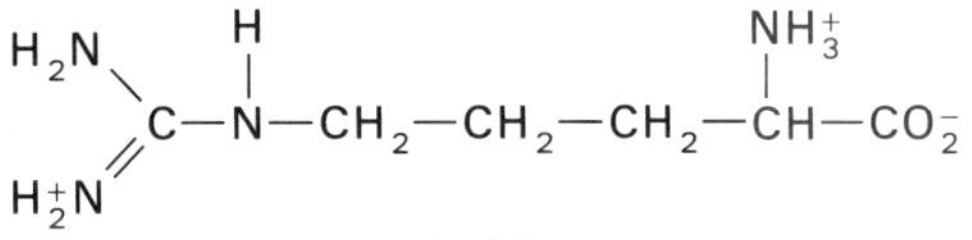

Arginine

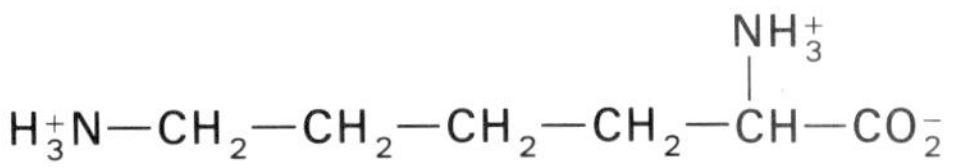

Lysine

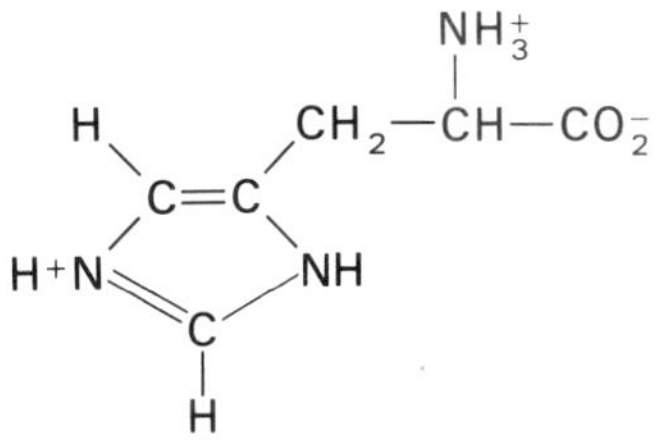

Histidine (predominantly positively charged below pH 6)

Some side chain groups are related to substances that have chemical properties which are interesting in their own right. For example, the side chain of histidine is the *imidazole* group, in which the nitrogen atom can use its lone pair of electrons to gain a proton (as a base) or can act as a *ligand* to bind metal ions. The side chain of tryptophan is the *indole* group, which is found in nature in a number of substances (see Section 12.3).

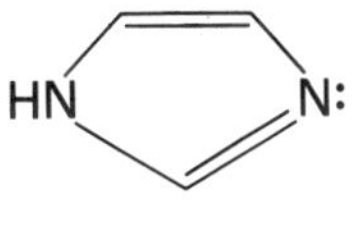

Imidazole

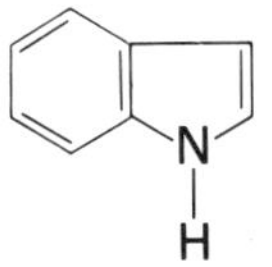

Indole

The side chain of arginine contains the *guanidinium* function, which is an organic nitrogen base. The guanidinium group is about 1000 times stronger as a base than NH_3. Finally, the side chain of tyrosine contains the *phenol* group, which can act as a weak proton donor.

Guanidinium ion
(conjugate acid of the base, guanidine)

Phenol

Phenoxide
anion

The nomenclature of amino acids is primarily empirical. A large number of these compounds were first isolated from natural sources during the 1800s, and many of the names refer to the original source from which they were isolated. Thus, asparagine was first isolated in 1806 from asparagus juice. Glycine is so called because in crystalline form it has a sweet taste (Greek *glykys*, "sweet").

In addition to the "fundamental 20" amino acids, there are over 150 less common amino acids known. Most of these are not associated with proteins and many are simple derivatives of the 20 common amino acids. Such amino acids may be metabolic intermediates or part of a nonprotein biomolecule. Table 2.3 gives the structures and functions of a few of these less common amino acids.

Importance of Amino Acids

Amino acids not only act as the building blocks of proteins but also serve as chemical precursors for many important nitrogen-containing compounds. For example, glycine is required for the biosynthesis of the heme group of hemoglobin. Tryptophan is the precursor of a family of substances important in the biochemistry of the nervous system. Tyrosine is the starting material for the biosynthesis of the skin pigment *melanin*. The biosynthesis of nitrogen-containing molecules is discussed in more detail in Chapter 12.

Higher animals cannot produce all of the individual amino acids from a simple nitrogen source and a carbon source as can bacteria. Therefore, certain amino acids must be taken in directly as part of the diet. The following are the essential amino acids in humans:

arginine*	lysine	tryptophan
histidine	methionine	valine
isoleucine	phenylalanine	
leucine	threonine	(* required by growing children only)

Table 2.3 Some Less Common Amino Acids.

Amino acid	Structure	Function/Occurrence
4-hydroxyproline (proline derivative)	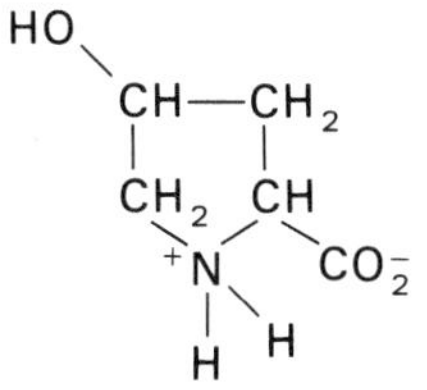 HO $CH—CH_2$ $CH_2\ CH$ $^+N\quad CO_2^-$ $H\quad H$	Found in the fibrous protein collagen.
Demosine (derived from four lysine molecules)	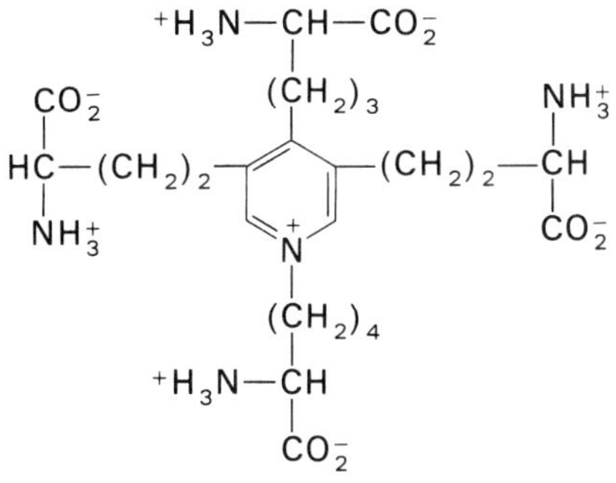 $^+H_3N—CH—CO_2^-$ $(CH_2)_3$ NH_3^+ CO_2^- $HC—(CH_2)_2—(CH_2)_2—CH$ $NH_3^+\quad CO_2^-$ N^+ $(CH_2)_4$ $^+H_3N—CH$ CO_2^-	Found in the fibrous protein elastin (connective tissue). The structure permits two-dimensional stretching in the complete protein.
β-alanine	$^+H_3N—CH_2CH_2—CO_2$	Part of the structure of the vitamin pantothenic acid.
Ornithine	NH_3^+ $^+H_3N—CH_2—CH_2—CH_2—CH—CO_2^-$	A biosynthetic intermediate in the formation of urea.
3,5-Diiodotyrosine	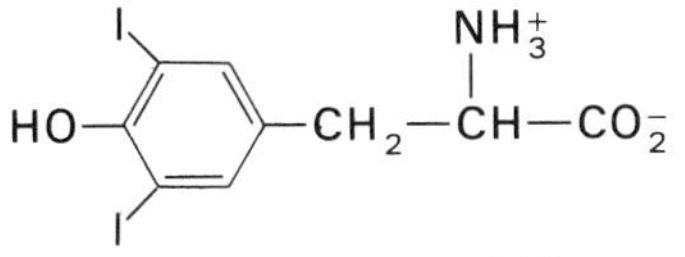 NH_3^+ $HO—\quad—CH_2—CH—CO_2^-$ (with I substituents)	Involved in iodine metabolism in the thyroid gland.
Canaline	NH_3^+ $^+H_3N—O—CH_2—CH_2—CH—CO_2^-$	A toxic amino acid isolated from the jack bean.

In well developed societies we don't commonly observe the consequences of a diet lacking in one or more of these essential amino acids. This is not the case for much of the world's population. In many Third World countries, children are fed on a diet consisting almost entirely of one grain such as corn, millet or plantain. These foods are low in protein and particularly deficient in lysine. This gives rise to the widespread deficiency disease, *kwashiorkor*. To combat this, new strains of cereal grains high in lysine are being developed. Also, in many areas of the world, synthetic amino acids are now being made available as a dietary supplement.

2.3

Stereochemistry of Amino Acids

In biological systems, the shape of a molecule is often crucially important, particularly the arrangement of groups around an asymmetric center. All amino acids in Table 2.2 except glycine can exist in two or more *nonsuperimposable mirror image forms*, or **enantiomers**.

For amino acids (and sugars), the *absolute configuration* of the molecule is referenced to the structure of D-glyceraldehyde. Figure 2.3 shows the relationship between the enantiomers of glyceraldehyde and of a representative amino acid, alanine. You can see the mirror image relationship between the enantiomers D- and L-alanine. All D-amino acids are stereochemically related to D-glyceraldehyde, as is D-alanine in this figure. All L-amino acids have absolute configurations corresponding to the arrangement of analogous groups in the structure of L-glyceraldehyde. In the case of amino acids synthesized in the laboratory by nonstereospecific reactions, an equimolar mixture of D and L forms is called a *racemic mixture*, or *racemate*. A racemic mixture of D- and L-alanine is represented as DL-alanine.

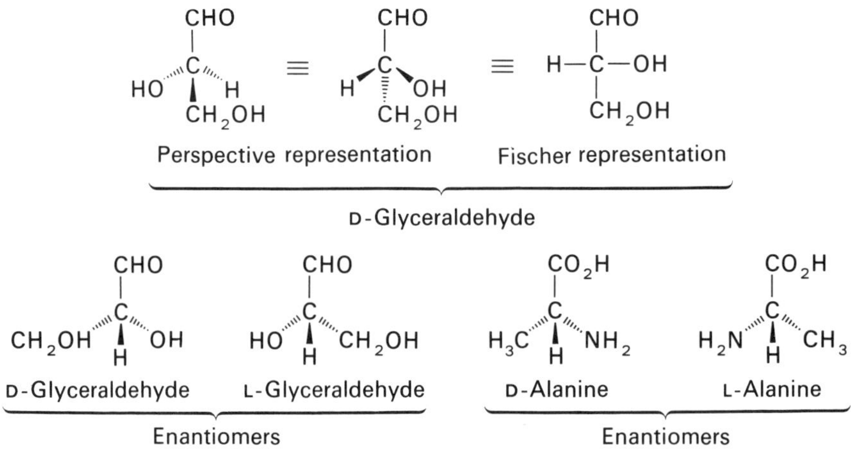

Figure 2.3

The absolute configuration of amino acids can be used to compare enantiomers of glyceraldehyde and alanine.

Only L-amino acids have been observed in proteins. The mirror image counterparts, D-amino acids, are not known to be associated with the metabolism of higher organisms. However, D-amino acids are important in specialized structures and in substances associated with lower forms of life. For example, the cell walls of bacteria (Chapter 4) involve D-alanine or other D-amino acids, depending on the particular organism. A number of antibiotics of bacterial origin include D-amino acids in their molecular structure. An example of this is the cyclic peptide gramicidin S, an antibiotic which has been isolated from certain strains of bacteria. The structure of gramicidin S is shown in Figure 2.4.

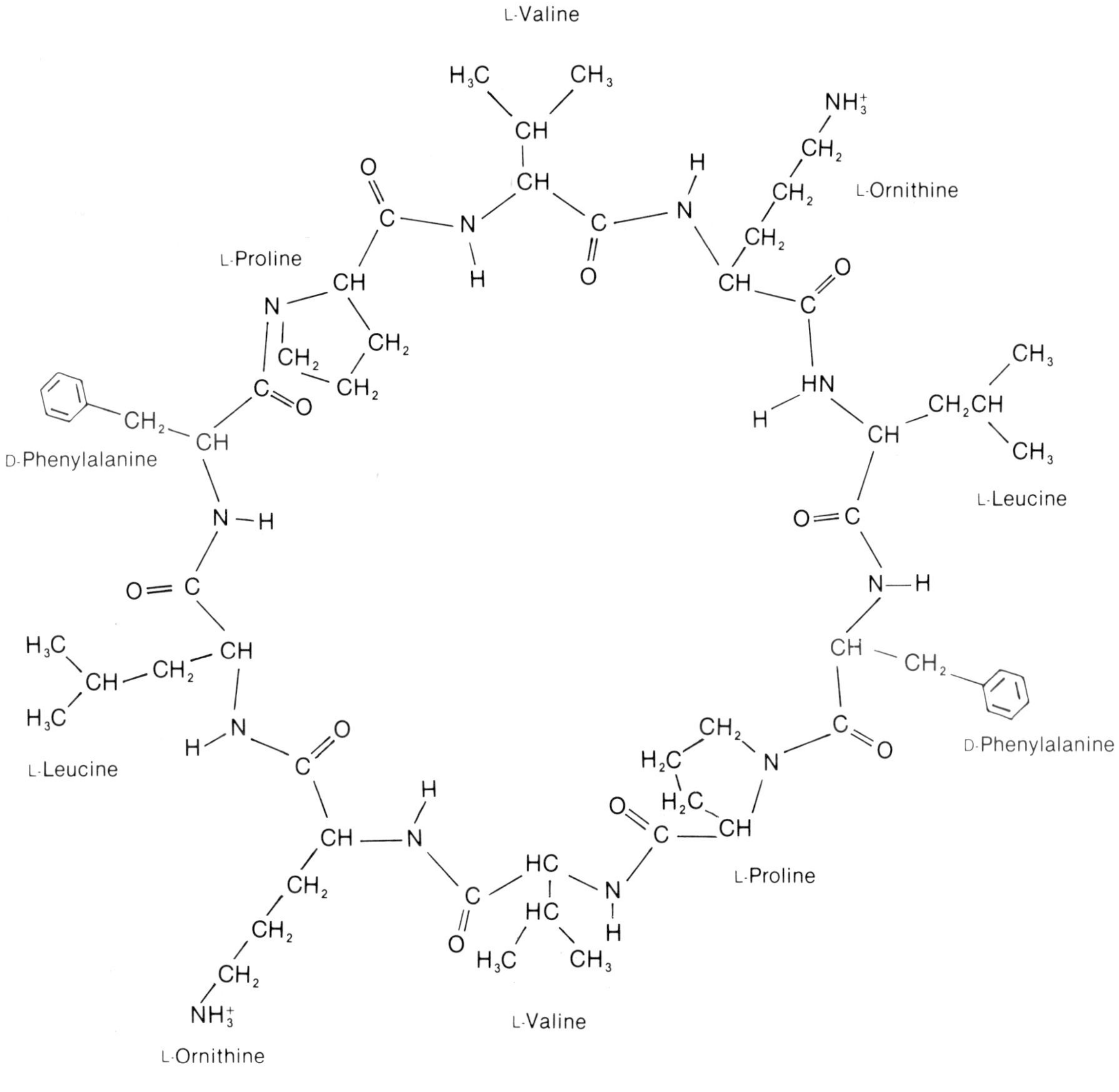

Figure 2.4

The structure of gramicidin S, an antibiotic of bacterial origin containing D-amino acids in its structure. Note also how proline is incorporated into the polypeptide chain.

2.4

Acids and Bases

Amino acids always contain at least one group that is a proton donor (acid) and at least one group that is a proton acceptor (base). Therefore, the acid–base behavior of amino acids is an important part of their chemistry. Before discussing this behavior we will briefly review some of the general chemistry of acids and bases.

The most useful general definition of an acid or a base for biochemical purposes is the Brønsted definition:

Acid: a proton donor; **base:** a proton acceptor

This definition permits us to consider as acid-base reactions a large number of processes which occur in living systems. A simple inorganic example is given in Eq. (1):

$$\underset{\text{acid}}{H^+Cl^-} + \underset{\text{base}}{Na^+OH^-} \longrightarrow H_2O + Na^+ + Cl^- \tag{1}$$

It is often more convenient to write this *neutralization reaction* between an acid and a base in ionic form:

$$OH^- + H^+ \longrightarrow H_2O$$

Since Cl^- and Na^+ ions do not participate in the reaction, we can consider them "spectator ions," and drop them from both sides of the chemical equation.

In any Brønsted acid-base reaction, an acid molecule donates a proton to a base molecule, producing a new *conjugate base* and a new *conjugate acid*. This is shown in Eq. (2) for the reaction between NH_3 and H_2O:

$$\underset{\text{base}}{\underset{\text{acid}}{HOH}} + NH_3 \rightleftharpoons OH^- + NH_4^+ \tag{2}$$

acid ←—— conjugate acid-base pair ——→ conjugate base

base ←—— conjugate acid-base pair ——→ conjugate acid

In aqueous solution, an acid will donate a proton to H_2O, yielding a hydrated proton species which can be represented as H_3O^+ (hydronium ion).

Water can also undergo an acid-base reaction with itself to produce H_3O^+ and OH^-. This reaction, shown in Eq. (3), is termed the self-dissociation of water:

$$HOH + HOH \rightleftharpoons H_3O^+ + OH^- \qquad (3)$$

The equilibrium constant for this process is called K_w, the ion product of water, and is given by

$$K_w = [OH^-][H_3O^+] = 1 \times 10^{-14} \quad \text{at } 25°C \qquad (4)$$

Note that the concentration of H_2O does not appear in the equilibrium expression. This occurs because the concentration of H_2O is so much greater than the concentration of the ions that it can be considered as unchanged after the ionization reaction, and we include it in the value of K_w. For *any* aqueous solution at 25°C, the ion product of water is equal to 10^{-14}. Since there are equal amounts of OH^- and H_3O^+ ions in pure water at 25°C, the concentration of these ions in pure water is 1×10^{-7} mole/liter each. In an acidic solution, the concentration of H_3O^+ is greater than $10^{-7} M$, and by Eq. (4), $[OH^-]$ must be less than $10^{-7} M$. For example, hydrochloric acid, HCl, is a strong acid: it dissociates completely in aqueous solution to yield H_3O^+ and Cl^- ions. When 2×10^{-3} moles of HCl are dissolved in a liter of H_2O, the resulting concentration of H_3O^+ is nearly equal to $2 \times 10^{-3} M$. This is much greater than $10^{-7} M$, so the solution is acidic.

If we wish to consider a series of solutions of differing acidity, we can express the concentration of H_3O^+ in two ways, exponential and logarithmic:

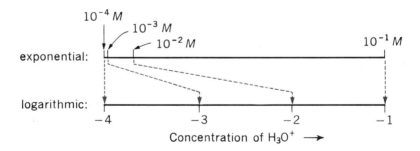

In many situations it is much less cumbersome to express the concentration of H_3O^+ in logarithmic form, giving rise to the pH notation:

$$pH = -\log[H_3O^+] \quad \text{or} \quad [H_3O^+] = 10^{-pH} \text{ moles/liter}$$

The relationship between $[H_3O^+]$, $[OH^-]$, and pH for a number of solutions is shown in Table 2.4 (p. 32).

Table 2.4 Hydronium Ion Concentration, Hydroxide Ion Concentration, and pH.

$[H_3O^+]$	$[OH^-]$	$log[H_3O^+]$	pH	
1	1×10^{-14}	0	0	acidic
1×10^{-2}	1×10^{-12}	-2	2	↑
2×10^{-3}	5×10^{-12}	-2.7	2.7	
1×10^{-5}	1×10^{-9}	-5	5	
1×10^{-7}	1×10^{-7}	-7	7	neutral
1×10^{-9}	1×10^{-5}	-9	9	
5×10^{-10}	2×10^{-5}	-9.3	9.3	
1×10^{-13}	1×10^{-1}	-13	13	↓
1×10^{-14}	1	-14	14	basic

The strength of an acid in water is determined by the extent to which the acid dissociates in solution to yield H_3O^+. We can therefore define a strong acid as one which dissociates completely, as shown by the example of nitric acid, HNO_3, in Eq. (5):

$$HNO_3 + H_2O \rightleftharpoons H_3O^+ + NO_3^- \quad (5)$$

initial conc: 0.1M

equilibrium conc: negligible   0.1M 0.1M

Similarly, a strong base, such as NaOH, dissociates completely when dissolved in water.

A weak acid or base is not completely dissociated in aqueous solution. Therefore, the H_3O^+ concentration in a solution of a weak acid does not equal the original concentration of the undissociated acid. A quantitative measure of the strength of an acid, HA, is given by the value of the *equilibrium constant* for the dissociation of the acid in aqueous solution (see also Chapter 6):

$$\underset{\text{acid}}{HA} + \underset{\text{base}}{HOH} \rightleftharpoons \underset{\substack{\text{conjugate} \\ \text{acid of } H_2O}}{H_3O^+} + \underset{\substack{\text{conjugate} \\ \text{base of HA}}}{A^-}$$

$$K_A \equiv \text{acid dissociation constant} = \frac{[H_3O^+][A^-]}{[HA]}$$

$$pK_A = -\log K_A$$

Table 2.5 Dissociation Constants of Some Weak Acids at 25°C.

System	K_A	pK_A
$CH_3{-}C(=O){-}OH + H_2O \rightleftharpoons H_3O^+ + CH_3{-}C(=O){-}O^-$ Acetic acid ⇌ Acetate ion	1.86×10^{-5}	4.73 (stronger)
$H_2CO_3 + H_2O \rightleftharpoons H_3O^+ + HCO_3^-$ Carbonic acid ⇌ Bicarbonate ion	4.52×10^{-7}	6.34
$H_2PO_4^- + H_2O \rightleftharpoons H_3O^+ + HPO_4^{2-}$ Dihydrogen phosphate ion ⇌ Monohydrogen phosphate ion	6.31×10^{-8}	7.20
$NH_4^+ + H_2O \rightleftharpoons H_3O^+ + NH_3$ Ammonium ion ⇌ Ammonia	5.7×10^{-10}	9.24 (weaker)

The extent to which HA dissociates is reflected in the value of K_A. Thus, K_A for a *strong* acid would be *much* greater than 1. The value of K_A for a *weak* acid would generally be *less* than one, as shown by the values of K_A for some weak acids listed in Table 2.5.

The value of K_A tells us something about the relative amounts of reactants and products at equilibrium in an acid dissociation reaction. For example, the value of K_A for acetic acid is 1.86×10^{-5}. This means that *at equilibrium*, the product of the concentrations of H_3O^+ and of acetate ion is approximately 50,000 times *smaller* than the concentration of undissociated acetic acid.

If we add a base to an acid, neutralization occurs. This can be described as the nearly complete reaction between H_3O^+ and a base. Some neutralization reactions are:

Acid		Base		
HNO_3 Nitric acid	+	NH_3 Ammonia	$\longrightarrow$	$NH_4^+NO_3^-$ Ammonium nitrate
HCl Hydrochloric acid	+	KOH Potassium hydroxide	$\longrightarrow$	$H_2O + K^+Cl^-$ Potassium chloride
$CH_3{-}C(=O){-}OH$ Acetic acid	+	$NaOH$ Sodium hydroxide	$\longrightarrow$	$H_2O + CH_3{-}C(=O){-}O^-Na^+$ Sodium acetate

Since the reaction between an acid and a base is stoichiometric, we can *titrate* an acid of unknown concentration by adding a base of known concentration until neutralization is achieved. If we perform this titration step by step and measure the pH of the solution after each addition of base by using a pH meter, we can construct a *titration curve* for the acid. The titration curve for acetic acid is shown in Figure 2.5.

We can analyze the titration curve in Figure 2.5 in terms of the four regions I, II, III, and IV shown in the figure.

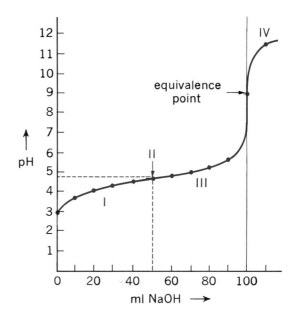

Figure 2.5

Titration curve of 100 mL of $0.1M$ CH_3COOH (acetic acid) with $0.1M$ NaOH

Region I In this part of the curve, the equilibrium concentration of acetate ion is less than the equilibrium concentration of the free acid, CH_3COOH. As we add hydroxide ion to the solution, the equilibrium reaction

$$H_2O + CH_3COOH \rightleftharpoons CH_3COO^- + H_3O^+$$

$$K_A = 1.87 \times 10^{-5} = \frac{[H_3O^+][CH_3COO^-]}{[CH_3COOH]}$$

shifts to the right as more H_3O^+ reacts with the added OH^- to form H_2O. As the ratio $[CH_3COO^-]/[CH_3COOH]$ increases due to the addition of base, the concentration of H_3O^+ *must* decrease in order for the value of the equilibrium expression to remain constant.

Region II Here we have added exactly one-half the amount of base needed to completely react with the amount of acid originally present. Thus, an *equimolar* (50–50) mixture of CH_3COOH and CH_3COO^- is produced. Application of the equilibrium expression yields

$$1.87 \times 10^{-5} = \frac{[H_3O^+][CH_3COO^-]}{[CH_3COOH]} = [H_3O^+] \quad \text{when } [CH_3COO^-] = [CH_3COOH]$$

Therefore, at the halfway point of neutralization of a weak acid by a strong base, $[H_3O^+] = K_A$, or pH = pK_A, as shown by the dotted line in Figure 2.5. In the part of the titration curve around region II, the pH of the solution changes slowly as we add more base. This is because the *ratio* of acid to conjugate base does not change rapidly in this part of the titration curve. We call this part of the curve the *buffer* region. A buffer is a solution containing a weak acid and its conjugate base which resists large changes in pH when a strong acid or a strong base is added to it. Blood serum and many other aqueous media in living systems are buffers that prevent large fluctuations in pH.

Region III   Here the equilibrium concentration of acetate ion is greater than the equilibrium concentration of free acetic acid. When we have added an equimolar amount of sodium hydroxide to the acetic acid, we have reached the *equivalence point* of the titration curve. The pH rises sharply here because there is insufficient undissociated acetic acid to provide any buffering.

Region IV   From this point on the pH of the solution is determined by the concentration of OH^-, which increases linearly with added NaOH.

This completes our discussion of some of the important aspects of acid-base chemistry. Now let's see how this applies to the properties of amino acids.

2.5

Acid-Base Behavior of Amino Acids

The physical properties of amino acids appear somewhat unusual when compared with carboxylic acids which lack an α-amino group. For example, compare the properties of alanine and lactic acid shown in Table 2.6. Lactic acid is an α-hydroxy acid in which the hydroxy group is incapable of gaining or losing a proton under ordinary conditions. Its physical properties are those of a polar carboxylic acid. Alanine, on the other hand, has physical properties that suggest an ionic substance, even in the solid state. For example, lactic acid is a liquid at room temperature, whereas alanine is a solid with a very high melting point for an organic compound of its molecular weight. Ethanol, which is not a good solvent for ionic substances, dissolves lactic acid well, but does not dissolve an appreciable amount of alanine.

Table 2.6 Some Physical Properties of Lactic Acid and the Amino Acid Alanine.

	$\underset{\displaystyle CH_3-CH-CO_2H}{\overset{\displaystyle NH_2}{\mid}}$ *Alanine*	$\underset{\displaystyle CH_3-CH-CO_2H}{\overset{\displaystyle OH}{\mid}}$ *Lactic acid*
Molecular weight	89.09	90.8
M. P. (°C) (racemate)	264–96 (with decomposition)	16.8
B. P. (°C) (racemate)	—	122 (15 mm Hg)
K_A (25°C) (COOH)	4.5×10^{-3}	1.4×10^{-4}
Solubility in ethanol (25°C)	0.0087 g/100 g solvent	soluble in all proportions

These points are all consistent with the fact that pure alanine in the solid state, and in a given pH range in solution, exists as a dipolar ionic form known as a **zwitterion**. The zwitterion results from the presence of both a proton-donating (acid) group and a proton-accepting (base) group in the amino acid molecule.

$$\Delta G^\circ = -7100 \text{ cal/mole}$$

In a solution of alanine, the free energy change of -7100 cal/mole on going from the neutral form to the zwitterion form indicates that the zwitterion is favored by a factor of over 100,000 to 1 (see Chapter 7).

The zwitterion has no *net* charge, as opposed to the charged ionic forms of amino acids we will discuss shortly. The pH at which the equilibrium concentration of amino acid zwitterions is a maximum is called the *isoionic pH*, or pI. This pH value is near or equal to the *isoelectric point*, defined as the pH value of an amino acid solution at which the amino acid (or protein) does not migrate in an electric field. The isoelectric point is an experimentally-determined quantity which depends on the nature of the buffer salts and other ions in the solution. The isoionic point, on the other hand, is an inherent property of the acid-base system and can be predicted if the titration curve of the system is known. While amino acids are generally soluble in water, the solubility of amino acids reaches a minimum in a solution having a pH value equal to the isoionic point, because the concentration of the least soluble form, the zwitterion, is at a maximum. Since amino acids exist as zwitterions in the solid state, dissolving an amino acid in pure water produces a solution whose pH value is approximately that of the isoionic point.

Figure 2.6

Diagram of electrophoresis. As the proteins (or amino acids) migrate toward the oppositely charged electrode, the ionic species having the greatest negative charge will migrate most rapidly toward the positive electrode, and vice versa. (At pH 8.5 most known proteins are negatively charged; that is, the isoelectric point of the protein in solution is less than pH 8.5.) Any species which is at its isoelectric point will remain at the origin. In actual practice, a dye can be used to render the separated protein spots visible.

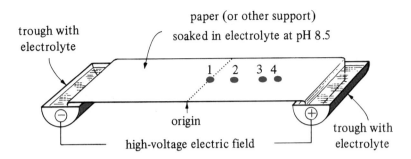

Knowing the isoionic point of an amino acid, or, in a more complicated context, a protein, can be useful for determining conditions for the crystallization of material by a technique known as *isoelectric precipitation*. This is because the solubility of an amino acid or a protein is a minimum at the isoionic pH, as we mentioned earlier. Knowing the isoionic point can also be useful for separating amino acids or proteins, since it permits us to predict the way in which the ions would migrate in an electric field. Separating substances on the basis of their electrical charge in this fashion is called *electrophoresis*. Figure 2.6 shows this process diagrammatically.

In Section 2.4 we discussed the titration curve of a simple organic acid, acetic acid. Titration curves are useful for visualizing the various ionic forms of an acid or a base as a function of pH. The titration curve for alanine is shown in Figure 2.7; it is typical of the behavior of amino acids which do not have

Figure 2.7

Titration curve for alanine. We can visualize the physical situation as the addition of a strong base to a solution of alanine which has been carefully acidified.

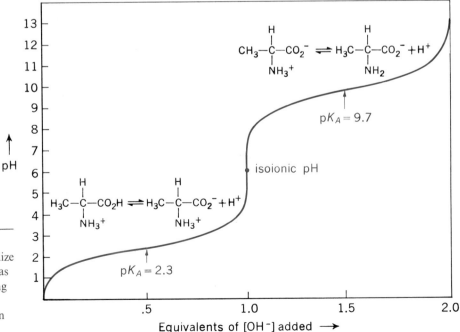

ionizable side chain groups. The curve in the figure indicates a dibasic acid which can lose a total of two protons in two separate steps. This means we must consider two acid-base equilibria, shown in Eq. (6):

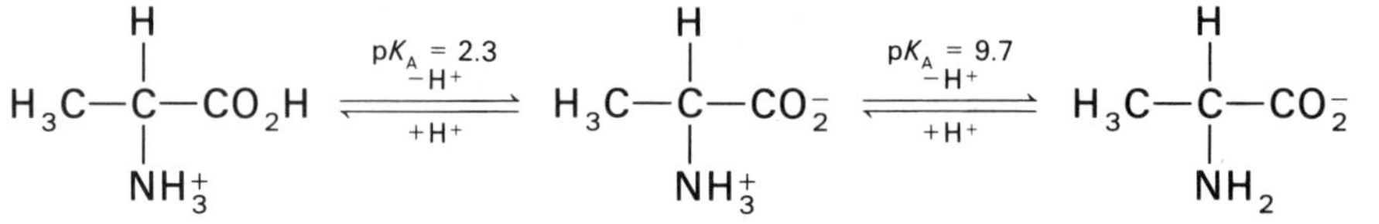

$$
\underset{\substack{H_3C-\overset{\displaystyle\overset{H}{|}}{\underset{\underset{NH_3^+}{|}}{C}}-CO_2H}}{} \underset{+H^+}{\overset{\substack{pK_A = 2.3 \\ -H^+}}{\rightleftharpoons}} \underset{\substack{H_3C-\overset{\displaystyle\overset{H}{|}}{\underset{\underset{NH_3^+}{|}}{C}}-CO_2^-}}{} \underset{+H^+}{\overset{\substack{pK_A = 9.7 \\ -H^+}}{\rightleftharpoons}} \underset{\substack{H_3C-\overset{\displaystyle\overset{H}{|}}{\underset{\underset{NH_2}{|}}{C}}-CO_2^-}}{} \quad (6)
$$

Net charge: +1   0   −1

pH at which species predominates: 1   6   11

Starting with the form having a net charge of +1, which exists in an acid solution, the addition of base (which reacts with H_3O^+) causes the first acid-base equilibrium to shift to the right. The carboxyl group is a stronger acid than the α-ammonium group, and thus loses its proton at a lower pH. The zwitterion form results from the complete neutralization of the carboxylic acid group. Further addition of base causes the loss of a proton from the α-ammonium group. Two pK_A values are evident: 2.3 for the carboxyl group, and 9.7 for the α-ammonium group. Physically, this means that at pH 2.3 there is a 50–50 mixture of cation and zwitterion forms; and at pH 9.7 there is a 50–50 mixture of zwitterion and anion forms. In Table 2.5 we saw that the pK_A of an aliphatic carboxylic acid such as acetic acid is about 4.8. Alanine has a carboxyl group with a pK_A value of 2.3, which means that it is over 300 times stronger as an acid than is acetic acid. This is due to the presence next to the carboxyl group of the positively charged α-ammonium group, which repels the carboxyl proton and makes its dissociation more favorable.

The acid-base behavior of side chain groups is important to proteins. Consider the acid-base properties of lysine, summarized in Eq. (7):

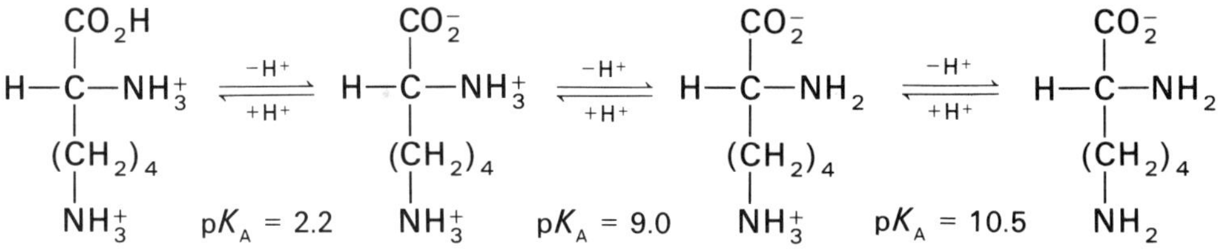

$$
\underset{\substack{\overset{\displaystyle CO_2H}{|} \\ H-\overset{|}{C}-NH_3^+ \\ \overset{|}{(CH_2)_4} \\ \overset{|}{NH_3^+}}}{} \underset{+H^+}{\overset{-H^+}{\rightleftharpoons}} \underset{\substack{\overset{\displaystyle CO_2^-}{|} \\ H-\overset{|}{C}-NH_3^+ \\ \overset{|}{(CH_2)_4} \\ \overset{|}{NH_3^+}}}{} \underset{+H^+}{\overset{-H^+}{\rightleftharpoons}} \underset{\substack{\overset{\displaystyle CO_2^-}{|} \\ H-\overset{|}{C}-NH_2 \\ \overset{|}{(CH_2)_4} \\ \overset{|}{NH_3^+}}}{} \underset{+H^+}{\overset{-H^+}{\rightleftharpoons}} \underset{\substack{\overset{\displaystyle CO_2^-}{|} \\ H-\overset{|}{C}-NH_2 \\ \overset{|}{(CH_2)_4} \\ \overset{|}{NH_2}}}{} \quad (7)
$$

              $pK_A = 2.2$ $pK_A = 9.0$ $pK_A = 10.5$

Net charge   +2   +1   0   −1

(Zwitterion)

When lysine is part of the covalent structure of a protein, only the side chain ammonium group is capable of acid-base behavior. The equilibria in Eq. (7) tell us that at the physiological pH of 7.3, the side chain of lysine will be positively charged.

The presence or absence of a charged group can have a profound effect on protein structure. Changing the pH of a protein solution can change the ionization state of ionizable amino acid side chain groups, causing dramatic and often irreversible changes in the structure of the protein. For example, the addition of any acid to milk causes precipitation of the milk protein casein. Table 2.7 shows the pK_A values of acidic and basic amino acid side chain groups. Note that only the imidazole group of histidine can act significantly as a buffer at physiological pH. Since the pK_A of imidazole is 6.0, the pH of the equilibrium mixture of imidazole/imidazolium ion would change most slowly with the addition of strong acid or strong base in this region of the titration curve.

Table 2.7 Amino Acid Side Chain Group pK_A Values. Note that in a given protein, the actual pK_A of a given group might be considerably different from the values given here, due to the sensitivity of these pK_A values to the local environment of the side chain group.

Amino acid	Ionizable side chain group	pK_A
Aspartic acid	$\beta\text{-}CO_2H \rightleftharpoons \beta\text{-}CO_2^- + H^+$	3.7
Glutamic acid	$\gamma\text{-}CO_2H \rightleftharpoons \gamma\text{-}CO_2^- + H^+$	4.3
Histidine	$HN{\diagdown}NH^+ \rightleftharpoons HN{\diagdown}N + H^+$	6.0
Cysteine	$-SH \rightleftharpoons -S^- + H^+$	8.3
Tyrosine	$\bigcirc\!-OH \rightleftharpoons \bigcirc\!-O^- + H^+$	9.1
Lysine	$-NH_3^+ \rightleftharpoons -NH_2 + H^+$	10.5
Arginine	$-NH-\underset{\underset{NH_2^+}{\|}}{C}-NH_2 \rightleftharpoons -NH-\underset{\underset{NH}{\|}}{C}-NH_2 + H^+$	13.2

2.6

Chemistry of Amino Acids

The chemistry of amino acids falls into two broad classes: the chemistry of the amino group, and the chemistry of the R group. The amino group reactions are important in the quantitative analysis of amino acids in protein hydrolyzates and body fluids. We will briefly discuss some of these reactions here and will discuss others later in the chapter.

Amino Group Reactions

Amino group reactions rely on the ability of the amino group to act as a nucleophile, where the lone electron pair of the amino nitrogen forms a bond with an electron-deficient center in the appropriate reagent. Amino acids and other amines can be oxidized using the mild oxidant ninhydrin, yielding a blue-colored product:

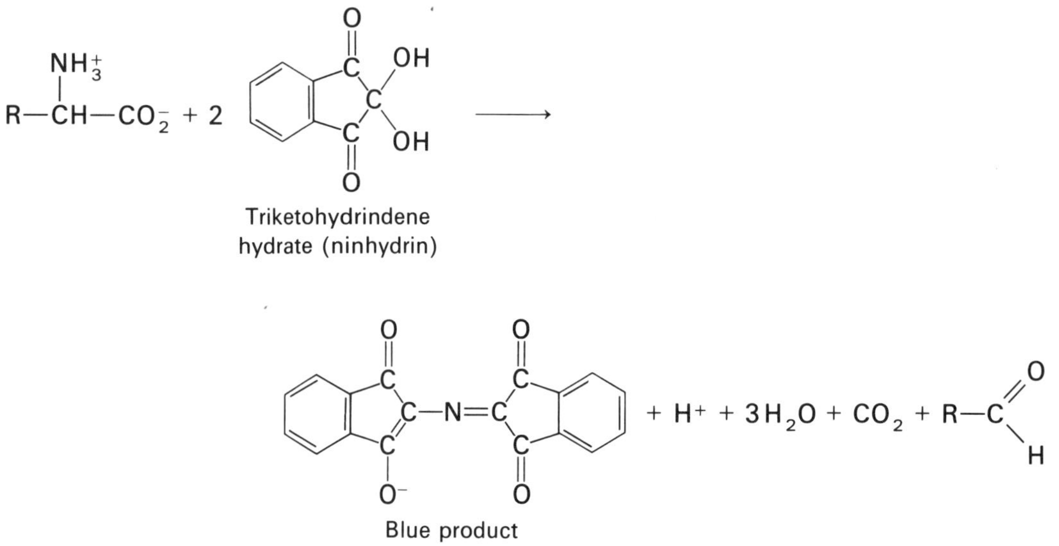

Triketohydrindene
hydrate (ninhydrin)

$$+ H^+ + 3H_2O + CO_2 + R-C \underset{H}{\overset{O}{\big\langle}} \qquad (8)$$

Blue product

Thus, the reaction of any amino acid with two equivalents of ninhydrin gives an intensely colored blue product. Proline and hydroxyproline, which are imino acids, produce a somewhat different product having a yellow color. This color reaction for amino acids is commonly used to render components of a mixture of amino acids visible when separated by electrophoresis, paper chromatography or ion exchange resins in an amino acid analyzer. The color reaction is also used to render fingerprints on porous surfaces visible. In this case, sufficient amino acids and other amines are released from the skin to result in a blue fingerprint when sprayed with ninhydrin reagent.

Two reagents which react with amino groups are Sanger's reagent (1-fluoro 2,4-dinitrobenzene; *FDNB*), and *dansyl chloride* (1-dimethyl-amino napthalene 5-sulfonyl chloride). The use of FDNB permits us to attach the intensely colored dinitrobenzene group to an amino acid, producing a colored dinitrophenyl (DNP) derivative, as shown in Eq. (9):

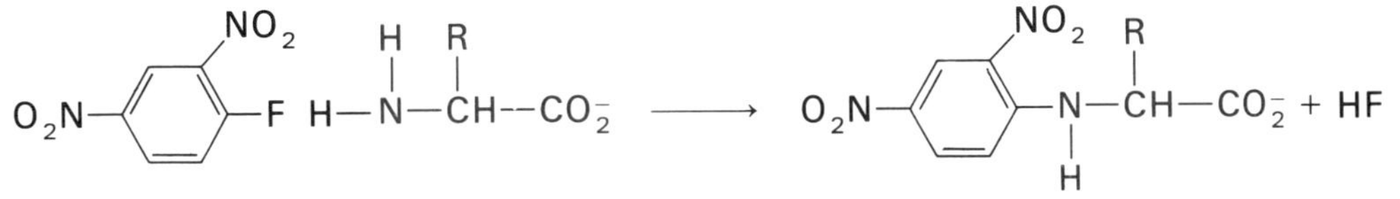

FDNB

DNP-amino acid
derivative (yellow–red)

$$\qquad (9)$$

Dansyl chloride possesses the advantage of producing amino acid derivatives which are intensely fluorescent. This permits us to detect minute amounts of amino acids due to the great sensitivity of fluorescence as an analytical technique. The reaction of dansyl chloride with an amino acid to produce a dansyl (DNS)-amino acid derivative is shown in Eq. (10):

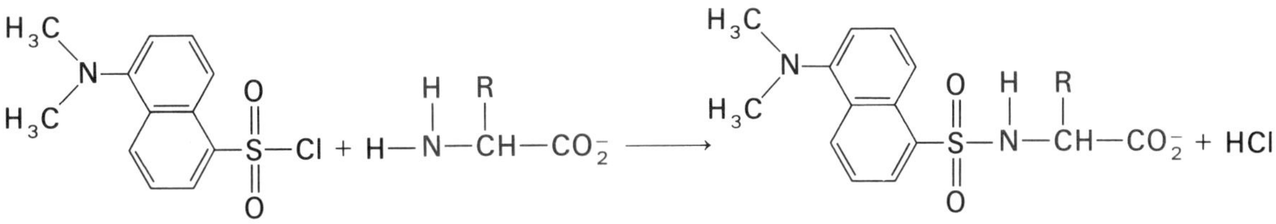

Dansyl chloride

DNS-amino acid
derivative

$$\qquad (10)$$

Both FDNB and dansyl chloride react with other amino groups, such as the amino group of the lysine side chain. These derivatives can be readily distinguished from those involving the α-amino group by chromatography.

R-Group Reactions

The reactions of the side chain groups in amino acids are generally typical of the chemical nature of each group. It is beyond the scope of this text to consider the chemistry of all R groups in detail. The interested student will find references at the end of this chapter particularly useful in this regard. However, the properties of the thiol group (—SH) of cysteine are worth considering in some detail, since they have a direct bearing on the overall structure of proteins.

An important redox reaction in biological systems involves the oxidation of two thiol groups to form a disulfide:

$$R\!-\!SH + HS\!-\!R + (\text{oxidant}) \;\rightleftharpoons\; R\!-\!S\!-\!S\!-\!R + (\text{oxidant} \cdot H_2)$$
Disulfide

In the case of the oxidation of the amino acid cysteine, the disulfide product is called *cystine*. While cystine is not important by itself, we can readily see how two parts of a protein chain, each containing the side chain of cysteine, can link together covalently for structural purposes, using this cystine type of disulfide bridge.

Home permanents are a practical example of this redox chemistry. The microscopic fibers that make up an individual hair strand are bound to each other by many disulfide links (see Section 2.12). The natural curl of a hair strand is largely a function of these disulfide links. The chemistry of a home permanent is summarized below.

1. React natural hair with a reductant to obtain hair with —S—S— links broken.

$$|—S—S—| \xrightarrow{\text{Reductant}} |—SH \quad HS—|$$

2. Set hair strands into desired form.

3. Treated hair is now oxidized to form new —S—S— links to maintain new overall strand shape.

$$|—SH \quad HS—| \xrightarrow{\text{Oxidant}} |—S—S—|$$

It is interesting that something as common as permanent waves actually involves some sophisticated protein chemistry.

2.7

Characteristics of Proteins

Proteins are sometimes referred to as **informational macromolecules**, because the amino acid sequence of a given protein reflects the genetic information contained in the base sequence of that portion of DNA which directed the biosynthesis of the protein. Each type of protein is characterized by:

1. **A specific chemical composition.** Each individual protein is a pure compound.

2. **A specific molecular weight.** All molecules in a given sample of a pure protein have the same molecular weight. This type of polymer is termed *monodisperse*. A random sample of a *polydisperse* polymer would contain molecules having a range of molecular weight values. Polydisperse macromolecules include various synthetic plastics as well as natural polysaccharides such as starch and cellulose. There is no direct genetic control of the size and sequence of such polymers, in contrast to an informational macromolecule such as a protein.

3. **A specific amino acid sequence.** The amino acid sequence of a given protein is genetically specified. However, small changes in the amino acid

sequence of a given protein can occur through a process known as *mutation*. Mutation can lead to desirable results if an "improved" protein is created through a process of natural selection at the molecular level. Mutation can also have disastrous results, as we will see when we consider sickle cell anemia later in this chapter.

There are two types of proteins, differentiated by the products obtained when the protein is hydrolyzed to its constituent monomeric units. These are simple proteins and conjugated proteins:

Simple protein: Amino acids *only*
Conjugated protein: Amino acids + nonprotein prosthetic group(s)

Table 2.8 summarizes some important types of conjugated proteins.

Table 2.8   Some Examples of Conjugated Proteins.

Class	Prosthetic group	Example
Nucleoproteins	Nucleic acid (DNA, RNA)	Viruses
Glycoproteins	Carbohydrate (sugar)	Egg albumin
Phosphoproteins	Phosphate	Casein (milk)
Hemeproteins	Iron-protoporphyrin	Hemoglobin
Lipoproteins	Phospholipid, cholesterol	Serum lipoproteins

The molecular weights of proteins range from an arbitrary minimum of about 5,000 to over 1,000,000. Proteins of molecular weight less than 5,000 are frequently referred to as *polypeptides*. There are a number of small polypeptides which are important in living systems and which have ten or fewer amino acid units. These are called *oligopeptides* (*oligo*, "a few"). We can often estimate the number of amino acid units in a given protein by the following relationship:

$$\frac{\text{molecular weight}}{120} = \text{number of amino acids}$$

(120 is the average molecular weight of an amino acid less 18, the molecular weight of water.)

At one time, proteins were classified according to their solubility, and it still is an important way of distinguishing proteins. The two major solubility categories are

1.   **Fibrous proteins.** These are insoluble in aqueous salt solution. The overall conformation of the protein molecule is generally either long fibers (α), as in the case of collagen (ligaments) and α-keratin (hair), or sheets (β), as in elastin. Such proteins constitute much of the connective tissue in higher organisms.

2. **Globular proteins.** These are, as the name implies, globular in conformation, and they are soluble in aqueous salt solution. Nearly all enzymes,
antibodies, protein hormones and transport proteins are globular proteins.

Some proteins are neither globular nor fibrous. These include myosin
(muscle fiber) and fibrinogen (blood clotting).

2.8

Nature of the Peptide Bond

The formation of a peptide bond from its constituent amino acids is thermodynamically unfavorable. The primary reason for this is that the formation of a
peptide (amide) link between two isolated amino acids requires that the reacting
groups first be converted from the more stable zwitterion forms to the un-ionized
(—COOH) and (—NH$_2$) forms. The unfavorable energetics of peptide bond
formation cause proteins to hydrolyze spontaneously. However, the reaction
is extremely slow in the absence of a catalyst.

The structure of the peptide bond was first elucidated by Linus Pauling in
the 1940s. A detailed diagram of the structure is shown in Figure 2.8. The
important features are:

1. The carbon-nitrogen peptide bond (1.32 Å) is shorter than the average
1.5 Å (1 Å = 10^{-8} cm) bond length of the C—N bond normally observed.
This indicates considerable *double bond character*, where the resonance
forms contributing to the overall peptide bond are

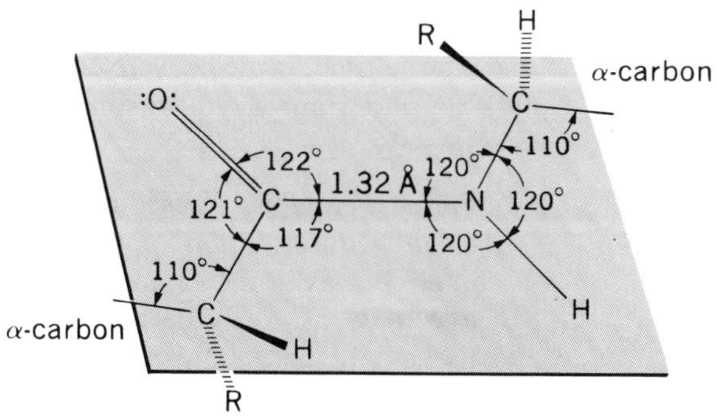

Figure 2.8

Planarity of the peptide
bond. The bond angles
about the carbonyl
carbon and the amide
nitrogen are all
approximately 120°. The
carbon–nitrogen bond
length is 1.32 Å. (From
W. H. Brown,
*Introduction to Organic
and Biochemistry*, 2nd
ed., Willard Grant Press,
Boston, 1976.)

2.   The high double bond character of the peptide bond requires that all atoms connected to the peptide carbon and nitrogen be *coplanar*, as shown in Figure 2.8.

We will soon see that these two features place considerable restraints on the types of regular structures a polypeptide can assume.

2.9

General Aspects of Protein Structure

Since proteins are central to the function of living organisms, the problem of determining structure from the level of the single peptide unit to the level of the overall three-dimensional protein structure represents an active area of scientific research. Each protein has a unique three-dimensional structure, determined by the sequence of amino acids in the polypeptide chain. To identify the three-dimensional structure of a protein, biochemists must perform four separate procedures.

1.   **Isolate and purify the protein.** Isolating one protein from the thousands present in a living cell presents a formidable task.

2.   **Determine the molecular weight and the amino acid composition.** Chemists may use several methods for determining the molecular weight of a large macromolecule. These include ultracentrifugation, disc gel electrophoresis, and gel permeation chromatography. The determination of the amino acid composition, the number and kind of amino acids present in the protein molecule, requires an instrument called an amino acid analyzer, which automatically separates and quantifies the amino acids present in a hydrolyzed protein.

3.   **Determine the amino acid sequence and the position of any disulfide bridges.** As we will see in the following section, this is primarily a chemical problem.

4.   **Determine the three-dimensional structure.** The physical method for doing this is called *X-ray crystallographic structure analysis* and requires a protein crystal about 0.1 mm on a side. Obtaining the protein crystal is often the most difficult step in the x-ray structure analysis of proteins.

We can divide the levels of organization of protein structure into four classes of ascending complexity. They are:

1.   **Primary structure.** This is simply the amino acid sequence in the polypeptide chain and the position of any disulfide bridges within the protein chain. The primary structure of a protein is maintained by covalent (peptide) bonds.

2. **Secondary structure.** This refers to the extent of regular local α-helical or β-pleated sheet structure associated with the overall structure of the protein. The secondary structure of a protein is maintained by hydrogen bonds between the carbonyl oxygens and amide nitrogens of the polypeptide chain.

3. **Tertiary structure.** This refers to the way in which the protein chain in globular proteins is bent and folded to form the overall three-dimensional structure of the protein molecule. The tertiary structure is maintained by interactions between amino acid R groups.

4. **Quaternary structure.** Many proteins exist as *oligomers*, or large molecules formed by the specific aggregation of identical or different subunits known as *protomers*. The arrangement of protomers in an oligomeric protein is known as the quaternary structure. Hemoglobin is a good example of such a protein: it exists as a tetramer containing two pairs of single-chain protomeric subunits. Proteins which have molecular weights in excess of 50,000 often involve two or more polypeptide chains.

2.10

Primary Structure of Proteins

In writing the amino acid sequence of a protein it is convenient to use three-letter abbreviations for the name of each amino acid. Table 2.9 summarizes both the three-letter abbreviation system and a more recently adopted one-letter code used by many researchers.

Figure 2.9 shows the covalent structure of an oligopeptide containing seven amino acid units, or *residues*. The amino acid in a protein chain with a free α-amino group is called the **N-terminal residue**. In Figure 2.9 this corresponds to the alanine residue. The amino acid with the free amino acid carboxyl group is called the **C-terminal residue**. This corresponds to the valine residue in Figure 2.9. It is customary to write out the primary structure of a protein

Figure 2.9

The covalent structure of a heptapeptide. All ionizable groups are shown in their ionized (physiological) form.

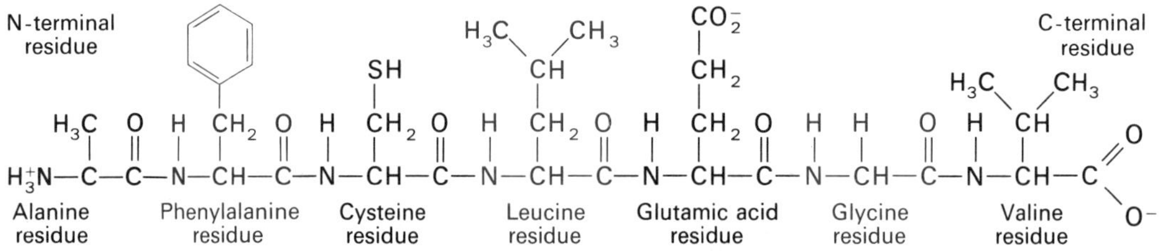

Table 2.9 Amino Acid Abbreviations.

Amino acid	Three-letter code	One-letter code
Alanine	Ala	A
Arginine	Arg	R
Asparagine	Asn	N
Aspartic acid	Asp	D
(Asn + Asp)*	Asx	B
Cysteine	Cys	C
Glutamine	Gln	Q
Glutamic acid	Glu	E
(Gln + Glu)*	Glx	Z
Glycine	Gly	G
Histidine	His	H
Isoleucine	Ile	I
Leucine	Leu	L
Lysine	Lys	K
Methionine	Met	M
Phenylalanine	Phe	F
Proline	Pro	P
Serine	Ser	S
Threonine	Thr	T
Tryptophan	Trp	W
Tyrosine	Tyr	Y
Valine	Val	V

* Both glutamine and asparagine easily hydrolyze to glutamic and aspartic acid, respectively. The published amino acid composition of a protein often states the sum of, for example, Glu and Gln.

proceeding from N-terminal to C-terminal left to right, as shown in the figure. Use of the three-letter symbols for amino acids makes it easier to represent the amino acid sequence of a protein; thus the heptapeptide shown in Figure 2.9 can be written

Ala-Phe-Cys-Leu-Glu-Gly-Val

End Group Analysis

A variety of chemical methods are available for determining the identity of the N-terminal and C-terminal amino acid residue, both of the entire protein chain and of each of the small peptide fragments produced by partial hydrolysis. Two N-terminal reagents were already discussed earlier in this chapter, FDNB and dansyl chloride. The application of either reagent is similar; Figure 2.10

shows the use of FDNB. The disadvantage of applying either FDNB or dansyl chloride is that for each application, the protein or peptide must be completely hydrolyzed to permit identification of the N-terminal residue.

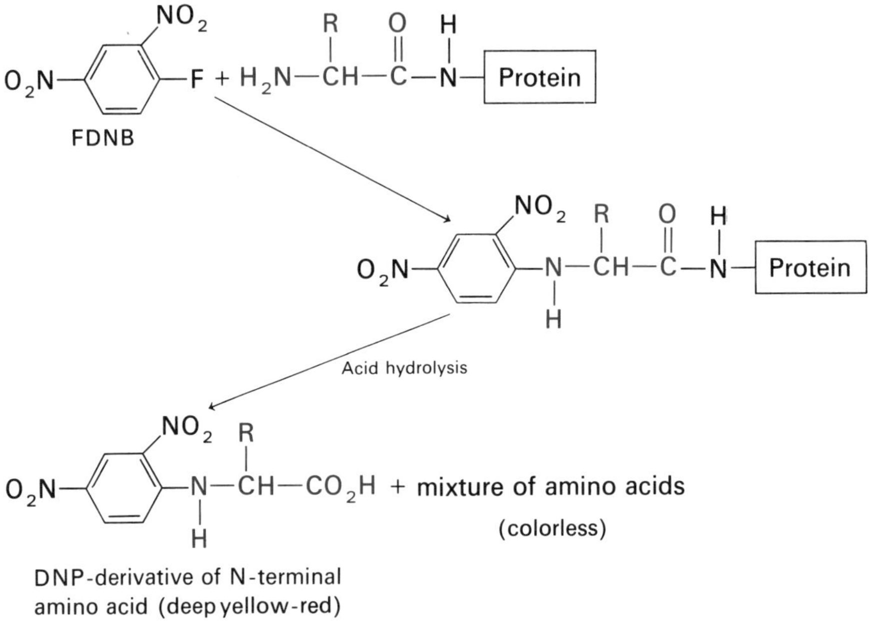

Figure 2.10

Use of FDNB to identify the N-terminal amino acid residue of a protein.

The *Edman reagent*, phenylisothiocyanate, has an advantage in that it lets us degrade the unreacted part of the protein chain step by step, one residue at a time. In theory, we can use this method to determine the entire sequence of a large peptide chain, but practically speaking it is not possible to sequence more than 40–50 residues in a given polypeptide using the Edman procedure. The procedure is shown in Figure 2.11. In the Edman method, the yields of the PTH-amino acid derivatives are nearly 100%, which is important in continuous sequencing. After each cycle, we separate and identify the N-terminal amino acid derivative by chromatography and repeat the procedure using the remainder of the protein or oligopeptide. The use of anhydrous trifluoroacetic acid permits us to cleave the derivatized N-terminal amino acid without the concurrent hydrolysis of the remaining polypeptide chain.

The best method for C-terminal analysis uses enzymes called **carboxypeptidases**. Carboxypeptidase A and B are pancreatic enzymes with differing residue specificity that catalyze the hydrolysis of the polypeptide chain from the C-terminal end. In general, the C-terminal residue is liberated most rapidly, followed by the next residue in from the C-terminal end of the chain, and so on. We can usually identify the last two or three amino acid residues from the C-terminal end of a protein chain by this method (see Problem **9** at the end of this chapter).

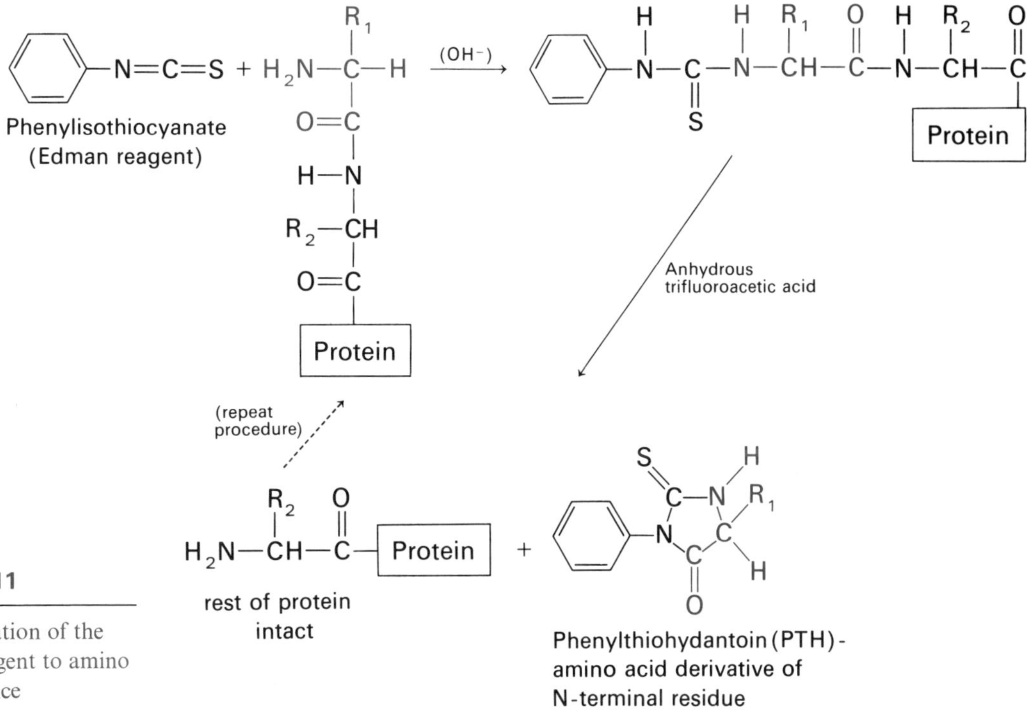

Figure 2.11

The application of the Edman reagent to amino acid sequence determination.   .

Partial Hydrolysis of Protein

Since we have methods for identifying the sequence of amino acids in small peptides (i.e., Edman method), it is important to be able to make fragments of large polypeptides in a specific or nonrandom manner. Partial acid hydrolysis of a protein is not generally suitable since it produces a random mixture of small peptide fragments.

There is a variety of enzymatic methods of cleavage that rely on the highly specific way in which a given enzyme reacts with a given compound or functional group (see Chapter 3). There are many protein-hydrolyzing enzymes (*proteases*) associated with digestive fluids of animals, as well as proteases from plants and bacteria. The specific action of all these proteases is dictated by the nature of the amino acid side chain groups associated with the peptide bond being hydrolyzed. The actions of some important protein-hydrolyzing enzymes are described in Table 2.10. Using each of these enzymes on a separate sample of a given protein provides us with unique mixtures of fragments. Parts of the amino acid sequences of these fragments overlap the amino acid sequences of other fragments produced by the other enzymes acting on the same protein. We can reconstruct the sequence of the entire protein chain by comparing the sequences of such peptide fragments. An example of this is shown in Figure 2.12, where we deduce the primary structure of a small polypeptide from its overlapping fragments.

Table 2.10 Enzymatic Methods for Specific Hydrolysis of Peptide Bonds in Proteins.

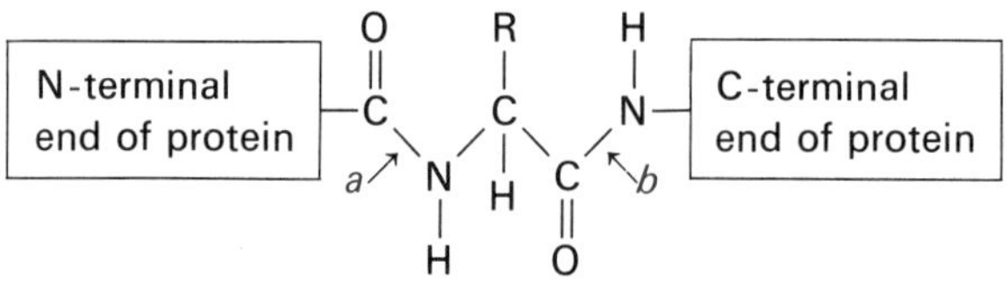

Enzyme	Cleaves at	R group preferred*	Source
Trypsin	b	Lys, Arg	Pancreas
α-Chymotrypsin	b	Tyr, Trp, Phe	Pancreas
Pepsin	a or b	Trp, Phe, Tyr, Met, Leu	Gastric mucosa
Papain	b	Arg, Lys, Gly	Papaya melon

* Some of these enzymes will cleave peptide bonds involving R groups other than those noted, but at a much slower rate.

Figure 2.12

Deduction of the sequence of a dodecapeptide using overlapping fragments whose sequences have been experimentally determined.

Fragments produced by treatment with trypsin	Gly-Phe-Met-Lys/Leu-Cys-Tyr-Lys/Ala-Arg/Ser-Leu
Fragments produced by treatment with chymotrypsin	Gly-Phe/Met-Lys-Leu-Cys-Tyr/Lys-Ala-Arg-Ser-Leu
Deduced polypeptide	Gly-Phe-Met-Lys-Leu-Cys-Tyr-Lys-Ala-Arg-Ser-Leu

2.11

Relationships between Primary Structure and Function

The sequence of amino acids gives a particular protein its unique identity and function. It is interesting to consider the question of how much the amino acid sequence of a protein can be changed before the function of that protein is lost or altered. In some cases large changes can occur without appreciably altering the function; in other cases, changing only one amino acid residue can effect a profound alteration in the properties of the protein.

Primary Structure and Species Variation

Let us first consider the primary structure of a protein isolated from several different species but exhibiting the same function in each. The protein *cytochrome c* is found in all animals, plants, and aerobic microorganisms. This protein functions as an electron carrier (see Chapter 9) and contains a heme prosthetic group in which the iron cycles between Fe^{3+} and Fe^{2+} as electrons are gained or lost. This is a fundamental energy-related process in all living things.

Since cytochrome c is found in a range of organisms from primitive to advanced, we consider it an ancient protein: one that existed from an early period in the evolution of living things. The amino acid sequence of cytochrome c has been determined for over 50 different organisms, ranging from yeast to humans. The differences in primary structure have been used as a basis for constructing family trees of living organisms which closely agree with those produced using classical morphological data. The use of protein primary structures has become a valuable tool for biologists studying the evolution of living organisms.

Cytochrome c in animals has 104 amino acids, while cytochrome c in lower forms of life has from 3 to 7 additional amino acids in the protein chain. Figure 2.13 shows the amino acid sequence of human heart cytochrome c. For all species studied so far, 27 of the amino acid residues in the primary structure of cytochrome c are absolutely invariant. These are the colored residues in the figure. Table 2.11 diagrams some of the differences between the types of cytochrome c isolated from various species.

```
     O   H                                              S—Heme—S
     ‖   |                                              |        |
CH₃C — N-Gly-Asp-Val-Glu-Lys-Gly-Lys-Lys-Ile-Phe-Ile-Met-Lys-Cys-Ser-Gln-Cys-His-Thr-Val-Gln-Lys-Gly-Gly-
                                  10                    14*       17        20
```

$$CH_3C$$

Lys-His-Lys-Lys-Thr-Gly-Pro-Asn-Leu-His-Gly-Leu-Phe-Gly-Arg-Lys-Thr-Gly-Glu-Ala-Pro-Gly-Tyr-Ser -Tyr-Thr-
                30   40

Ala-Ala-Asn-Lys-Asn-Lys-Gly-Ile-Ile-Trp-Gly-Glu-Asp-Thr-Leu-Met-Glu-Tyr-Leu-Glu-Asn-Pro-Lys-Lys-Tyr-Ile-
50   60   70

Pro-Gly-Thr-Lys-Met-Ile-Phe-Val-Gly-Ile-Lys-Lys-Lys-Glu-Glu-Arg-Ala-Asp-Leu-Ile -Ala-Tyr-Leu-Lys-Lys-Ala-
            80   90   100

Thr-Asn-GlyCOOH

---

* In one species studied so far this is an alanine residue.

Figure 2.13

The primary structure of human heart cytochrome c, showing the locations of the residues found to be the same for cytochrome c in all forms of life studied so far.

The difference between cytochrome c in yeast and in man, for example, is far less than is implied in Table 2.11, because in many cases the differences in primary structure involve substituting for an amino acid at a given position in the cytochrome c of yeast, an amino acid which has similar properties in the cytochrome c of man. This replacement of one amino acid residue in a protein chain with a different one which has the same properties (polar, apolar, positively charged, or negatively charged) is termed *conservative replacement*. For

Table 2.11 Comparative Differences in Amino Acid Sequences at Positions in the Primary Structure of Cytochrome c Isolated from Various Species.

Species difference	Differences in primary structure
Man ↔ Monkey	1 position is different
Man ↔ Horse	12 positions are different
Man ↔ Rattlesnake	14 positions are different
Mammals ↔ Chicken	10–15 positions are different
Mammals ↔ Tuna	17–21 positions are different
Vertebrates ↔ Yeast	43–48 positions are different

example, in human cytochrome c there is an isoleucine residue (apolar) at position 57, counting in from the N-terminal end. In cytochrome c from cotton seed, we find a valine (apolar) residue at position 57. Both are apolar, and even though there is a difference, the apolar properties of position 57 are conserved.

Cytochrome c is generally considered to be an ancient protein in which extensive alteration of the primary structure has occurred even though certain crucial amino acid residues have been retained. However, this variability in primary structure is not true for all proteins. *Histones*, for instance, are strongly basic proteins that are intimately associated with the structure of DNA in the chromosomes of eucaryotic cells. Comparing the protein histone IV from calf thymus with that from pea seedlings, one discovers that only 2 out of a total of 102 amino acid residues are different and these 2 are conservative replacements. The fact that the plant and animal lines probably diverged about 2 billion years ago means that every residue in this protein is important to its function. Any changes caused by mutation would impair the function of histone IV and would thus be eliminated through natural selection. This also suggests that histone IV is a protein that completed its molecular evolution many millions of years ago.

Primary Structure and Genetic Defects

The process of mutation (see Chapter 13), expressed in many instances as the replacement of an amino acid residue at a given position in a protein, is a continuously occurring phenomenon. As described here, mutation can cause the replacement of an amino acid residue with one having similar properties (conservative replacement) or different properties (nonconservative replacement). In some cases such a replacement is beneficial and is, in fact, evolution at the molecular level. In other cases, a nonconservative replacement can have unfortunate consequences. The effects of mutation on the function of hemoglobin is a well-studied case in point.

Hemoglobin is an oxygen–transporting protein found in the blood of vertebrates, including humans. It is a tetramer consisting of 2 α chains and 2 β chains bound together to form the functional hemoglobin molecule (see Figure 2.24). An α chain of hemoglobin has 141 amino acid residues and a β chain has 146 residues. Mutations have commonly been observed in both of these chains. Over 150 mutant hemoglobins have been identified and it is reasonable to assume that many times that number exist. Many of the amino acid substitutions seen in human hemoglobins cause little or no alteration in the properties of the protein. Table 2.12 shows some amino acid substitutions in hemoglobin that do have harmful consequences.

Table 2.12   Some harmful amino acid replacements in human hemoglobin. All of the replacements shown are nonconservative.

Hemoglobin type	Location in chain	Normal residue	Mutant residue
Hb S	$\beta 6$	Glu	Val
Hb C	$\beta 6$	Glu	Lys
Hb I	$\alpha 16$	Lys	Glu
Hb M $_{Boston}$	$\alpha 58$	His	Tyr
Hb M $_{Saskatoon}$	$\beta 63$	His	Tyr

The first group of 3 hemoglobins in the table is characterized by the fact that low oxygen concentrations (venous blood) promote the crystallization of the defective hemoglobin into needle-like crystals. This changes the shape of the red blood cells from the normal biconcave disc shape to a sickle or crescent shape. These distorted cells are more fragile and tend to rupture in the capillaries, leading to a condition called *sickle cell anemia*.

In sickle-cell hemoglobin, represented as Hb S, the α-chains are unaffected. However, in the β-chain of normal hemoglobin there is a glutamic acid residue at position 6 (counting from the N-terminal end), while in the β-chain of Hb S a valine residue appears at position 6. This nonconservative change (from negative at pH 7 to hydrophobic) is sufficient to bring about a major change in the properties of the protein.

An individual who inherits sickle cell hemoglobin from both parents rarely reaches adulthood without medical care, while an individual inheriting sickle cell hemoglobin from only one parent is said to have sickle cell *trait* and is able to lead a relatively normal life. High incidences of sickle cell anemia are observed in populations living in locations where virulent forms of malaria are present. Such regions include parts of Africa, India and Malaysia. Because the malarial parasite cannot develop well in the blood of persons having sickle cell trait, these people have an evolutionary advantage over those not protected from malaria. This is a case where natural selection has chosen the lesser of two evils: sickle cell trait rather than malaria.

The second group of abnormal hemoglobins in Table 2.12 is characterized by the following mutation: an amino acid residue involved with the heme prosthetic group is replaced by an amino acid residue of different chemical properties. In normal hemoglobin, the heme iron is in the Fe^{2+} state required for its function as an oxygen carrier. Here, the replacement near the heme iron appears to promote the oxidation of the Fe^{2+} by O_2 to yield Fe^{3+}. Hemoglobin in the Fe^{3+} state is termed methemoglobin and is unable to transport oxygen. The genetic condition where either the α chains or the β chains are in the Fe^{3+} state is termed *methemoglobinemia*.

Primary Structure and Functional Differentiation

One other example of the relationship of primary structure to protein function is worth mentioning, since it illustrates an important point concerning the evolution of proteins. Just as species have evolved from simple cells to complex organisms, a concurrent evolution of protein structure and function at the molecular level has taken place to meet the increasing demands for specialization.

Consider the neurohypophyseal (posterior part of the pituitary gland) hormones *oxytocin* and *vasopressin*. Table 2.13 shows the primary structure and properties of each of these two oligopeptide hormones. Only two differences are readily apparent, at positions 3 and 8. Nevertheless, these differences are sufficient to bring about a large change in physiological response. Peptides such as vasopressin and oxytocin probably evolved from a *common ancestral protein*, as evidenced by the extensive **homology** of the amino acid sequences apparent from the similarity in their primary structures. Many sets of different proteins exhibit such sequence homology, causing biologists to hypothesize that they evolved by divergent paths from a common ancestral protein.

Table 2.13   Comparison of the Primary Structures and Functions of Bovine Oxytocin and Bovine Vasopressin.

	Oxytocin	*Vasopressin*
Effect on diuresis:	no effect	inhibits
Effect on blood pressure:	lowers	elevates
Effect on lactation:	stimulates ejection of milk	not much effect
Effect on uterine contractions:	stimulates*	stimulates

Primary structures

```
               ┌──────S—S──────┐
Oxytocin:  Cys-Tyr-Ile-Gln-Asn-Cys-Pro-Leu-Gly-NH₂**

               ┌──────S—S──────┐
Vasopressin: Cys-Tyr-Phe-Gln-Asn-Cys-Pro-Arg-Gly-NH₂**
```

* This hormone is often administered to women in labor during childbirth to stimulate contractions of the uterus.

** The notation Gly-NH_2 means that the C-terminal carboxyl group is present as the *amide* derivative in the native oligopeptide.

2.12

Secondary Structural Features of Proteins

As a result of investigations in the 1930s and 1940s by Linus Pauling, R. B. Corey, and others, Pauling and Corey predicted the occurrence of the α-helix as one type of regular structure in proteins. This prediction has since been supported by direct experimental evidence. The α-helix, along with similar helical structures, forms a major structural part of fibrous proteins such as α-keratin in hair, as well as of polymers in some of the amino acids such as poly-L-alanine. It is also a feature of the secondary structure of globular proteins where portions of the protein chain exist in a helical conformation. The structural features of the α-helix are shown in Figure 2.14.

Some amino acids can cause breaks or bends in the regular α-helix structure. For example, proline and hydroxyproline (in collagen) cannot participate in the regular geometry of the α-helix due to the fixed geometry of the cyclic proline molecule. In globular proteins, the presence of proline usually causes the polypeptide chain to bend or fold. Disruption of the regular helix

Figure 2.14

Folding of the extended polypeptide chain into a regular α-helix. Shown here is the formation of a right-handed α-helix from a linear polypeptide chain.

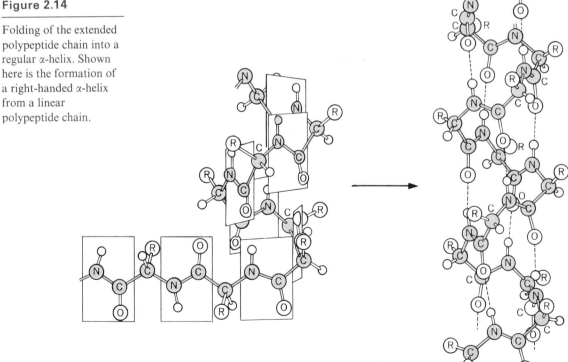

Figure 2.14 (cont.)

A right-handed α-helix showing some of the dimensions of the α-helix structure. (After C. B. Anfinsen, *The Molecular Basis of Evolution*, John Wiley and Sons, New York, 1959.)

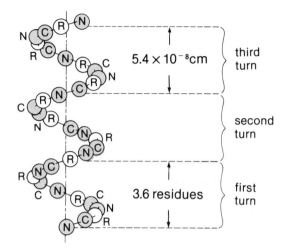

5.4×10^{-8} cm } third turn

} second turn

3.6 residues } first turn

Major features of the α-helix

1.   The helical structure is stabilized by intra-chain hydrogen bonds between the carbonyl oxygens and the amide hydrogens. In the α-helix each $\diagdown$C═O is hydrogen-bonded to the fourth $\diagdown$N—H down the chain. The α-helix offers nearly ideal geometry for this type of hydrogen bonding.

2.   There are 3.6 amino acid residues per turn of the helix. Each turn extends the coiled chain by 5.4×10^{-8} cm.

3.   The R groups all extend out from the cylindrical surface of the α-helix in the same direction relative to the surface. This minimizes steric interactions between side chain groups.

4.   The planar character of the peptide group is maintained.

5.   A right-handed α-helix is more stable for L-amino acids. Right-handed means that if you turn the helix clockwise, it screws downward, and is thus analogous to the right-handed threads in common wood and machine screws. Mixed D- and L-amino acids do not produce a stable α-helix.

structure can also occur when there are several adjacent amino acid residues with the same charge. In this case, disruption of the structure occurs through electrostatic repulsions. Figure 2.15 illustrates the transition from an α-helix structure in poly-L-lysine at pH 12 to a *random coil* structure, formed when the amino groups of the lysine residues gain protons after the polymer is placed in a pH 7 medium.

In addition to being an element of the secondary structure of globular proteins, the α-helix is the characteristic feature of such fibrous proteins as the

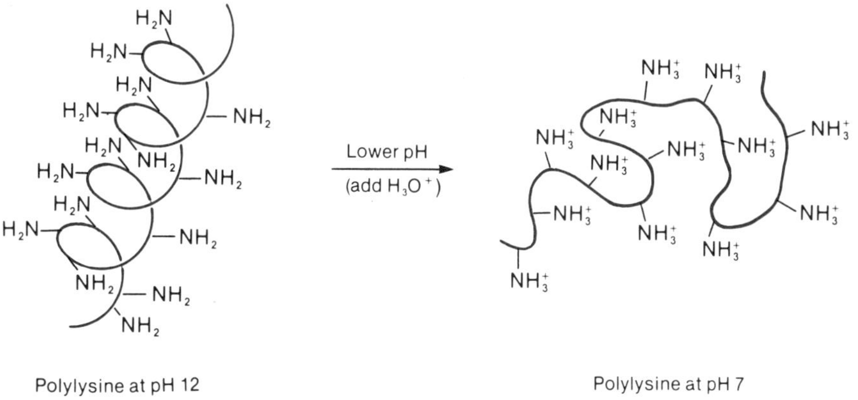

Figure 2.15

The transition from α-helix to random coil in polylysine. At pH 12 all —NH₂ groups in the molecule are uncharged, giving it an α-helix structure. At pH 7 the —NH₂ groups have each gained a proton and become positively charged; repulsion among these groups twists the molecule into a random (unstructured) coil.

Polylysine at pH 12

Lower pH
(add H₃O⁺)

Polylysine at pH 7

keratins of hair, wool, and fingernails. In the case of hair, three α-keratin chains intertwine as shown in Figure 2.16, forming a *protofibril*. Nine protofibrils surround two protofibrils ("9 + 2") to form a *microfibril*. Hundreds of microfibrils cohere to form a *macrofibril*. Each hair strand is made up of dead cells cemented together; within each cell, many macrofibrils are packed together in a direction parallel to the axis of the hair strand. The α-helices are bound together by disulfide cross-links between cysteine residues on neighboring keratin chains. Hard α-keratins such as horn or fingernails have many disulfide cross-links; the α-keratins of skin, on the other hand, are low in sulfur content and are thus flexible and elastic.

Figure 2.16

The supracoiling of three α-helices in hair and wool forms a protofibril.

It is interesting to note that the "9 + 2" arrangement of protofibrils in hair is found in other fibrous proteins in nature. One such case occurs in bacterial flagella. Another instance is in sperm flagella, shown in the electron micrograph in Figure 2.17.

In addition to the α-helix, Pauling and Corey proposed another regular structure called the *β-pleated sheet structure*, shown in Figure 2.18. Both the α-helix and the β-pleated sheet structures easily accommodate the planar peptide group. However, while the α-helix conformation is stabilized by *intra*-chain hydrogen bonds, the β-pleated sheet conformation is stabilized by *inter*-chain hydrogen bonding between the ＼N—H and ＼C=O groups on adjacent chains. In the pleated sheet structure, the side chain groups in a given polypeptide chain extend alternatively up and down with respect to the plane of the sheet and perpendicular to it. Adjacent polypeptide chains in a β-pleated sheet structure may run in a parallel direction (from N-terminal to C-terminal), or in an antiparallel direction as shown in Figure 2.18.

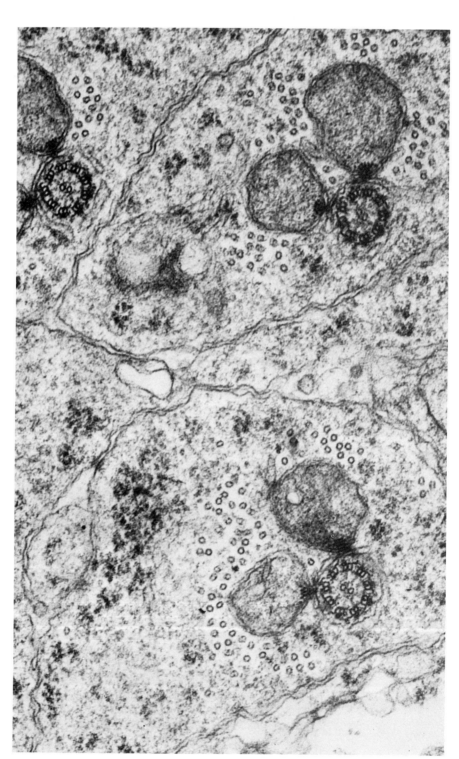

Figure 2.17

Electron micrograph of a
section through
developing spermatids in
the testes of the fruit fly,
Drosophila Paulistorium,
shown 20,000 ×. Note
the 9 + 2 arrangement
of microtubular
structures that form the
flagellum. The large
bodies adjacent to the
flagella are mitochondria.
(Electron micrograph
courtesy of Dr. R. Peter
Kernaghan.)

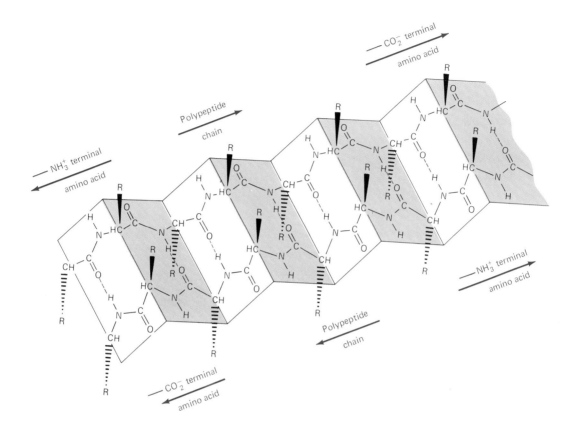

Figure 2.18

β-Pleated sheet
conformation showing
two polypeptide chains
running antiparallel.
Hydrogen bonding
between chains is
indicated by dotted lines.
(From W. H. Brown and
J. McClarin, *Introduction
to Organic and
Biochemistry*, 3d ed.,
Willard Grant Press,
1981.)

One protein which exhibits extensive pleated sheet structure is silk. The protein chain of silk predominantly contains repeating hexapeptide units consisting of

-(Gly-Ser-Gly-Ala-Gly-Ala)-

Successive β-pleated sheets in silk are arranged as shown in Figure 2.19. The layers of sheets in the silk structure are held together by relatively weak noncovalent forces involving the side chain groups. This gives the silk fiber great flexibility, while the pleated sheet nature of the covalent backbone (no helical structure) gives the fiber great resistance to stretching.

In the enfolding of the polypeptide chain of globular proteins, the β-pleated sheet structure may form part of the secondary structure of the protein. In this case, short segments of parallel or antiparallel parts of the polypeptide chain can hydrogen bond together in a manner analogous to the more extensive β-pleated sheet structures discussed here.

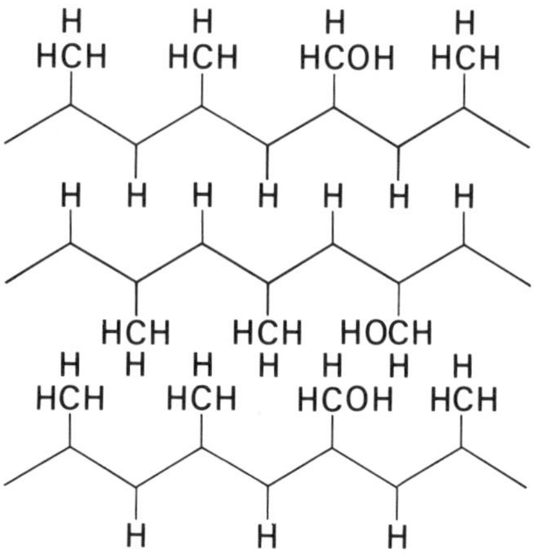

Figure 2.19

A schematic representation of three parallel polypeptide chains in the β-pleated sheet conformation of silk. This is a side-on view of the type of β-pleated sheet structure shown in Figure 2.18.

A Case in Point: Collagen, a Structural Protein

Over 30% of the protein material in the human body consists of the structural protein collagen. It comprises more than half of the total material in teeth, bone, tendons, skin, cartilage and the cornea of the eye. Collagen is a remarkable material in that it has a high tensile strength, an unusual structure, and it contains hydroxylysine and hydroxyproline, amino acids that occur in few other proteins. One common collagen-derived substance is *gelatin*. When collagen is boiled, its structure is permanently destroyed and gelatin results. Because of the large amount of hydrophilic ("water-loving") side chains present in gelatin, it forms gels in aqueous solution.

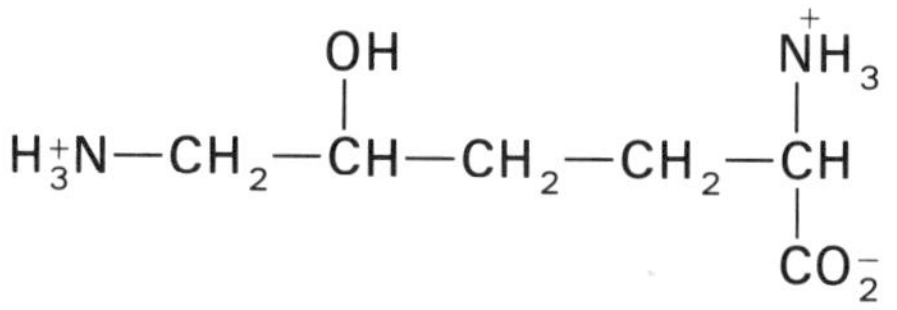

Hydroxylysine 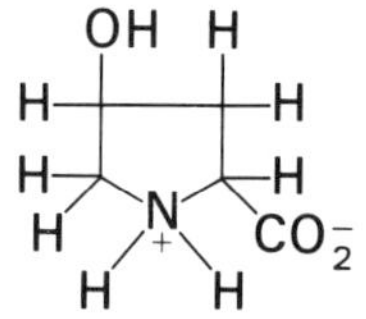   Hydroxyproline

The structure of collagen is made up of three levels:

1. The covalent backbone consists of individual protein chains with a molecular weight of approximately 100,000 each. The most abundant amino acid residue is glycine, accounting for 33% of the total amino acid residues present. Proline is also abundant (12%), and also present are the unusual amino acids, hydroxyproline and hydroxylysine.

2. Three chains combine to form a triple helix in the secondary structure. This triple helix is the basic structural unit of collagen and is called *tropocollagen*. It is a rod 15 Å in diameter and 3000 Å long. In the tropocollagen helix, the three strands are hydrogen bonded to each other via

the peptide —NH groups of the glycine residues and the peptide —C=O groups on other chains. This is a distinctly different helical structure from the α-helix.

3. The tropocollagen units are covalently linked to form a bundle or sheaf called a collagen *microfibril*. The collagen microfibril may be formed into parallel bundles in the case of tendon formation, or in sheets, like fibers forming paper, in the case of skin.

Figure 2.20 shows the relationship between tropocollagen, the collagen microfibril, and some of the structures containing this remarkable protein. The covalent cross-links between the tropocollagen units involve the side chains of the lysine residues. The degree of cross-linking determines the rigidity of the

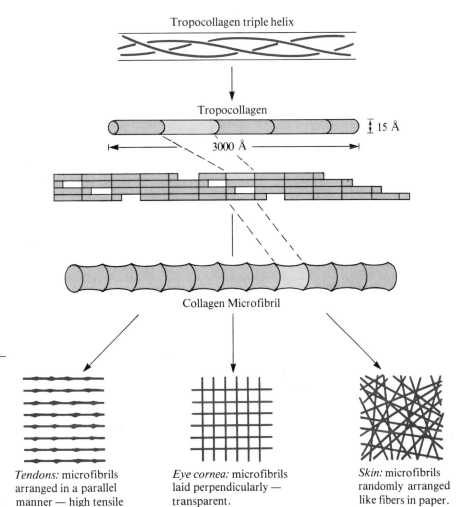

Figure 2.20

The assembly of collagen showing the collagen triple helix combined to form tropocollagen units, which are in turn grouped to create collagen microfibrils. Collagen microfibrils form the structural basis for many tissues.

Tendons: microfibrils arranged in a parallel manner — high tensile strength.

Eye cornea: microfibrils laid perpendicularly — transparent.

Skin: microfibrils randomly arranged like fibers in paper.

collagen structure. Interference with the cross-linking process produces a condition known as *lathyrism* which causes collagen-containing structures to weaken and become easily deformed. The most obvious deformation is seen in abnormal curvature of the spine and other load-bearing bone structures. The cross-linking process appears to be continuous, and it is thought that the stiffening of skin and other tissues associated with the process of aging is due to an increasing amount of cross-linking in these collagen-containing tissues. The hydroxylysine residues of tropocollagen appear to play a role in the formation of collagen fibers. While it is presently unclear just what this role is, it is known that vitamin C (ascorbic acid) is required for the formation of hydroxylysine; the vitamin C deficiency disease, *scurvy*, is characterized by a breakdown of blood vessels and skin structure due to improperly formed collagen.

2.13

Tertiary Structure of Proteins

Before 1957, no three-dimensional structures of proteins were known. In the period 1961–63, J. C. Kendrew reported the structure of the oxygen storage protein myoglobin and M. F. Perutz reported the three-dimensional structure of hemoglobin. For this work Kendrew and Perutz received the Nobel Prize in Chemistry in 1962. Currently there are over 50 proteins whose three-dimensional structures have been determined by X-ray crystallography.

Figure 2.21 shows the three-dimensional structure of myoglobin together with that of the β-chain of hemoglobin, for purposes of comparison. Note that there is considerable structural homology between myoglobin and the individual chains of hemoglobin. This is another example of proteins that have evolved from a common ancestral protein. The myoglobin structure is unusual in that it has a very high (77%) α-helix content and no cysteine residues. Compare this with the structure of the enzyme α-chymotrypsin, shown in Figure 3.17 in Section 3.8. In myoglobin we find eight segments of α-helix forming the sides of a box which encloses the heme prosthetic group.

In the three-dimensional structure of globular proteins, nearly all polar R groups are on the surface, in contact with the aqueous (polar) environment. The interior of the protein is predominantly composed of hydrophobic R groups. This relative location of polar and hydrophobic R groups is found in most globular proteins. Hydrophobic amino acid residues tend to favor the interior of the protein structure where the interaction with solvent (H_2O) is highly restricted. *Hydrophobic interactions* between hydrophobic R groups in the interior of a given protein are a major factor in the stabilization of the three-dimensional structure of the protein in an aqueous environment. This effect is related to the tendency of oil to coalesce and form a separate phase when mixed with water. In addition to hydrophobic interactions, other factors

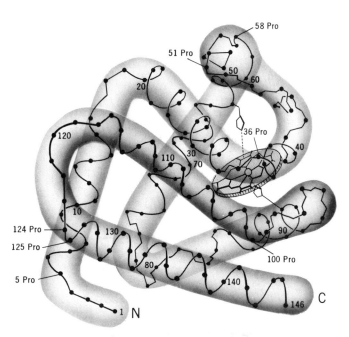

β-chain of hemoglobin

Figure 2.21

Three-dimensional structure of the β-chain of hemoglobin and of myoglobin. The heme group in each is shown as a disc-shaped unit. Note that many of the folds in the protein chain occur at proline (Pro) residues. (Figure from M. F. Perutz, "The Hemoglobin Molecule," copyright © 1964 by Scientific American, Inc. All rights reserved.)

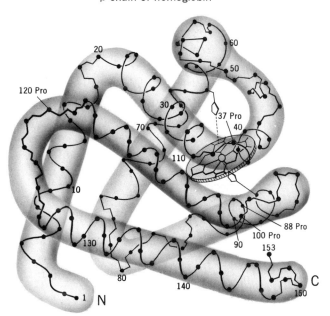

Myoglobin molecule

Figure 2.22

Forces of interaction
which help stabilize the
secondary and tertiary
structure of proteins:
(a) ionic interactions;
(b) hydrogen bond
    interactions;
(c) hydrophobic
    interactions;
(d) covalent disulfide
    links.

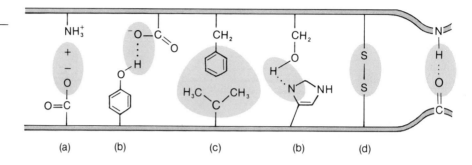

contribute to the stability of the tertiary structure of a protein. These supple-
mentary factors are summarized in Figure 2.22.

It is important to recognize that the structure of a protein is not maintained
by any single factor, but rather by a combination of factors. The disruption of
any given set of stabilizing interactions usually brings about a major alteration
in the tertiary structure of the protein. We can distinguish between two states
of a protein: the **native** state and the **denatured** state. A native protein exists
within a three-dimensional structure that enables it to perform its biological
function. If we create a reversible or irreversible structural change in a protein
which causes the protein to lose all or part of its biological function, we say the
protein has been denatured. There are several ways to denature proteins,
including:

1. heat treatment;
2. large pH changes;
3. chemical treatment;
4. mechanical treatment.

Frying an egg causes the ovalbumin to denature. Whipping egg whites
also brings about denaturation of the ovalbumin. Adding lemon juice to milk
denatures and precipitates the milk protein casein. Adding a detergent to most
proteins in solution causes a disruption of the hydrophobic interactions in the
interior of the protein molecule, and thus causes a change in the overall protein
structure.

Multi-Chain Proteins

Many proteins whose tertiary structures are composed of more than one
polypeptide chain are formed from single-chain precursors. Generally such a
single-chain precursor has little or no previous biological activity, and is
activated by the cleavage of one or more peptide links in the polypeptide chain.
This results in a multi-chain protein with two or more shorter chains in its tertiary
structure. The protein hormone, *insulin*, is a good example of this type.

Insulin acts to stimulate the cellular intake of glucose and fatty acids from
the blood. Insulin deficiency leads to a condition called *diabetes mellitus* which

Figure 2.23

The relationship
between insulin and
proinsulin from beef
pancreas. The enzyme
catalyzed cleavage of the
inactive single-chain
precursor, proinsulin,
results in the formation
of the active two-chain
insulin molecule.

must be controlled by daily injections of bovine insulin. Insulin is a two-chain
protein produced in the pancreas from the single-chain protein *proinsulin*.
Figure 2.23 shows the relationship between these two proteins. The inactive
precursor, proinsulin, contains 84 amino acid residues. A chain 33 amino acid
residues long is removed from the proinsulin chain between residues 30 and 64
and discarded. The two short chains thus produced are already joined by
disulfide links in the original tertiary structure of proinsulin. These two chains
constitute the active insulin molecule.

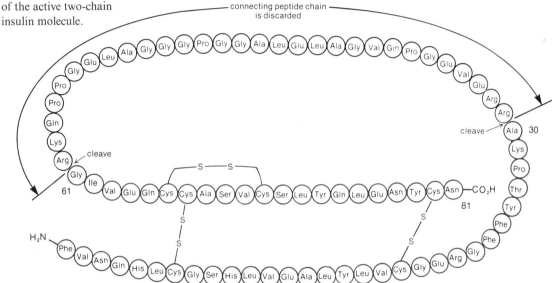

Proinsulin (inactive); one chain

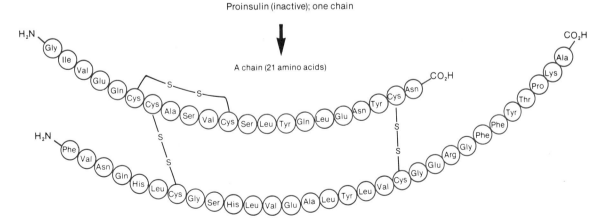

A chain (21 amino acids)

B chain (30 amino acids)

Insulin (active); two chains

At first glance this method of producing active proteins may appear to be an inefficient use of energy and materials. This is not the case, however, because the formation of multi–chain proteins in this manner ensures that the correct tertiary structure is created in the final product. The probability that a correct structure would result from the linking of two separately-formed protein chains is small. For example, if insulin chains are separated using the chemistry discussed earlier in this chapter for the cleavage of disulfide links (p. 42), and then recombined, only a small fraction of the chains recombine in such a way that biologically-active insulin is formed. Many other proteins also appear to be formed from inactive precursors, as we shall see in the case of the enzymes trypsin and chymotrypsin in Chapter 3.

2.14

Quaternary Structure of Proteins

The quaternary structure of an oligomeric protein is maintained by the same kinds of non-covalent forces that stabilize the tertiary structure of any protein. Many proteins have been shown to be oligomeric, but interestingly few, if any, contain an odd number of protomers. Table 2.14 lists a few of the many known examples.

Table 2.14 Some Oligomeric Proteins and Their Characteristics.

Protein	Molecular weight	Number of protomers	Function
Hemoglobin	64,500	4	Oxygen transport (blood).
α-Amylase	97,000	2	Starch hydrolysis (saliva).
Lactate dehydrogenase	150,000	4	Enzyme of anaerobic glycolysis (Chapter 8).
Catalase	232,000	4	Enzyme catalyzes the breakdown of hydrogen peroxide.
Tobacco mosaic virus (nucleoprotein)	40,000,000	2130	Plant virus. The protomers are arranged in an ordered fashion around a central nucleic acid molecule.

Oligomeric proteins may consist of identical subunits; of non-identical subunits arranged in fixed amounts in a prescribed manner as in hemoglobin; or of non-identical subunits arranged in a prescribed manner but in varying amounts. Let's consider an example of each of these latter two types of oligomeric proteins.

Hemoglobin: Invariant Quaternary Structure

Hemoglobin is the oxygen-carrying protein in red blood cells. Each liter of blood contains about 150 grams of hemoglobin. Normal adult hemoglobin consists of a tetramer composed of 2 α chains and 2 β chains arranged in a tetrahedral manner. Figure 2.24 shows how the protomers of hemoglobin are arranged to form the quaternary structure of the protein. The contact faces in the quaternary structure of hemoglobin are bound by charged group interaction, hydrogen bonding and hydrophobic interactions between hydrophobic amino acid residues.

The quaternary structure of hemoglobin is closely connected with its unusual properties as an oxygen carrier. Figure 2.25 shows the relationship

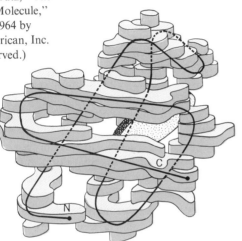

Hemoglobin α-chain

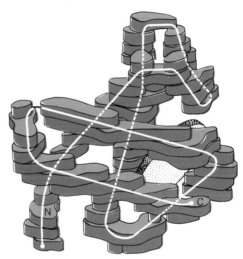

Hemoglobin β-chain

Top view

Side view

Figure 2.25

Oxygen saturation curves for hemoglobin and myoglobin under physiological conditions. The horizontal axis shows the concentration of oxygen the protein is in equilibrium with, in terms of partial pressure, and the vertical axis shows the relative amount of protein-bound oxygen for a given pO_2. In the physiological range of oxygen pressures, hemoglobin can exchange much more oxygen than myoglobin.

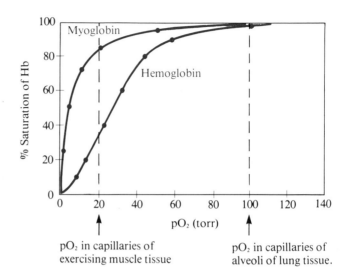

pO₂ in capillaries of exercising muscle tissue

pO₂ in capillaries of alveoli of lung tissue.

between oxygen pressure (concentration) in blood and oxygen binding for both hemoglobin and the oxygen-storage protein myoglobin. The pressure of oxygen in red blood cells leaving the lungs is about 100 torr while the pressure in the capillaries in active muscle tissue is only about 20 torr, because muscle cells are consuming oxygen. The difference in saturation percentage on going from $pO_2 = 100$ torr to $pO_2 = 20$ torr for the myoglobin curve in Figure 2.25 is smaller than the same change for the hemoglobin curve. This means that myoglobin is capable of gaining and releasing much less oxygen than hemoglobin. The binding curve for hemoglobin indicates that it is able to "load" oxygen efficiently at higher oxygen concentrations and "unload" oxygen efficiently at lower oxygen concentrations. The major structural difference between hemoglobin and myoglobin is that myoglobin is a monomer and hemoglobin is a tetramer (see Figure 2.21).

The *sigmoid* shape of the O_2 binding curve for hemoglobin is indicative of *cooperative binding*, which means that the binding of an oxygen molecule by each unit in the quaternary structure of hemoglobin makes it progressively easier for the remaining units to bind oxygen. It is known that the tertiary structure of each hemoglobin protomer undergoes a change when oxygen is bound. This change is transmitted through the quaternary structure to the other protomeric units and makes it easier for them to bind oxygen. Figure 2.26 shows one model for this cooperative effect.

Figure 2.26

A model for the stepwise binding of oxygen to hemoglobin. Binding of one molecule in one subunit of the quaternary structure increases the binding ability of an adjacent subunit.

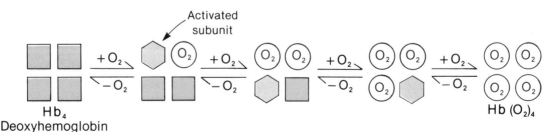

Hb₄
Deoxyhemoglobin

Hb (O₂)₄

There is considerable evidence that oxygen binding in hemoglobin is accompanied by an alteration in quaternary structure. It was observed 40 years ago that crystals of deoxyhemoglobin disintegrated when they were allowed to interact with oxygen. More recently, X-ray crystal structure comparison of oxyhemoglobin and deoxyhemoglobin has shown considerable structural differences. The origin of these changes in quaternary and tertiary structure lies in the properties of the heme iron in oxy- and deoxyhemoglobin. In hemoglobin, the Fe^{2+} is coordinated with the porphyrin ring of the heme group (see Figure 3.5) and a histidyl residue of the protein. Figure 2.27 shows the changes that occur when oxygen is gained or lost by a hemoglobin protomer. When oxygen is bound by hemoglobin, the heme iron interacts with the oxygen. As a result, the heme iron's covalent radius decreases and it is pulled into the plane of the porphyrin ring. Even though the iron ion moves only 0.74 Å during the change from oxy- to deoxyhemoglobin, this shift results in a large alteration in the tertiary and quaternary structure of the hemoglobin molecule. Some of the atoms in the protein chain actually move as much as 5.7 Å as a result of this seemingly minor shift.

Other things that affect the quaternary structure of hemoglobin will cause a corresponding shift in the oxygen binding curve. One such substance is *2,3-diphosphoglycerate* or *2,3-DPG*. This physiologically important substance is present in normal red blood cells in roughly equimolar amounts with hemoglobin. In the presence of 2,3-DPG, the amount of oxygen lost by hemoglobin on going from alveolar (lungs) blood to muscle capillaries is greater than that

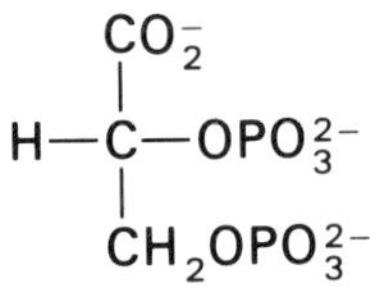

Figure 2.27

The origin of the conformational changes in hemoglobin upon binding of oxygen. Interaction with oxygen causes the Fe^{2+} to become smaller and be pulled into the plane of the porphyrin ring. The histidyl residue coordinated with the iron is pulled along with it, thus other parts of the protein molecule undergo conformational changes to a greater or lesser degree. This change is transmitted through the quaternary structure of hemoglobin to other protomers and enhances their ability to bind oxygen.

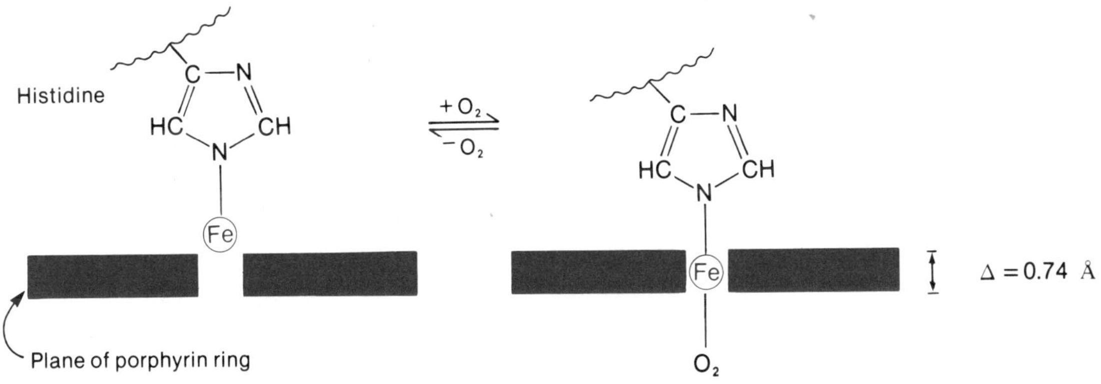

$$\begin{array}{c} CO_2^- \\ | \\ H-C-OPO_3^{2-} \\ | \\ CH_2OPO_3^{2-} \end{array}$$

2,3-diphosphoglycerate

lost by hemoglobin alone. This means that 2,3-DPG enables oxyhemoglobin to

Histidine

Plane of porphyrin ring

$\Delta = 0.74$ Å

"unload" oxygen more completely than unassisted hemoglobin can. It is now known that 2,3-DPG binds in the central cavity of deoxyhemoglobin (see Figure 2.24) but not in the central cavity of oxyhemoglobin. The reason for this is that the expansion of the tertiary structure of hemoglobin on binding oxygen renders this cavity too small for 2,3-DPG to fit. Binding of 2,3-DPG by deoxyhemoglobin shifts the equilibrium

$$HbO_2 \rightleftharpoons Hb + O_2 \qquad (11)$$

to the right by stabilizing one of the products.

The binding of 2,3-DPG plays a crucial role in the transport of oxygen across the placenta from maternal to fetal circulation. Fetal hemoglobin, called hemoglobin F, differs from adult hemoglobin in that the two β subunits found in adult hemoglobin are replaced by two different subunits designated as γ subunits. Hemoglobin F (HbF) binds oxygen more strongly than hemoglobin A (adult). Recently it has been discovered that the reason for the higher oxygen affinity of HbF is that it binds 2,3-DPG less strongly, and is thus less stabilized (see Eq. 11). Because the deoxy form of HbF is not as stable as adult hemoglobin, the oxygenated form of HbF is more favored. Figure 2.28 shows the effect the different 2,3-DPG binding affinities have on the oxygen-binding curves of fetal and adult hemoglobin under physiological conditions.

The change in the quaternary structure of hemoglobin on going from deoxyhemoglobin to oxyhemoglobin affects its acid-base properties as well. The alteration in quaternary structure causes the environments of certain ionizable groups of hemoglobin to change. This results in a net change in acid-base properties as shown in Eq. 12.

$$DeoxyHb(H^+)_n + O_2 \rightleftharpoons OxyHb \cdot (O_2) + nH^+ \qquad (12)$$

Figure 2.28

Oxygen-binding curves of fetal hemoglobin (HbF) and adult hemoglobin (HbA) showing the greater oxygen affinity of HbF *in the presence of 2,3-DPG*. The greater oxygen affinity of HbF causes oxygen to pass from maternal to fetal circulation across the placenta.

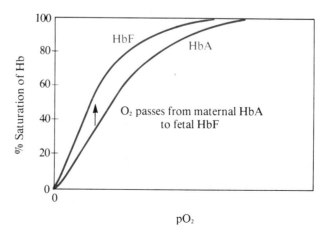

In actively-metabolizing tissues such as exercising muscle, considerable amounts of H^+ ions are produced. The increased amount of H^+ causes the equilibrium in Eq. 12 to shift to the left, making more O_2 available to the tissues. This change in oxygen affinity by hemoglobin as a function of $[H^+]$ is called the *Bohr Effect*.

Isozymes: Proteins with a Variable Quaternary Structure

The arrangement of non-identical subunits in varying amounts within a fixed structure gives rise to the existence of different forms of the same protein within the same organism. In the case of enzymes (catalytic proteins) the multiple forms are called **isozymes**. In many cases, as we shall see in Chapter 3, the properties of the isozyme forms need not be the same. The enzyme lactate dehydrogenase (LDH) illustrates isozymes very well. Lactate dehydrogenase catalyzes the reaction:

$$\underset{\text{Lactate anion}}{CH_3-\overset{\overset{\displaystyle OH}{|}}{CH}-CO_2^-} + \underset{\text{Oxidant}}{A} \rightleftharpoons \underset{\text{Pyruvate anion}}{CH_3-\overset{\overset{\displaystyle O}{||}}{C}-CO_2^-} + \underset{\substack{\text{Reduced}\\\text{oxidant}}}{AH_2}$$

LDH is a tetramer whose quaternary structure is shown in the remarkable electron micrograph of Figure 2.29. There are two distinct types of protomers: M (muscle) subunits and H (heart) subunits. You might predict that, given a tetramer with a choice of two subunits in each position, five different combinations of the two protomers could be constructed:

M_4	M_3H	M_2H_2	MH_3	H_4
M M	M M	M M	M Ⓗ	ⒽⒽ
M M	M Ⓗ	ⒽⒽ	ⒽⒽ	ⒽⒽ

This is, in fact, the case. M and H subunits differ in net charge so the 5 isozyme forms differ in overall net charge. This permits us to use electrophoresis to separate the 5 forms of LDH, as shown in Figure 2.30. In humans, the LDH isozymes are richer in M subunits in muscle and liver tissue and richer in H subunits in heart muscle. This fact is important in diagnosing diseases of the liver and heart and distinguishing one from another. In both cases, LDH is released from damaged cells and added to the LDH already present in the blood. Electrophoresis of blood serum LDH from a patient with hepatitis (liver disease) would show abnormally high levels of M_4 and M_3H forms. Electrophoresis of LDH from a patient with some form of heart muscle

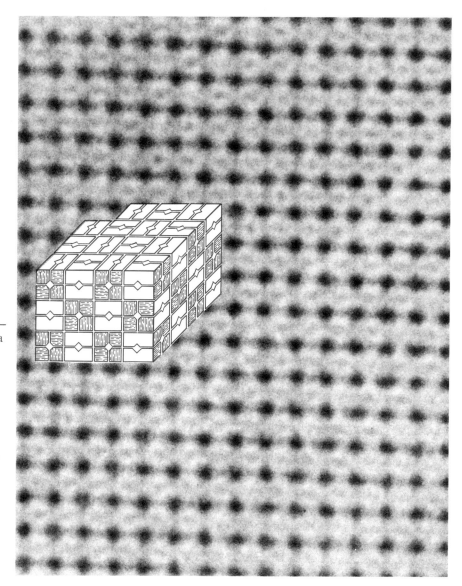

Figure 2.29

Electron micrograph of a thin section of a single crystal of lactic dehydrogenase, magnification 1,120,000. The superimposed drawing represents the molecular crystal structure of this enzyme, showing how the protomers are arranged in the tetrameric quaternary structure of lactic dehydrogenase. (Courtesy of Dr. L. W. Labaw, National Institutes of Health.)

degeneration would show an abnormal abundance of H_4 and H_3M isozymes. Serum protein concentrations are sensitive indicators of disease and are extremely useful in medical diagnosis. We will see more examples of this in Chapter 3.

The subject of protein chemistry is an exciting part of biochemistry. We have mentioned a few areas of current interest in our discussion of protein structure. For example, the determination of the primary structure of the same protein taken from different species is becoming a taxonomic basis for inter-

Figure 2.30

Electrophoresis of the lactate dehydrogenase isozymes. If a purified preparation is used as shown below, a reagent such as ninhydrin can be used to render the separated spots visible. The intensity of the spots depends on the amount of protein present. When a sample of blood serum is analyzed for lactate dehydrogenase isozymes, a selective staining procedure is used to visualize the LDH isozymes in the presence of many other serum proteins.

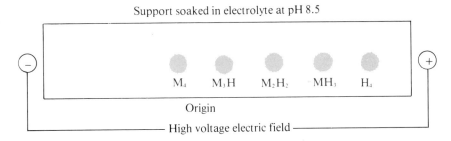

Support soaked in electrolyte at pH 8.5

M_4 M_3H M_2H_2 MH_3 H_4

Origin

High voltage electric field

relating species of organisms. The idea of "biochemical individuality" is becoming more apparent as the detailed properties are determined of such proteins as hemoglobin, taken from different individuals. Establishing homologies between different proteins having the same or similar functions, in terms of both amino acid sequence and overall structure, has contributed considerably to our understanding of evolutionary processes at the molecular level. The study of the catalytic proteins, enzymes, as we will see in Chapter 3, has already yielded much information on the molecular affairs of living organisms.

Suggestions for Further Reading

Dickerson, R. E. and Geis, I. *The Structure and Action of Proteins.* Menlo Park, Cal.: W. A. Benjamin, Inc., 1969. An eminent scientist and a leading scientific illustrator have collaborated to produce an overview of protein molecular biology and biochemistry which is unusually well done.

Kopple, K. D. *Peptides and Amino Acids.* Menlo Park, Cal.: W. A. Benjamin, Inc., 1966. This monograph goes into the details of amino acid and peptide chemistry in a readable and useful manner at the undergraduate level.

Light, A. *Proteins: Structure and Function.* Englewood Cliffs, N. J.: Prentice-Hall, Inc., 1974. This well written and beautifully clear survey of protein chemistry is the book to consult as a first step beyond the level of this chapter.

Stryer, L. *Biochemistry.* San Francisco: W. H. Freeman & Co., 1975. This comprehensive text has an excellent detailed discussion of hemoglobin and collagen.

Scientific American

Allison, A. C. "Sickle Cells and Evolution." August 1956.

Dayhoff, M. O. "Computer Analysis of Protein Evolution." July 1969.

Dickerson, R. E. "The Structure and History of an Ancient Protein." April 1972.

Doty, Paul. "Proteins." September 1957.

Edelman, G. M. "Structure and Action of Antibodies." August 1970.

Kendrew, J. C. "The Three-Dimensional Structure of a Protein Molecule." December 1961.

Perutz, M. F. "The Hemoglobin Molecule." November 1964.

————. "Hemoglobin Structure and Respiratory Transport." December 1978.

Zuckerkandl, E. "The Evolution of Hemoglobin." May 1965.

Problems

1. For each of the terms below, provide a brief definition or explanation.

   Acid   Peptide bond
   Amino acid   Primary structure
   Base   Protein
   Buffer   Quaternary structure
   Globular protein   Secondary structure
   α-Helix   Tertiary structure
   Informational macromolecule   Zwitterion

2. Complete each of the reactions given below, showing the structures of reactants and products.

   (a) Ninhydrin + alanine =   (c) Dansyl chloride + glycine =
   (b) FDNB + lysine =   (d) Phenyl isothiocyanate + histidine =

3. Refer to Figure 2.5 and the discussion that follows this figure.

   (a) What is $[H_3O^+]$ in a liter of solution containing 0.1 mole of acetic acid and 0.1 mole of acetate anion?

   (b) Sufficient solid NaOH is added to the solution in part (a) above to add 0.05 moles of OH^- to the solution. Assuming the volume of the solution does not change, what are the new concentrations of acetic acid, acetate anion and H_3O^+?

   (c) What is the pH of the solution in part (a) of this question; part (b)? Compare this pH change with what you would observe if you added 0.05 moles of NaOH to a solution of pure H_2O.

4. Sketch the titration curve of serine, given $pK_{\alpha\text{-}NH_3^+} = 9.2$ and $pK_{COOH} = 2.4$. In each pH range, show the principal species in equilibrium.

5. Table 2.13 presents the primary structures of the peptide hormones oxytocin and vasopressin.

   (a) Draw the structures of the ionizable groups in each peptide.
   (b) Predict which peptide has the higher isoionic point.

6. Write out reactions for the following experiments, and draw structures where appropriate.

   (a) How would you quantitatively determine the presence of a primary amine functional group?
   (b) How would you shorten a tripeptide chain by one unit only?

7. Given the polypeptide

$$\text{Lys-Ala-Lys-Asp-Leu-Cys-Pro-Tyr-Val-Glu-His}$$
$$1234567891011$$

   give the number(s) of the amino acid residue(s) which:

   (a) Have nonpolar side chains.
   (b) Have negatively charged side chains at pH 7.

(c) Can form covalent cross-links with similar groups in other parts of a polypeptide chain.

(d) Would react with FDNB.

(e) Have polar but uncharged side chains at pH 7.

8. D. P. Botes recently reported the primary structure of one of the toxic principles from the venom of the Cape Cobra (*Naja nivea*) [*J. Biol. Chem.*, **246**, 7383 (1971)]. Residues 1–33 are given below:

(N-terminal) Ile-Arg-Cys-Phe-Ile-Thr-Pro-Asp-Val-Thr-Ser-Glu-Ala-Cys-Pro-
              1   2   3   4   5   6   7   8   9 10 11 12 13 14 15

Asp-Gly-His-Val-Cys-Tyr-Thr-Lys-Met-Trp-Cys-Asp-Asn-Phe-Cys-Gly-Met-Arg
 16   17 18 19 20 21 22 23 24 25 26 27 28 29 30 31 32 33

Referring to this amino acid sequence, give the number(s) of the residues involved in the following situations:

(a) The residues involved in peptide links cleaved by α-chymotrypsin.

(b) The residues involved in peptide links cleaved by trypsin.

(c) The residues which react with phenylisothiocyanate (Edman reagent).

(d) The residues where a fold in the peptide chain might occur.

9. Kamen *et al.* have reported the primary structure of a 27-residue diheme peptide isolated from the photoanaerobe *Chromatium* [*J. Biol. Chem.*, **273**, 3083 (1962)]. Treatment of the peptide with pepsin under the conditions used by these workers produced three large peptide fragments, A, B, and C. The data below was obtained from studying the action of carboxypeptidase on fragments A and B.

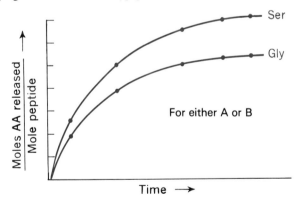

What does the data indicate concerning the amino acid (AA) sequence at the C-terminal end of fragment A or B?

10. A heptapeptide has the following amino acid composition:

Val(1), Lys(1), Leu(2), Phe(1), Gly(2)

Reaction of the above heptapeptide with FDNB and subsequent hydrolysis in 6*M* HCl gave the DNP derivative of valine plus six other amino acids. Hydrolysis of another sample of the heptapeptide with trypsin yielded two fragments: a dipeptide containing lysine and valine and a pentapeptide fragment which gave the DNP derivative of glycine on treatment with FDNB and subsequent hydrolysis. Hydrolysis with chymotrypsin gave two fragments, one of which was a dipeptide. Treatment of the original heptapeptide with carboxypeptidase liberated glycine most rapidly. Write out the amino acid sequence of this peptide.

11. Given below are the sequences of some of the fragments used by Sanger to deduce the amino acid sequence of insulin. Using these, deduce the proper sequence for the octapeptide of which they are a part. (None of the residues given appears more than once in the octapeptide.)

Phe-Val	Cys-Gly	His-Leu
Glu-His	Val-Asp	Asp-Glu
Leu-Cys	Phe-Val-Asp	Glu-His-Leu

12. Write out the structures for three of the possible tripeptides containing Tyrosine, Lysine, and Cysteine (one of each). Using short (3-letter) notation, give the sequences of the remaining possible tripeptide.

13. Refer to Figure 2.25 on page 68 which shows the oxygen saturation curves for hemoglobin and myoglobin.

(a) Reading from the graph as best as you can, determine how much oxygen in terms of percentage of saturation of hemoglobin is carried from the lungs ($pO_2 = 100$ torr) to active muscle tissue ($pO_2 = 20$ torr).

(b) Estimate the amount of O_2 carried by myoglobin under the conditions described in part (a).

(c) Assuming there are 150 grams of hemoglobin per liter of blood and the molecular weight of hemoglobin is 64,500 g/mole, how many moles of O_2 per liter of blood corresponds to 100% saturation of hemoglobin?

(d) Convert your answers in parts (a) and (b) to moles of O_2 per liter. In the case of myoglobin, assume a "blood" made up of 150 grams of myoglobin per liter (Mol. wt. = 16,900).

14. The table below shows some of the 150 or so known amino acid replacements in human hemoglobins.

Abnormal H_b (designation)	Normal residue and position in chain	Replacement
	In alpha chain	
(a) I	16, Lys	Asp
(b) M_{Boston}	58, His	Tyr
(c) $G_{Chinese}$	30, Glu	Gln
	In beta chain	
(d) S	6, Glu	Val
(e) Zurich	63, His	Arg
(f) Köln	98, Val	Met

Which of these correspond to nonconservative changes in amino acid side chain character? Which correspond to conservative changes? In each case *briefly* indicate the reason for your answer.

CHAPTER 3

Enzymes

One of the most dramatic functions of proteins is that of enzyme activity. Enzymes mediate, control, and catalyze the chemical activities of a living cell. Although the action of enzymes has been studied for over 100 years, the manner in which an enzyme protein molecule functions is still not completely understood.

Practically speaking, enzyme technology has existed for centuries, as in wine-making. In fact, much of our early understanding of enzyme action resulted from the work of Pasteur, Kühne, Buchner, and others investigating the action of yeast. Kühne was the first to introduce the term *enzyme* (Greek: "in yeast"), in 1878. Enzymes are a common part of our daily lives; for example, you can buy enzymes to tenderize meat. Cheese was and still is made by treating milk with an enzyme called *rennin*, obtained from the lining of a calf's stomach. Fermentation has always been an important art.

The protein nature of enzymes was not demonstrated conclusively until 1926, when Sumner prepared crystals of the enzyme urease and showed them to be protein. Since then over 150 enzymes have been crystallized and over 1000 have been isolated in varying degrees of purity. This is not a large number when you consider that even a simple organism such as *E. coli* has about 3000 different proteins. In cells of higher animals, the number of different proteins may range in the millions. In many ways, the study of enzymes has provided a bridge between the biological and physical sciences. Investigations of the structure, action, and properties of enzymes will occupy chemists, biologists, and physicists for many years to come.

As we will see, enzymes possess a number of unique properties as catalysts which are crucial to the operation of a cell. Before we learn more about enzymes, let's discuss what we mean by the term "catalyst."

3.1

Nature of Catalysis

Enzymes enhance reaction rates in a selective and efficient manner. In considering the nature of a chemical reaction, we should ask two fundamental questions:

1. "To what extent does the reaction occur?" (*Thermodynamics*)
2. "How fast does the reaction proceed?" (*Kinetics*)

The first question refers to the position of equilibrium: that is, the ratio of products to reactants. It is important to recognize the dynamic nature of chemical equilibria. For example, consider the system

$$A + B \underset{k_r}{\overset{k_f}{\rightleftharpoons}} C + D$$

This system is at equilibrium when the rate of formation of C and D equals the rate of the back reaction: that is, the formation of A and B.

$$k_f[A][B] = k_r[C][D]$$

If we take the ratio of the forward and reverse rate constants, we obtain

$$\frac{k_f}{k_r} = \frac{[C][D]}{[A][B]} = K_{equilibrium}$$

This important result emphasizes the dynamic nature of chemical equilibrium, in which the forward and reverse reactions take place concurrently. Usually we are interested in the formation of products from reactants,

$$A + B \xrightarrow{k_f} C + D$$

In considering the second question, the kinetics of this reaction, it is convenient to visualize the changes in potential energy of the system by using the reaction coordinate diagram shown in Figure 3.1. In this figure, the products

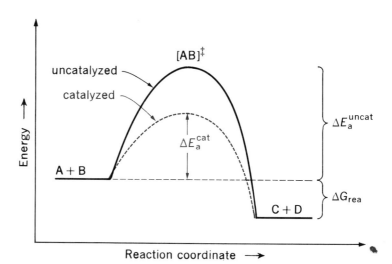

Figure 3.1

Reaction coordinate diagram showing the effect of a catalyst on activation energy.

exhibit a lower potential energy than the reactants, so the equilibrium favors the product side of the reaction. The energy barrier of the reaction is given by E_a, *the activation energy.* This is the energy needed to sufficiently activate the reactants to form the activated complex $[AB]^{\ddagger}$. In the presence of a catalyst, the reaction proceeds along the dashed curve in Figure 3.1, where the energy barrier is lower. Thus, more reactant molecules contain enough energy to form the activated complex, increasing the reaction rate. Note that lowering E_a affects the forward *and* reverse reactions in equal measure. Therefore, *a catalyst does not affect the position of chemical equilibrium* (K_{eq}); it affects only the forward and reverse *rates.* An energetically unfavorable reaction will not yield more product by the use of a catalyst, because ΔG_{rea} is unaffected by the catalyst, as is apparent from Figure 3.1.

Enzymes are biological catalysts. Like simple catalysts, enzymes provide an alternate reaction pathway (mechanism) that is lower in activation energy, and hence faster. A good example is the enzyme *catalase*, which catalyzes the reaction

$$2H_2O_2 \longrightarrow O_2 + 2H_2O$$

This is an important reaction physiologically, since hydrogen peroxide is a toxic product of certain reactions in metabolism and must be broken down rapidly. Table 3.1 lists several E_a values for this reaction under various conditions. The tremendous rate enhancements observed with enzymes as catalysts cannot be explained completely on the basis of lowered E_a; there are other factors involved which we will consider later.

Table 3.1 Activation Energy for Decomposition of H_2O_2. (Since rate depends on an exponential function of E_a, small differences in E_a reflect large differences in reaction rate.)

System	E_a (kcal/mole)	
Uncatalyzed	18.0	
Fe(II) (aq), catalyst	13.0	rate difference $\sim 10^{10}$
Pt (s), catalyst	12.0	
Catalase, biocatalyst	5.0	

As catalysts, enzymes are unique compared to simple inorganic or organic catalysts. The special catalytic properties of enzymes include the following points:

1. Enzymes enhance reaction rates under ordinary (physiological) conditions of pressure, temperature, and pH. This is rarely the case with other catalysts.

2. Enzymes function with an unusually high degree of selectivity or specificity with respect to the reactants acted upon and the reaction type catalyzed. Thus, competing reactions and side reactions are not observed in enzyme catalysis.

3. Enzymes provide tremendous reaction rate increases compared with ordinary catalysts.

These three points make the living state of matter possible.

3.2

Enzyme Classification and Nomenclature

Of the many types of proteins discussed in Chapter 2, the largest and most specialized class is the enzymes. Since the biosynthesis of proteins is genetically governed, enzymes, by conducting the biochemical functions of the cell, serve as agents for the expression of genetic information in the cell.

Because the chemical affairs of a cell are multiple and diverse, there are many different types of enzymes. Originally, scientists classified enzymes according to the reaction catalyzed, if a rational scheme of nomenclature was applied at all. The chemical substance that interacts with an enzyme as a primary reactant is termed the **substrate** of that enzyme. Consider the following:

$$H_2O + H_2N-\overset{\overset{\displaystyle O}{\|}}{C}-NH_2 \xrightarrow{\text{Urease}} 2NH_3(aq) + CO_2(g)$$

Urea

Enzyme: urease
Substrate: urea (primary reactant; H_2O is taken for granted)
Trivial name: urea + ase = urease
Systematic name: urea amidohydrolase

This example illustrates a problem which has faced all scientists—that of nomenclature. The trivial or common name of most enzymes is of the form "substrate-ase," or "substrate-reaction type-ase." Often the trivial name or names of a particular enzyme are not very descriptive.

The systematic naming of enzymes is currently based on a set of rules known as the International Enzyme Commission, or IEC, system, adopted by the International Union of Biochemistry. In the IEC system, each enzyme has a highly descriptive (and often long) systematic name. In addition, many enzymes have an accepted trivial name for everyday use. Thus, catalase is the accepted

Table 3.2   IEC Classification of Enzymes.

1. Oxidoreductases
(Oxidation-reduction reactions)

1.1 Acting on $-\overset{\mid}{C}H-OH$

1.2 Acting on $-\overset{\mid}{C}=O$

1.3 Acting on $-CH=CH-$

1.4 Acting on $-\overset{\mid}{C}H-NH_2$

1.5 Acting on $-\overset{\mid}{C}H-NH-$

1.6 Acting on NADH; NADPH

2. Transferases
(Transfer of functional groups)
2.1 One-carbon groups
2.2 Aldehydic or ketonic groups
2.3 Acyl groups
2.4 Glycosyl groups
2.7 Phosphate groups
2.8 S-containing groups

3. Hydrolases
(Hydrolysis reactions)
3.1 Esters
3.2 Glycosidic bonds
3.4 Peptide bonds
3.5 Other C—N bonds
3.6 Acid anhydrides

4. Lyases
(Addition to double bonds or
the reverse of that reaction)

4.1 $-\overset{\mid}{C}=\overset{\mid}{C}-$

4.2 $-\overset{\mid}{C}=O$

4.3 $-\overset{\mid}{C}=N-$

5. Isomerases
(Isomerization reactions)
5.1 Racemases

6. Ligases
(Formation of bonds with
ATP cleavage)
6.1 C—O bonds formed
6.2 C—S bonds formed
6.3 C—N bonds formed
6.4 C—C bonds formed

trivial name for the enzyme "$H_2O_2 : H_2O_2$ oxidoreductase." In this text, we will refer to enzymes by their accepted trivial names.

The IEC system is summarized in Table 3.2. As you can see, there are six major classes of enzyme-catalyzed reactions. Within each class are subcategories based on the nature of the functional group acted on by the enzyme. The numerous subcategories suggest the large diversity of chemical reaction types catalyzed by enzymes. It would be appropriate here to consider some specific examples that illustrate the variety and scope of enzyme chemistry.

1.   Oxidoreductases (Oxidation-reduction)

Oxidoreductases function in the general format shown in Figure 3.2. The scheme indicates that some part of the molecular structure of the oxidoreductases can exist in either an oxidized or a reduced form. The oxidoreductase cycles between its oxidized and reduced forms as successive substrate molecules

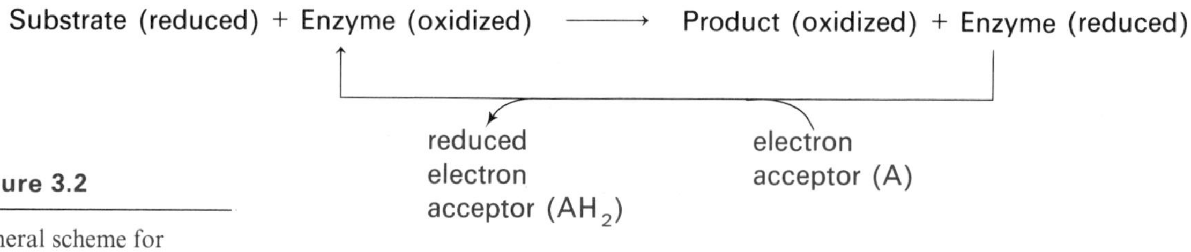

Substrate (reduced) + Enzyme (oxidized) $\longrightarrow$ Product (oxidized) + Enzyme (reduced)

reduced
electron
acceptor (AH$_2$)

electron
acceptor (A)

Figure 3.2

General scheme for oxidoreductase action. The curved arrow indicates that the "Enzyme (reduced)" reacts with an electron acceptor (A) to yield (AH$_2$) and "Enzyme (oxidized)." This "Enzyme (oxidized)" reacts with the substrate again to create more product. Note that in this case, two electrons (2e$^-$) are transferred from the substrate to the electron acceptor, along with two protons (2H$^+$).

are oxidized. The nature of the electron acceptor determines which of two types of oxidoreductases we are considering, dehydrogenases or oxidases.

Dehydrogenases are redox enzymes which do not use oxygen as an electron acceptor. Biological oxidations generally involve the removal of two electrons from the substrate by a process called dehydrogenation, an example of which is shown below:

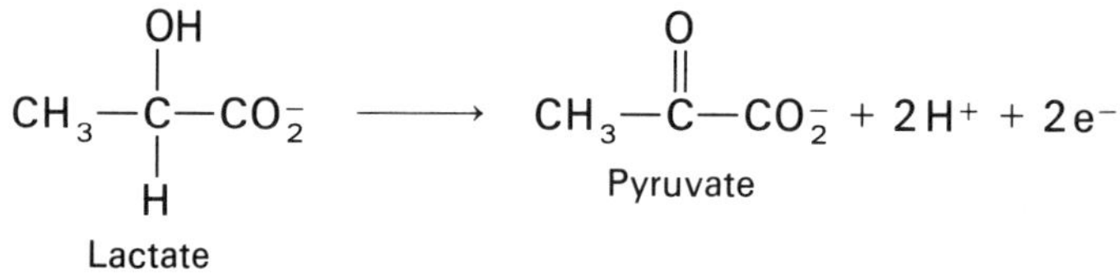

$$CH_3-\underset{\underset{H}{|}}{\overset{\overset{OH}{|}}{C}}-CO_2^- \longrightarrow CH_3-\overset{\overset{O}{\|}}{C}-CO_2^- + 2H^+ + 2e^-$$

Lactate   Pyruvate

Here two electrons are removed from the secondary alcohol functional group of lactic acid to form the ketone functional group of pyruvic acid.

Oxidases are redox enzymes for which oxygen, O$_2$, serves as an electron acceptor. The reaction catalyzed by the enzyme L-amino acid oxidase illustrates this:

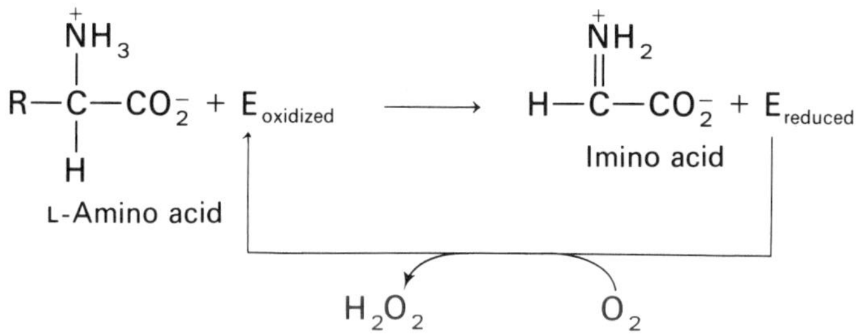

$$R-\underset{\underset{H}{|}}{\overset{\overset{+}{\overset{NH_3}{|}}}{C}}-CO_2^- + E_{oxidized} \longrightarrow H-\overset{\overset{+}{\overset{NH_2}{\|}}}{C}-CO_2^- + E_{reduced}$$

L-Amino acid   Imino acid

H$_2$O$_2$ O$_2$

Other types of oxidoreductases include *oxygenases*, which incorporate oxygen directly into the substrate molecule, and *peroxidases*, which use H$_2$O$_2$ as an electron acceptor. We can illustrate the action of peroxidases by the reaction catalyzed by catalase:

$$H_2O_2 + H_2O_2 \xrightarrow{\text{Catalase}} O_2(g) + 2H_2O$$

| Substrate | Electron acceptor | Oxidized substrate | Reduced electron acceptor |

2.   Transferases (Group Transfer Reactions)

Reactions in which a group moves from one position to another are important in biochemistry. Thus, we find a wide variety of transferase enzymes. We categorize them primarily by the nature of the group transferred. As an illustration, let us consider the action of two of the traditional types of transferases: transaminases and kinases.

Transaminase (aminotransferase) enzymes catalyze the reversible transfer of an amino group from an amino acid to a keto acid. Such an equilibrium is:

$$X-\overset{\overset{\textstyle O}{\|}}{C}-CO_2^- + Z-\overset{\overset{\textstyle \overset{+}{N}H_3}{|}}{\underset{\underset{\textstyle H}{|}}{C}}-CO_2^- \underset{\text{Transaminase}}{\rightleftharpoons} X-\overset{\overset{\textstyle \overset{+}{N}H_3}{|}}{\underset{\underset{\textstyle H}{|}}{C}}-CO_2^- + Z-\overset{\overset{\textstyle O}{\|}}{C}-CO_2^-$$

α-keto acid α-amino acid

This type of amino transfer reaction is very important in the biochemistry of amino acids.

Certain transaminases are also useful in the diagnosis of particular diseases. For example, heart tissue is rich in the enzyme glutamic oxaloacetic transaminase (GOT), which catalyzes the reaction

$$^-O_2C-CH_2-\overset{\overset{\textstyle \overset{+}{N}H_3}{|}}{CH}-CO_2^- + {}^-O_2C-CH_2CH_2-\overset{\overset{\textstyle O}{\|}}{C}-CO_2^- \overset{\text{GOT}}{\rightleftharpoons}$$

Aspartate α-ketoglutarate

$$^-O_2C-CH_2-\overset{\overset{\textstyle O}{\|}}{C}-CO_2^- + {}^-O_2C-CH_2CH_2-\overset{\overset{\textstyle \overset{+}{N}H_3}{|}}{CH}-CO_2^-$$

Oxaloacetate   Glutamate

Liver tissue is rich in a similar transaminase, but pyruvic acid is involved instead of oxaloacetic acid. In certain diseases, such as infectious hepatitis, the tissues undergo degeneration due to the infection. These transaminases are then released into the blood. Therefore, an elevated serum level of these transaminases is diagnostic of this type of infection. Enzyme chemistry is becoming an increasingly important tool in medical diagnosis.

Kinases are transferase enzymes which play an important role in transferring energy from one system to another in the form of a "high-energy phosphate bond." The molecule adenosine triphosphate (ATP) acts as a "medium of exchange" for phosphate bond energy in living systems. In this capacity, ATP is generally a reactant in a kinase-catalyzed reaction. An example is the reaction of pyruvate kinase, which is a step in glycolysis (see Chapter 8):

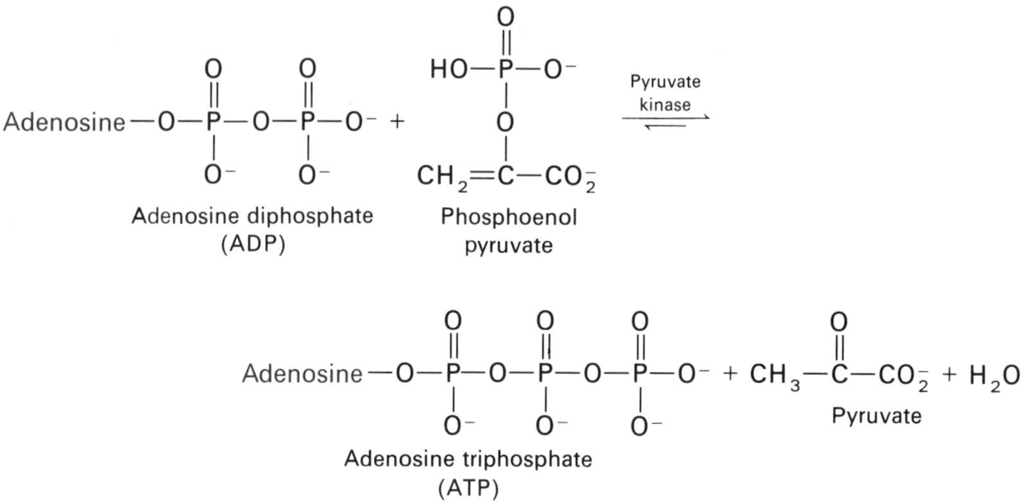

$$\text{Adenosine}-\text{O}-\overset{\overset{\text{O}}{\|}}{\underset{\underset{\text{O}^-}{|}}{\text{P}}}-\text{O}-\overset{\overset{\text{O}}{\|}}{\underset{\underset{\text{O}^-}{|}}{\text{P}}}-\text{O}^- \; + \quad \begin{array}{c}\overset{\text{O}}{\|} \\ \text{HO}-\text{P}-\text{O}^- \\ | \\ \text{O} \\ | \\ \text{CH}_2=\text{C}-\text{CO}_2^-\end{array} \quad \xrightarrow{\overset{\text{Pyruvate}}{\text{kinase}}}$$

Adenosine diphosphate   Phosphoenol
(ADP)   pyruvate

$$\text{Adenosine}-\text{O}-\overset{\overset{\text{O}}{\|}}{\underset{\underset{\text{O}^-}{|}}{\text{P}}}-\text{O}-\overset{\overset{\text{O}}{\|}}{\underset{\underset{\text{O}^-}{|}}{\text{P}}}-\text{O}-\overset{\overset{\text{O}}{\|}}{\underset{\underset{\text{O}^-}{|}}{\text{P}}}-\text{O}^- \; + \; \text{CH}_3-\overset{\overset{\text{O}}{\|}}{\text{C}}-\text{CO}_2^- \; + \; \text{H}_2\text{O}$$

Adenosine triphosphate   Pyruvate
(ATP)

In this case, a phosphoryl group ($-\text{PO}_3^{2-}$) is reversibly transferred from phosphoenol pyruvate to ADP.

3. Hydrolases (Hydrolysis Reactions)

We usually classify hydrolases on the basis of the bond hydrolyzed. This classification is summarized in Figure 3.3. In Chapter 2 we discussed a few peptidases such as trypsin and chymotrypsin in connection with determining a protein's primary structure.

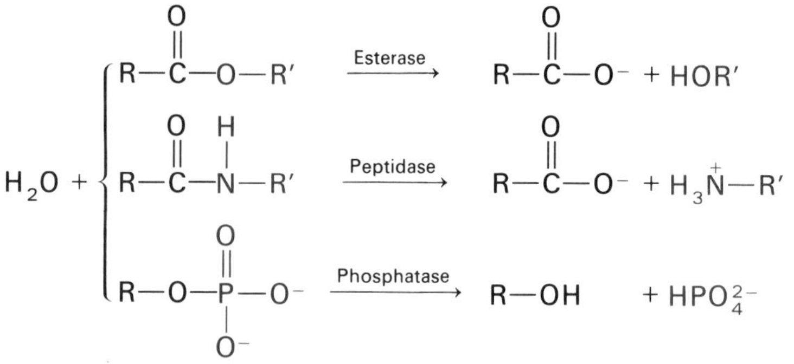

$$\text{H}_2\text{O} + \begin{cases} \text{R}-\overset{\overset{\text{O}}{\|}}{\text{C}}-\text{O}-\text{R}' & \xrightarrow{\text{Esterase}} & \text{R}-\overset{\overset{\text{O}}{\|}}{\text{C}}-\text{O}^- + \text{HOR}' \\[2ex] \text{R}-\overset{\overset{\text{O}}{\|}}{\text{C}}-\overset{\overset{\text{H}}{|}}{\text{N}}-\text{R}' & \xrightarrow{\text{Peptidase}} & \text{R}-\overset{\overset{\text{O}}{\|}}{\text{C}}-\text{O}^- + \text{H}_3\overset{+}{\text{N}}-\text{R}' \\[2ex] \text{R}-\text{O}-\overset{\overset{\text{O}}{\|}}{\underset{\underset{\text{O}^-}{|}}{\text{P}}}-\text{O}^- & \xrightarrow{\text{Phosphatase}} & \text{R}-\text{OH} \quad + \text{HPO}_4^{2-} \end{cases}$$

Figure 3.3

Some reactions catalyzed by hydrolases.

4. Lyases (Addition of Groups Across a Double Bond or the Reverse of That Reaction)

An example of a lyase is the hydration of the double bond of fumaric acid by the enzyme fumarase:

$$H_2O + \quad \underset{\text{Fumarate}}{\overset{-O_2C}{\underset{H}{\diagup}} C = C \overset{H}{\underset{CO_2^-}{\diagdown}}} \quad \overset{\text{Fumarase}}{\rightleftarrows} \quad \underset{\text{Malate}}{-O_2C-\overset{H}{\underset{H}{C}}-\overset{OH}{\underset{H}{C}}-CO_2^-}$$

As with many enzyme systems, we must consider the reverse reaction in some cases. Thus, pyruvic acid decarboxylase is a lyase because we can consider it a catalyst for the reverse of the reaction where acetaldehyde is added across the carbon-oxygen double bond in CO_2, even though, in practice, the reaction proceeds irreversibly as written because CO_2 is evolved:

$$CH_3-\overset{O}{\overset{||}{C}}-C\overset{\diagup O}{\diagdown_{O^-}} \quad \xrightarrow{H^+} \quad CH_3-\overset{O}{\overset{||}{C}} + \overset{O}{\underset{||}{C}} \\ \qquad\qquad\qquad\qquad\qquad\quad H \quad O$$

5. Isomerases (Equilibrium between Two Isomers)

An example of an isomerase is the reaction catalyzed by alanine racemase, an enzyme found in bacteria:

$$\text{L-Alanine} \quad \overset{\text{Alanine racemase}}{\rightleftarrows} \quad \text{D-Alanine}$$

Some isomerases can catalyze sugar interconversions (Chapter 4). There are also *cis-trans* isomerases, one of which, retinene isomerase, is directly involved in the biochemistry of vision.

6. Ligases (Formation of Bonds with ATP Cleavage)

These enzymes catalyze reactions that form chemical bonds. We often call such enzymes *synthetases*. Many of the important molecules in biological systems, such as proteins, polysaccharides, and nucleic acids, contain fundamental chemical linkages that are thermodynamically unstable (see Chapters 1 and 6). Therefore, the formation of a peptide bond, for example, is an energy-requiring process. The *driving force* for such ligase-catalyzed reactions is generally the exergonic (energy-releasing) removal of a phosphoryl or pyrophosphoryl group from ATP. We can illustrate this by the enzyme D-alanylalanine synthetase of bacteria:

$$\underset{\text{D-Alanine}}{CH_3-\overset{\overset{+}{N}H_3}{\underset{|}{CH}}-CO_2^-} + \underset{\text{D-Alanine}}{H_3\overset{+}{N}-\overset{CH_3}{\underset{\underset{CO_2^-}{|}}{CH}}} + ATP \quad \rightleftarrows \quad \underset{\text{D-Alanylalanine}}{H_3\overset{+}{N}-\overset{H_3C}{\underset{H}{C}}-\overset{O}{\overset{||}{C}}-\overset{H}{\underset{}{N}}-\overset{CH_3}{\underset{}{CH}}-CO_2^-} + ADP + HPO_4^{2-}$$

3.3

Enzyme Proteins and Cofactors

Within the class of proteins that show catalytic activity we find considerable diversity of overall form. Many enzymes are simple proteins; that is, complete hydrolysis yields only a mixture of amino acids. α-Chymotrypsin and trypsin are examples of enzymes which are simple proteins. Many enzymes, when subjected to complete hydrolysis, yield other groups or substances in addition to amino acids. Such *conjugated protein* enzymes are common. The format of a conjugated protein is summarized below.

Protein part (inactive) + Nonprotein part = *Holoenzyme* (active)
(apoenzyme) $\begin{cases} \text{prosthetic group;} \\ \text{coenzyme;} \\ \text{metal ion cofactor} \end{cases}$

Coenzymes and prosthetic groups actually refer to the same thing, although we can make an arbitrary distinction between the two: an ordinary coenzyme is *loosely* bound by the apoenzyme and can be reversibly dissociated; a coenzyme which is *strongly* bound to the apoenzyme is called a *prosthetic group*. Often, a prosthetic group is covalently linked to the apoenzyme. The nature of the coenzyme dictates the type of chemical process catalyzed by the holoenzyme. The coenzyme molecule is situated at a location in the tertiary structure of the enzyme known as the **active center** or active site. The active center is that part of the enzyme which *binds* the substrate molecule and brings about, by virtue of the groups located at the active center, whatever reaction the enzyme catalyzes.

Metal Ion Cofactors

Metal ions are crucial components of all living organisms. Certain metal ions may be required for stabilizing protein structure by charge–charge interactions. Ions such as Na^+, K^+ and Ca^{2+} may be necessary in metabolic processes involving transport of materials across membranes. Metal ions may also be intimately involved in the catalytic action of enzyme molecules. Scientists are only beginning to understand the roles metal ions play in living systems. As inorganic chemists continue to study the action of these ions in organic structures, a new discipline of bioinorganic chemistry is developing.

Table 3.3 summarizes the biochemical significance of some metal ions. Metal ions function primarily because they are able to bind with groups associated with either protein structure or non-protein prosthetic groups or both. The Fe^{2+} of myoglobin and hemoglobin is an example of the last case, where the metal ion is not only coordinated with the four nitrogens of the heme porphyrin ring, but is also coordinated to an imidazole nitrogen of a protein histidyl residue. Let's consider some metal ions in more detail.

Table 3.3 Metal Ion Cofactors in Enzymes. As the inorganic chemistry of living systems becomes better understood, more metal ions will be added to those in this table, and the functions of those already listed may be expanded.

Ion	Function
Na^+	Principal extracellular cation: may be important in controlling the activity of certain enzymes.
K^+	Principal intracellular cation: required for nerve impulse transmission; stabilizes structure in some proteins; required for activity in some enzymes.
Mg^{2+}	Essential for the activity of many enzymes; complexes readily with many protein groups and often stabilizes subunits in oligomeric proteins. Also complexes with ATP.
Ca^{2+}	Required for activity in some enzymes; required for nerve impulse transmission; calcium phosphate forms the hard structure in bones and teeth.
Mn^{2+}, Zn^{2+}	Required for activity in some enzymes.
Fe^{2+}, Fe^{3+}	Fe^{2+} is essential in oxygen transport, as in hemoglobin; both ions are essential in heme and non-heme iron redox proteins and oxidoreductases.
Co^{2+}	Essential in vitamin B_{12}-requiring enzymes.
Cu^+, Cu^{2+}	Essential for oxygen transport in marine organisms; essential metal ion in some oxidoreductases.
$Mo(VI)$	Required for certain oxidoreductases; required for nitrogen fixation.
Se (?)	A highly toxic metal recently found to be required in trace amounts by certain enzymes.
Cr^{3+}	In trace amounts it may be required for normal cellular absorption of glucose.
V (?)	Certain marine animals have a cellular energy economy based on vanadium instead of iron.
Ni^{2+}	Required in certain enzymes such as urease; the first enzyme to be crystallized by Sumner in 1926.

Ordinarily there is a more than sufficient amount of essential metal ions in a normal human diet; metal ion deficiency diseases have most often been observed in animals on laboratory diets. For example, rats living on purified casein (milk protein) died due to degeneration of liver cells. It was found that this could be prevented by adding minute amounts of selenium to the diet.

Although this element is highly toxic, the experiment proved that selenium is required in trace amounts. In some areas of the world, dietary factors may lead to a deficiency in certain metal ions. In some parts of the Middle East, poor growth and development caused by zinc deficiency have been observed in locations where unleavened bread forms a major part of the diet. This kind of cereal product is rich in *phytic acid* which forms strong complexes with Zn^{2+} and thus makes the zinc ion unavailable for absorption. When dough is acted on by yeast during the production of *leavened* bread, enzymes in the yeast break down the phytic acid, and the Zn^{2+} is released and can be absorbed.

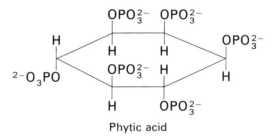

Phytic acid

Iron is a crucial element in human nutrition. In most body tissues, its total concentration in all forms is about $3 \times 10^{-4}M$, but in red blood cells, the concentration is $2 \times 10^{-2}M$. Iron is carefully conserved in the body. Since the average lifetime of a normal human red blood cell is about 126 days, about 1/126 or 0.85% of these are broken down daily. This corresponds to about 25 milligrams of iron. Almost all of this iron is reused since a normal male adult absorbs only about 1 milligram of iron a day from the diet. Certain physiological conditions such as pregnancy and menses lead to a condition of anemia, and the iron in the diet must then be supplemented by soluble forms such as ferrous sulfate. Iron is stored principally in the liver and spleen, where it is combined with a protein called *ferritin*. This protein contains 23% iron by weight. In addition to ferritin, the transport of iron in the blood requires a special protein called *transferrin*.

3.4

Vitamins and Coenzymes

Higher animals and many microorganisms lack the ability to biosynthesize certain substances that are necessary for normal metabolism. These organic molecules which are needed in small amounts in the diet are called vitamins. Vitamins as chemical substances essential to human nutrition were not discovered until the early 1900s when the Polish chemist Casimir Funk isolated a substance obtained from rice polishing that was effective in curing the dietary deficiency disease *beriberi*. This substance was an amine (thiamine) and Funk proposed the theory that the four recognized dietary deficiency diseases of that time, *scurvy*, *rickets*, *pellagra* and *beriberi*, all resulted from a deficiency of specific substances in the diet which he termed "vitamines." We now know that this theory is correct even though not all vitamins are amines. Each of the vitamins associated with the deficiency diseases noted above has been isolated and synthesized. Indeed, the production and sale of synthetic vitamins is now a multimillion dollar business.

Vitamins may be either polar molecules which are *water soluble* or non-polar molecules *soluble in fat solvents*. Most of the water-soluble vitamins serve as building blocks for coenzymes. They provide that part of the coenzyme molecule that the animal cannot make itself from simple precursors. Thus, many vitamin deficiency diseases cause an insufficiency of a given coenzyme and the resulting inactivity of the family of enzymes that requires the coenzyme.

Before we discuss the actual chemistry of coenzymes, let's consider some of the general features of fat-soluble and water-soluble vitamins.

Water-Soluble Vitamins

Ascorbic acid (vitamin C) is nutritionally essential for primates and guinea pigs; other species of animals are able to biosynthesize it. In Chapter 2 we saw that vitamin C is vital in the formation of the structural protein collagen. A diet deficient in ascorbic acid leads to a condition called *scurvy*, characterized by sore gums, loose teeth, breakdown of capillaries, hemorrhages, painful joints and slow healing of wounds. Scurvy was once the scourge of sailors and explorers who had to exist for extended periods on diets of dried and salted food. It was not until 1757 that it was recognized that fresh fruits and vegetables protected people from scurvy. This gave rise to the inclusion of lime or lemon juice in the sailors' daily ration. In normal adults, a daily intake of about 50 milligrams of ascorbic acid is sufficient.

Ascorbic acid

Biotin is a nutritionally essential factor found in grains, eggs, liver and yeast. No specific disease can normally be ascribed to biotin deficiency in the diet because in most animals intestinal bacteria synthesize sufficient biotin to meet the animal's needs. Biotin deficiency can be produced artificially in certain animals. For example, feeding rats a diet containing large amounts of raw egg whites results in loss of hair, dermatitis and loss of muscular coordination. This

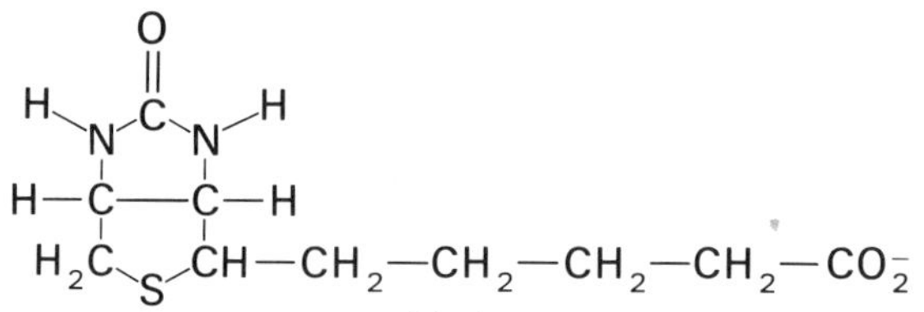

Biotin

is due to the presence of a glycoprotein in raw egg white called *avidin* which binds biotin almost irreversibly.

$$\text{Avidin} + \text{Biotin} \rightleftharpoons \text{Avidin} \cdot \text{Biotin} \qquad K_{eq} = 10^{15}$$

The reaction between avidin and biotin results in an artificially induced biotin deficiency.

Cobalamin (vitamin B_{12}) is obtained mainly from meat, particularly kidney, liver and brain. At least 50% of the vitamin B_{12} in normal adults is produced by intestinal bacteria which are capable of biosynthesizing this complex substance (see Figure 3.9). Only minute amounts of this vitamin are usually needed in the body, the normal concentration of B_{12} in the blood being about $2 \times 10^{-10} M$. Vitamin B_{12} is also noteworthy because it contains Co(III) covalently bound to the organic part of the vitamin.

Vitamin B_{12} deficiency leads to a condition called *pernicious anemia*. This gives rise to the formation of immature and fragile red blood cells and irreversible degradation of the central nervous system. In one form of the disease, a protein factor essential for the intestinal absorption of B_{12} is lacking. This usually strikes only older people and can be controlled by periodic injections of vitamin B_{12}. Symptoms of vitamin B_{12} deficiency have also been observed in strict vegetarians, whose blood levels of vitamin B_{12} are sometimes only 1/5 to 1/2 of normal.

Folic acid is essential for cell growth and multiplication. Good sources for this vitamin include leafy green vegetables (*fol*iage), liver, yeast and wheat germ. Folic acid deficiency is widespread in the world because of poor diet, disease or both. Since folic acid is essential in nucleic acid biosynthesis (see Section 12.3), new genetic material, hence, new cells, cannot be produced in the absence of folic acid. A deficiency in folic acid first shows up as an anemia because red blood cells are replaced more rapidly than other cells in the body.

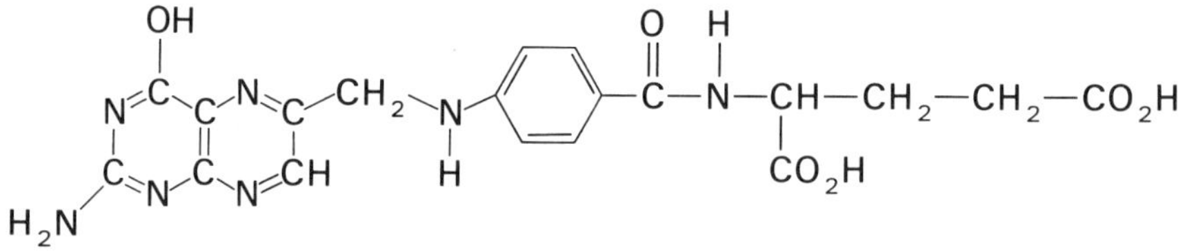

Folic acid

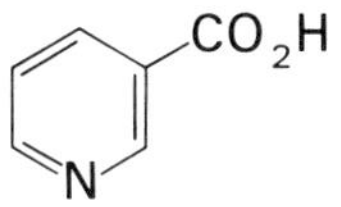

Niacin, or nicotinic acid, is a member of the vitamin B_2 complex and is found in many plant and animal sources. Niacin deficiency results in a disease called *pellagra*. At one time pellagra was a widespread disease in the United States, with over 100,000 cases reported annually during 1910–1936. Pellagra is characterized by severe dermatitis, diarrhea and eventually death if untreated by proper diet. Niacin can be biosynthesized to some extent in humans from the amino acid tryptophan. Unfortunately, pellagra still occurs in regions where corn is the major food, a staple low in both niacin and tryptophan, an essential amino acid.

Pantothenic acid is a dietary precursor for coenzyme A (see Figure 3.8). It is produced by green plants and most microorganisms but not by higher animals. Meats, eggs and milk are important dietary sources. Because coenzyme A is a central molecule in metabolism, a deficiency of pantothenic acid would have grave results. However, since this substance is widespread in foods, such a deficiency is not ordinarily observed in humans.

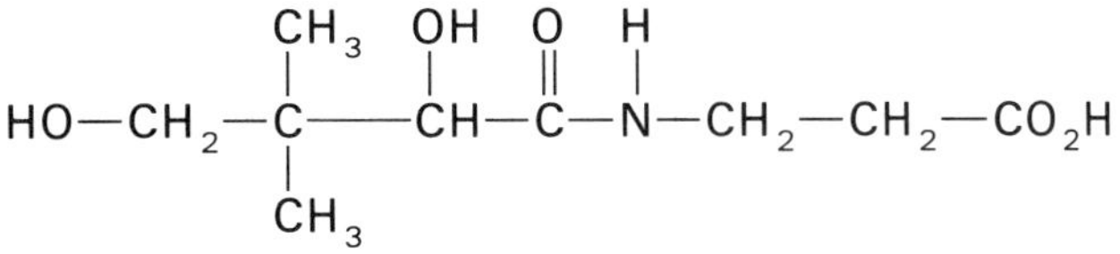

Nicotinic acid

Pantothenic acid

Pyridoxine (vitamin B_6) is found in a wide range of foods and a deficiency of this vitamin in normal humans is rarely observed.

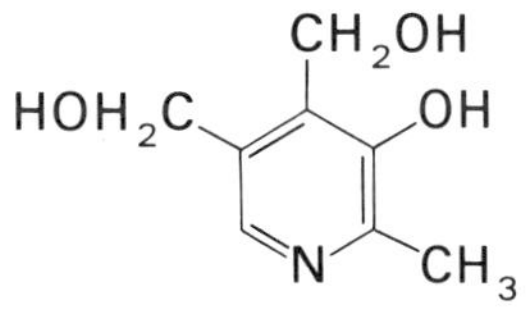

Pyridoxine

Riboflavin (vitamin B_2) is found in all biological material and riboflavin deficiency, by itself, does not cause any human disease. Plants and most

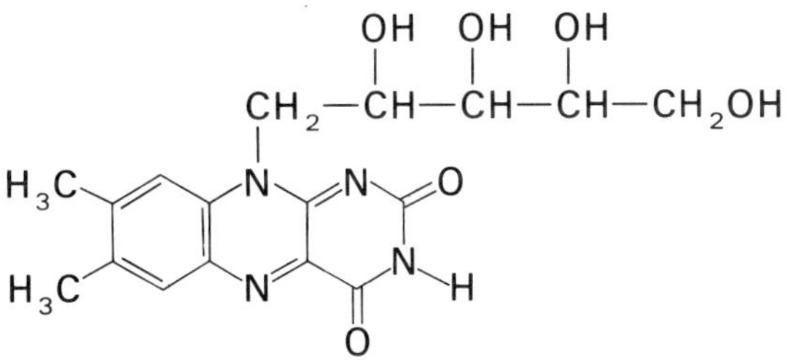

Riboflavin

microorganisms can biosynthesize riboflavin but higher animals must gain this vitamin from the diet. In the United States, milk and eggs are important sources of riboflavin.

Thiamine (vitamin B_1) is required in the diet of all animals except ruminants. Historically, thiamine deficiency has been widespread in parts of Asia where polished (white) rice is a major part of the diet. The outer layers of grain seeds are rich in thiamine. These layers are removed when grain is milled, thus polished rice is almost devoid of thiamine, even though the whole grain is a good source of the vitamin. Thiamine deficiency gives rise to the disease *beriberi*, which can take two forms: "wet" or cardiovascular beriberi, and "dry" or neurological beriberi. Without proper diet, both can be fatal. Because thiamine is easily soluble in water and unstable when subjected to prolonged high temperature, excessive cooking of foods ordinarily rich in thiamine, such as beans, can destroy this vitamin.

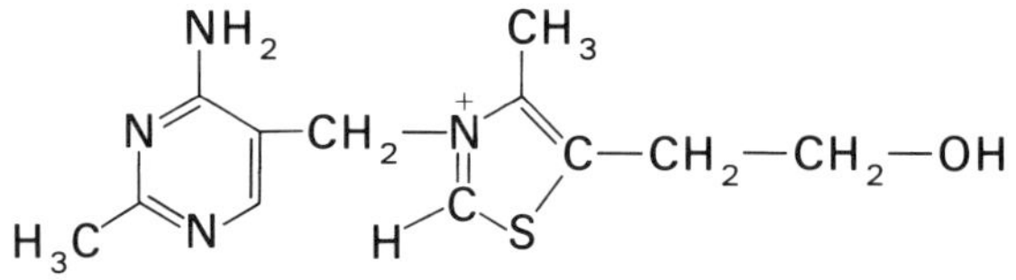

Thiamine

Fat-Soluble Vitamins

Vitamin A (retinol) is required in the diet of all higher animals. Good sources of vitamin A include leafy green vegetables, yellow vegetables and fish liver oils. Vitamin A, unlike most other vitamins, can be stored in the body in relatively large amounts, particularly as part of the lipid material in the liver. When the primary alcohol group of retinol is oxidized, the aldehyde *retinal* is produced. Retinal is a central molecule in the biochemistry of vision, as we shall see in Chapter 6. Because vitamin A is needed to make retinal, an early indicator of vitamin A deficiency is night-blindness.

Except for the process of vision, the metabolic roles of vitamin A are not well understood. Other results of vitamin A deficiency reflect its importance in metabolism. These include dry skin, hardening of the skin, eye inflammation, poor skeletal growth and poor resistance to infection.

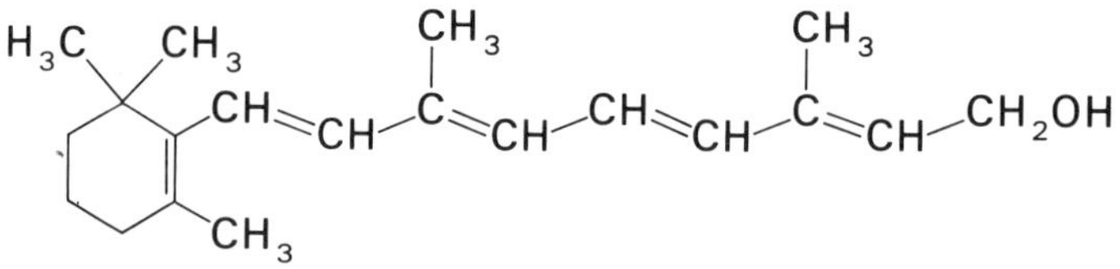

Vitamin A (retinol)

Vitamin D (calciferol) is principally involved in the control of calcium metabolism. Vitamin D deficiency results in the disease *rickets* in which growing bones in children are soft, easily deformed and poorly calcified. Under ordinary conditions, humans are able to produce sufficient vitamin D through the following reaction:

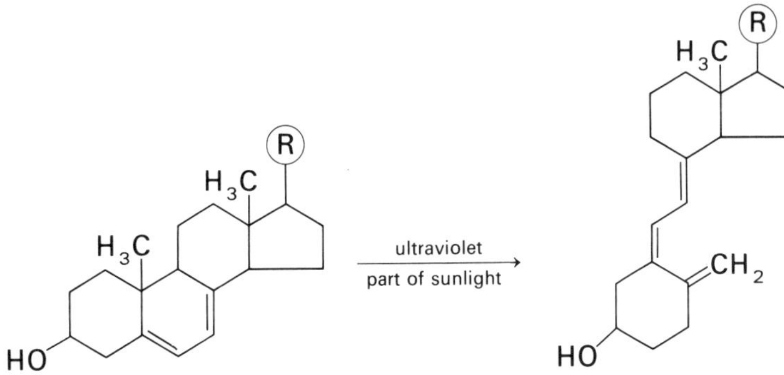

7-dehydrocholesterol
R = C_8H_{17}

Cholecalciferol
(vitamin D_3)

Irradiation of another steroid, ergosterol, yields ergocalciferol (vitamin D_2). The two forms of vitamin D differ only in the number of carbon atoms in the R group shown above.

Vitamin D deficiencies can arise in children living in northern latitudes where exposure to sunlight is limited. The major dietary sources of vitamin D are fish liver oils and vitamin D-enriched milk. Excessive amounts of vitamin D in the diet can be toxic, causing decalcification of bone and deposition of calcium salts under the skin and in the kidneys.

Vitamin E (α-tocopherol) appears to act as an antioxidant to protect unsaturated fatty acids from oxidation. There is also considerable evidence that it is needed for normal cell membrane structure and function. Vegetable oils such as wheat germ, safflower, cottonseed and corn oil are good sources of vitamin E. Vitamin E deficiency can lead to a weakening of red blood cell membranes, resulting in hemolysis.

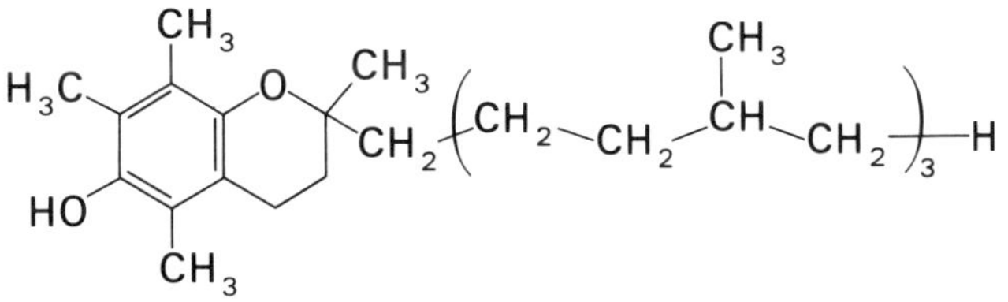

Vitamin E

Vitamin K is required for the formation of prothrombin, a protein essential for the blood-clotting process. There are two forms of this vitamin: K_1 which is found in plant tissue and K_2 which is produced by intestinal

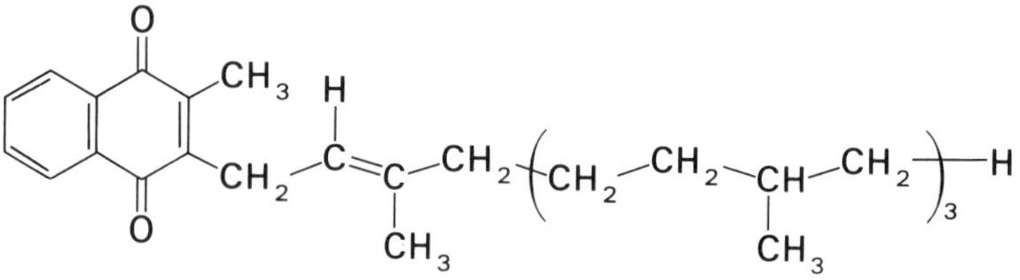

Vitamin K$_1$
(Vitamin K$_2$ has 2 double bonds in the side chain)

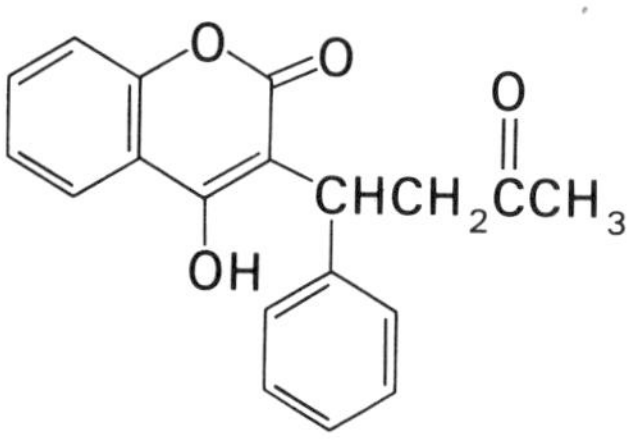

Warfarin

bacteria. In normal individuals, vitamin K deficiency is rare. However, prolonged antibiotic therapy can cause a deficiency by killing the intestinal flora and thus destroying a favorable symbiosis between bacteria and man.

Vitamin K deficiency increases the time required for blood to clot. In such a case, even a minor cut or bruise can become serious due to hemorrhaging. This condition can also be caused by certain compounds which interfere with vitamin K metabolism and prevent its action in the blood-clotting process. These compounds include *warfarin*, a substance used as rat poison and in the treatment of diseases involving excessive blood clotting. One such disease is *thrombophlebitis*, where clots form in the veins and obstruct normal circulation in the extremities, particularly in the legs.

Vitamins are clearly important constituents of the diet. Some vitamins such as niacin, folic acid, biotin, vitamin B$_{12}$ and vitamin K can be supplied either wholly or partially by intestinal bacteria. Table 3.4 lists a set of recommended dietary allowances for the vitamins. These allowances may vary considerably depending on age, sex and physiological conditions.

Table 3.4 A listing of recommended adult dietary allowances for the vitamins.

Vitamin	Approximate daily allowance
Ascorbic acid	60 mg
Biotin	0.3 mg
Cobalamin (B$_{12}$)	0.006 mg
Folic acid	0.4 mg
Niacin	20 mg
Pantothenic acid	10 mg
Pyridoxine (B$_6$)	2 mg
Riboflavin	1.7 mg
Thiamine (B$_1$)	1.5 mg
Vitamin A	1.3 mg
Vitamin D	0.02 mg
Vitamin E	10–15 mg*
Vitamin K	0.001 mg*

* Estimated values

3.5

Coenzymes and Chemical Reactions

Many of the chemical reactions associated with the living state of matter are brought about by enzymes that require a coenzyme for activity. In many cases the isolated coenzyme can bring about a similar reaction, usually with much poorer efficiency, at a slower rate, and with no selectivity. The study of the chemistry of coenzymes is interesting since it gives us a better appreciation of the molecular nature of biological processes. Thus it will be useful to consider these substances in more detail. Table 3.5 briefly summarizes the functions and sources of a number of coenzymes.

Table 3.5   Vitamins and Coenzymes.

Coenzyme	Vitamin precursor	Function
Nicotinamide adenine dinucleotide (NAD^+)	Nicotinic acid (Niacin)	Redox
Nicotinamide adenine dinucleotide phosphate ($NADP^+$)	Nicotinic acid (Niacin)	Redox
Flavin adenine dinucleotide (FAD)	Riboflavin	Redox
Flavin mononucleotide (FMN)	Riboflavin	Redox
Lipoic acid	None	Redox
Heme	None	Redox
Thiamine pyrophosphate (TPP)	Thiamine	Oxidative decarboxylation
Pyridoxal phosphate (Py)	Pyridoxine (Vitamin B_6)	Transaminations, racemizations
Coenzyme A (CoASH)	Pantothenic acid	Acyl group transfer
Tetrahydrofolic acid (THF)	Folic acid	One-carbon group transfer
Biotin	Biotin	(CO_2) transfer
5'-Deoxyadenosyl cobalamin (coenzyme B_{12})	Cobalamin (Vitamin B_{12})	Group transfer

Redox Coenzymes

We can divide biological oxidations into two broad types: hydrogen-transfer reactions and electron-transfer reactions. The primary distinction between the two types is that, in the case of hydrogen-transfer reactions, two electrons and two protons are removed from the substrate, while in the case of electron-transfer reactions, only electrons are removed from the substrate. We will first consider hydrogen carriers.

NAD^+ and $NADP^+$

The structure of *N*icotinamide *A*denine *D*inucleotide (NAD^+) and of the closely related *N*icotinamide *A*denine *D*inucleotide *P*hosphate ($NADP^+$) is shown below. (In the older literature you will find NAD^+ referred to as DPN, diphosphopyridine nucleotide, and $NADP^+$ labeled TPN, triphosphopyridine nucleotide.)

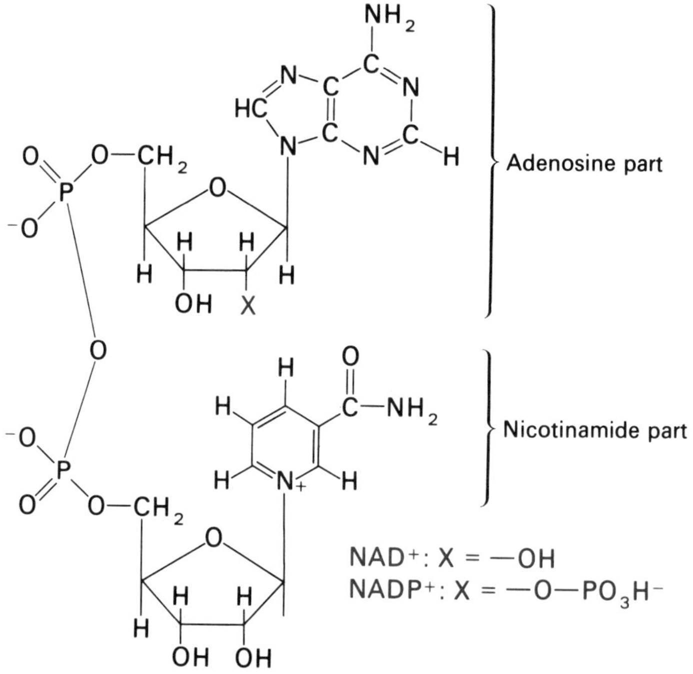
Adenosine part
Nicotinamide part
NAD+: X = —OH
NADP+: X = —O—PO3H−

The adenosine part of NAD^+ seems to often be involved in binding the coenzyme to the apoenzyme. The redox-active part of NAD^+ (or of $NADP^+$) is the nicotinamide portion, which is shown below in its oxidized and reduced forms. (We use the symbol R to denote the rest of the NAD^+ molecule.)

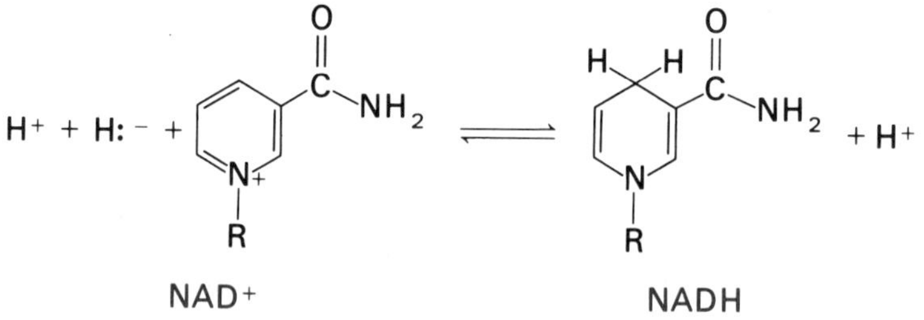
H+ + H:− +
NAD+
NADH
+ H+

Figure 3.4

Action of the NAD^+-requiring enzyme lactate dehydrogenase (LDH). In the overall reaction, the oxidoreductase cycles between oxidized and reduced forms as successive substrate molecules are oxidized and successive electron acceptor molecules are reduced. Thus, a small amount of enzyme can bring about the conversion of a considerably larger amount of reactant into product.

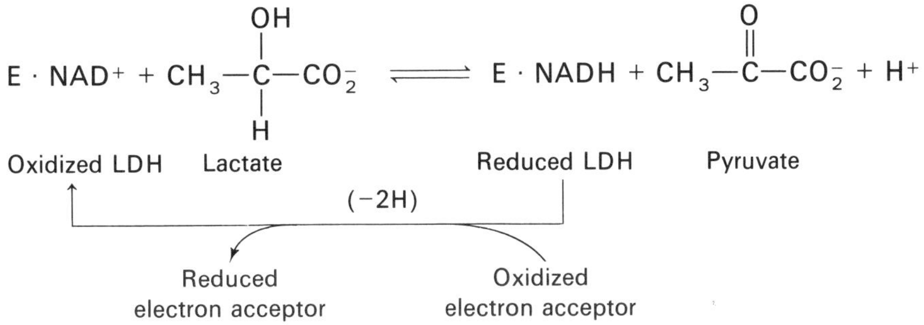

The action of this coenzyme can be illustrated by the reaction of lactate dehydrogenase, as shown in Figure 3.4. The holoenzyme is represented as $E \cdot NAD^+$. Note the participation of an electron acceptor, discussed earlier in this chapter. Since NADH does not react with oxygen in enzymatic systems, an organic molecule or another redox enzyme must carry out the oxidation of $E \cdot NADH$ so that it can oxidize the next substrate molecule.

The two coenzymes NAD^+ and $NADP^+$ act in an identical manner as far as redox activity is concerned. However, there is a differentiation of function between the two on the subcellular level. For example, in liver tissue, most of the cellular NADP is in the reduced form (NADPH) to provide reducing equivalents ($H^+ + e^-$) for biosynthesis. On the other hand, most of the cellular NAD^+ is in the oxidized form to extract reducing equivalents from "food" molecules. We will discuss the roles of NAD^+ and $NADP^+$ in more detail in the chapters on metabolism.

FAD and FMN

The flavin coenzymes (Latin: *flavus*, "yellow") are riboflavin-derived coenzymes which act as hydrogen carriers in biological systems. The structures of FAD, flavin adenine dinucleotide, and FMN, flavin mononucleotide, are:

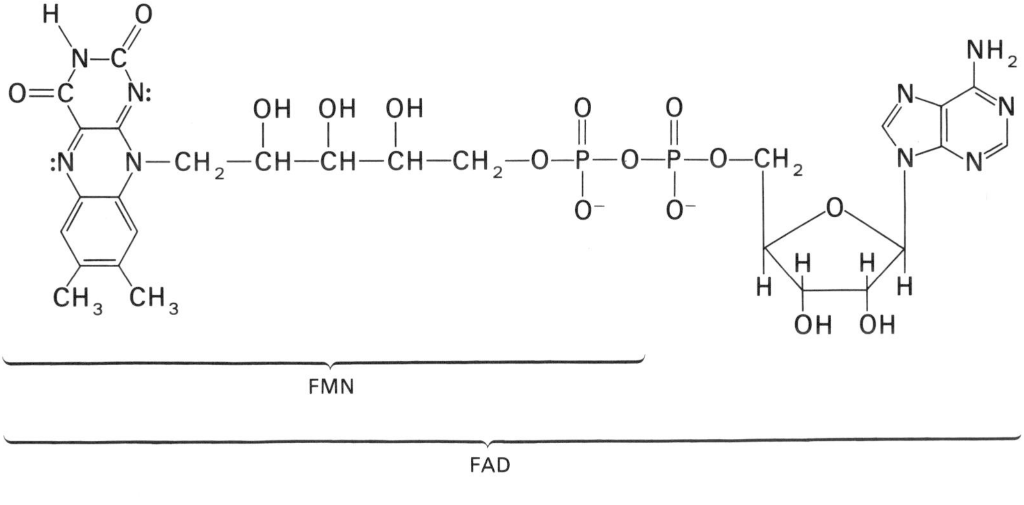

Both FAD and FMN contain the same redox-active part in the form of the isoalloxazine portion, shown below in oxidized and reduced form:

Oxidized
(yellow)

Reduced
(pale yellow)

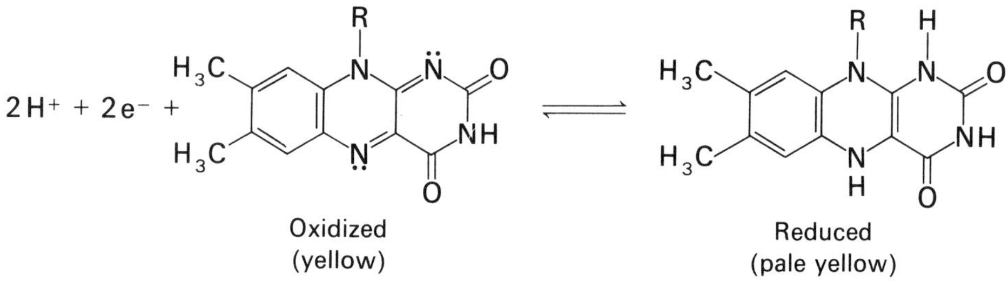

There is considerable versatility in the action of flavin enzymes due to the chemical properties of the reduced isoalloxazine. The way reoxidation of reduced flavoproteins occurs depends on the properties of the particular apoenzyme. Table 3.6 lists various ways in which reduced flavin enzymes can react with an electron acceptor.

Table 3.6 Flavoenzymes and Electron Acceptors.

Enzyme type	Electron acceptor	Product of reaction with acceptor
Oxidase	O_2	H_2O_2
Oxygenase	O_2 + organic molecule (X)	H_2O + X—H_2
Dehydrogenase	Organic molecule (X)	X—H_2

Both flavin enzymes and NAD enzymes are closely involved in those metabolic processes that break down "food" molecules, such as glucose, by oxidation reactions, thus providing chemical energy for the cell. We shall see in Part II how biological oxidations provide the energy input needed to sustain the living state.

Other Hydrogen Carriers

Lipoic acid is a hydrogen carrier that can exist in two oxidation states due to the chemistry of the sulfhydryl-disulfide redox pair. The structures of the oxidized and reduced forms of lipoic acid are

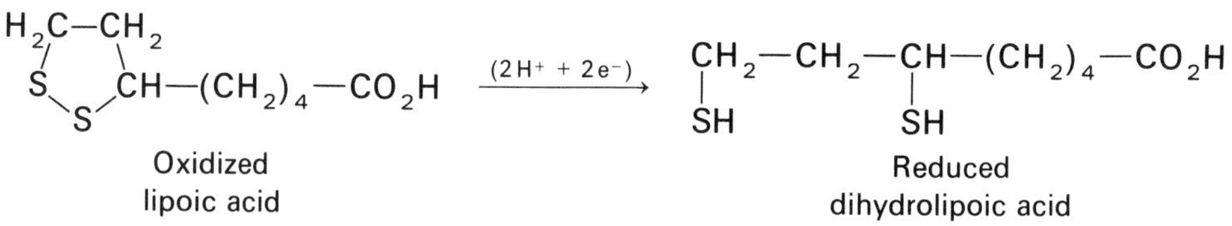

Oxidized
lipoic acid

Reduced
dihydrolipoic acid

Lipoic acid can act as a hydrogen carrier because of the redox properties of the —SH group.

It is interesting to note that, as was the case with most coenzymes, the discovery of lipoic acid involved a laborious extraction and purification of the coenzyme from a natural source. In isolating lipoic acid, Gunsalus and his colleagues extracted 10 tons of beef liver to obtain a final yield of 30 milligrams of lipoic acid!

Another sulfhydryl compound important in biological oxidations is *glutathione*, or *γ*-L-glutamyl-L-cysteinylglycine. Glutathione acts as a biological hydrogen carrier due to its sulfhydryl group, and is widely distributed in nature.

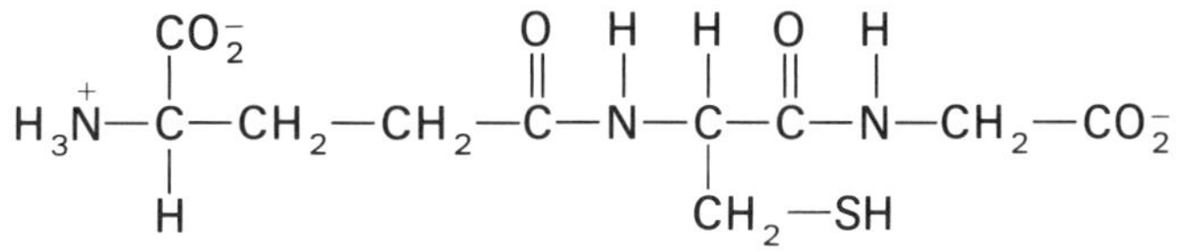

Glutathione

Quinones are aromatic compounds that can exist in different oxidation states. They occur naturally, as in the cases of coenzyme Q, which is important in respiration (see Chapter 9), and vitamin K, which is a key factor in blood clotting. The general scheme for the oxidation-reduction of such quinone coenzymes is

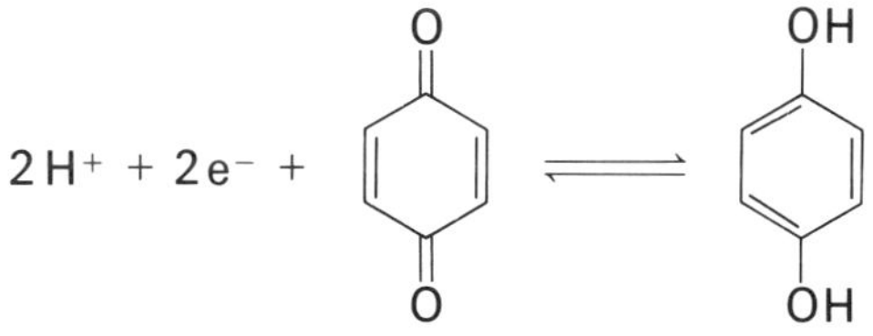

Quinone (oxidized)   Hydroquinone (reduced)

Electron-carrying coenzymes represent a class of redox-active materials that plays a vital role in generating energy in living systems. Generally, a metal ion is involved which can exist in two (or more) oxidation states. The metal ion is associated with a ligand or binding group, together making up the coenzyme. While metal ions such as Cu^+/Cu^{2+} are important in nature, the iron-containing electron carriers are the best known. We will consider only the heme coenzymes in this section and reserve mention of others for the appropriate context later.

Heme Coenzymes

All heme coenzymes contain the *porphyrin* macrocyclic ring system shown in Figure 3.5. The basic porphyrin system shown occurs in nature with various substituents at the numbered positions on the ring. For example,

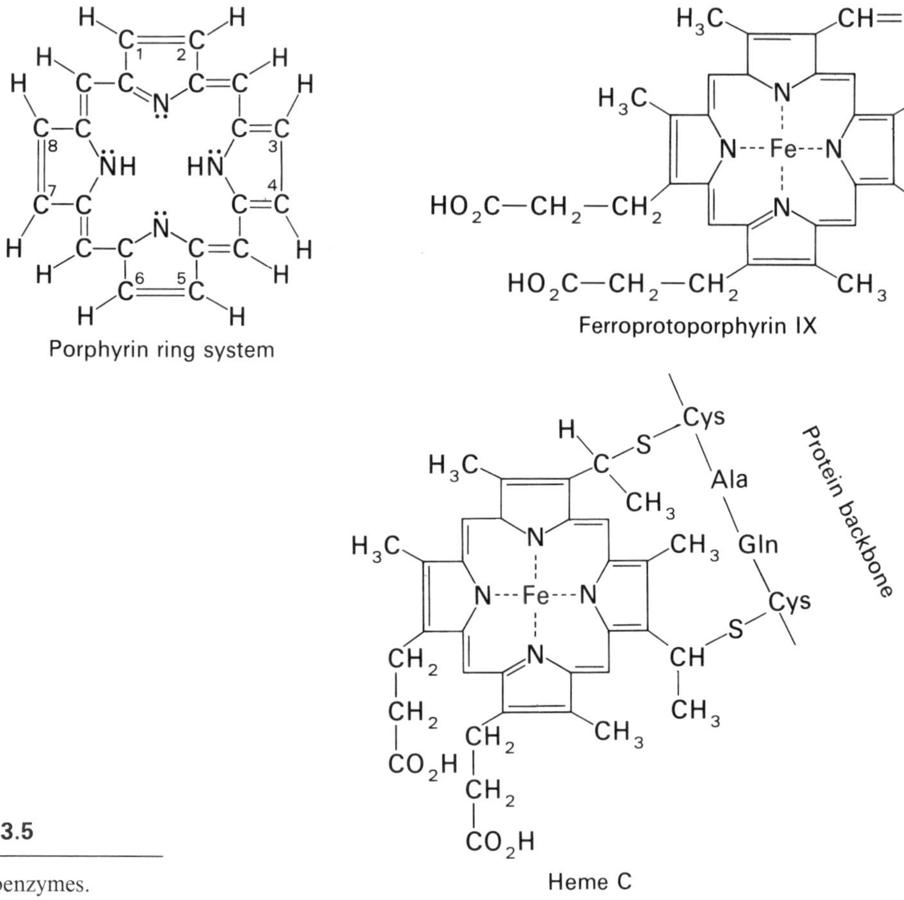

Porphyrin ring system

Ferroprotoporphyrin IX

Figure 3.5

Heme coenzymes.

Heme C

Figure 3.5 pictures the heme prosthetic group associated with such proteins as hemoglobin, myoglobin, catalase, and b-type cytochromes. It is termed *ferro-protoporphyrin IX* because in an iron–porphyrin system containing four methyl, two vinyl, and two propanoic acid substituents, there are fifteen possible isomers, one of which is a naturally-occurring isomer arbitrarily designated *IX*.

In hemoglobin and myoglobin, which are both oxygen carriers, iron is in the $+2$ oxidation state. In the enzyme catalase the iron is normally in the $+3$ oxidation state. In the electron-transferring cytochromes, the iron alternates between the $+2$ and $+3$ oxidation states as an electron is gained or lost. Such ring systems are excellent chelating agents; that is, they readily bind metal ions by forming coordinate covalent bonds between the electron-rich nitrogen atoms in the ring and the electron-deficient metal ion. In fact, we can prepare stable complexes of porphyrins with most transition-metal ions.

Other types of heme prosthetic groups have different ring substituents. This variability with respect to the porphyrin part of heme systems gives rise to four broad classes of heme coenzymes, including ring systems corresponding to

Heme A, Heme a_2, ferroprotoporphyrin IX (as in Figure 3.5), and Heme C. The Heme C prosthetic group of cytochrome c is also shown in Figure 3.5. It is interesting to note that in this case the Heme C prosthetic group is linked to the apoprotein by covalent thio-ether bonds between the ring and a portion of the cytochrome c protein chain. In addition to the four classes, the plant pigments chlorophyll a and chlorophyll b contain similar but *not* identical ring systems, where the metal ion is Mg^{2+} rather than a transition-metal ion (see Chapter 10).

The **cytochromes** are heme proteins that can transfer electrons due to the ability of the bound iron to undergo the process:

$$Fe^{3+} + e^- \rightleftharpoons Fe^{2+}$$

The ease with which reduction occurs is expressed as the redox potential (measured in volts). Cytochromes are often arranged in living systems in a sequential **electron transport chain**. In such a chain, cytochromes which are better electron donors when reduced pass electrons to cytochromes which are poorer electron donors when reduced. In the mitochondrion, this process is known as the respiratory chain (Chapter 9). Reducing equivalents (electrons) feed into the respiratory chain from NADH and FADH coenzymes. These electrons pass from a higher potential energy to a lower energy and ultimately reduce O_2 to H_2O. In the process, energy is conserved in the form of the high-energy phosphate bond in ATP. The redox potential of the iron in a given cytochrome is determined largely by the structure of the protein, which provides an environment for the heme coenzyme that is unique for each cytochrome.

The remaining coenzymes we will consider form an important part of the biochemical basis of a living cell. The chemistry of these coenzymes illustrates a good part of the chemistry of the entire cell.

Thiamine Pyrophosphate

Thiamine by itself is able to catalyze a number of reactions involving carbonyl compounds. Not surprisingly, then, enzymes which require thiamine pyrophosphate are also involved in carbonyl reactions. The structure of thiamine pyrophosphate is

Thiamine pyrophosphate

The active part of this coenzyme is the sulfur-containing *thiazole ring*. The thiamine pyrophosphate prosthetic group forms a covalent intermediate with the substrate. This reaction involves the thiazole ring of the coenzyme and an

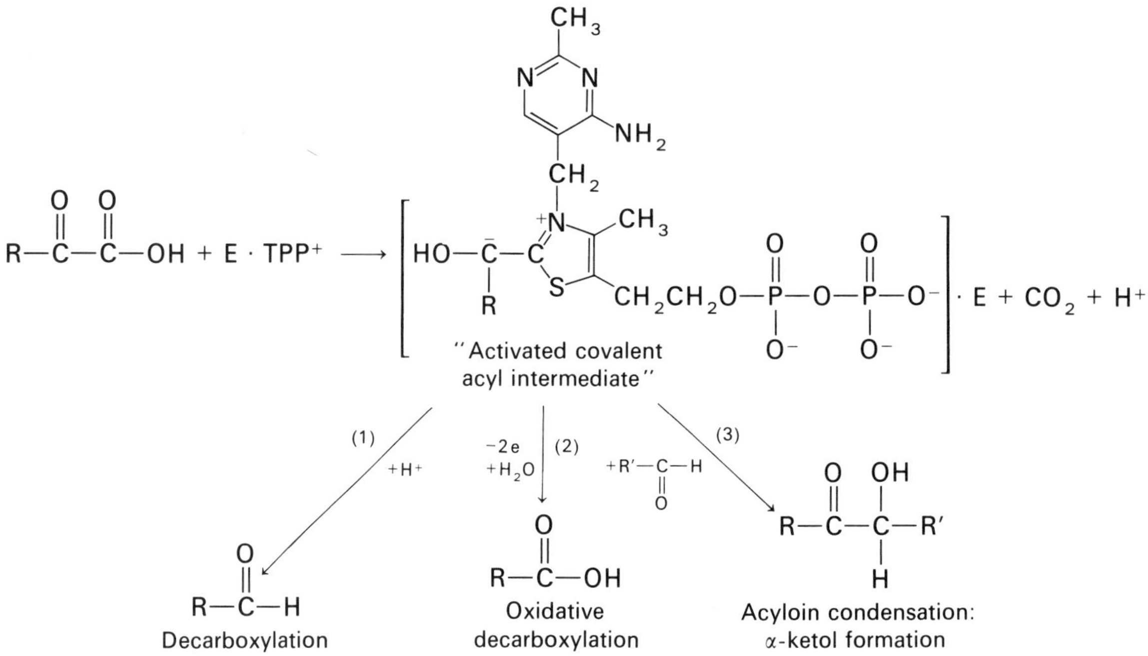

Figure 3.6

Basic scheme for reactions catalyzed by thiamine pyrophosphate-requiring enzymes (E · TPP). The structure of the covalent substrate-thiazole intermediate is shown.

acyl group derived from the α-keto acid by elimination of CO_2. The process is shown in Figure 3.6. The oxidative decarboxylation reaction (reaction 2 in Figure 3.6) is very important in the breakdown of glucose into CO_2 and H_2O.

Pyridoxal Phosphate

Pyridoxal enzymes are closely associated with amino acid metabolism. Such enzymes include transaminases, racemases, and certain decarboxylases. In these enzymes the required coenzyme is pyridoxal phosphate, whose structure is shown below. Its activity as a coenzyme resides in the ability of the aldehyde function of pyridoxal to form a *Schiff base intermediate* with the amino group of, for example, an amino acid substrate.

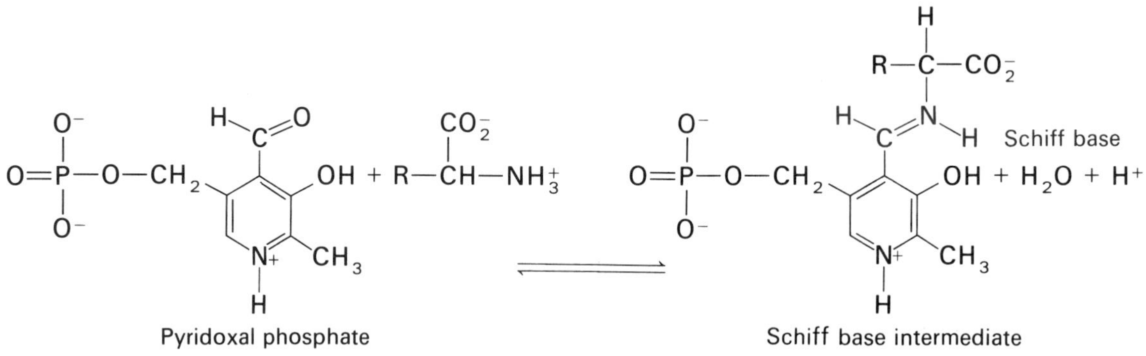

During the formation of the Schiff base intermediate, electron density shifts from the amino acid to the pyridine ring. This alters the properties of the amino acid part of the Schiff base intermediate and causes decarboxylation, racemization, or transamination to take place, depending on the nature of the particular pyridoxal enzyme. Figure 3.7 summarizes some of the reactions involving amino acids which are catalyzed by pyridoxal phosphate-requiring enzymes.

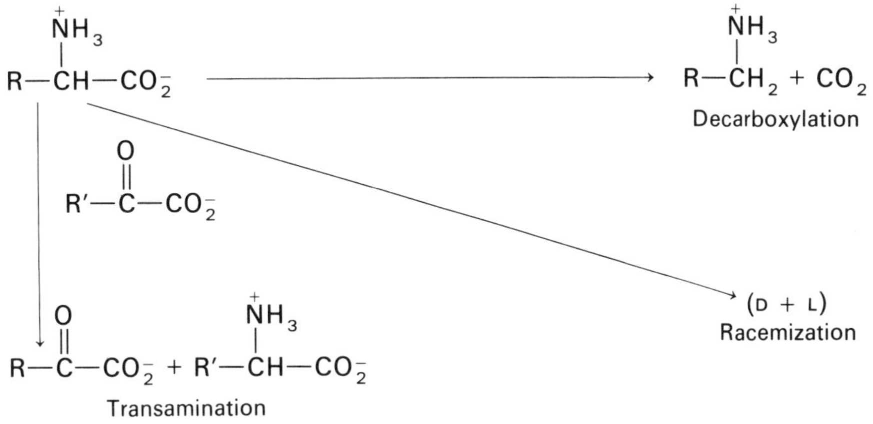

Figure 3.7

Some reactions catalyzed by pyridoxal enzymes.

Coenzyme A

There are many situations in molecular biology and biochemistry where reactions of the following format are important:

$$R-C \overset{O}{\underset{OH}{\big\langle}} + H-Y \rightleftharpoons R-C \overset{O}{\underset{Y}{\big\langle}} + H_2O$$

This type of reaction is termed an acylation reaction, and, as given here, the equilibrium for such a system lies far to the left. However, in the laboratory we would not use a carboxylic acid as an acyl group donor, but would use instead a more reactive compound:

$$R-C \overset{O}{\underset{Cl}{\big\langle}} + H-Y \rightleftharpoons R-C \overset{O}{\underset{Y}{\big\langle}} + HCl$$

Here the reaction favors products (i.e., goes to completion). Reactive acyl donors such as acetyl chloride are not suitable for the chemical affairs of a living cell. Therefore, the cell requires a biological acyl group carrier to act as a source of reactive acyl groups. This is the function of coenzyme A, shown in Figure 3.8.

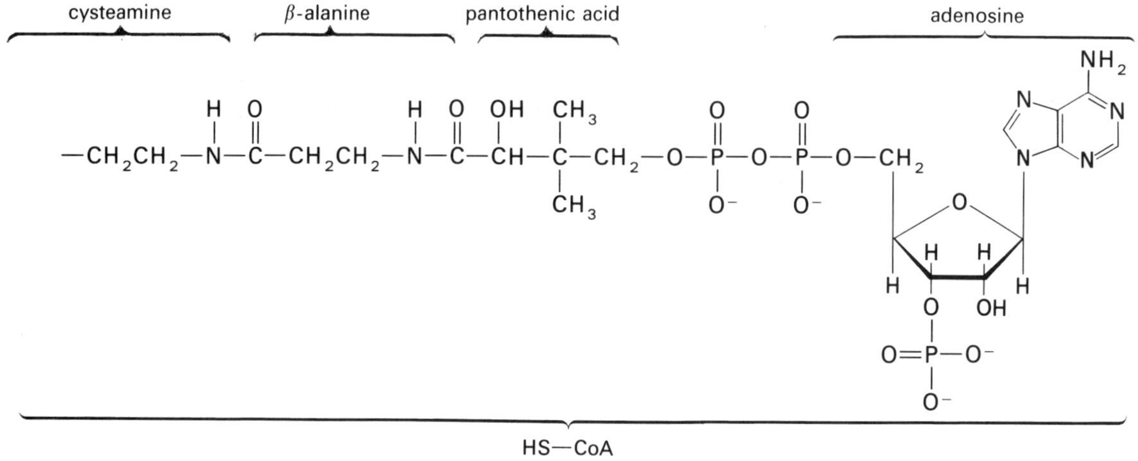

Figure 3.8

Coenzyme A.

The active part of the coenzyme A molecule is the sulfhydryl group. Thus, coenzyme A is often represented as CoASH or HSCoA. The overall form of CoASH-mediated acyl group transfer reactions is

$$\underset{\text{Carboxylic acid}}{R{-}CH_2CO_2^-} + HSCoA \xrightarrow[\text{circumstances}]{\text{Driving force depends on}} \underset{\text{Acyl-CoA}}{R{-}CH_2\overset{\overset{\displaystyle O}{\|}}{C}{-}S{-}CoA} + H_2O \longrightarrow \left[\begin{array}{c}\text{Variety of}\\ \text{acylation}\\ \text{reactions}\end{array}\right]$$

Note that the thiol-ester linkage in the acyl-CoA is analogous to an oxygen-ester function. Thiol-esters, however, are more reactive than their oxygen counterparts. We will see later, in Chapter 9, that the thiol-ester of acetic acid and Coenzyme A, acetyl CoA, is of particular importance.

Tetrahydrofolate

The transfer of one carbon fragment, such as a $—CH_3$, $—\overset{\overset{\displaystyle O}{\|}}{C}—H$ or $—CH_2OH$ group, represents an important step in biosynthesis. Such reactions generally involve enzymes which require tetrahydrofolic acid (THF) as a coenzyme (Chapter 12).

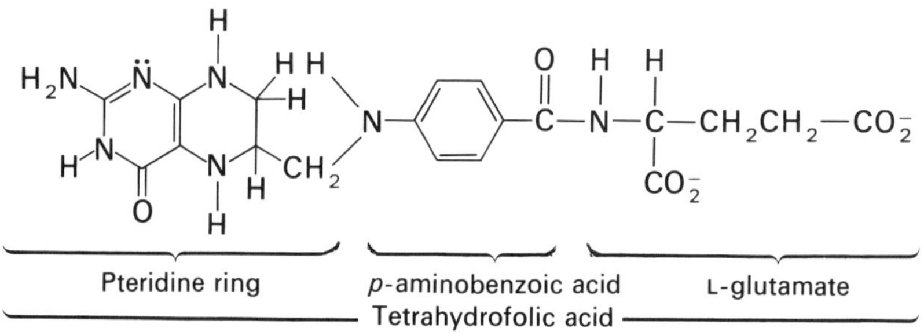

It is the pteridine ring in THF that directly takes part in the action of the coenzyme. THF is the coenzyme required in the biosynthesis of methane. It is also, as we shall see, associated with the action of the sulfa drugs.

Biotin

Many reactions in living systems involve the fixation of CO_2. In most cases enzyme-bound biotin acts as a carboxyl group carrier. The reaction of carbohydrate metabolism shown below illustrates the type of reaction catalyzed by biotin-requiring enzymes:

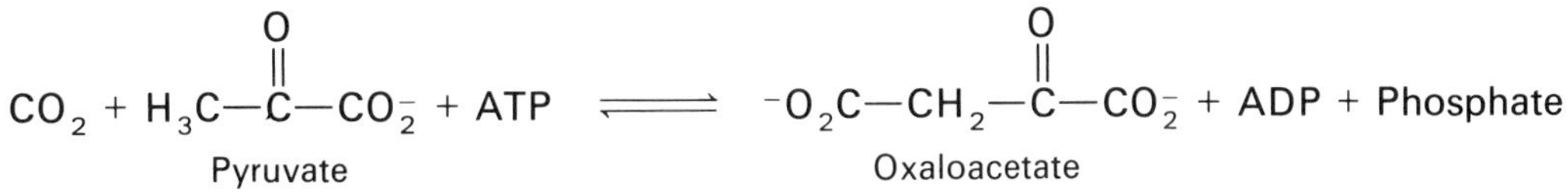

$$CO_2 + H_3C-\overset{\overset{O}{\|}}{C}-CO_2^- + ATP \rightleftharpoons {}^-O_2C-CH_2-\overset{\overset{O}{\|}}{C}-CO_2^- + ADP + Phosphate$$

$$\underset{\text{Pyruvate}}{} \qquad\qquad\qquad \underset{\text{Oxaloacetate}}{}$$

This particular reaction is catalyzed by pyruvate carboxylase and requires the phosphate bond energy of one mole of ATP to provide the driving force for the reaction. In tissues, biotin is found covalently bonded to proteins via the amide link shown below:

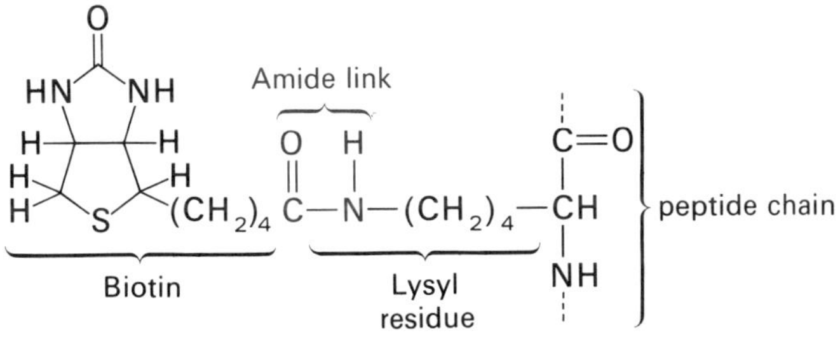

5'-Deoxyadenosyl Cobalamin (Vitamin B$_{12}$)

The nature of vitamin B_{12} is quite similar to the heme coenzymes. A close look at the structure of vitamin B_{12}, shown in Figure 3.9, reveals that the *corrin* ring system shown here is different from the porphyrin ring of the heme coenzymes. Also, Co(III) has replaced iron as the required metal ion. A general scheme for the rearrangement reactions catalyzed by B_{12}-requiring enzymes is

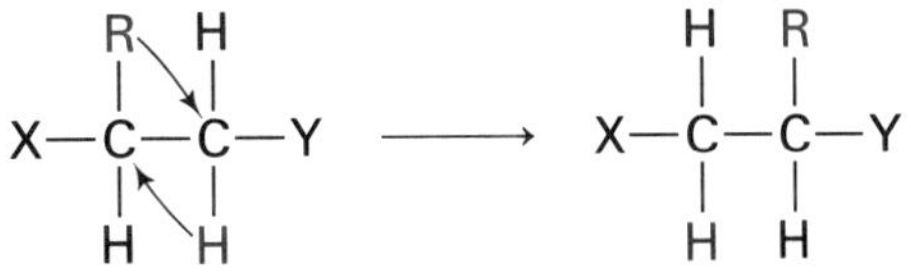

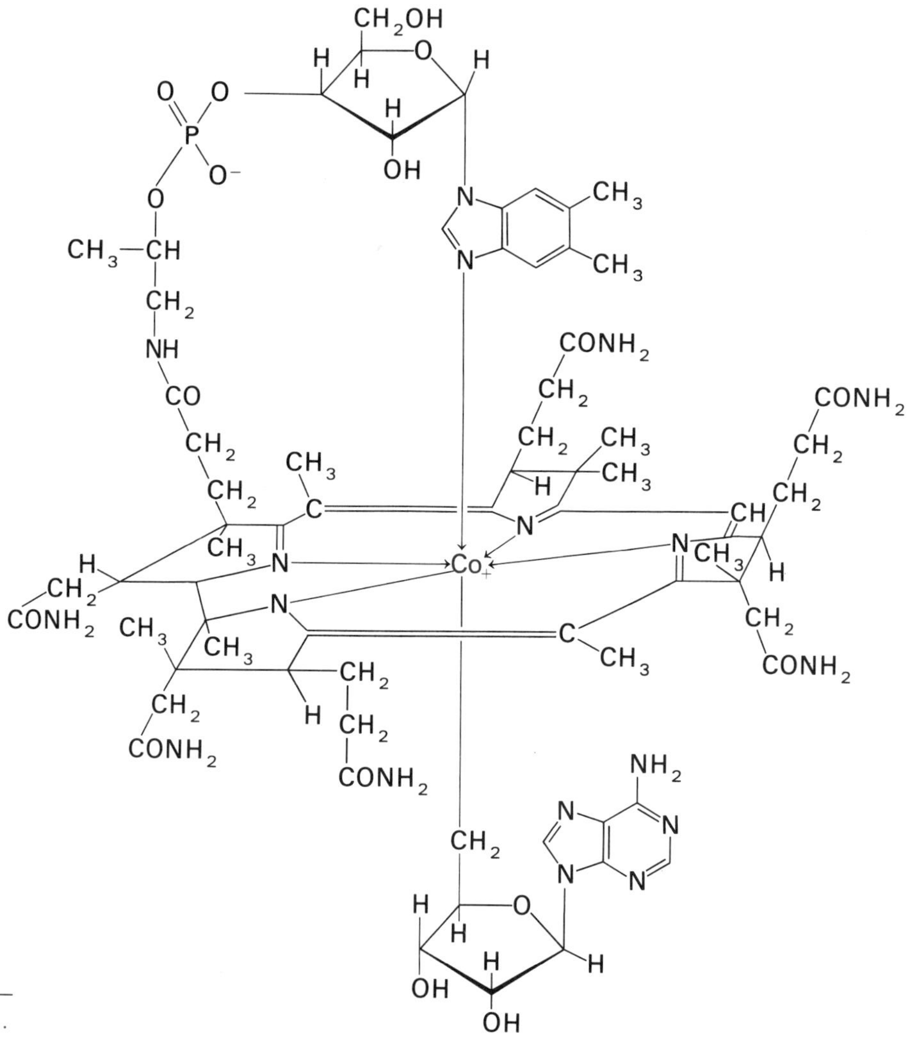

Figure 3.9

Structure of vitamin B_{12}.

Higher animals, as well as some microorganisms, have an absolute requirement for vitamin B_{12}. The biosynthesis of this vitamin by soil, water, and intestinal bacteria is an example of how higher animals are dependent on lower forms of life. It has been estimated that about 40% of the daily requirement of vitamin B_{12} in humans is provided by intestinal flora.

One important class of biochemical reactions remains to be discussed. This is the transfer of phosphoryl groups from a high-energy phosphoryl group donor to a phosphoryl group acceptor. In living systems, high-energy phosphoryl group donors are generated by the biochemical breakdown, or catabolism, of food molecules. Ultimately, as we shall see in Part II, the high-energy phosphoryl groups collect in the form of the major phosphoryl group donor in the cell, adenosine triphosphate (ATP):

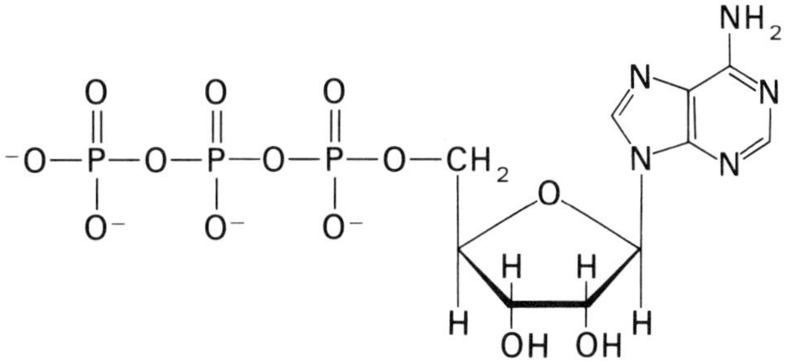

Adenosine triphosphate (ATP)

We can illustrate the general form of such a reaction with a hydroxy compound serving as a phosphoryl group acceptor, as shown below. This example is catalyzed by a kinase. We could consider the coenzyme to be ATP, but the distinction is not clear since we might also consider ATP to be a cosubstrate. The central role of ATP in cellular energetics will be discussed in detail in Chapter 7.

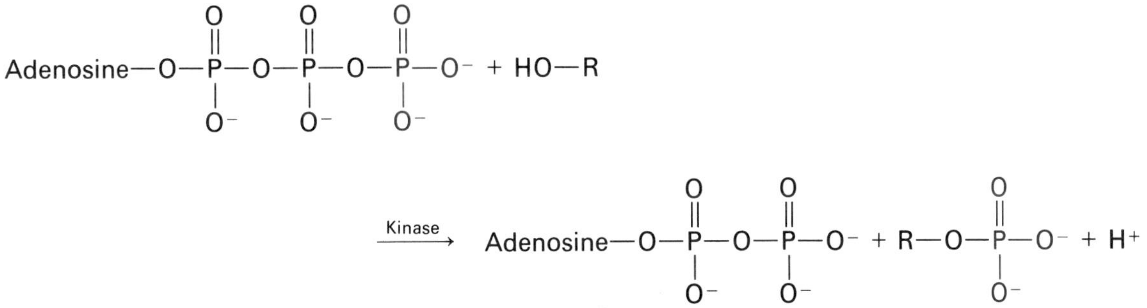

3.6

Mechanism of Enzyme Action

Since enzymes are catalysts, the study of the means by which enzymes bring about the transformation of substrate into product centers on the effect of various factors on the *rate* of an enzyme reaction. As with any reactant, it is important to define a set of conventions regarding enzyme activity or concentration. Molar concentration units are not always appropriate for enzymes because the molecular weight of the protein involved may not be known. Table 3.7 summarizes some of the conventions we use regarding enzyme activity.

Table 3.7   Accepted Conventions for the Expression of Enzyme Activity

1 activity unit	= amount of enzyme causing the transformation of 1.0×10^{-6} moles of substrate per minute at 25°C, under optimal conditions.
Specific activity	= number of units/mg protein
Turnover number	= number of substrate molecules acted on per unit of time and per active center

In purifying an enzyme, an increase in specific activity when going from step to step indicates that we are removing impurities. Ordinarily, we start with a mixture of proteins derived from a natural source. Since the total amount of protein is large, the specific activity of any given enzyme in such a mixture will be relatively low. Because purification involves separating the protein in question from the other proteins present in the mixture, successive purification steps increase the percentage of the enzyme in relation to the total mixture.

Many factors influence the rate of an enzyme reaction. Among the most important are the concentrations of substrate and enzyme. Some other key factors are temperature, pH, ionic strength, and the presence of inhibitors. In fact, anything that affects the tertiary structure of the enzyme protein will affect the rate of the enzyme reaction.

Effects of Temperature, pH, and Enzyme Concentration

In many systems the effect of temperature on an enzyme reaction is similar to the behavior shown in Figure 3.10. The reaction rate increases with rising temperatures, as we would predict. However, the curve begins to fall off at high temperatures and eventually the enzyme loses all activity as the protein becomes denatured by heat. Many enzymes function optimally in the temperature range of 25–37°C.

Figure 3.10

Effect of temperature on enzyme activity.

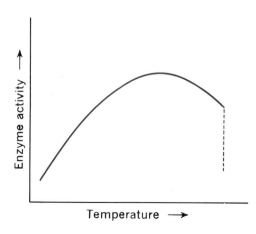

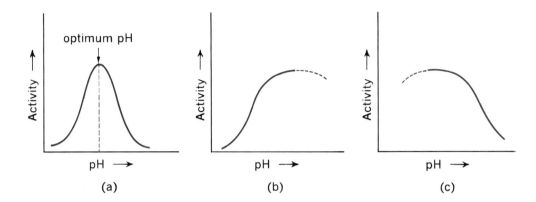

(a)   (b)   (c)

Figure 3.11

Effect of pH on enzyme
activity.

The effect of pH on an enzyme reaction is complicated by several factors which may be concurrent. Thus, we might see behavior corresponding to Figures 3.11(a), (b), or (c), depending on the nature of the enzyme. We often observe the behavior shown in Figure 3.11(a), where the pH reaches an optimum. The reaction rate falls off on either side of the pH optimum for any combination of three possible reasons:

1. The enzyme protein may become denatured by high or low pH extremes.
2. The enzyme may require ionizable amino acid side chain groups, which may be active in only one ionization state.
3. The substrate may gain or lose protons and be reactive in only one charge form.

While many enzymes are found to be most active around pH 7, some enzymes are maximally active at either high or low pH [Figure 3.!1(b), (c)], depending on the physiological context in which the enzyme operates. One example is the enzyme pepsin, the proteolytic enzyme of gastric juice, which has an optimum pH around 2.

The effect of enzyme concentration on the rate of an enzyme–catalyzed reaction is generally as shown in Figure 3.12. The rate increases linearly with

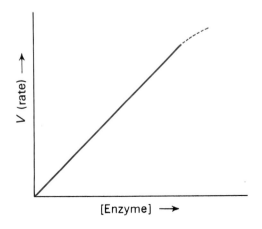

Figure 3.12

Effect of enzyme
concentration on reaction
rate.

increasing enzyme concentration as long as the concentration of enzyme is much less than that of the substrate. This is usually the case under physiological conditions.

Effect of Substrate Concentration

In the late 1800s and early 1900s, a number of scientists quantitatively studied the effect of substrate concentration on the rate of various enzyme reactions. As you can see from Figure 3.13, the rate of the enzyme-catalyzed reaction first increases with increasing substrate concentration. As we raise the substrate concentration still further, however, we achieve a limiting, or maximum, rate. Further augmentation of substrate concentration has no additional effect on the reaction rate. This phenomenon is called *saturation kinetics*. At the substrate concentrations that produce the maximum reaction rate, we can consider the catalyst to be "saturated with substrate." Saturation kinetics is observed in cases where a binding step is required for a reaction to proceed. In the case of an enzyme-catalyzed reaction, this means that the enzyme must first bind the substrate before the chemical reaction involving the substrate takes place. This corresponds to the general form:

Enzyme + Substrate $\rightleftharpoons$ Enzyme-Substrate Complex $\longrightarrow$ Enzyme + Products

This is the *Michaelis–Menten* mechanism for enzyme reactions which will be discussed in more detail in the Section 3.7. As Figure 3.13 shows, the concentration of substrate needed to give $\frac{1}{2}$ the maximum reaction rate ($\frac{1}{2} V_{max}$) is called K_M, the *Michaelis constant*. The Michaelis constant is related to the equilibrium constant for the dissociation of the enzyme-substrate complex into enzyme and substrate. A small value for K_M means that an enzyme-substrate complex is very stable: the enzyme has a high affinity for the substrate. Conversely, a large value for K_M means that the enzyme has a low affinity for the substrate: the equilibrium favors the enzyme + substrate, not the enzyme-substrate complex. In the next section, we will learn how we can measure K_M as we discuss the treatment of enzyme kinetics in more detail.

Figure 3.13

The effect of substrate concentration on the reaction rate of an enzyme-catalyzed reaction: saturation kinetics. Enzyme-catalyzed reactions increase in rate with growing substrate concentration until a limiting maximum rate, V_{max} is attained. The concentration of substrate needed to give a rate equal to $\frac{1}{2}V_{max}$ is labeled K_M, the Michaelis constant, and is a measure of enzyme-substrate affinity.

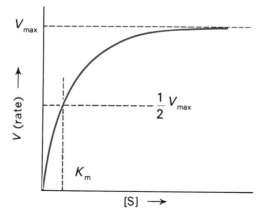

3.7

Enzyme Kinetics

The form of the curve in Figure 3.13 is a hyperbola, which for a chemical system suggests that the rate depends on the equilibrium

$$xy \rightleftharpoons x + y$$

where $[xy]$ and $[x]$ are constant.

In 1913, Michaelis and Menten proposed the following mechanism to account for the behavior of enzyme reactions:

$$\underset{(x)}{E} + \underset{(y)}{S} \rightleftharpoons \underset{(xy)}{ES} \xrightarrow{k_{cat}} E + Product$$

where E = free enzyme, ES = enzyme-substrate complex, and S = substrate, and $[S] \gg [E]$ and $[ES]$. The rate of appearance of the product, V, is given by the expression

$$V = k_{cat}[ES] \tag{1}$$

Since the concentration of the enzyme-substrate complex, ES, is governed by the equilibrium with E and S, it is clear that the Michaelis–Menten mechanism gives us the observed hyperbolic velocity/substrate curve. However, the form of the rate expression given in Eq. (1) is not informative, since we generally cannot measure $[ES]$ directly. The question then becomes: what is an expression for V in terms of quantities we can measure?

Consider the equilibrium between ES, E, and S expressed as a dissociation reaction:

$$ES \underset{k_r}{\overset{k_f}{\rightleftharpoons}} E + S \tag{2}$$

We can define an equilibrium constant, K_s, which is a quantitative measure of the affinity between the enzyme and the substrate:

$$K_s = \frac{[E][S]}{[ES]} = \frac{k_f}{k_r} \tag{3}$$

If we have a "good" substrate, K_s will be small; the equilibrium concentration of ES will be large compared to that of the free enzyme, reflecting high enzyme-substrate affinity. Since the reaction rate depends directly on the concentration of ES, by Eq. (1), the rate will be faster than it would be for a poor substrate. A "poor" substrate will yield a larger value of K_s, a lower equilibrium concentration of ES, and a correspondingly slower rate.

The total amount of enzyme present is a quantity we can measure: it is the amount of enzyme originally added to the reaction mixture. By a simple consideration of conservation of matter, the total enzyme concentration, $[E]_0$, is given by Eq. (4):

$$[E]_0 = [E] + [ES] \tag{4}$$

By manipulating Eqs. (3) and (4) we can obtain an expression for [ES] in terms of quantities we can measure:

$$K_s = \frac{[E]_0[S] - [ES][S]}{[ES]}; \quad \text{thus } [ES] = \frac{[E]_0[S]}{K_s + [S]} \tag{5}$$

If we insert the result of Eq. (5) into Eq. (1), we get the Michaelis–Menten rate expression:

$$\text{rate} = V = \frac{k_{cat}[E]_0[S]}{K_s + [S]} \tag{6}$$

Let's stop and see if this result fits the experimental observations by applying Eq. (6) to two limiting cases.

Case 1   [S] is much larger than K_s. In Eq. (6), $K_s + [S] \cong [S]$. So

$$V \cong \frac{k_{cat}[E]_0[S]}{[S]} = k_{cat}[E]_0 \equiv V_{max}$$

Physically this means that the enzyme is saturated with substrate. By Le Châtelier's principle, the equilibrium between E, ES, and S shifts almost completely toward ES. Thus, a limiting rate, V_{max}, is predicted at high substrate concentrations.

When $V = V_{max}/2$, it can be shown that $[S] = K_s$ (see Problem 5).

Case 2   [S] is much smaller than K_s. In Eq. (6), the term $K_s + [S] \cong K_s$. Therefore, under these conditions, Eq. (6) becomes

$$V = \frac{k_{cat}[E]_0[S]}{K_s}$$

This is a first-order rate expression with respect to substrate. Experimentally, the rate varies linearly with substrate concentration in the initial part of the velocity/substrate curve.

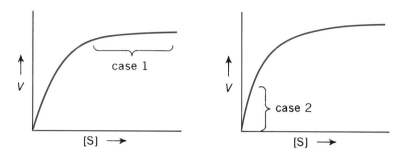

Since the terms K_s and k_{cat} define the nature of a given enzyme–substrate system, it is useful to measure these quantities experimentally. We can linearize the Michaelis–Menten rate equation to facilitate the analysis of experimental data used in determining K_s and k_{cat}. There are several ways to do this. The most direct way is to take the reciprocal of both sides of Eq. (6), which gives rise to the Lineweaver–Burk expression given in Eq. (7):

$$1/V = 1/k_{cat}[E]_0 + (K_s/k_{cat}[E]_0)(1/[S])$$

or

$$1/V = 1/V_{max} + (K_s/V_{max})(1/[S]) \tag{7}$$

Equation (7) defines a straight line. A plot of $1/V$ vs $1/[S]$ is called a *Lineweaver–Burk plot*, or a double reciprocal plot. This is shown in Figure 3.14.

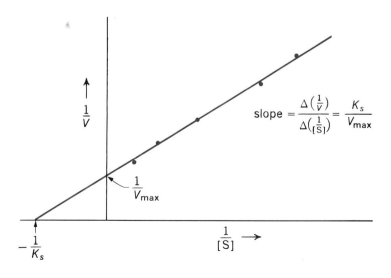

Figure 3.14

Lineweaver–Burk plot.

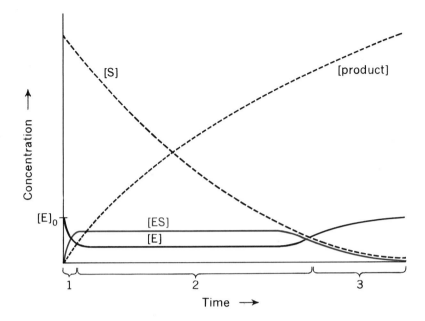

Figure 3.15

Variation of [E], [ES], and [S] in an enzyme-catalyzed reaction.

It is important to note that in deriving the Michaelis–Menten rate expression we have assumed that S, E, and ES are all in equilibrium with each other. If this were absolutely true, there could be no formation of product; for if product were formed, the system would not be at equilibrium. The actual physical situation might be better visualized by considering the concentrations of the various participants in the enzyme reaction as functions of time. This is summarized in Figure 3.15. We can define three periods with respect to the time course of the enzyme reaction:

1. Pre-steady state period.

2. Steady state period, where [E] and [ES] do not change. This constitutes the major part of the reaction.

3. Post-steady state period.

The only complication added is that the formation of product must be taken into account. We handle this by defining the quantity K_M, the Michaelis constant; the true equilibrium constant K_s is often difficult to measure experimentally. Referring back to Eq. (3), we can define K_M by Eq. (8):

$$K_M = \frac{k_f + k_{cat}}{k_r} \qquad (8)$$

The quantity determined experimentally by rate measurements and Lineweaver–Burk plots is K_M. For many enzyme-substrate systems, the quantity k_{cat} in

Eq. (8) is small compared with k_f. Thus, K_M and K_s are often nearly equal. In any case, the form of the Michaelis–Menten rate expression is the same whether we use K_M or K_s.

At the time Michaelis and Menten proposed their mechanism, there was no direct evidence for an actual enzyme-substrate complex. It was not until 1943 that the first experimental evidence for an enzyme–substrate intermediate was reported by Chance for the enzyme peroxidase. Since then, one or more ES intermediates or complexes have been shown to exist for many enzymes.

In considering the effect of substrate concentration on an enzyme that requires a coenzyme, we must be sure that the coenzyme concentration does not limit the rate. In general, we can consider the coenzyme to be a second substrate, since it must bind to the apoenzyme before binding to the substrate:

$$E + \text{Coenzyme} \rightleftharpoons E \cdot \text{Coenzyme}$$

$$E \cdot \text{Coenzyme} + S \rightleftharpoons E \cdot \text{Coenzyme} \cdot S \longrightarrow E \cdot \text{Coenzyme} + \text{Product}$$

This scheme pertains to many enzymes. If we hold the substrate concentration constant and vary the coenzyme concentration, we can determine a "K_M" for the coenzyme in a manner analogous to that for the substrate, shown in Figure 3.14. In the case of coenzyme binding, the reciprocal of the coenzyme concentration would be the y-axis of the plot, and the binding constant obtained would be for the dissociation of the apoenzyme–coenzyme complex (holoenzyme).

3.8

Enzyme Specificity

The interaction of small molecules with large macromolecules is a major phenomenon in molecular biology. The binding of a molecule by a macromolecule is fundamental to the action of antibodies, hormones, toxins, odor and taste perception, and the action of enzymes. Enzymes as catalysts exhibit a high degree of specificity with respect to the substrate. That is, the shape of the substrate molecule is important in ensuring that not only does the substrate bind to the enzyme, but the part of the substrate molecule acted upon is presented in the proper orientation with respect to the active groups of the enzyme.

The Active Center

There is a location in the three-dimensional structure of an enzyme where the substrate is bound and acted upon. This is the *active center* or active site. Figure 3.16 pictures this idea for a hypothetical enzyme. It is important to

Figure 3.16

A schematic representation of an active center. The color part of the substrate represents the bond to be broken. The numbered R groups represent amino acid side chain groups in the primary structure of the protein. Note how the active center can consist of segments of the protein chain which are widely separated in the amino acid sequence of the protein.

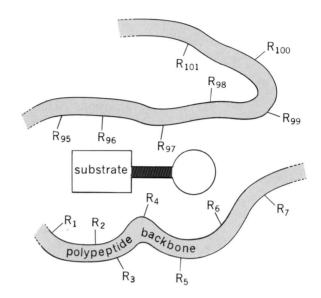

note that catalytically important groups of the enzyme may be widely separated in the primary structure of the protein. Their juxtaposition at the active site results from the folding of the polypeptide chain in the tertiary structure of the protein. The location of the amino acid side chain (R) groups of the enzyme protein is precisely specified by that part of the DNA strand (gene) coded for the amino acid sequence of the protein. Some of the R groups may take part in substrate binding, while others may have a catalytic function.

A three-dimensional representation of an enzyme is given in Figure 3.17 for the case of α-chymotrypsin. It is clear that only a small region of the protein participates in catalysis. The remainder of the structure is necessary mainly to organize and stabilize the active center.

Over the years, several hypotheses have been advanced to explain the function of the active center and enzyme catalysis. In 1894, Emil Fischer recognized the close stereochemical relationship between an enzyme and its substrate. Fischer proposed that an enzyme related to a substrate like a lock to a key. This lock-and-key hypothesis accounts for enzyme specificity, but implies a rigid active center. The analogy is therefore an oversimplification because we now have considerable experimental evidence for conformational changes in many enzymes during the course of the enzyme-substrate interaction. The evidence suggests that the enzyme's active center may adapt itself to the substrate during the formation of the ES complex. This structural or conformational adaptation in active center geometry, induced by the binding of a substrate molecule, would bring to bear R groups of the enzyme in the most efficient orientation for binding and catalysis. This is the induced–fit hypothesis of Koshland (1959). The essential features of the induced–fit mechanism for enzyme action are summarized in Figure 3.18.

Figure 3.17

A schematic drawing representing the conformation of the polypeptide chains in α-chymotrypsin. [From P. B. Sigler, *et al.*, *J. Mol. Biol.*, **35**, 143 (1968), with permission.] Note that there are three chains (A, B, and C) in this protein covalently linked by disulfide bridges. The active center region is shown in color, and the essential histidine and serine residues are indicated.

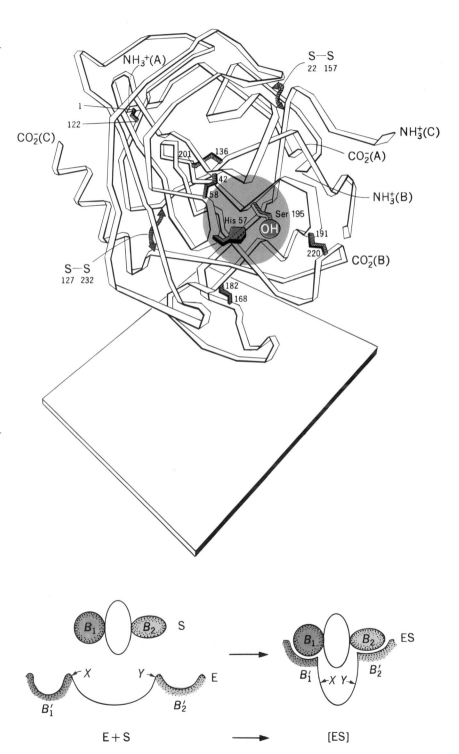

Figure 3.18

A representation of the induced-fit mechanism. The binding of the substrate through the interactions of groups B_1 and B_2 with groups B_1' and B_2' causes a change in the shape of the enzyme, so that catalytic groups X and Y in the active center of the enzyme are brought to the proper position relative to the substrate molecule. (From W. P. Jencks, *Catalysis in Chemistry and Enzymology*, McGraw-Hill, New York, 1969, used with permission.)

Table 3.8 Groups Associated with the Active Centers of Enzymes.

Group	Structure	Function
Catalytic groups		
Coenzymes (prosthetic groups)	—	Determine the nature of the enzyme reaction
Imidazole (histidine)	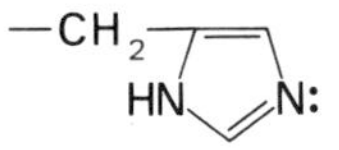 $-CH_2$ (imidazole ring, HN, N:)	Nucleophilic catalyst Base catalyst (Binds positively charged groups)
Hydroxyl (serine, tyrosine)	$-CH_2-\ddot{O}-H$	Nucleophilic catalyst (Binds positively charged groups)
Thiol (cysteine)	$-CH_2-\ddot{S}-H$	
Binding and positioning groups		
Carboxyl (glutamic acid, aspartic acid)	$-CH_2-C\underset{O^-}{\overset{O}{\big\langle}}$	Binds positively charged groups
Ammonium (arginine, lysine)	$-CH_2-NH_3^+$	Binds negatively charged groups
Alkyl (hydrophobic amino acids)	(hydrocarbon)	Bind hydrophobic groups

In addition to a possible coenzyme or metal ion cofactor, or both, a number of amino acid side chain groups are associated with the function of the active centers of enzymes. Some of these groups are listed in Table 3.8.

Types of Enzyme Specificity

We can distinguish between several types of enzyme specificity. Some enzymes act upon one and only one substance. Other enzymes act upon a given type of functional group or linkage, and still others accept only one optical isomer of an optically active substrate. Each specific case depends on the fact that a substrate molecule is three-dimensional in nature and must, therefore, have a specified relationship to the shape and chemical properties of the enzyme active center.

Absolute specificity   There are many enzymes which act upon one substance and no known others. An example of such a system is fumarase, an enzyme of the Krebs cycle (Chapter 9). Fumarase catalyzes exclusively the hydration of the double bond of fumarate, yielding L-malate, and is thus classed as a lyase:

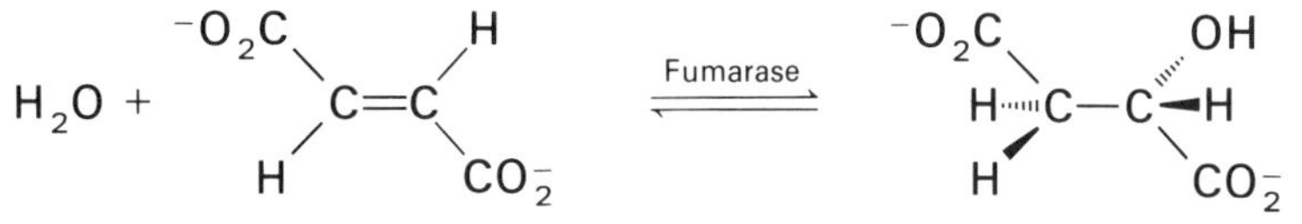

The structures of a number of nonsubstrates of fumarase are given in Figure 3.19.

Substances which are not substrates for a given enzyme may fail to interact with the enzyme for either of two reasons. In one case, the shape of the molecule or the nature of the functional groups may not permit the substrate to bind to the active center. In the second case, the substrate may bind to the active site, but be unable to undergo the reaction catalyzed by the enzyme. In this instance the substance may act as an inhibitor, since it interferes with the binding of a substrate molecule. This is an example of a competitive inhibitor (see Section 3.9).

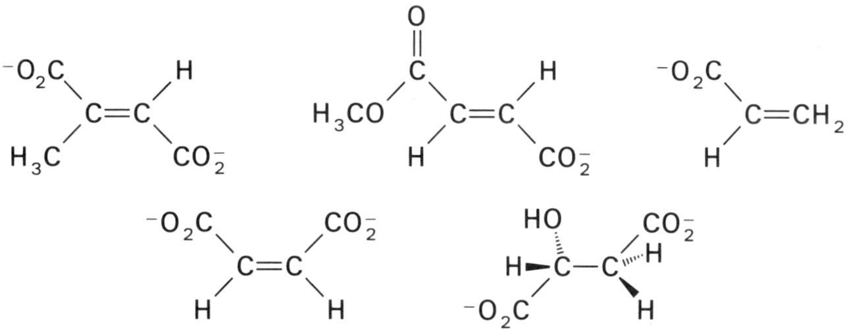

Figure 3.19

Some molecules which appear similar to fumarate and L-malate yet do not act as substrates for the enzyme fumarase.

Group-specific enzymes   Enzymes which act upon a given type of functional group are common. In fact, some of these have already been discussed in Chapter 2 in connection with determining the primary structure of proteins. Enzymes such as trypsin, chymotrypsin, and pepsin are group specific in that they catalyze the hydrolysis of a peptide bond between a number of different amino acid pairs. Table 3.9 gives examples of group-specific hydrolase enzymes.

Table 3.9 Group Specificity in Hydrolyzing Enzymes.

Type	Linkage acted upon	Examples
Phosphatases	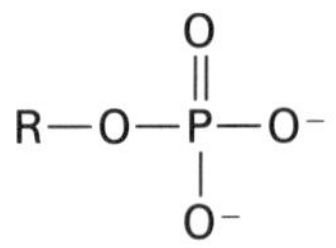 $R-O-\overset{\overset{O}{\|\|}}{\underset{\underset{O^-}{\|}}{P}}-O^-$	Acid phosphatase Alkaline phosphatase
Esterases	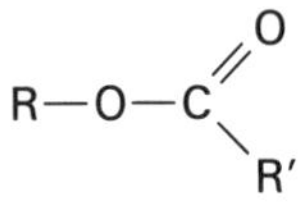 $R-O-C\overset{\diagup O}{\diagdown R'}$	Lipase Acetylcholinesterase
Proteases (peptidases)	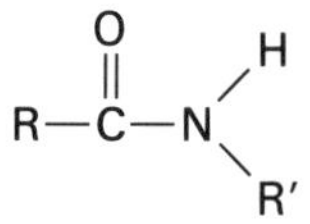 $R-\overset{\overset{O}{\|\|}}{C}-N\overset{\diagup H}{\diagdown R'}$	Trypsin α-Chymotrypsin Pepsin

Stereospecific enzymes Many enzymes interact with optically active substances where only one of the enantiomers reacts as a substrate. Such enzymes are said to be *stereospecific*. Group-specific enzymes may exhibit stereospecificity as well. The structural requirements of a substrate for reacting with group-specific enzymes are:

1. Reactive linkage or functional group } nonstereospecific
2. Binding or orienting group
3. Additional binding or orienting group} stereospecific

An illustration of a substrate meeting the requirements for a non–stereospecific enzyme is acetylcholine. Acetylcholine is hydrolyzed by the action of acetylcholinesterase, a crucial reaction in the biochemistry of nerve impulse transmission:

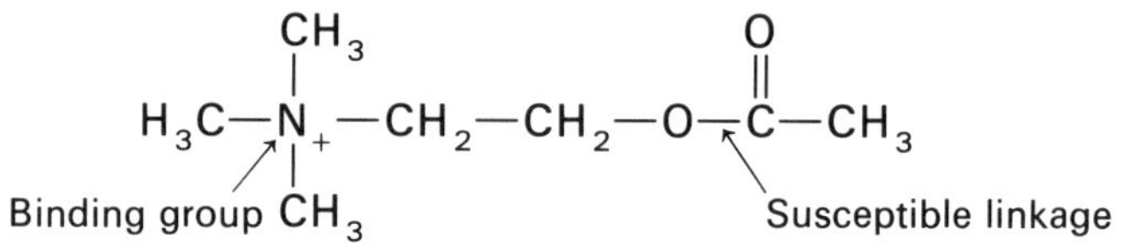

$$H_3C-\overset{\overset{CH_3}{\|}}{\underset{\underset{CH_3}{\|}}{N^+}}-CH_2-CH_2-O-\overset{\overset{O}{\|\|}}{C}-CH_3$$

Binding group   Susceptible linkage

Acetylcholine

The action of a stereospecific enzyme requires three points of interaction with a substrate molecule. Figure 3.20 shows how stereospecificity can arise, using an amino acid as an example. In this hypothetical case, the groups of the enzyme, represented by Ⓐ, B̄, and Ĉ, would be either catalytic or binding functions essential to the action of the enzyme.

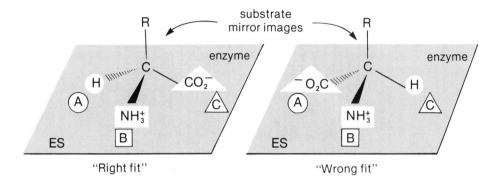

"Right fit"   "Wrong fit"

Figure 3.20

Interaction of the D and L isomers of an optically active substrate with a stereospecific enzyme.

The concept of enzyme specificity plays a vital role in the functioning of a cell. In the soluble cytoplasm, many enzyme-catalyzed reactions occur simultaneously in the same reaction medium. No interference among the various reactions takes place except in a highly specified way, involving some enzymes associated with the control of the reaction pathways taking place. We will consider the control of cellular reaction rate shortly.

3.9

Inhibition of Enzyme Action

Many substances act to inhibit, or decrease, the rate of enzyme reactions. Some inhibitors are potent poisons; others are valuable chemotherapeutic agents. We can distinguish two major types of enzyme inhibitors, differentiated by the mode of action. These are competitive inhibitors, which "compete" with the substrate for the active center, and non–competitive inhibitors, which affect the enzyme in such a way as to allow the substrate to bind to the active center but not to react to form product.

Competitive Inhibition

A substance which acts as a competitive inhibitor for a given enzyme is often a "look-alike," or more fundamentally, a "feel-alike" molecule with respect to a substrate molecule. Thus, the examples of nonsubstrates for fumarase given in Figure 3.19 are all competitive inhibitors of that enzyme. We can rework this reaction into a Michaelis–Menten format as

$$E + S \; \rightleftharpoons \; ES \text{ (active)} \; \longrightarrow \; E + Product \qquad (9)$$

$$E + I \; \rightleftharpoons \; EI \text{ (inactive)} \qquad (10)$$

In this scheme, the inhibitor I competes with S for the active site. By Le

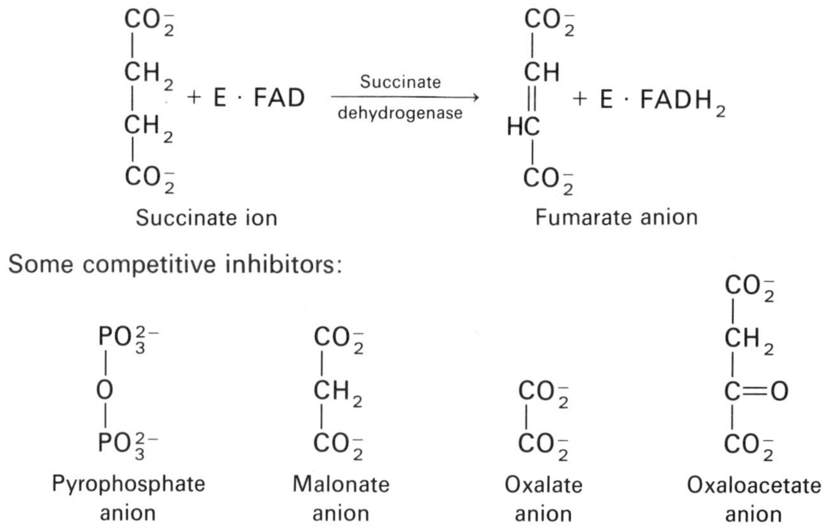

$$\underset{\text{Succinate ion}}{\begin{matrix} CO_2^- \\ | \\ CH_2 \\ | \\ CH_2 \\ | \\ CO_2^- \end{matrix}} + E \cdot FAD \xrightarrow[\text{dehydrogenase}]{\text{Succinate}} \underset{\text{Fumarate anion}}{\begin{matrix} CO_2^- \\ | \\ CH \\ \| \\ HC \\ | \\ CO_2^- \end{matrix}} + E \cdot FADH_2$$

Some competitive inhibitors:

Figure 3.21

Competitive inhibitors of succinate dehydrogenase. Note the resemblance of the inhibitors to the actual substrate of this enzyme.

$$\underset{\substack{\text{Pyrophosphate} \\ \text{anion}}}{\begin{matrix} PO_3^{2-} \\ | \\ O \\ | \\ PO_3^{2-} \end{matrix}} \qquad \underset{\substack{\text{Malonate} \\ \text{anion}}}{\begin{matrix} CO_2^- \\ | \\ CH_2 \\ | \\ CO_2^- \end{matrix}} \qquad \underset{\substack{\text{Oxalate} \\ \text{anion}}}{\begin{matrix} CO_2^- \\ | \\ CO_2^- \end{matrix}} \qquad \underset{\substack{\text{Oxaloacetate} \\ \text{anion}}}{\begin{matrix} CO_2^- \\ | \\ CH_2 \\ | \\ C{=}O \\ | \\ CO_2^- \end{matrix}}$$

Châtelier's principle, if [S] increases, the equilibrium concentration of ES must increase at the expense of EI. Thus, we can reverse the effect of a competitive inhibitor by simply increasing the substrate concentration. Another enzyme of the Krebs cycle (Chapter 9) provides a good illustration of the "look-alike" nature of competitive inhibitors. Figure 3.21 shows the succinate dehydrogenase system together with several competitive inhibitors.

As in the case of the binding of the substrate to an enzyme, we can define a dissociation constant for the equilibrium shown in Eq. (10). This is the *inhibition constant* $K_I = [E][I]/[EI]$. As with K_M, a *small* value for K_I reflects enzyme-inhibitor affinity and thus more complete inhibition of the enzyme. We can determine the value of K_I for a given enzyme-inhibitor system by performing rate experiments with and without the inhibitor present and then constructing the corresponding Lineweaver–Burk plots, as shown in Figure 3.22. Note that V_{max} remains unaffected by the presence of a competitive inhibitor while K_M *increases* in value by the factor $(1 + [I]/K_I)$.

Noncompetitive Inhibition

The noncompetitive inhibitor affects the enzyme either at a region other than the active center, or at the active center but without affecting the binding of substrate. Usually a noncompetitive inhibitor does not "look like" the substrate. The Lineweaver–Burk plot for the case of noncompetitive inhibition appears as in Figure 3.23. In the case of noncompetitive inhibition, K_M is unaffected: the substrate is free to bind to the active center. However, V_{max} is decreased by the factor $1/(1 + [I]/K_I)$. The breakdown of ES into E + Product is interfered with by a noncompetitive inhibitor, so its effect cannot be reversed by the addition of more substrate. As you can see from a comparison of Figures 3.22 and 3.23, competitive and noncompetitive inhibition are readily distinguished experimentally.

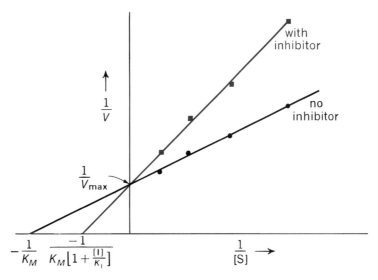

Figure 3.22

Lineweaver–Burk plots
for an enzyme subject to
the action of a
competitive inhibitor.

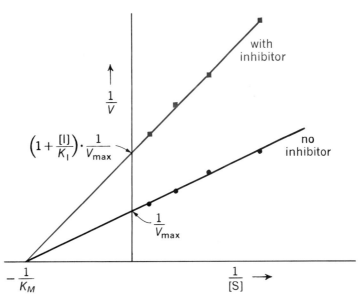

Figure 3.23

Lineweaver–Burk plot
for the case of
noncompetitive
inhibition.

Many heavy-metal ions act as noncompetitive inhibitors. For example, Ag^+, Hg^{2+}, and Pb^{2+} often inhibit enzymes in this fashion by forming mercaptide salts with sulfhydryl groups of the enzyme protein:

$$Ag^+ + R—SH \rightleftharpoons AgSR + H^+$$

Enzymes that require a metal ion cofactor are noncompetitively inhibited by complexing agents which remove the metal ion from the enzyme. For example, cyanide and sulfide ions are both highly toxic due to their ability to form strong complexes with transition-metal ions such as Fe^{2+}.

Some noncompetitive inhibitors bind to the enzyme at locations other than the active center. In doing so, the inhibitor induces a change in the tertiary or quaternary structure of the protein which disrupts the active center. This causes a decrease in activity, as represented in Figure 3.24.

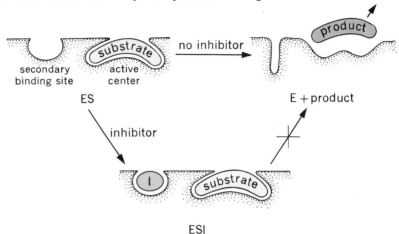

Figure 3.24

Noncompetitive inhibition by an inhibitor binding to a region of the enzyme other than the active center.

Enzyme Inhibitors as Poisons

Many substances inhibit enzymes by either affecting the nature of the coenzyme or by irreversibly affecting a group at the enzyme active center. For example, nerve gases are of the general structure shown:

General structure for organophosphorus nerve gases

Di-isopropylphospho-fluoridate

Such compounds combine irreversibly with an essential serine —OH group at the active center of acetylcholinesterase, preventing the transmission of nerve impulses. Substances like di-isopropylphospho–fluoridate appear to act preferentially on serine residues at the active centers of *serine hydrolases* such as acetylcholinesterase; other serine residues of the enzyme are not affected.

Enzymes requiring metal ions are generally inhibited by substances that can form strong complexes with the metal ion. For example, oxalate anion, $C_2O_4^{2-}$, forms insoluble oxalate salts with Ca^{2+} and Mg^{2+} ions, both essential ions for many enzymes. Spinach and rhubarb leaves contain significant amounts of oxalate ion and people eating excessive amounts of these foods may develop symptoms of oxalate poisoning. This is particularly true for rhubarb.

Both cyanide ion, CN^-, and sulfide ion, S^{2-}, exert their toxic effect in a similar way. They both form strong complexes with the Fe^{3+} ion of the heme protein cytochrome oxidase, a key enzyme in respiration and the only one that reacts directly with oxygen (see Chapter 9). Because the pathway by which oxygen is utilized is blocked, the effect is the same as suffocation. Actually, sulfide ion is the more toxic of the two. We often encounter this ion in the chemistry lab as H_2S, hydrogen sulfide, a gas that smells like rotten eggs. Fortunately, we are able to smell H_2S in concentrations far below the toxic level.

Carbon monoxide, CO, is an odorless gas that competes with O_2 for the oxygen binding sites on hemoglobin. Carbon monoxide is able to displace oxygen from hemoglobin because it binds over 200 times more strongly to hemoglobin than does O_2. There is evidence that CO only partially replaces the oxygen in oxyhemoglobin but in doing so, the remaining O_2 molecules are more strongly bound, thus making them less available to respiring tissues.

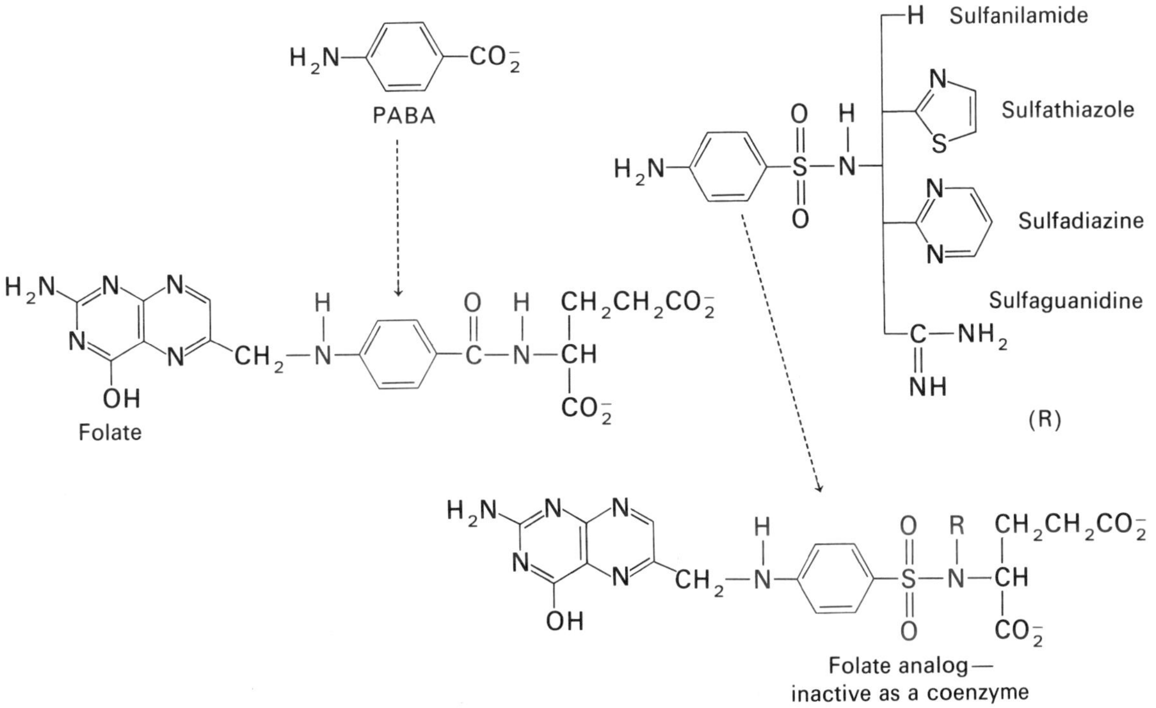

Figure 3.25

Structure and action of sulfa drugs. Bacteria using sulfa drugs to biosynthesize folic acid die of folic acid deficiency because the sulfa-containing analog created is not active as a coenzyme.

Enzyme Inhibitors as Drugs

Many important drugs work on the principle that all higher animals and many microorganisms lack the ability to synthesize certain essential organic compounds (metabolites). Such drugs are called *antimetabolites*. For example, many bacteria require para amino benzoic acid (PABA) to biosynthesize the coenzyme folic acid. Sulfa drugs are similar enough to PABA that bacteria are able to biosynthesize an *inactive* folic acid analog from sulfa drugs, as shown in Figure 3.25. The purpose of the sulfa drugs is to artificially induce a folic acid deficiency disease in harmful bacteria.

Antibiotics are substances produced by organisms that are toxic to other organisms. For example, the fungus *Penicillium notatum* produces *penicillin*, the first known antibiotic. Penicillin has become one of the most important of all drugs because it is so effective against many different bacteria. A number of natural penicillins are produced by different strains of Penicillium in which the

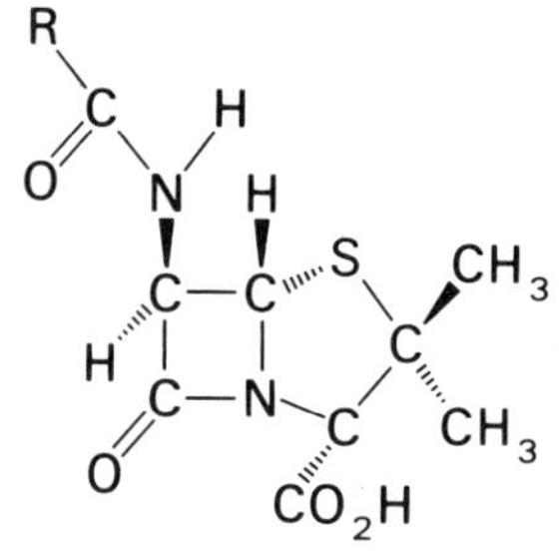

Penicillin

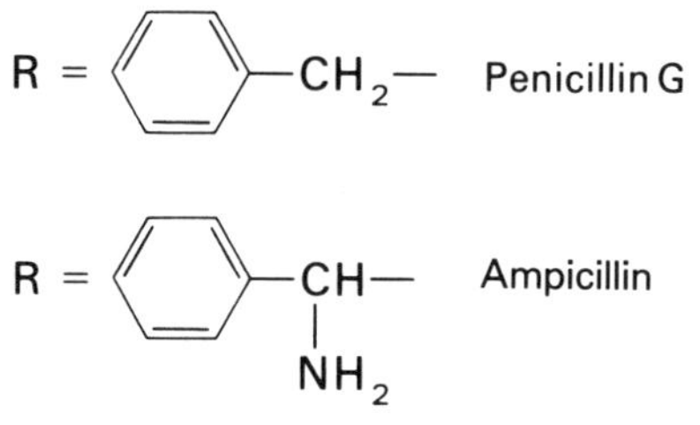

R group in the structure above can vary. Penicillin kills growing bacteria by acting as a noncompetitive inhibitor of a key enzyme involved with the biosynthesis of the bacterial cell wall (see Chapter 4). Without completed cell walls for protection, the bacteria rupture and die. There are other important forms of penicillin in which a natural penicillin is synthetically modified to produce an antibiotic with more desirable properties than the original parent compound. One such semi-synthetic penicillin is ampicillin (R = aminobenzyl) which has a broader spectrum of action than penicillin G.

Another interesting family of antibiotics is based on rifamycin, an antibiotic produced by the mold, *Streptomyces mediterranei*. Rifampicin is a closely related substance which has turned out to be an efficient drug for the

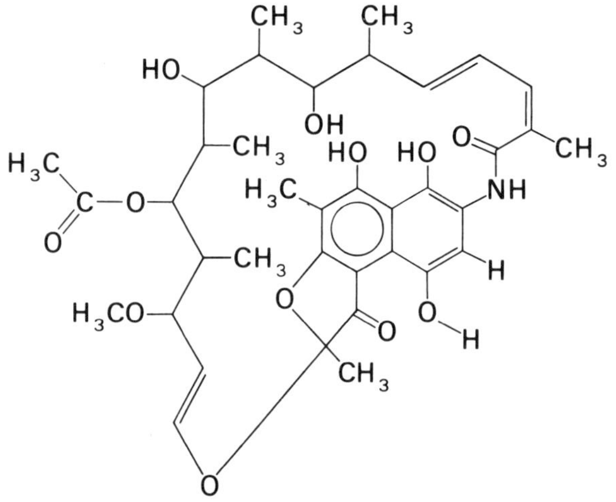

Rifampicin

treatment of tuberculosis. Rifampicin is a powerful inhibitor of the action of bacterial RNA polymerase $(K_1 = 2 \times 10^{-8}M)$, a key enzyme in protein biosynthesis. This prevents growing bacteria from synthesizing RNA and consequently proteins, thus leading to their death (see Chapter 13).

There are many antibiotics in common use today. The mode of action may differ for each family of antibiotics, as the two examples given here illustrate. Table 3.10 gives the names and modes of action for a few important antibiotics.

Table 3.10   Some antibiotics and their modes of action.

Antibiotic	Mode of Action
Penicillin (family)	Inhibits bacterial cell wall synthesis.
Tetracycline (family)	Inhibits bacterial protein synthesis by blocking the binding of aminoacyl transfer RNA molecules to the ribosomes (see Chapter 13).
Streptomycin (family)	Interferes with bacterial protein synthesis.
Tyrocidin	Disrupts the permeability of bacterial membranes to ions.
Amphotericin	Disrupts the proton permeability in the membranes of fungi.

There are other ways in which substances can inhibit enzymes. Some enzymes are subject to inhibition by certain molecules which act neither competitively nor noncompetitively, in that both K_M and V_{max} are affected. In addition, a certain kind of inhibition involving a special class of regulatory enzymes forms the basis of the self-regulation of metabolic pathways. This type of allosteric, or feedback inhibition, is considered in the next section.

3.10

Control of Enzyme Action

The chemistry of a living cell is largely sequential in nature. We can compare the cell's metabolic pathways to the branches of a tree, since in many cases the control of an entire pathway may involve regulating the activity of only one key enzyme (the "trunk" of the tree).

There are basically two ways to control the rate of an enzyme reaction. The first one regulates the concentration of active enzyme by either controlling the rate at which the enzyme protein is produced or by affecting the conversion between active and inactive enzyme forms. Control of enzyme reactions by regulating the rate of protein synthesis is a crude system because of the time lag involved in starting and stopping protein biosynthesis. The second method controls the activity of a enzyme through feedback inhibition. This type of regulation is capable of rapid responses to changes in demand for a given metabolite.

Feedback Control

Feedback control is present in all living organisms and exemplifies the high degree of efficiency associated with the utilization of materials and energy in living systems. During our discussion of metabolism we will discuss numerous examples of feedback control (see Parts II and III).

The concept of feedback control of enzyme reactions was first recognized in the late 1950s by Umbarger. The system he studied comprised the biosynthesis of isoleucine from threonine. These reactions are shown in Figure 3.26. The overall form of regulation by end-product inhibition is apparent from this example. The controlled step in a reaction sequence is catalyzed by an enzyme which possesses, in addition to the usual catalytic properties, susceptibility to inhibition or activation by other substances (modulators) that serve to control

Figure 3.26

Biosynthesis of isoleucine from threonine, showing control by feedback inhibition of the first step in the pathway through the end product isoleucine.

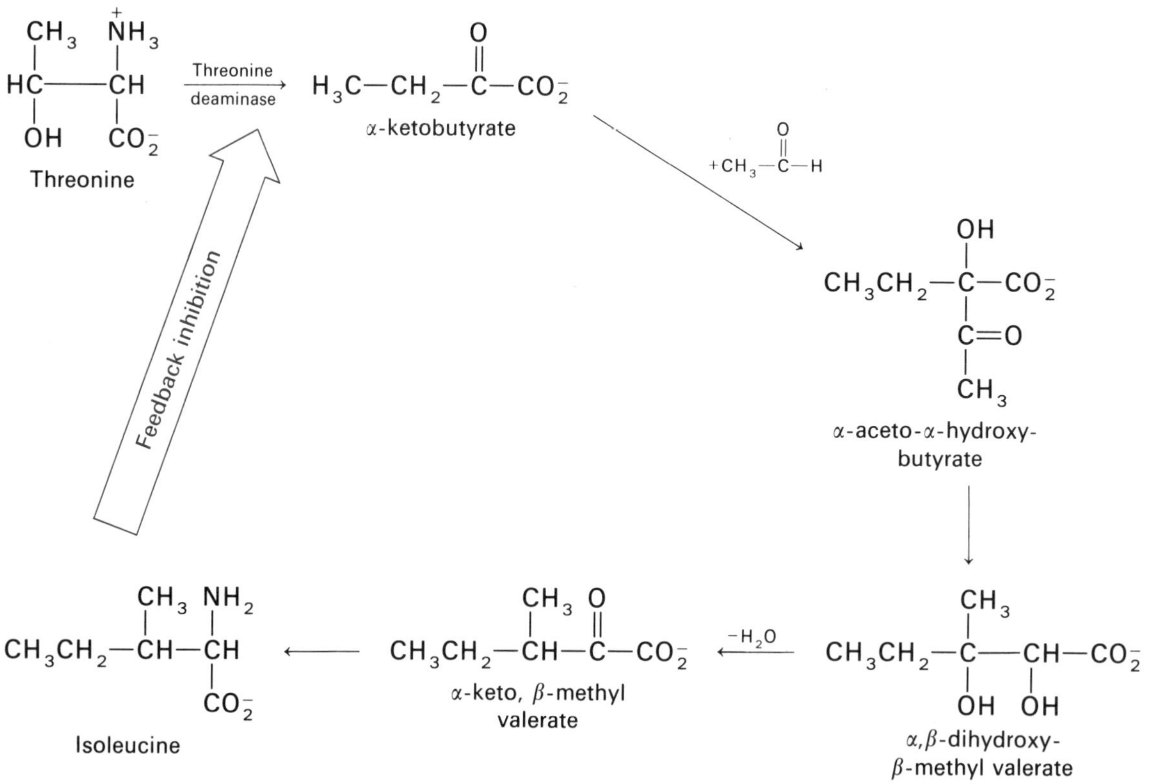

the rate of the reaction pathway. Such an enzyme is termed a **regulatory** or **allosteric** enzyme (Greek: *allo*, "other"; *steric*, "site"). The inhibition observed with regulatory enzymes in the presence of a specific modulator is neither purely competitive nor noncompetitive. In the case of isoleucine biosynthesis the regulatory enzyme threonine deaminase controls the rate of synthesis of isoleucine and thus its concentration in the cell. Isoleucine, in turn, inhibits threonine deaminase, and thus regulates its own concentration level. Isoleucine is called a *negative modulator*, or effector, of the regulatory enzyme threonine deaminase.

A familiar analogy for feedback control is shown in Figure 3.27. You can readily see how the level of product (in this example, water) controls the rate of its input, and thus its level.

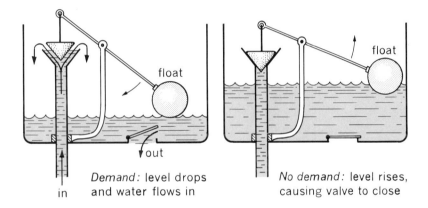

Figure 3.27

Feedback control: the level of material controls its rate of input into the system.

Regulatory enzymes can be subject to both positive (activation) and negative (inhibition) effects. An activator, or positive modulator, increases the catalytic action of a regulatory enzyme. In many cases, the substrate acts as an activator by binding first to a site other than the active site. By doing this, the substrate binds more easily to the enzyme. This is known as a **cooperative effect** and produces a *sigmoid* velocity/substrate curve, characteristic of most regulatory enzymes, as shown in Figure 3.28. A regulatory enzyme modulated by its own substrate is termed a *homotropic* system. A regulatory enzyme modulated by a substance other than substrate is termed a *heterotropic* system. A third class of regulatory enzymes, perhaps the most common, is the mixed *heterotropic-homotropic* system. If more than one substance affects the activity of a regulatory enzyme, it is said to be multivalent. In some cases as many as nine modulators have been observed to regulate the activity of a given regulatory enzyme.

Working in concert, positive and negative modulators of multivalent regulatory enzymes make possible the close control and integration of sugar, fat, and protein metabolism. The effects of positive and negative modulators on regulatory enzymes are summarized in Figure 3.28. In the graph of Figure

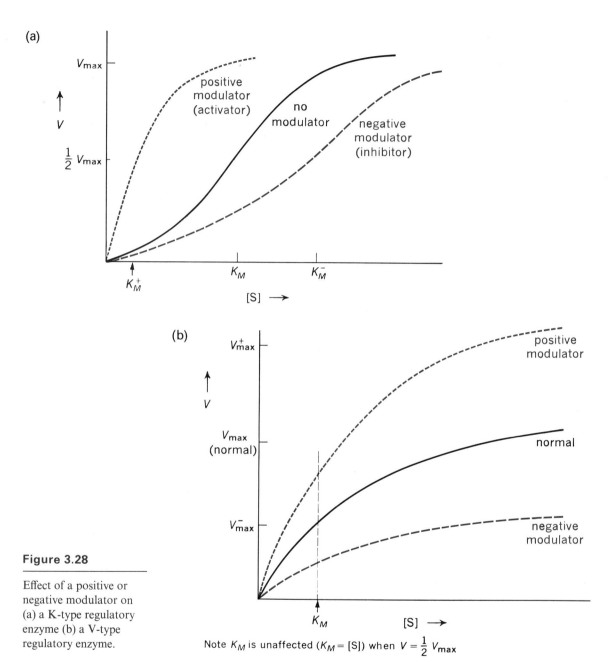

Figure 3.28

Effect of a positive or negative modulator on (a) a K-type regulatory enzyme (b) a V-type regulatory enzyme.

Note K_M is unaffected ($K_M = $ [S]) when $V = \frac{1}{2} V_{max}$

3.28(a), the effect of the activator causes an apparent decrease in K_M, thus increasing the rate. This is characteristic of allosteric systems known as K-systems. The effect of the negative modulator increases the value of K_M; thus a higher substrate concentration is needed to produce a given reaction rate. Figure 3.28(b) shows a plot of V vs [S] for a V-system. Here, the modulator affects $V_{\max}$ and not K_M. The two types of behavior summarized in Figure 3.28 are characteristic of the two major classes of regulatory enzymes.

The action of regulatory enzymes has been explained by the model shown in Figure 3.29. Another means of visualizing the action of a regulatory enzyme with only one subunit was given in Figure 3.24. Binding the modulator to a binding site in a regulatory subunit somehow causes a conformational effect felt at the active center of the catalytic unit. The fact that these enzymes are usually oligomeric, very large, and easily susceptible to denaturation tends to support this theory. Researchers have observed that regulatory enzymes show a considerable degree of conformational flexibility, making long-range interactions between regulatory sites and a catalytic site plausible. The work of Gerhart and others, using the regulatory enzyme *aspartate transcarbamylase* provides direct experimental support for this idea. They were able to destroy the regulatory properties of the enzyme by dissociating the subunits through chemical treatment. In such a case, the classical velocity/substrate curve was obtained from the action of the unregulated catalytic subunit.

While allosteric control can be demonstrated *in vitro* in isolated cell extracts, the second major form of control of enzyme action requires living cells in order to be demonstrated. This method is genetic control of enzyme action.

Figure 3.29

A model for a regulatory enzyme, showing the long-range effect of binding the modulator on the action of the catalytic site.

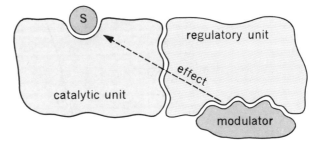

Genetic Control

Genetic control of enzyme action involves the **induction** or **repression** of the biosynthesis of the enzyme protein. This process controls the rate in a coarse manner, usually on an all-or-nothing basis, by changing the amount of enzyme present. Since proteins have a definite lifetime, ranging from minutes to months, there is a constant turnover, or formation and breakdown of enzymes in the cell. Due to the heavy investment of energy and materials entailed in the biosynthesis of proteins, many enzymes must be synthesized on demand. This is particularly the case for bacteria, which are at the mercy of the environment. Thus, bacteria have evolved metabolic control mechanisms that permit a high degree of economy in operation. Due to their relative simplicity, bacteria offer excellent systems for the study of induction and repression of enzymes.

Enzymes ordinarily present in a cell at a constant level are called *constitutive* enzymes. **Inducible** or **adaptive** enzymes are not ordinarily present in the cell, but are synthesized in response to the presence or absence of certain substrates. An extensively studied inducible enzyme system is the utilization of lactose in *E. coli*. In the absence of the sugar lactose, used as a carbon source, three key enzymes for the utilization of lactose are not present in the cell. Since maltose and glucose normally serve as carbon sources for protein synthesis in *E. coli* (Chapter 4), the biosynthesis of these key lactose-utilizing enzymes is said to be *repressed* or "turned off" in the absence of lactose. Of the three proteins, the enzyme β–galactosidase is normally assayed in demonstrating this effect. If we take *E. coli* cells from a glucose medium, wash them, and then place them in a lactose medium, we observe a lag period followed by the resumption of growth. During the lag period, in response to the presence of the *inducer*, in this case lactose, the cellular concentration of β-galactosidase is observed to increase rapidly. Figure 3.30 illustrates the time course of β-galactosidase induction. We will learn how cells regulate the synthesis of adaptive enzymes in Chapter 13.

Figure 3.30

Plot of the increase and decrease with time of β-galactosidase biosynthesis on the addition and removal of β-galactosidase inducers. (Redrawn from J. D. Watson, *Molecular Biology of the Gene*, Benjamin, 1970.)

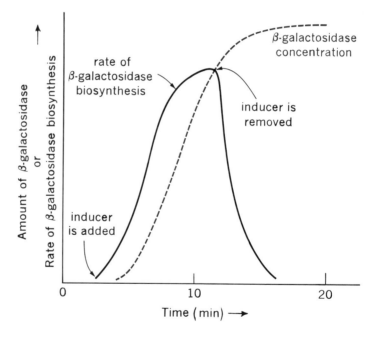

Enzyme Induction and Environmental Stress Many species have the ability to alter their metabolism in response to some environmental stress. For example, DDT-resistant insects detoxify the insecticide by producing the enzyme, DDT-dehydrochlorinase, when exposed to DDT. An interesting and dramatic example of a family of enzymes induced by environmental stress is the response of fish and certain other marine animals to oil pollution.

A number of fat soluble substances including insecticides, aromatic hydrocarbons (from petroleum) and steroid hormones induce the formation of certain oxidative enzymes in the livers of animals. These enzymes are called mixed-function oxidases (MFO) and catalyze the reaction:

$$A—H + 2e^- + 2H^+ + O_2 = A—OH + H_2O$$

where $A—H$ is an aromatic hydrocarbon. The induction of MFO is important to marine animals exposed to petroleum because the aromatic hydrocarbons in the oil are the most toxic and most water-soluble part of the petroleum, and therefore the most harmful to marine life. The hydroxylation reaction shown above is a crucial first step in the detoxification of ingested aromatics. For example, the level of liver MFO in brown trout living in an oil-polluted pond has been found to be over 10 times higher than the level of MFO in brown trout living in a pollution-free pond. Trout are just one of the many higher animals which can produce enzymes to enable them to deal with change in their environment, either natural or man-made.

Other Controls of Enzyme Action

In the preceding sections, we showed how an organism can effect control of enzyme activity in response to the presence or absence of a given metabolite (genetic control) or in response to a change in demand for a given metabolite (allosteric control). Other means of controlling enzyme activity can reflect the physiological role of the particular enzyme. The best example of this is the action of lactate dehydrogenase (LDH).

We discussed the nature of isozymes in Chapter 2, using the enzyme lactate dehydrogenase as an example. It turns out that LDH is intimately connected with the control of glycolysis at the organism level in higher animals. LDH is tetrameric and can exist as a mixture of five combinations (isozymes) of two types of monomeric protein chains: H (heart) chains and M (muscle) chains. Isozymes rich in H chains, such as H_4 and H_3M isozymes, are associated with aerobic tissues such as heart muscle, where lactate is removed from circulation, oxidized to pyruvate, and then further oxidized to CO_2 and H_2O. The H_4 isozyme of LDH is inhibited by pyruvate, thus favoring the reaction

$$\text{Lactate} \xrightarrow{H_4—LDH} \text{Pyruvate} + 2e^- + 2H^+$$

In skeletal muscle, periods of intense activity can deplete the tissues of oxygen and lead to the accumulation of lactic acid as the product of anaerobic glucose catabolism. In such "anaerobic" tissues, LDH takes the form of isozymes rich in M chains: M_4 and M_3H. Unlike H_4, the M_4 isozyme is not inhibited by pyruvate, and thus converts it to lactate rapidly:

$$\text{Pyruvate} + 2e^- + 2H^+ \xrightarrow{M_4—LDH} \text{Lactate}$$

What is important here is that a differentiation occurs in the properties of LDH, depending on the physiological role of the particular type of tissue in which the enzyme functions.

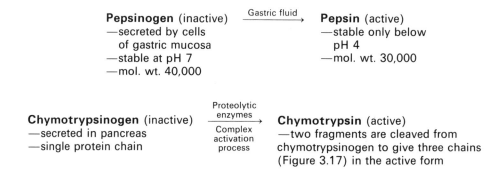

Pepsinogen (inactive) $\xrightarrow{\text{Gastric fluid}}$ **Pepsin** (active)
—secreted by cells   —stable only below
  of gastric mucosa   pH 4
—stable at pH 7   —mol. wt. 30,000
—mol. wt. 40,000

Chymotrypsinogen (inactive) $\xrightarrow[\substack{\text{Complex}\\\text{activation}\\\text{process}}]{\substack{\text{Proteolytic}\\\text{enzymes}}}$ **Chymotrypsin** (active)
—secreted in pancreas   —two fragments are cleaved from
—single protein chain   chymotrypsinogen to give three chains
                                               (Figure 3.17) in the active form

Figure 3.31

Zymogen activation.

One other control mechanism which we should discuss relies on the biosynthesis of **zymogens**, inactive precursors for a given protein. Many enzymes are made in an inactive form. This is particularly true for the proteolytic enzymes associated with digestion. Figure 3.31 illustrates this for the cases of the proteases pepsin and α-chymotrypsin. It is important to bear in mind that if pepsin were produced directly, it would become denatured by the pH 7 conditions of the gastric mucosal cells in which the enzyme is synthesized. By the initial production of the inactive precursor pepsinogen, which is stable at pH 7, followed by the subsequent activation of pepsinogen by the action of gastric juice, the cell can produce under ordinary conditions an enzyme which is stable only below pH 4. In the case of α-chymotrypsin, not only is the enzyme initially produced in a form which will not begin to degrade the cell in which it was made, but another possible problem is avoided. The protein α-chymotrypsin consists of three chains held together by disulfide bridges (Figure 3.17). The probability for error in the assembly of three separate chains to make an active enzyme is likely to be large. However, by first making an inactive precursor with the disulfide bridges in place, and then removing two small segments during the course of activation, the cell produces a three-chain protein with almost no chance for error.

In addition to the control mechanisms just discussed, the interchange between active and inactive enzyme forms by covalent modification represents another important regulation device. A good example of this device is the control of the activity of glycogen phosphorylase, a key enzyme in the breakdown of the storage polysaccharide, glycogen. As shown in Figure 3.32, this enzyme

Figure 3.32

Enzyme control by covalent modification. The control of the activity of glycogen phosphorylase involves the conversion of the inactive form of the enzyme to the active form by the reaction of hydroxyl groups with ATP to produce phosphate-ester groups. The regulatory process for this enzyme also involves other factors as well (see Section 12.5).

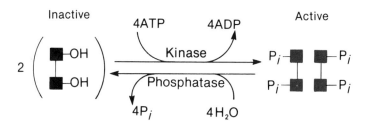

is converted from an inactive to an active form by the phosphorylation of serine hydroxyl groups in the inactive form. The activation process also involves a major change in the quaternary structure of glycogen phosphorylase from a dimer to a tetramer. The reaction is discussed in detail in Section 12.5. Here we will see how this activation process is, itself, under hormonal control.

3.11

Applications of Enzymes

Because of the unique properties of enzymes, increasing use is being made of these catalysts in many areas of industrial and clinical biochemistry. In 1975, the value of the enzymes isolated for various industrial and biomedical purposes was over $51 million in the U.S. Let's look at some applications of enzymes in more detail.

Enzymes in Clinical Diagnosis

Enzymes can be used in clinical diagnosis in either of two ways: (1) as indicators of disease; (2) as test reagents for detecting metabolite concentrations. In the first case, the presence of abnormally high levels of an enzyme in the blood often indicates the breakdown of cells in a diseased organ. For example, hepatitis causes the breakdown of liver tissue and the release of liver enzymes into the blood. Because the liver is the seat of nitrogen metabolism, transaminases are particularly abundant. The enzymes, glutamic oxaloacetic transaminase (GOT) and glutamic pyruvic transaminase (GPT) are key, and elevated concentrations of these in the blood are early indicators of possible hepatitis. One widely used clinical test for serum GOT uses the following reactions:

$$\overset{\overset{+}{N}H_3}{\underset{|}{^-O_2CCH_2CHCO_2^-}} + \overset{\overset{O}{\parallel}}{^-O_2CCH_2CH_2CCO_2^-} \xrightarrow{(GOT)} \overset{\overset{O}{\parallel}}{^-O_2CCH_2CCO_2^-} + \overset{\overset{+}{N}H_3}{\underset{|}{^-O_2CCH_2CH_2CHCO_2^-}} \quad (1)$$

Aspartate α-ketoglutarate Oxaloacetate Glutamate

$$\text{Oxaloacetate} + \text{NADH} + \text{H}^+ \xrightarrow[\text{dehydrogenase}]{\text{Malate}} \overset{\overset{OH}{|}}{^-O_2CCH_2CHCO_2^-} + \text{NAD}^+ \quad (2)$$

Malate

Reaction (1) involves no change in light absorption, but in reaction (2), NADH absorbs light at 340 nm while NAD^+ does not. In this *coupled assay*, the amount of GOT present in the original serum sample is proportional to the rate at which light absorption at 340 nm decreases, and hence the rate at which NADH decreases. Table 3.11 lists some other enzymes commonly assayed as indicators of disease.

Table 3.11 Some of the major enzymes used as indicators of disease.

Enzyme assayed	Organ affected or disease
acid phosphatase	cancer of the prostate
alkaline phosphatase	liver, bone
amylase	pancreas
glutamic oxaloacetic transaminase (GOT)	liver, heart
glutamic pyruvic transaminase (GPT)	liver, heart
lactate dehydrogenase (LDH)	liver, red blood cells
creatine phosphokinase (CPK)	brain, muscle, heart

Enzymes are also used as test reagents to measure the amount of specific metabolites in a mixture. Enzyme specificity permits the clinical chemist to measure one substance in the presence of many others that might interfere with an ordinary non-enzymatic test. One of the most widely used enzyme reagents is the glucose oxidase assay for glucose in the blood or urine. The reactions used in this coupled assay are as follows:

$$\text{Glucose} + O_2 + H_2O \xrightarrow[\text{oxidase}]{\text{Glucose}} \text{Gluconic acid} + H_2O_2 \qquad (1)$$

$$\underset{\text{(colorless)}}{H_2O_2 + o\text{-Dianisidine}} \xrightarrow{\text{Peroxidase}} \underset{\text{(brown)}}{\text{oxidized Dianisidine} + H_2O} \quad (2)$$

In reaction (1) above, one mole of H_2O_2 is formed per mole of glucose present. Reaction (2) uses the H_2O_2 formed in (1) to produce a colored oxidized product. The intensity of the brown color is proportional to the amount of glucose present. For testing outside of a lab environment, glucose test paper is available, consisting of paper strips impregnated with the two enzymes and the color-forming compound.

Enzymes and Food Technology

The use of enzymes as industrial catalysts is currently undergoing rapid development due to two unique properties of enzymes: (1) they are highly specific and efficient; (2) they function at or near room temperature and are therefore highly energy-efficient. For example, corn starch is converted into corn syrup by the following reactions:

$$\text{Starch} \xrightarrow{\alpha\text{-amylase}} \text{Dextrins} \xrightarrow{\text{Glucoamylase}} \text{Glucose} \xrightarrow[\text{isomerase}]{\text{Glucose}} \text{Fructose}$$

The end product is a mixture of glucose and fructose which is sweeter than glucose alone. In the last step of this process, the glucose solution passes through a reactor containing the enzyme glucose isomerase bonded to an inert support. Product emerges from the exit of the reactor as a pure product mixture of glucose and fructose. In 1974 over 1 billion pounds of high fructose corn syrup were produced in this manner.

In reducing our dependence on petroleum, cellulase enzymes are being used to convert waste cellulose from wood, paper and agricultural waste to ethanol, a liquid fuel. In another case, enzyme technology is being applied to a pollution problem arising from liquid whey, a by-product of the cheese industry that contains lactose. Until recently whey was simply released into the nearest river, but now immobilized lactase is being used to convert whey into sugar syrup, for sweetening or fermentation into alcohol. Table 3.12 summarizes some of the industrial uses of enzymes.

Table 3.12 Selected industrial uses of purified enzymes.

Enzyme	Action	Selected uses
Amylases	Hydrolyze starch	Glucose formation from starch for sweetening and fermentation
Invertase	Hydrolyzes cane sugar to glucose and fructose	Manufacture of invert sugar for candy
Microbial proteases	Hydrolyze proteins	Detergent additive, meat tenderizer, bread-baking
Papain	Hydrolyzes protein	Meat tenderizer
Rennin	Hydrolyzes peptides	Curdles milk in making cheese
Glucose oxidase	Oxidizes glucose in presence of O_2	Removal of O_2 from food products; used for glucose test in diagnosis
Glucose isomerase	Converts glucose to fructose	Production of high-fructose (sweeter) corn syrup

We have discussed some of the properties of enzymes as catalytic proteins. In Parts II and III of this text we will see how the interrelated processes of metabolism proceed. Each metabolic step or reaction uses a catalyst—an enzyme. Before we can begin our discussion of metabolism, we must discuss the three remaining classes of biomolecules: carbohydrates, nucleic acids and lipids.

Suggestions for Further Reading

Lehninger, A. L. *Biochemistry*. 2d ed. New York: Worth Publishers, Inc., 1975. Chapters 8 and 9 contain a good treatment of enzymes at a more advanced level.

Metzler, D. E. *Biochemistry—The Chemical Reactions of Living Cells*. New York: Academic Press, 1977. This is a well-documented and thoroughly indexed comprehensive text. Chapters 6 and 7 feature an in-depth discussion of enzyme chemistry. Chapter 8 contains a good discussion of coenzymes.

Skinner, K. J. "Enzyme Technology—A Special Report." *Chemical and Engineering News*, August 18, 1975, pp. 22–41. An interesting and readable discussion of the enzyme industry in an easily available weekly publication.

Scientific American

Changeux, Jean-Pierre. "The Control of Biochemical Reactions." April 1965.

Koshland, D. E., Jr. "Protein Shape and Biological Control." October 1973.

Mosbach, K. "Enzymes Bound to Artificial Matrices." March 1971.

Neurath, H. "Protein-Digesting Enzymes." December 1964.

Phillips, D. C. "The Three-Dimensional Structure of an Enzyme Molecule." November 1966.

Problems

1. For each of the terms given below, provide a brief definition or explanation.

Activation energy	Conjugated protein	Michaelis constant
Active center	Enzyme	Saturation kinetics
Allosteric	Hydrolase	Specificity
Antimetabolite	Inducible enzyme	Substrate
Catalyst	Isomerase	Transferase
Coenzyme	Ligase	Vitamin
Competitive inhibitor	Lyase	Zymogen

2. Enzymes often have two names—an official IEC name, and a more commonly used "trivial name." A number of reactions are given below in terms of functional groups or compounds. For each reaction, give the overall IEC class of the enzyme which would catalyze the reaction, as well as the trivial name for the class. For example:

$$
\begin{array}{ccc}
\underset{\displaystyle \underset{\text{OH}}{|}}{\overset{\displaystyle \overset{\text{H}}{|}}{\text{C}-\text{C}-\text{C}}} & \longrightarrow & \text{C}-\overset{\displaystyle \overset{\text{O}}{\|}}{\text{C}}-\text{C}
\end{array}
\quad
\begin{array}{l}
\text{oxidoreductase; dehydrogenase} \\
\text{(or oxidase)}
\end{array}
$$

(a) $R\!-\!\!\bigcirc\!\!-\!H \longrightarrow R\!-\!\!\bigcirc\!\!-\!OH$

(b) ATP + D-glucose → ADP + D-glucose 6-phosphate

(c) glucose 6-phosphate → glucose 1-phosphate

(d)

$$R-\overset{\overset{\displaystyle O}{\|}}{C}-R + R'-\overset{\overset{\displaystyle H}{|}}{\underset{\underset{\displaystyle H}{|}}{C}}-NH_2 \longrightarrow R-\overset{\overset{\displaystyle NH_2}{|}}{\underset{\underset{\displaystyle H}{|}}{C}}-R + R'-\overset{\displaystyle C}{\underset{\displaystyle H}{\overset{\displaystyle \nearrow O}{\diagdown}}}$$

(e)

$$H_2O + R-C\overset{\displaystyle \nearrow O}{\underset{\displaystyle O-R'}{\diagdown}} \longrightarrow R-CO_2H + R'OH$$

3. Give a specific example for each of the following enzymes, including the commonly used (trivial) name of the enzyme; the reaction catalyzed, showing representative substrate(s); and a reasonable source from which the enzyme can be isolated. (You may have to go to the library to do this.)

   (a) A peptidase   (b) A transferase   (c) A dehydrogenase   (d) An isomerase

4. For each of the following vitamin-deficiency diseases, give the name and structure of the vitamin involved.

   (a) Beriberi   (b) Pellagra   (c) Scurvy   (d) Rickets

5. Prolonged antibiotic therapy can cause deficiencies in certain vitamins by killing intestinal flora. List the vitamin deficiencies that could be caused in this way.

6. Give the structures of the amino acid side chain groups that could be involved in metal ion cofactor binding. Assume the ionic forms that predominate at pH 7.3.

7. Give the name and structure of the coenzyme that would most likely be required by an enzyme catalyzing each of the following reactions. In some cases, more than one coenzyme might be appropriate.

   (a) $^-O_2C-CH_2CH_2-CO_2^- \longrightarrow$

   $$^-O_2C-CH=CH-CO_2^- + 2H^+ + 2e^-$$

   (b) $CH_3\overset{\overset{\displaystyle OH}{|}}{C}H-CO_2^- + O_2 \longrightarrow CH_3\overset{\overset{\displaystyle O}{\|}}{C}-CO_2^- + H_2O_2$

   (c) $CH_3\overset{\overset{\displaystyle O}{\|}}{C}-CO_2^- + H^+ \longrightarrow CH_3\overset{\overset{\displaystyle O}{\|}}{C}-H + CO_2$

   (d) $CH_3-\overset{\overset{\displaystyle \overset{+}{N}H_3}{|}}{C}H-CO_2^- + H^+ \longrightarrow CH_3CH_2\overset{+}{N}H_3 + CO_2$

   (e) $CH_2=CH-CH_2\overset{\overset{\displaystyle O}{\|}}{C}-SCoA + 2H^+ + 2e^- \longrightarrow$

   $$CH_3CH_2CH_2\overset{\overset{\displaystyle O}{\|}}{C}-SCoA$$

8. Show the products for the hydrolysis of the linkages shown in Table 3.9.

9. Using the reasoning presented in Figure 3.20, show how pyruvate dehydrogenase can produce *only* L-lactate from the optically inactive pyruvate:

Pyruvate ion → L-lactate ion

$$V = \frac{V_{max}[S]}{K_M + [S]}$$

10. Derive two expressions which linearize the simple Michaelis-Menten expression shown in the margin. In the laboratory, are the expressions you derive really equivalent? Why?

11. Show that $V = (\frac{1}{2})V_{max}$ when $[S] = K_M$.

12. How can you experimentally distinguish between competitive and noncompetitive inhibition of enzyme action?

13. The enzyme L-amino acid oxidase catalyzes the following reaction:

Using L-leucine as a substrate at pH 8.0 and 35°C, the following data was obtained:

[Leucine] (moles/liter)	0.00065	0.00086	0.00106	0.00136	0.00187	0.00287
$V \left(\dfrac{moles}{liter} \cdot \dfrac{1}{min}\right)$	1.56×10^{-4}	1.91×10^{-4}	2.16×10^{-4}	2.50×10^{-4}	2.85×10^{-4}	3.30×10^{-4}

Using this data, construct a Lineweaver–Burk plot and determine the values of K_M and V_{max}.

Under the same conditions, experiments were performed in the presence of $5 \times 10^{-4}\ M$ sodium benzoate. Construct a Lineweaver–Burk plot with the following data:

[Leucine] (moles/liter)	0.00106	0.00136	0.00187	0.00287
V in the presence of inhibitor $\left(\dfrac{moles}{liter} \cdot \dfrac{1}{min}\right)$	1.45×10^{-4}	1.74×10^{-4}	2.11×10^{-4}	2.67×10^{-4}

Referring to Figures 3.22 and 3.23, determine the value of K_I. What is the nature of the inhibition by benzoate ion?

14. How do sulfa drugs act as antimetabolites?

15. In Chapter 2 we learned about the manner in which hemoglobin binds oxygen. How do the oxygen binding properties of hemoglobin serve as a model for a positive modulator acting on a homotropic allosteric K-type regulatory enzyme?

16. Why is allosteric regulation of a metabolic process a fine control mechanism, as opposed to genetic control (repression) which is a coarse control?

4

Carbohydrates

The family of compounds called carbohydrates includes substances which are extremely abundant and important in the biological world. Well over half the organic carbon in the world is in the form of carbohydrates. The most common of these compounds is cellulose, the structural carbohydrate of plants. It has been estimated that over 10^{11} tons of cellulose are produced and broken down in the biosphere each year.

We often think of carbohydrates as substances such as glucose, which many heterotrophs use as a primary source of energy. In Chapters 7, 8 and 9 we will see how such organisms use carbohydrates to obtain chemical energy. However, carbohydrates can also serve a structural purpose—not only in plants, but also in the cell walls of bacteria and in the matrix, or ground substance, surrounding cells in the tissues of higher organisms.

There are three broad classes of carbohydrates: **monosaccharides**, **oligosaccharides**, and **polysaccharides**. We can completely hydrolyze both polysaccharides and oligosaccharides, producing monosaccharides, and further hydrolysis does not yield any molecules smaller than monosaccharides. Oligosaccharides are polymers composed of from two to six monosaccharide units. Polysaccharides such as starch and cellulose contain thousands of monosaccharide units joined by hydrolyzable covalent linkages. We will consider the chemistry of monosaccharides first, since they form the basis for the more complex carbohydrates.

4.1

Monosaccharides

A monosaccharide is either a *polyhydroxy aldehyde* or a *polyhydroxy ketone*, with the overall empirical formula $(CH_2O)_n$. From this formula you can readily see how the designation "carbohydrate" arose. If we subtract the elements of H_2O from the empirical formula for a monosaccharide, we are left with just carbon. The nomenclature of the monosaccharides depends on the number of carbon atoms, as shown in Table 4.1.

Table 4.1 Nomenclature of Monosaccharides.

Number of carbons	Family name
3	Triose
4	Tetrose
5	Pentose
6	Hexose
7	Heptose

In the biological world pentoses and hexoses are most abundant, although trioses, tetroses, and heptoses also play key roles in the metabolism of plants and animals. Monosaccharides are further separated into two categories, **aldoses** and **ketoses**, depending on the nature of the functional group: aldehyde or ketone. The two simplest monosaccharides, glyceraldehyde and dihydroxyacetone, illustrate this categorization:

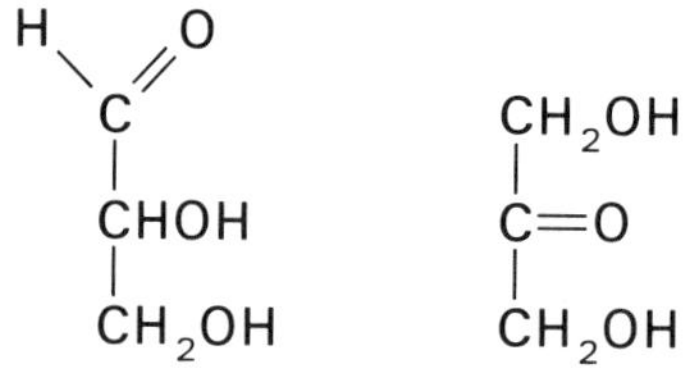

$$
\begin{array}{cc}
\overset{\displaystyle H\diagdown\diagup O}{\underset{|}{C}} & CH_2OH \\
CHOH & C=O \\
| & | \\
CH_2OH & CH_2OH
\end{array}
$$

Glyceraldehyde,   Dihydroxyacetone,
an *aldo*triose   a *keto*triose

Note that while dihydroxyacetone is a symmetrical molecule, the second carbon atom of glyceraldehyde is asymmetric, so glyceraldehyde exists in two non-superimposable mirror-image forms. All monosaccharides (and thus higher sugars) except dihydroxyacetone contain one or more asymmetric carbons.

Monosaccharides: Stereochemistry

Much of our initial understanding of stereochemistry was due to the work of Emil Fischer on the structure of sugars. By 1884, when Fischer first published his investigations into sugar chemistry, four hexoses, one pentose, and three disaccharides had been isolated, purified, and characterized. Much of the early structural work done by Fischer and others was based on the hexose *glucose*, $C_6H_{12}O_6$, the central monosaccharide of both the plant and animal worlds. On the basis of chemical evidence, Fischer showed that glucose (and other aldohexoses) contains four asymmetric carbon atoms, which he labeled as follows:

$$
\underset{1}{CHO}-\underset{2}{\overset{*}{CHOH}}-\underset{3}{\overset{*}{CHOH}}-\underset{4}{\overset{*}{CHOH}}-\underset{5}{\overset{*}{CHOH}}-\underset{6}{CH_2OH}
$$

Figure 4.1

Fischer conventions for planar representation of stereoisomers. We arrange the model of the molecule with the carbon chain vertical so that the CH$_2$OH or equivalent group is at the bottom, the CHO or equivalent group is at the top, and all the H and OH groups extend toward the front. The model is then flattened to produce a planar representation. We can rotate the Fischer representation only in the plane of the paper. Inverting the representation, as in flipping a pancake, produces a completely different stereoisomer.

$$\text{CHO} \quad \text{H}-\overset{\cdot}{\text{C}}-\text{OH} \equiv \text{H}\overset{\text{CHO}}{\diamond}\text{OH} \equiv \text{H}\overset{\text{CHO}}{\diamond}\text{OH} \equiv \text{H}-\overset{\text{CHO}}{\underset{\text{CH}_2\text{OH}}{\text{C}}}-\text{OH}$$

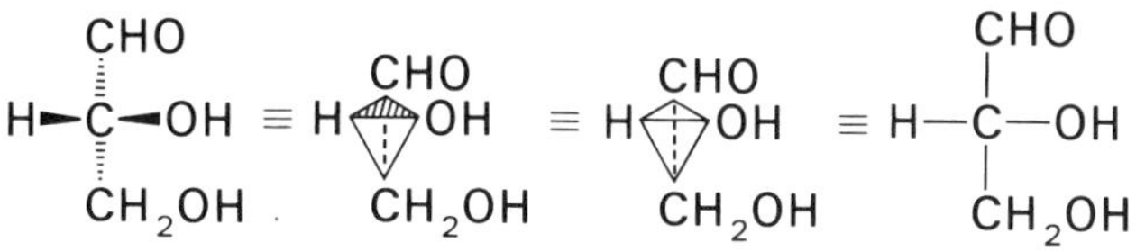

D-Glyceraldehyde

Two mirror-image forms, or enantiomers, correspond to *each* asymmetric center. Thus, for the *aldohexoses*, we find 2^4 possible stereoisomers. Fischer established a set of conventions for representing these different stereoisomers, illustrated in Figure 4.1 for the aldotriose D-glyceraldehyde. The configuration shown in Figure 4.1 was arbitrarily chosen by Fischer as the reference configuration for all D-sugars. The configuration for monosaccharides with more than three carbons is based on the relative configuration of the asymmetric center farthest from the carbonyl group. Figure 4.2 shows the configurations for a tetrose, a pentose, and a hexose. Fischer chose L-malic acid as the parent compound for the L-sugars:

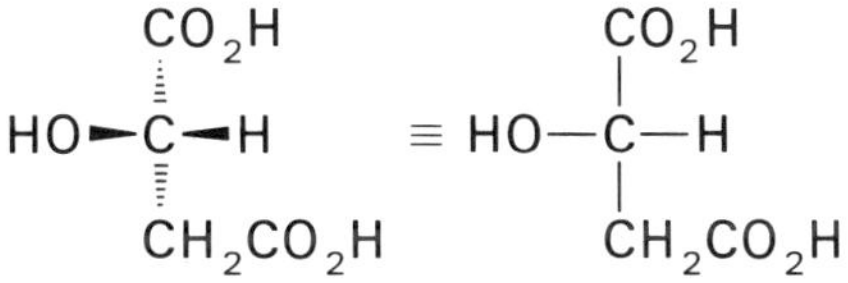

$$\text{CO}_2\text{H} \quad \text{HO}-\overset{\cdot}{\text{C}}-\text{H} \equiv \text{HO}-\overset{\text{CO}_2\text{H}}{\underset{\text{CH}_2\text{CO}_2\text{H}}{\text{C}}}-\text{H}$$

L-Malic acid

The structures for sugars of the D-configuration were determined by painstaking synthesis from D-glyceraldehyde, using reactions in which the stereochemistry of the reactant was conserved. D-Sugars are universally important in living systems; their mirror-image L-counterparts are rarely found in nature. Some L-sugars are found in the cell wall structures of certain bacteria.

Figure 4.3 shows some important D-sugars in their Fischer representations. A number of pairs of sugars differ only in the configuration at one asymmetric carbon atom. Compounds related in this way are called *epimers*.

Figure 4.2

Representations of D-sugar configurations.

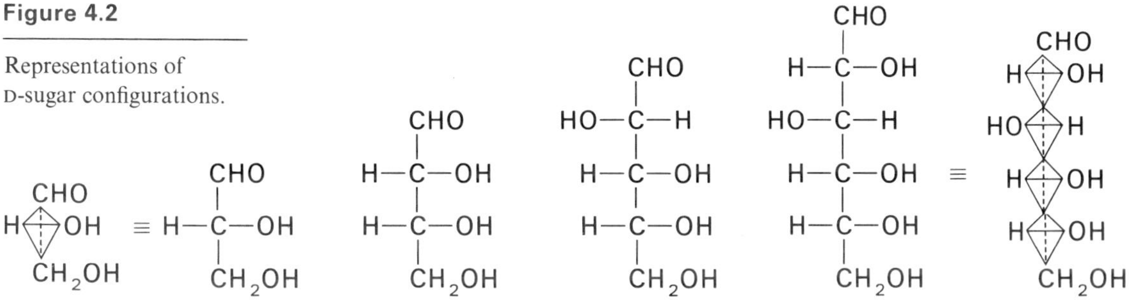

D-Glyceraldehyde   D-Erythrose   D-Arabinose   D-Glucose

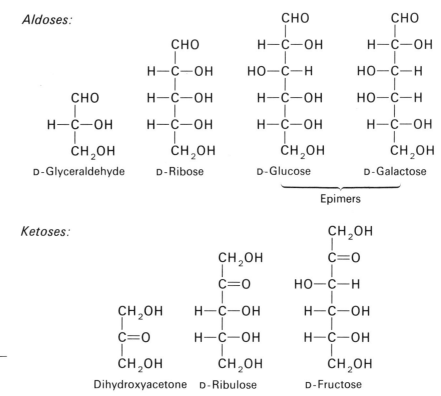

Figure 4.3

The structures of some important D-sugars.

D-Glucose and D-galactose are epimers, since they differ only at carbon 4. The interconversion of epimers is an important process in carbohydrate metabolism.

Monosaccharides: Cyclic Structures

Aldehydes and ketones react with alcohols to form hemiacetals and hemiketals, respectively (see Appendix B). Because monosaccharides have either an aldehyde or ketone group plus alcohol groups, hemiacetal or hemiketal formation can occur internally to yield a cyclic structure. Due to bond angle

$$R—OH + R'—C{\overset{O}{\overset{\parallel}{}}}H \rightleftharpoons R'—\overset{OH}{\underset{OR}{C}}—H$$

Alcohol   Aldehyde   Hemiacetal

strain, 5- and 6-membered ring structures are favored for the sugars. Figure 4.4 shows the formation of a cyclic hemiacetal for the case of D-glucose. In this case, two distinct forms are produced depending on whether the alcohol group reacts

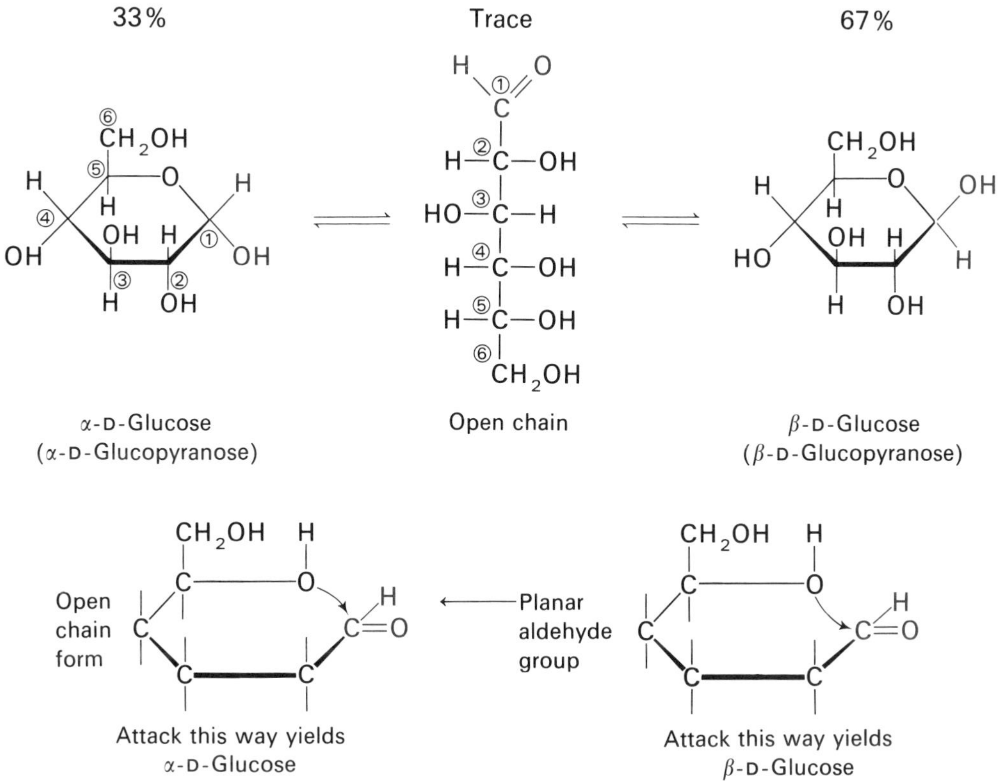

33%   Trace   67%

α-D-Glucose   Open chain   β-D-Glucose
(α-D-Glucopyranose)   (β-D-Glucopyranose)

Attack this way yields   Attack this way yields
α-D-Glucose   β-D-Glucose

Figure 4.4

Anomeric forms of
D-glucose, given in the
Haworth projection
representation. These
forms differ only in the
configuration of the
hemiacetal carbon.

with the carbonyl carbon from above or below the plane of the carbonyl group. These two forms, designated α and β, are related as *anomers*. In solution, these two anomeric forms are in equilibrium with the open chain form.

When pure α-D-glucose, β-D-glucose or natural D-glucose is dissolved in water, the resulting solution is the same equilibrium mixture of α, β and open-chain forms shown in Figure 4.4. The interconversion of α and β forms is called *mutarotation* and is catalyzed by enzymes called *mutarotases*.

The six-membered ring structure resulting from internal hemiacetal formation in aldohexoses is called a *pyranose* ring after the heterocycle *pyran*. Ketoses can undergo similar ring formation through an analogous *hemiketal* structure. Aldopentoses, ketohexoses, and, in some cases, aldohexoses can also form five-membered ring cyclic forms known as *furanose* structures.

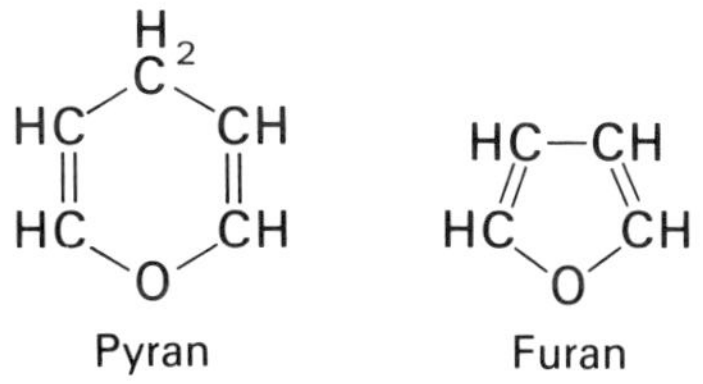

Pyran   Furan

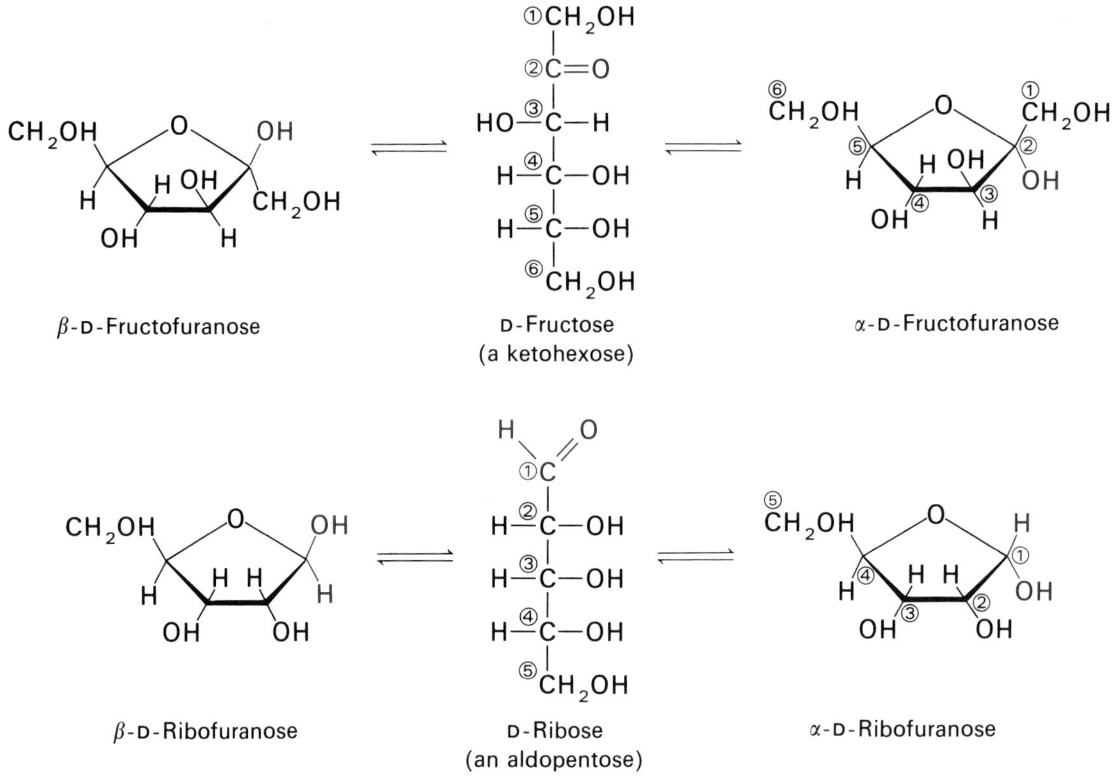

β-D-Fructofuranose D-Fructose (a ketohexose) α-D-Fructofuranose

β-D-Ribofuranose D-Ribose (an aldopentose) α-D-Ribofuranose

Figure 4.5

Cyclic forms of some monosaccharides.

The pyranose form is preferred by most aldohexoses. The furanose form predominates for most ketohexoses and aldopentoses. Figure 4.5 gives some examples of cyclic structures of sugars other than D-glucose.

The perspective representations of the cyclic forms are called *Haworth projection* formulas. This is a convenient way of noting the differences in configuration of each position in various sugars. Such cyclic forms are not planar, however, as the Haworth projections imply. As in the case of cyclohexane, the pyranose ring can exist in a chair or a boat form, depending on the steric properties of the individual sugar. Glucose, for example, favors the chair conformation in both anomeric forms:

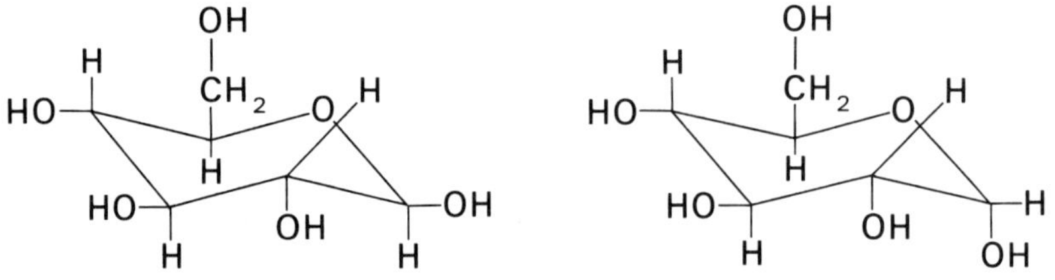

Chair form of β-D-Glucopyranose

Chair form of α-D-Glucopyranose

In this text, we will use the Haworth formulas wherever they are appropriate. You should bear in mind that such representations are only approximations of the true structure of the molecule.

Monosaccharides: Reactions

Aldoses are *reducing sugars*, meaning that the free aldehyde function of the open chain form is capable of being oxidized to a carboxylic acid group. Ketoses are not readily oxidized under the mild conditions in which aldoses oxidize. This difference forms the basis for a variety of identification tests, particularly for glucose, which, as an aldohexose, is a reducing sugar. Two of these tests are:

Tollen's Test

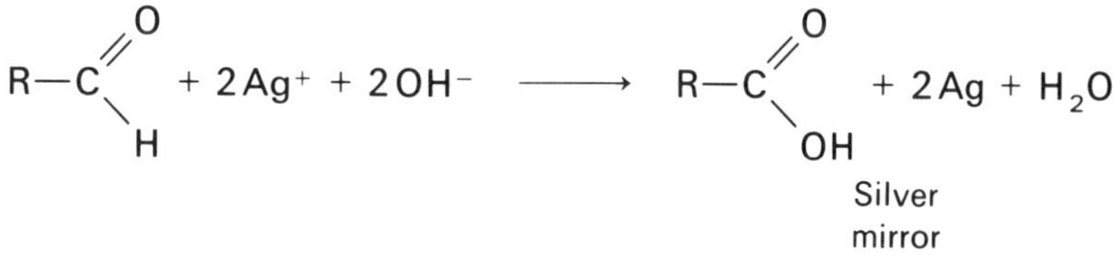

$$R-C\overset{O}{\underset{H}{\big<}} + 2\,Ag^+ + 2\,OH^- \longrightarrow R-C\overset{O}{\underset{OH}{\big<}} + 2\,Ag + H_2O$$

Silver
mirror

Fehling's (Benedict's) Test

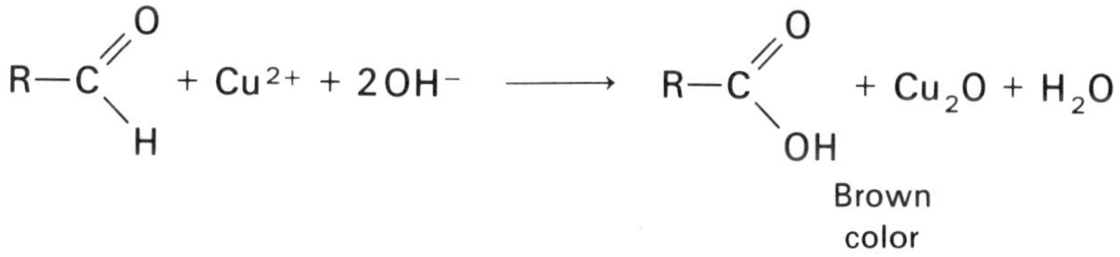

$$R-C\overset{O}{\underset{H}{\big<}} + Cu^{2+} + 2\,OH^- \longrightarrow R-C\overset{O}{\underset{OH}{\big<}} + Cu_2O + H_2O$$

Brown
color

Chemical oxidation of aldoses generally produces *aldonic* acids. In some cases, the aldonic acid forms an internal ester, or *lactone*:

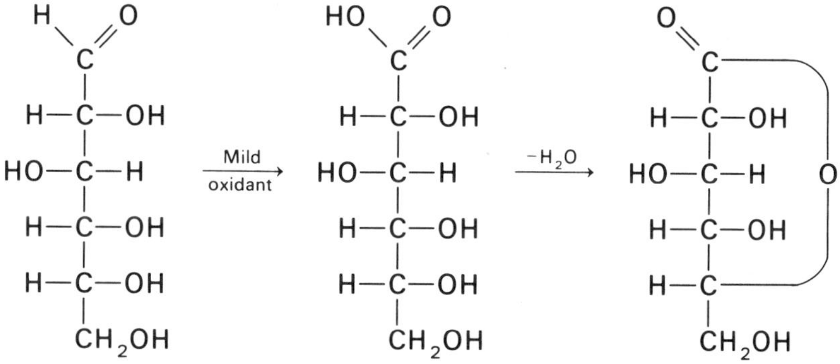

D-Glucose $\xrightarrow[\text{oxidant}]{\text{Mild}}$ D-Gluconic acid $\xrightarrow{-H_2O}$ δ-Gluconolactone

Some aldonic acids are strong organic acids and some of their salts are used in medicine. For example, calcium gluconate is a stable and biologically acceptable form for administering calcium in cases of deficiency. L-Ascorbic acid, or vitamin C, is a lactone derived from a hexonic acid containing a double bond between carbons 2 and 3. While the physiological role of L-ascorbic acid remains to be clarified completely, it most likely involves the redox properties of L-dehydroascorbic acid.

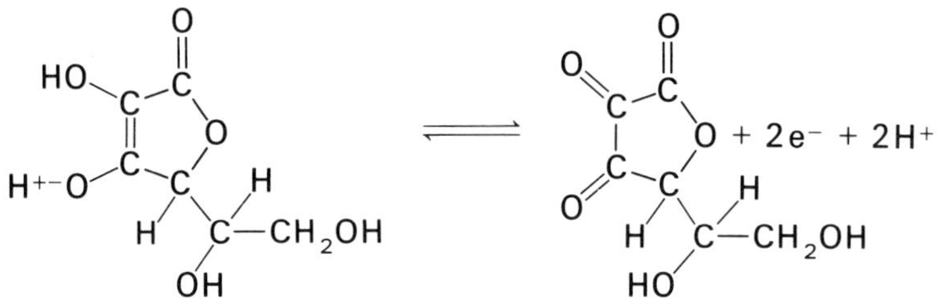

L-Ascorbic acid   L-Dehydroascorbic acid

When the number 6 carbon of a hexose is oxidized by an enzyme, a *uronic acid* is produced. For example, D-glucose yields D-glucuronic acid, a component of the polysaccharide hyaluronic acid.

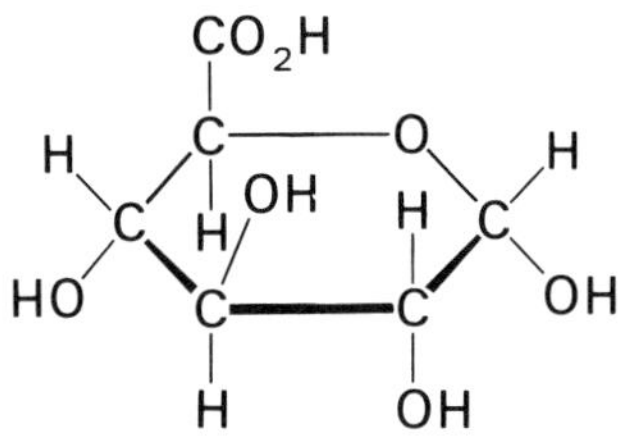

α-D-Glucuronic acid

The alcohol functions of sugars can exhibit reactions such as ester and ether formation, which are characteristic of the alcohol functional group. In addition to forming esters with carboxylic acids, sugars form esters with inorganic oxy-acids as well. Nitrocellulose is a poly-nitro ester of the polysaccharide cellulose. More important physiologically are the phosphate esters of sugars. Several sugar-phosphate esters are important intermediates in metabolism, as we shall see in Chapter 8. At physiological pH, the sugar-phosphoric acid esters are anionic due to ionization of the remaining phosphoric acid protons.

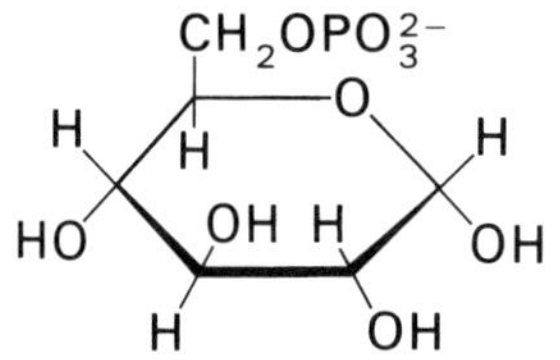

α-D-Glucose 6-phosphate 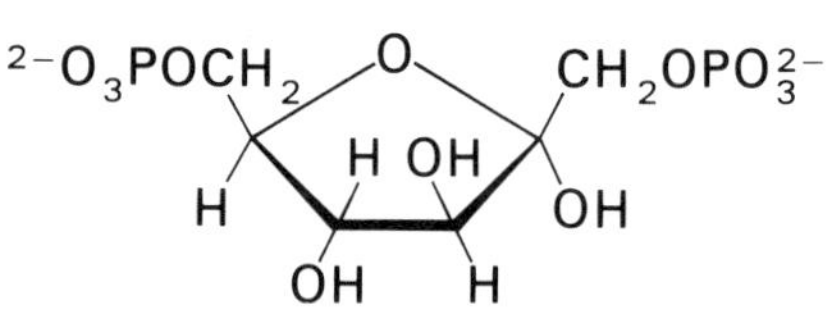   α-D-Fructose 1,6-diphosphate

An important reaction involving the alcohol functions in sugars is the formation of an acetal derivative at the anomeric —OH group (see Appendix B). This reaction forms the basic covalent link in oligo- and polysaccharides. We will consider this aspect of monosaccharide chemistry shortly.

Monosaccharides: Other Types

In addition to the biologically important D-aldoses and D-ketoses, there are other types of monosaccharides, which we can consider as derivatives of those sugars already mentioned.

The *amino sugars* are a class of monosaccharides where an —NH$_2$ group replaces a hydroxyl group in a simple monosaccharide, usually at the number 2 carbon. Amino sugars are commonly found in nature as *N*-acetyl derivatives, which are uncharged at physiological pH. One important example is *N*-acetyl-D-glucosamine. We shall see later in this chapter that this particular amino sugar is a component of the polysaccharide *chitin*, which is a major component of the exoskeletons of insects and crustaceans.

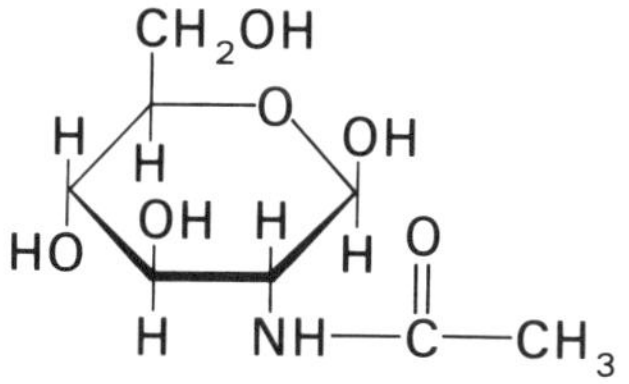

N-acetyl-β-D-Glucosamine 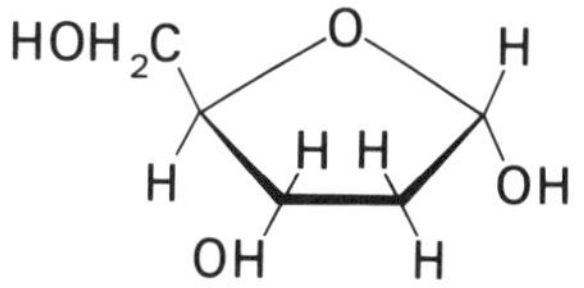   2-deoxy-α-D-Ribose

The *deoxysugars* result from the replacement of an —OH group in a sugar by an —H, by, for example, selective reduction. Probably the most important deoxysugar is 2-deoxyribose, the carbohydrate constituent of DNA.

4.2

Oligosaccharides

The most common oligosaccharides are disaccharides, composed of two monosaccharide units (identical or different) joined by a hydrolyzable *glycoside* link. A number of trisaccharides and higher oligosaccharides are found in nature, but we will confine our attention to the common disaccharides maltose, sucrose, and lactose.

The glycoside link formed between two monosaccharide units is chemically identical to an acetal (in the case of aldoses) or a ketal (in the case of ketoses). An acetal is formed from the reaction of a hemiacetal and an alcohol:

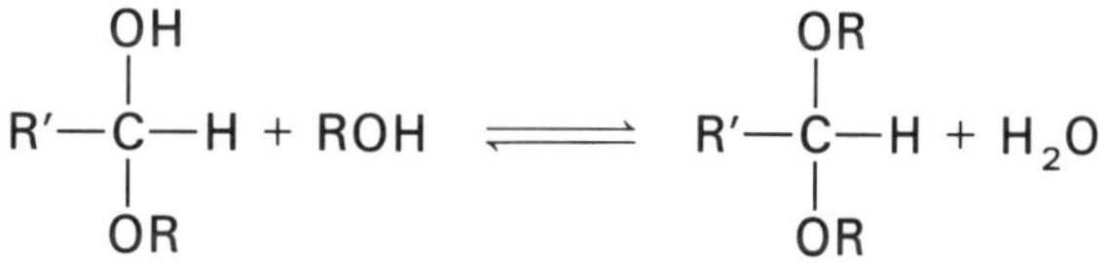

Hemiacetal   Alcohol   Acetal

Hemiketals react with alcohols in a similar way to yield ketals (see Appendix B). Because monosaccharides in their cyclic forms contain both hydroxyl (alcohol) groups and a hemiacetal or hemiketal group, it is possible to apply the chemistry just discussed to learn how two or more monosaccharide units can be joined together.

Disaccharides are glycosides formed when one monosaccharide reacts as an alcohol and another reacts as a hemiacetal (or hemiketal). The simplest disaccharide is probably *maltose*, which on hydrolysis yields a solution of pure D-glucose. Equation (1) shows how maltose forms from two glucose units:

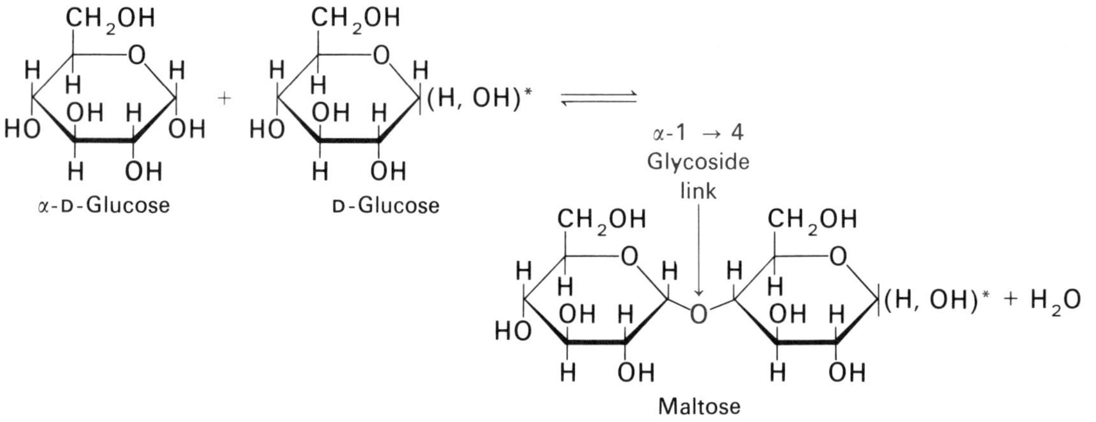

$$\alpha\text{-D-Glucose} + \text{D-Glucose} \rightleftharpoons \text{Maltose} + H_2O \qquad (1)$$

α-D-Glucose

D-Glucose

α-1 → 4 Glycoside link

Maltose

* The designation (H, OH) means that the anomeric position of this monosaccharide is free to mutarotate.

The glycoside linkage shown in Eq. (1) is termed an α-1 → 4 link because it forms from the anomeric —OH of one D-glucose unit in the α configuration and the number 4 —OH group of the other D-glucose unit. Since an unsubstituted anomeric carbon still remains in maltose, it is free to mutarotate; that is, it can exist in any of the three equilibrium forms α, β, or open chain. Thus, maltose is a reducing sugar.

Cellobiose is a disaccharide which we can consider to be the basic repeating unit in cellulose:

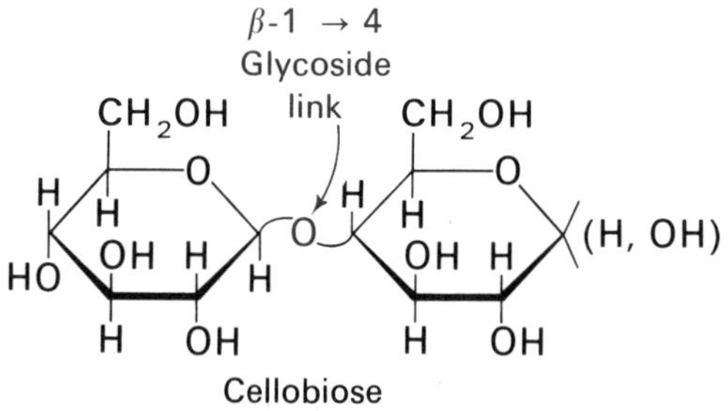

β-1 → 4 Glycoside link

Cellobiose

Looking at the structure of cellobiose, it is apparent that the disaccharide is formed from the anomeric —OH of one unit in the β configuration and the number 4 —OH of the other unit. This type of glycoside link is a β-1 $\rightarrow$ 4 link. (You should bear in mind that the "S"-shaped bond is not a true representation of the shape of a β-1 $\rightarrow$ 4 link, merely a convenience to make it readily distinguishable from the α counterpart.)

Sucrose (cane sugar) is the disaccharide we probably have the most direct experience with. Since the glycoside link forms from the anomeric hydroxyls of both monosaccharide units, sucrose is not a reducing sugar. Thus, it does not undergo mutarotation. Sucrose can be hydrolyzed either enzymatically or chemically to produce an equilibrium mixture of glucose and fructose that is sweeter, pound for pound, than sucrose. This mixture is called invert sugar because the hydrolysis is accompanied by an inversion in optical rotation from clockwise (dextrorotary) to counterclockwise (levorotary). Honey is a naturally occurring form of predominantly invert sugar.

Lactose, or milk sugar, is found only in milk. It yields D-galactose and D-glucose upon hydrolysis.

Sucrose
(α-D-Glucopyranosyl-β-D-Fructofuranoside)

β-1 $\rightarrow$ 4
Glycoside
link

Lactose

4.3

Polysaccharides

The bulk of the carbohydrate molecules in nature exist in the form of high-molecular-weight polysaccharides, which are used either for structural purposes or for the storage of chemical energy. Since the basic covalent connection between monomeric units in a polysaccharide is a glycoside link, polysaccharides are also called *glycans*. There is a vast range of variation in the components and structural properties of polysaccharides. They can differ in the nature of the monosaccharides which make up the repeating units, in the chain length, and in the existence and extent of branching. In general, a pure sample of a particular polysaccharide contains molecules of varying degrees of polymerization—that is, of molecular weight. In contrast to proteins, where the sequence of amino acids, and thus the molecular weight, is precisely specified, polysaccharides are

Table 4.2   Examples of Polysaccharide Types.

Name	Source	Monosaccharide unit(s)	Repeating linkages	Form
Homopolysaccharides				
Laminaran	Seaweeds (*laminaria*)	D-glucose	α-1 $\rightarrow$ 3	Linear
Inulin	Jerusalem artichoke	D-fructose	α-1 $\rightarrow$ 4	Linear
Cellulose	Plants	D-glucose	β-1 $\rightarrow$ 4	Linear
Amylose	Plants (starch)	D-glucose	α-1 $\rightarrow$ 4	Linear
Amylopectin	Plants (starch)	D-glucose	α-1 $\rightarrow$ 4; α-1 $\rightarrow$ 6	Branched
Glycogen	Animals	D-glucose	α-1 $\rightarrow$ 4; α-1 $\rightarrow$ 6	Branched
Chitin	Arthropods, fungi	*N*-acetyl-D-glucosamine	β-1 $\rightarrow$ 4	Linear
Heteropolysaccharides				
Gum arabic	Acacia trees	D-galactose L-arabinose L-rhamnose D-glucuronic acid	Not fully known	Branched
Blood group substances	Humans	D-galactose D-glucosamine L-fucose	Not known	Branched
Hyaluronic acid	Higher animals	D-glucuronic acid *N*-acetyl-D-glucosamine	β-1 $\rightarrow$ 3 β-1 $\rightarrow$ 4	Linear

said to be *polydisperse*, because a given sample contains polysaccharide molecules of various chain lengths. *Homopolysaccharides* produce a solution of only one type of monosaccharide upon complete hydrolysis. *Heteropolysaccharides* produce solutions containing two or more types of monosaccharides. Heteropolysaccharides containing amino sugars are referred to as *mucopolysaccharides*. Table 4.2 lists a few polysaccharides, illustrating the wide range of types of these macromolecules.

Homopolysaccharides

Many homopolysaccharides serve as carbohydrate reserves for the organism in which they occur. For example, laminaran, the reserve carbohydrate of the seaweed *laminaria*, can make up as much as 50% of the dry weight of the seaweed frond, depending on the season. We are certainly familiar with corn starch and potato starch, which are reserve carbohydrates for these two plants.

Starches fall into two categories: *amylopectin* and *amylose*. Amylose is a linear macromolecule; amylopectin is branched. Molecular weights of starches can range into the millions. The basic repeating unit for both amylose and amylopectin is the maltose unit:

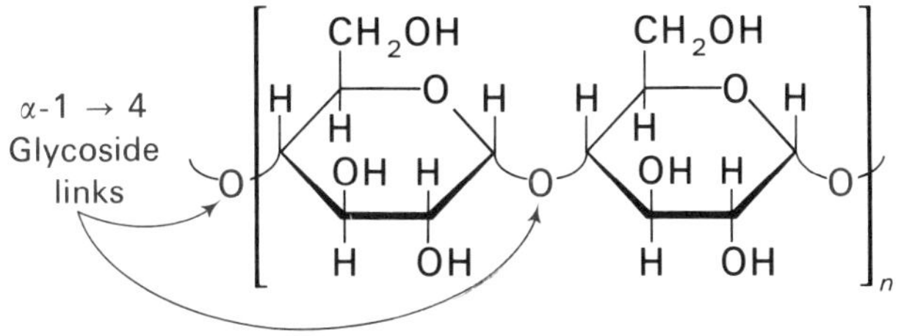

Starch molecules have two distinct ends: a nonreducing end with a free number 4—OH group, and a reducing end with a free anomeric —OH group. Branching in amylopectin takes place approximately once every 25 glucose units in the polysaccharide chain. The point of branching occurs at the formation of an α-1 → 6 linkage, shown in Figure 4.6. *Glycogen*, or muscle sugar, is a branched-chain reserve carbohydrate produced in animals. It is basically identical to amylopectin except that glycogen is more highly branched.

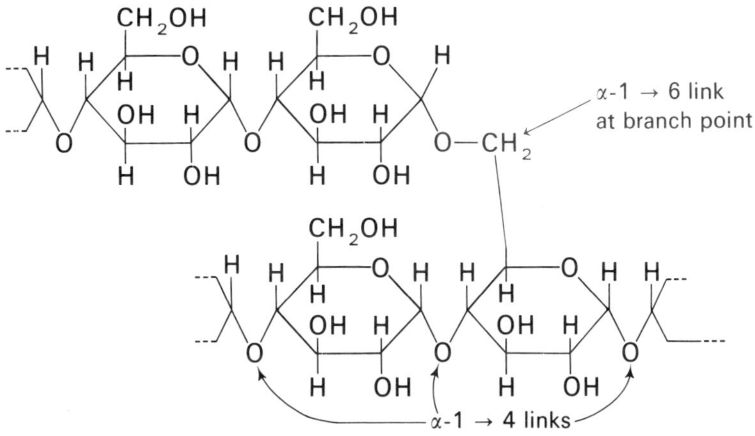

Figure 4.6

Mode of branching in amylopectin and in glycogen.

The enzymatic hydrolysis of starch is an important process in digestion. We can divide starch-hydrolyzing enzymes into two classes: α-amylases of saliva and pancreatic juice, and β-amylases of germinating seeds. The correct term for an α-amylase is α-1,4-glucan,4-glucanohydrolase. This name describes the action of the enzyme, illustrated in Figure 4.7. The action of β-amylase, or

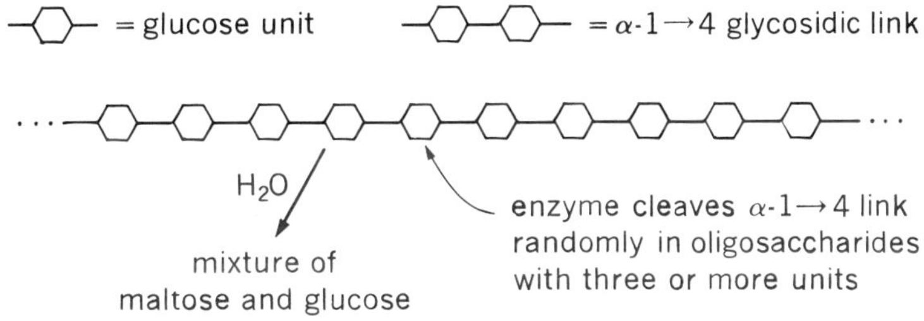

Figure 4.7

Action of α-amylase on amylose.

more specifically α-1,4-glucan maltohydrolase, is more specific in that the enzyme cleaves off successive maltose units, working in from the nonreducing end. Thus, the hydrolysis of *amylose*, as catalyzed by β-amylase, yields a solution of maltose. The action of β-amylase on an *amylopectin* is affected by the presence of the α-1 → 6 links at the branch points, which are not hydrolyzable by the enzyme. Thus, hydrolysis ceases when the enzyme reaches a branch point, as shown in Figure 4.8. The polysaccharide remaining in such a solution, after the incomplete hydrolysis of amylopectin by a β-amylase, is called a *dextrin*. Dextrins are low-to-medium molecular weight branched polysaccharides, and are commonly used as a base for the paper paste used by schoolchildren.

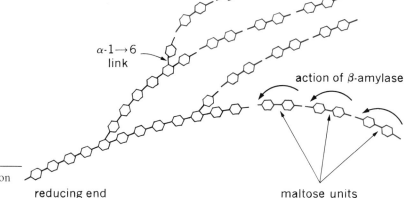

Figure 4.8

Action of β-amylase on amylopectin.

Iodine in aqueous solution with starch becomes adsorbed by the highly-hydrated starch molecules, producing a blue-black color. The color depends on the average size of the molecule. We can follow the hydrolysis of a starch by observing the color changes in the presence of iodine:

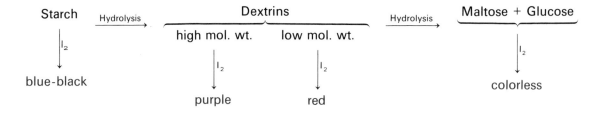

Cellulose and chitin are two structural polysaccharides which deserve attention due to their abundance in nature. Cellulose and starch produce the same solution of D-glucose upon complete hydrolysis. However, a mouthful of white bread produces a sweet taste due to hydrolysis of the starch by the salivary amylases, while a mouthful of cotton, which is pure cellulose, elicits no sweet taste. The structural difference between cellulose and amylose may seem minor, but it is enough to render cellulose unhydrolyzable, and thus indigestible by α-amylases in human digestive fluids. Whereas amylose is α-1 $\rightarrow$ 4 linked, cellulose is β-1 $\rightarrow$ 4 linked. Cellulose is a linear high-molecular-weight homopolysaccharide whose basic repeating unit, cellobiose, was shown in Section 4.2.

Chitin is similar to cellulose except that it is a mucopolysaccharide composed of β-1 $\rightarrow$ 4 linked N-acetyl-D-glucosamine units:

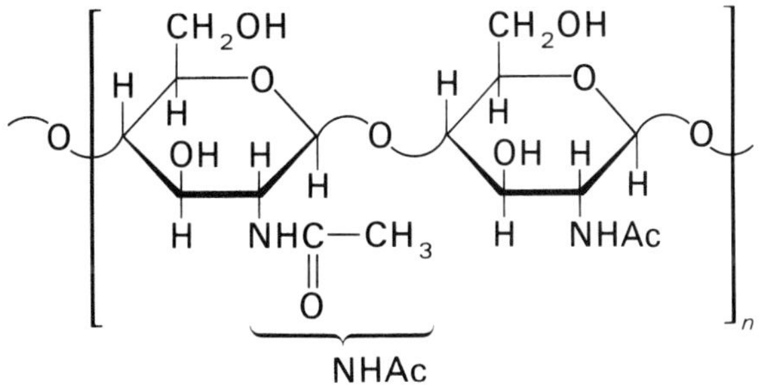

Repeat unit of chitin

Chitin is generally strongly resistant to hydrolysis, although the enzyme *chitinase* accomplishes the job easily. Chitin forms the basis for the durable material of mold spore coats and of the exoskeletons of insects and crustaceans.

Heteropolysaccharides

A wide variety of polysaccharides contain two or more monosaccharide units. Many vegetable gums contain as many as four different monosaccharide units. Some vegetable gums, such as agar and carageenan, are important thickening agents in food technology.

Hyaluronic acid is a heteropolysaccharide which is an important component of the connective tissue of animals. The structure of the hyaluronic acid repeating unit is

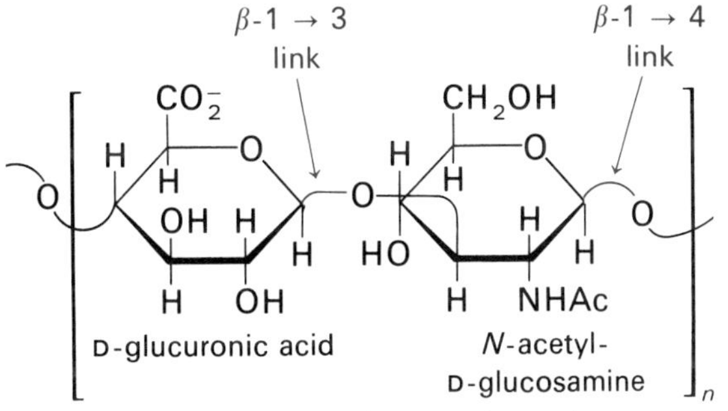

Hyaluronic acid repeating unit

At physiological pH, the hyaluronic acid molecule becomes thoroughly hydrated due to the large number of anionic carboxylate functional groups. This produces a gel, or thick fluid, depending on the size of the molecule. Hyaluronic acid in tissues is generally associated with protein material. It serves as an important component of the ground substance or matrix which binds cells together in connective tissue, and is present also in the lubricating synovial fluid of joints and in the vitreous humor of the eye. Many snake and bee venoms contain a "spreading factor" consisting of an enzyme, hyaluronidase, that hydrolyzes hyaluronic acid and thus facilitates spreading of the toxic substances in the venom throughout the tissue.

Complex Heteropolysaccharides

Polysaccharides are intimately involved in the biochemistry of cell surfaces. In the cells of higher animals, heteropolysaccharides, glycoproteins, and lipo-polysaccharides (polysaccharides covalently linked to lipids) are associated with the immunological properties of the cells. Complex heteropolysaccharides are associated with the antigenic properties of certain bacteria; the outer "slimy" coat, or *capsule* in others; and the basic covalent structure of the bacterial cell wall.

4.4

Bacterial Cell Walls

The bacterial cell wall is a highly structured protective coat surrounding the cell membrane of bacteria and it possesses a considerable degree of mechanical strength. This strength is vital to a free-living organism such as a bacterium, which must be able to exist in a hypo-osmotic environment where an unprotected cell would absorb water and burst due to hypotonic lysis. We can divide bacteria into two broad classes: Gram-positive and Gram-negative, depending on the behavior of the cells during the Gram staining procedure. The differences between the cell walls of the two classes of bacteria account for the different staining behavior. Table 4.3 summarizes the properties of the cell walls of Gram-positive and Gram-negative bacteria.

Table 4.3 Properties of Cell Walls in Gram-positive and Gram-negative Bacteria.

Bacteria	Constituents of cell wall	Physical properties of cell wall
Gram-positive	Pure carbohydrates and protein	Rigid, hard
Gram-negative	Carbohydrates and protein, lipids, lipoproteins, lipopolysaccharides	Flexible, soft

Although Gram-positive and Gram-negative bacteria differ in the complexity of the overall cell coat, the covalent framework of the cell wall is basically the same in both types of organisms. This covalent framework consists of a single bag-shaped molecule that encompasses the cell membrane and is called a peptidoglycan, or *murein* (Latin: *murus*, "wall"). While details vary from one species to another, the basic structure is the same. Thus, the repeat unit of bacterial cell wall peptidoglycan shown in Figure 4.9 is representative. The nature of the tetrapeptide varies from one species to another. What is important is that the tetrapeptide includes at least one free carboxyl function and at least one free amino function, permitting cross-linking of parallel heteropolysaccharide-peptide chains with short oligopeptide bridges to form the overall

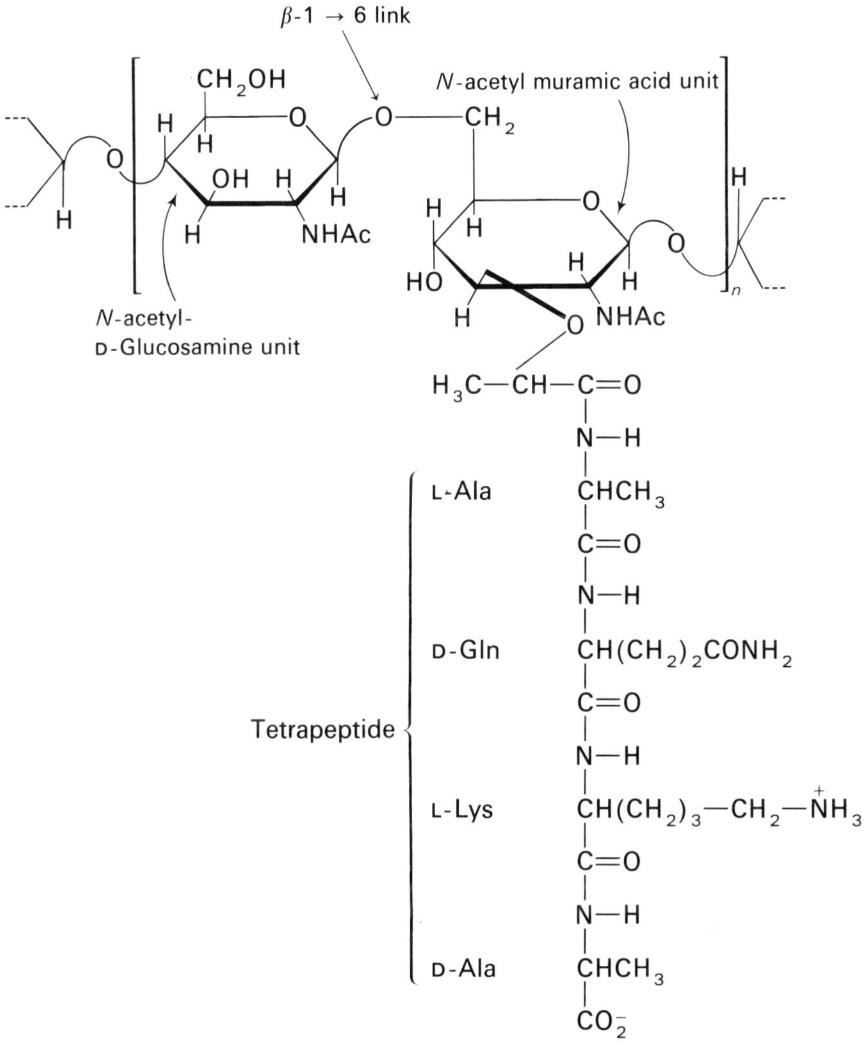

Figure 4.9

Typical repeat unit of bacterial cell wall peptidoglycan.

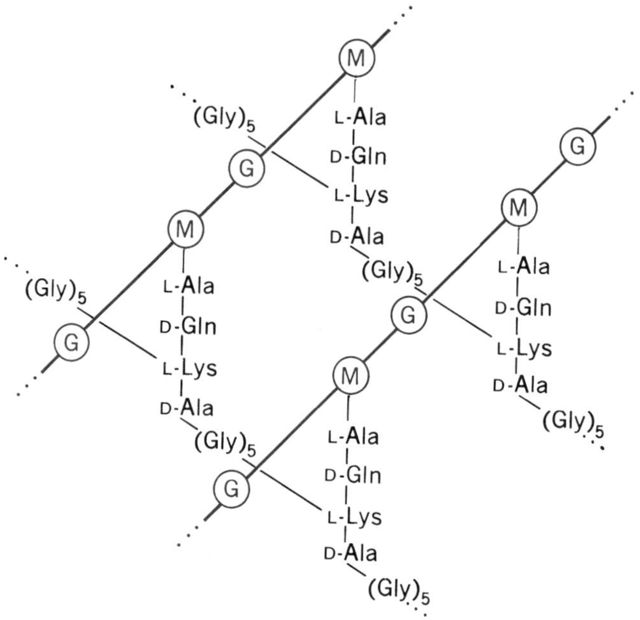

Figure 4.10

Portion of the
peptidoglycan of *S.
aureus*, showing cross-
linking of adjacent
heteropolysaccharide-
tetrapeptide chains by
pentaglycine peptide
bridges. G = *N*-acetyl-
D-glucosamine,
Ⓜ = *N*-acetyl muramic
acid.

cell wall peptidoglycan. The arrangement of the peptidoglycan components in the cell wall of *Staphylococcus aureus* is shown in Figure 4.10 and illustrates how the cross-linked cell wall structure is produced. The overall result is a peptidoglycan network surrounding the cell.

The biosynthesis of the cell wall takes place outside the cell membrane. Certain antibiotics interfere with this process. Penicillin, for example, inhibits the enzyme that catalyzes the final cross-linking between the pentapeptide bridge and the D-alanine residue of the heteropolysaccharide-tetrapeptide backbone.

In addition to the peptidoglycan framework, bacterial cell walls contain a number of carbohydrate-containing *accessory substances*, such as the teichoic acids of Gram-positive organisms and the lipopolysaccharides of Gram-negative bacteria. The functions of many of these accessory substances remain unclear.

4.5

Glycoproteins

In Chapter 2 we saw that there are a number of classes of conjugated proteins. *Glycoproteins* are one such class in which the carbohydrate part is covalently linked to the polypeptide chain. We have seen in this chapter that polysaccharides are not, by themselves, informational macromolecules, in that there is no genetically-specified sequence of monomer units as is the case with proteins. Glycoproteins, however, do permit the inclusion of carbohydrate units into an informational macromolecule. Because many glycoproteins in higher animals occur on the cell surfaces, the informational role of glycoproteins is the basis for such effects as cell–cell recognition; tissue rejection in organ transplants; and blood typing for matching blood donors and recipients. Table 4.4 summarizes some glycoproteins and their functions.

Table 4.4   Some Glycoproteins and Their Sources and Functions.

Glycoprotein	Source	Function
Antifreeze protein	Antarctic fish	Prevents blood plasma from freezing
Mucins	Salivary and gastric mucous secretion	Viscous lubricants
Collagen	Bone, connective tissue	Structural (see Chapter 2)
Fibrinogen	Blood	Blood clot formation
Immune globulins	Blood	Basis of antibody system
Follicle-stimulating hormone (FSH)	Pituitary gland	Controls estrogen production at beginning of reproductive cycle

Blood Group Substances

Blood group substances are glycoproteins that differ among individuals, according to their blood type. Individuals are classified into four main blood types; A, B, AB and O. Blood serum from individuals with type O blood contains antibodies that cause the clumping and precipitation of red blood cells from type A, B and AB donors. Blood donations from type A, B and AB individuals are mutually incompatible in the same way. It is noteworthy that neither A, B nor AB serum contain antibodies for type O erythrocytes. This is why blood type O individuals are sometimes called "universal donors."

The molecular basis for blood types is the erythrocyte surface protein, *glycophorin*. Part of this protein protrudes from the erythrocyte membrane and contains many oligosaccharide units covalently linked to serine and threonine hydroxyl groups of the polypeptide chain. The parts of the oligosaccharide units that actually interact with the binding sites of the antibody proteins are called the *blood group antigens*. These consist of 3 or 4 monosaccharide units as shown in Figure 4.11. It is remarkable that the substitution of just one *N*-acetyl group for one hydroxyl group of one monosaccharide unit can mean the difference between type A and type B blood. In the case of type AB individuals, the erythrocytes contain both A and B blood group antigens.

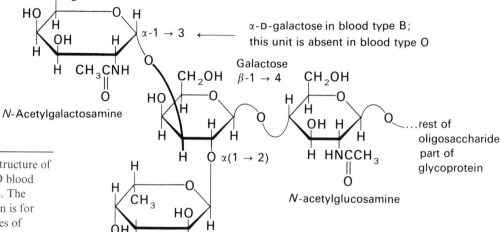

α-D-galactose in blood type B; this unit is absent in blood type O

Galactose β-1 → 4

N-Acetylgalactosamine

N-acetylglucosamine

...rest of oligosaccharide part of glycoprotein

α-L-fucose

Type A blood group antigen

Figure 4.11

The covalent structure of the A, B and O blood group antigens. The structure shown is for one of two types of oligosaccharide chains. One has the β-1 → 4 linkage between the galactose and N-acetyl glucosamine unit as shown, and the other has a β-1 → 3 linkage at this point.

Suggestions for Further Reading

Morrison, R. T. and Boyd, R. N. *Organic Chemistry*. Boston: Allyn and Bacon, 1973. This standard organic chemistry text contains two excellent chapters on the organic chemistry of carbohydrates.

White, A.; Handler, P.; Smith, E.; Hill, R. and Lehman, I. *Principles of Biochemistry*. 6th ed. New York: McGraw-Hill, 1978. Includes several good chapters on the biochemistry of carbohydrates and cell walls.

Scientific American

Albersheim, P. "The Walls of Growing Plant Cells." April 1975.
Kretchmer, N. "Lactose and Lactase." October 1972.
Sharon, N. "The Bacterial Cell Wall." May 1969.
———. "Glycoproteins." May 1974.
———. "Lectins." June 1977.

Problems

1. For each of the terms given below, provide a brief definition or explanation.

Aldose	Glycan	Murein
Amino sugar	Glycoprotein	Mutarotation
Anomer	Glycoside link	Pentose
Deoxysugar	Hexose	Pyranose
Furanose	Ketose	

2. Draw the structures of the product(s) of the following reactions:

(a) D-glucose $+ 2\,Ag^+ + 2\,OH^- \longrightarrow$

(b) [product from (a)] + dehydrating agent $\longrightarrow$

(c) cellulose $\xrightarrow[\text{hydrolysis}]{\text{complete}}$ (d) amylose $\xrightarrow[\text{hydrolysis}]{\text{complete}}$

(e) α-D-galactose $\xrightarrow{\text{mutarotation}}$

3. The monosaccharides D-glucose and D-fructose react to form a disaccharide. Draw the structures of all the possible disaccharides that can be produced using α- and β-1 → 4 and 1 → 6 links.

4. Sugar hydroxyl groups react with methyl iodide in the presence of Ag_2O to form methyl *ethers* by a process known as methylation:

$$R\text{—OH} + I\text{—CH}_3 \xrightarrow{Ag_2O} R\text{—O—CH}_3 + HI$$
$$\text{Ether}$$

This is a useful tool for carbohydrate chemists working on structure determination. For example, a pure oligosaccharide containing six glucose units can be exhaustively methylated with CH_3I and Ag_2O and then completely hydrolyzed to yield only 2,3,6-tri-O-methylglucose:

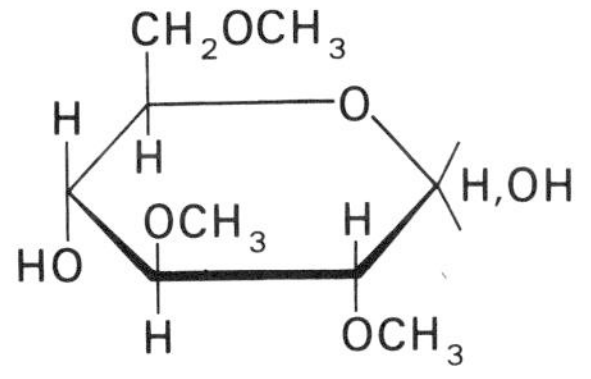

The original pure oligosaccharide could not be oxidized by silver ion or cupric ion. Give the structure of the oligosaccharide, assuming α-glycoside links.

5. Using structures, show why sucrose is not a reducing sugar.

6. We have been representing a β-1 → 4 glycoside link, as in cellulose, using an "S"-shaped notation. This is, of course, not a true representation of how the bond actually looks. Using Haworth projection formulas, draw the actual structure of a penta-saccharide fragment of cellulose.

7. Draw the structures of all the products resulting from the complete hydrolysis of

(a) Hyaluronic acid
(b) Peptidoglycan unit (shown in Figure 4.9).

8. Refer to Figure 4.11:

(a) Draw the structures of the blood group antigens for types B and O blood groups.
(b) Using the number 1 carbon of α-D-glucose as a point of attachment, show how an oligosaccharide chain could be covalently attached to a serine and to a threonine residue of a polypeptide.

CHAPTER 5

Nucleotides and Nucleic Acids

The nucleic acids *deoxyribonucleic acid* (*DNA*) and *ribonucleic acid* (*RNA*) provide the chemical basis for information transfer within all cells. Nucleic acids are polymer chains composed of monomeric units called *nucleotides*. Nucleic acids are informational macromolecules—each nucleic acid has a unique nucleotide sequence analogous to the unique amino acid sequence of a given protein. In this chapter we will learn about the monomeric units of DNA and RNA and how these units are joined to form a nucleic acid strand. In doing this, we will also see that some of the nucleotides and their derivatives are important biomolecules in their own right. Finally, we will examine the covalent structure of DNA and RNA, features of the three-dimensional structure of these molecules, and the combination of nucleic acids with proteins to form such structures as ribosomes and viruses.

In Chapter 13 we will build on the material covered in this chapter and examine the way nucleic acids function in the transfer of information in the cell.

5.1

Mononucleotides

Mononucleotides are the recurring building blocks of all nucleic acids. The three components of a mononucleotide can be isolated and identified as shown in Eq. (1):

$$\text{Mononucleotide} \xrightarrow[\text{hydrolysis}]{\text{Complete}} \begin{cases} \text{Heterocyclic nitrogen bases} \\ \text{Five-carbon sugar} \\ \text{Phosphate} \end{cases} \tag{1}$$

Two of these components are already familiar to us. Phosphate shows up as a key inorganic substance in cellular affairs. The five-carbon sugar component is either D-*ribose* in the case of RNA or *2-deoxy* D-*ribose* in the case of DNA. Just as L-amino acids constitute the building blocks of proteins, the pentose part of nucleic acids is present as the D-optical isomer only.

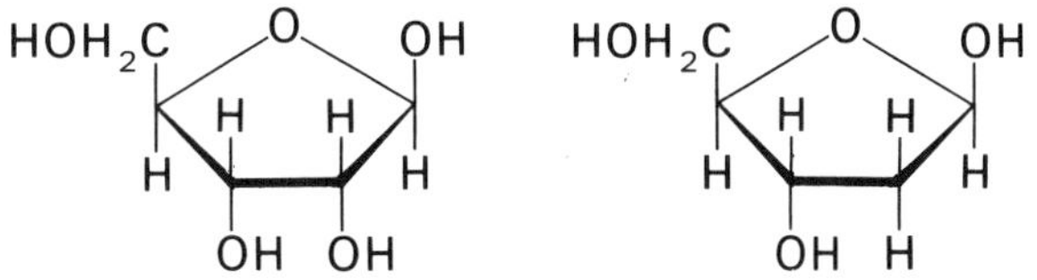

D-Ribose   2-deoxy D-ribose

The heterocyclic nitrogen bases fall into two broad families: purines and pyrimidines.

Purine bases: Adenine and Guanine Figure 5.1 shows how the two purine bases adenine and guanine are related to the basic purine nucleus. Adenine is already familiar to us as the base associated with ATP, ADP, and AMP. Guanine and adenine occur in both DNA and RNA isolated from natural sources.

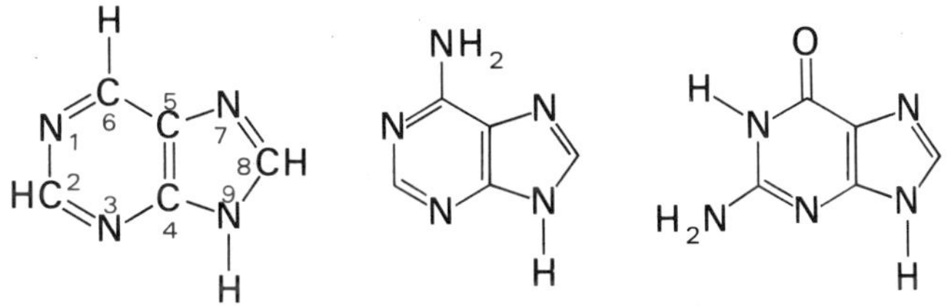

Figure 5.1

The purine bases adenine and guanine.

Purine   Adenine (A)   Guanine (G)

Pyrimidine bases: Cytosine, Uracil, and Thymine Figure 5.2 shows how the structures of cytosine, uracil, and thymine are related to the pyrimidine nucleus. Cytosine occurs in both DNA and RNA; thymine is found primarily in DNA; and uracil is found only in RNA.

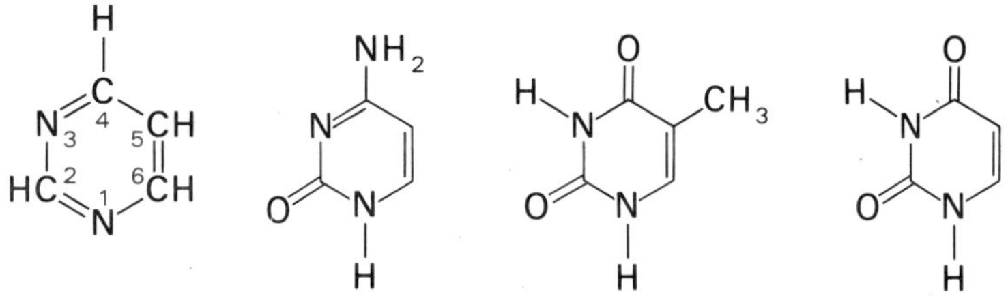

Figure 5.2

The pyrimidine bases cytosine, thymine, and uracil.

Pyrimidine   Cytosine (C)   Thymine (T)   Uracil (U)

Although Adenine, Guanine, Cytosine, Thymine, and Uracil account for the major bases found in the mononucleotide units of DNA and RNA, certain types of RNA contain other *minor bases*. These minor bases are generally derivatives of the major bases, as we can readily see from the structures of the minor bases shown in Figure 5.3.

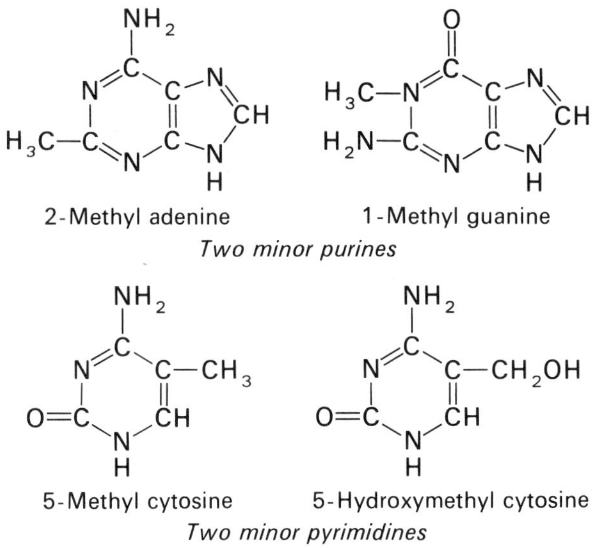

Figure 5.3

Some minor bases found in RNA.

2-Methyl adenine 1-Methyl guanine
Two minor purines

5-Methyl cytosine 5-Hydroxymethyl cytosine
Two minor pyrimidines

Properties of Purine and Pyrimidine Bases

Each of the purine and pyrimidine bases can exist in two or more *tautomeric* forms in chemical equilibrium. The structures given in Figures 5.1, 5.2 and 5.3 represent the major tautomeric forms for each base at pH 7. The tautomers of uracil are shown in Figure 5.4.

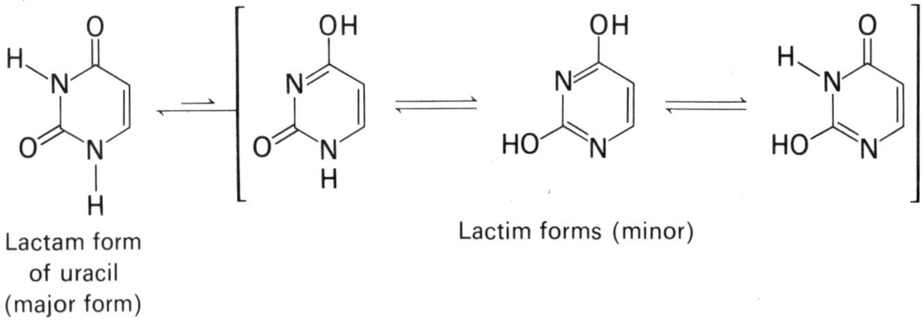

Figure 5.4

Tautomeric forms of uracil. Purine and pyrimidine bases exist as multiple forms in self-equilibrium called *proton tautomers*. The base structures used throughout this text are the predominant forms that exist at pH 7.

Lactam form
of uracil
(major form)

Lactim forms (minor)

The purine and pyrimidine bases of nucleic acids contain functional groups that permit hydrogen bonding. Because of the molecular structural features of purine and pyrimidine bases, certain hydrogen-bonded pairs of bases are much more energetically favorable than others. The selectivity of hydrogen bonding between bases is shown in Figure 5.5. We will see in Chapter 13 that this hydrogen bonding interaction or **base pairing** is crucial to the structure and function of nucleic acids.

The nitrogen heterocyclic ring system of the purine and pyrimidine bases absorbs ultraviolet light strongly in the 250–280 nm region. This property provides a useful means for detecting and quantifying nucleic acids and nucleotides spectrophotometrically.

Figure 5.5

Hydrogen bonding between purine and pyrimidine bases (base pairing). Due to differences in molecular structure, certain pairs of bases hydrogen bond together much more strongly than others. Note that adenine forms much stronger hydrogen bonds with thymine than with other moles of adenine.

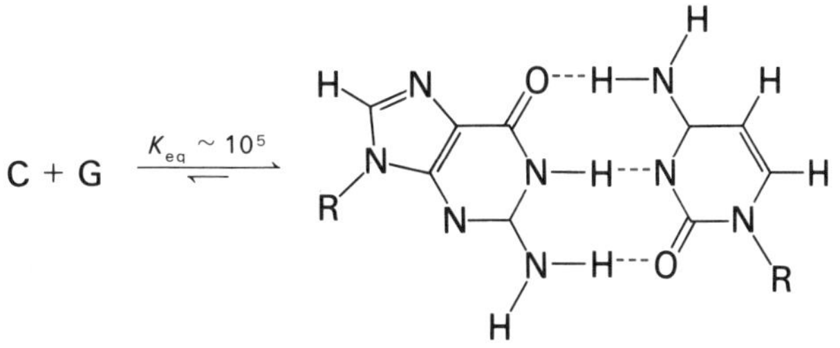

Three hydrogen bonds

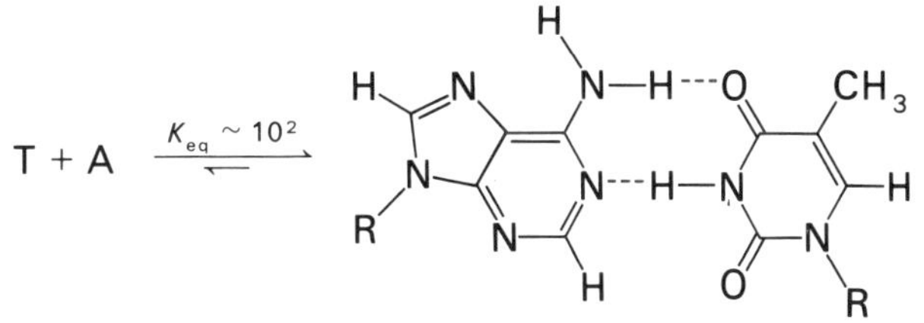

Two hydrogen bonds

$$A + A \underset{}{\overset{K_{eq} \sim 3}{\rightleftharpoons}} A{\cdots}A$$

Nucleosides and Nucleotides

A pentose and a pyrimidine or purine base can combine to form an N-glycosyl derivative called a **nucleoside**. Examples of some nucleoside structures are shown in Figure 5.6. While nucleosides are not ordinarily found in free form in the cell, their phosphoric acid esters are important metabolites.

Mononucleotides are esters of nucleosides in which the oxyacid, phosphoric acid, has formed an ester link with one of the pentose —OH groups. The phosphoric acid part of the nucleotide structure is a strong dibasic acid having pK_a values of approximately 1.0 and 6.0. This means that at a physiological pH of 7, the nucleotides present are predominantly of the dianion form. Therefore, from now on we will refer to these compounds as phosphate esters

Figure 5.6

The structures of some nucleosides, including two possible pentoses. The figure shows the reaction of a pentose (β-D-ribose) with a nitrogen base (adenine) to form a nucleoside (adenosine). Nucleosides normally occur as phosphate esters rather than in the form shown here. Note that the pentose ring positions are given with primes (e.g., (2′)) to differentiate sugar ring positions from those of the purine or pyrimidine ring.

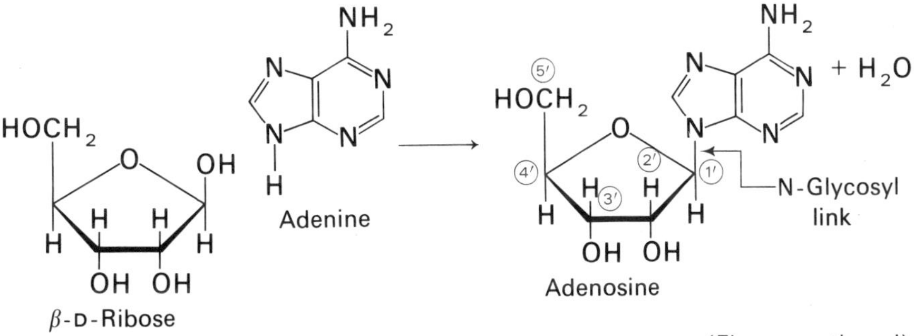

(Figure continued)

Ribonucleosides **Deoxyribonucleosides**

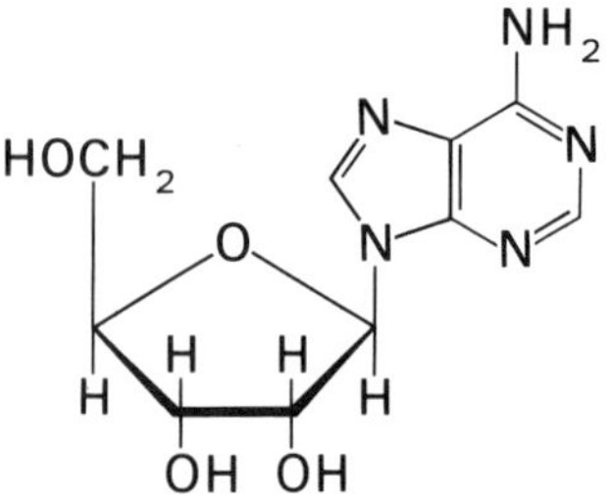
Adenosine

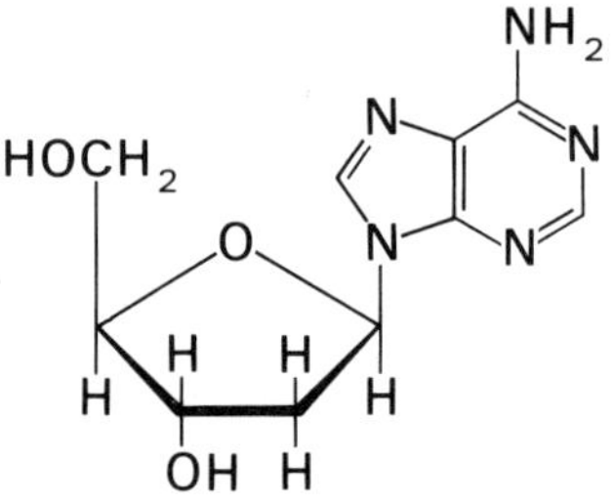
Deoxyadenosine

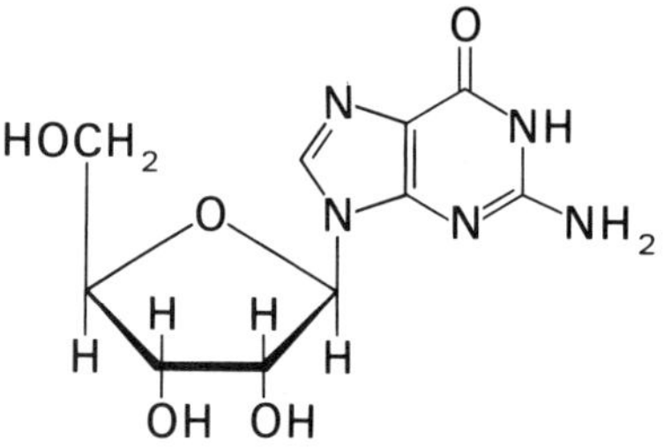
Guanosine

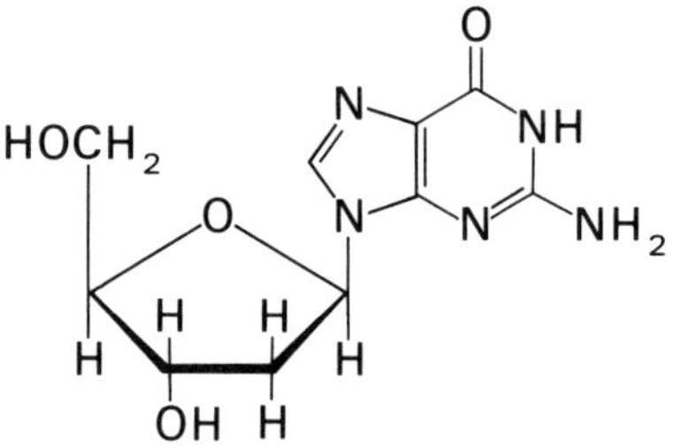
Deoxyguanosine

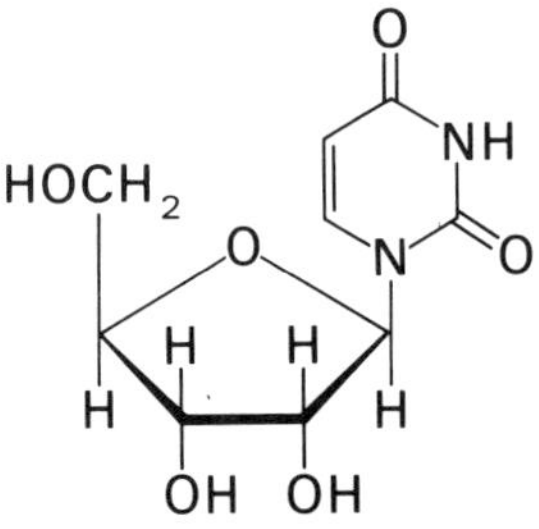
Uridine *

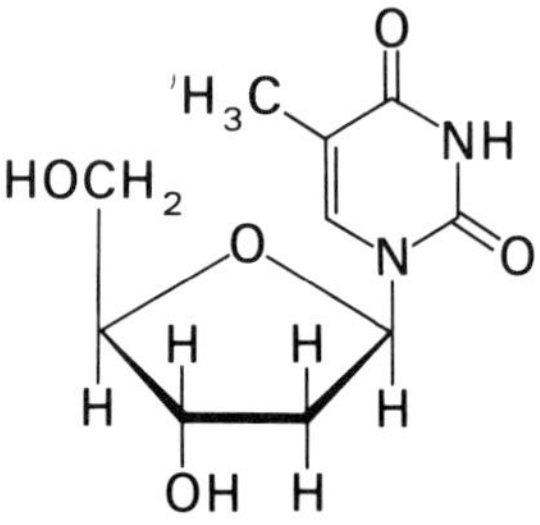
Deoxythymidine *

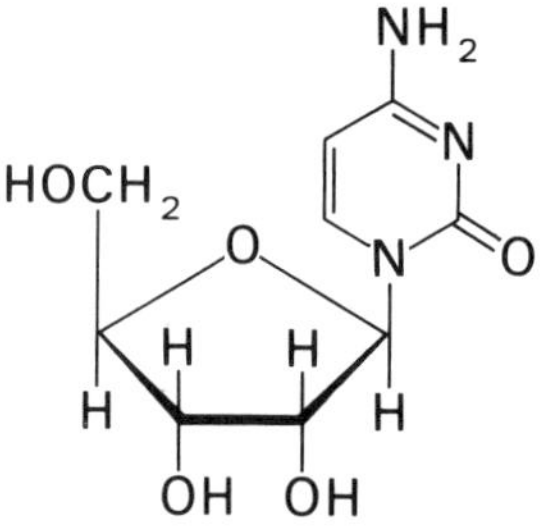
Cytidine

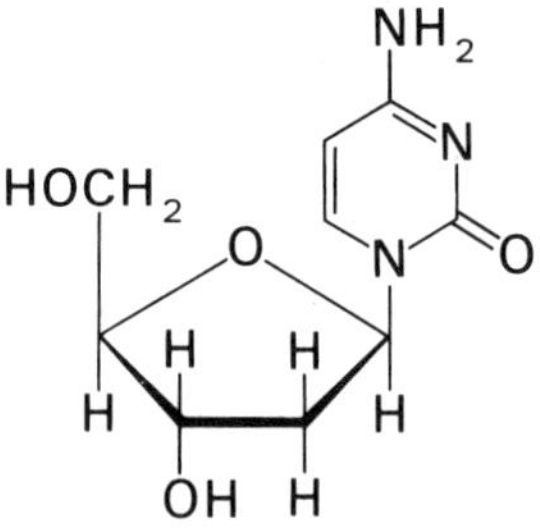
Deoxycytidine

Figure 5.6 (cont.)

The structures of some nucleosides.

* Remember that uracil is found in RNA and thymine in DNA. Both bases have similar hydrogen-bonding properties.

and represent them in the form $R-O-PO_3^{2-}$. Figure 5.7 shows some of the adenine ribonucleoside monophosphates (adenine ribonucleotides) known to occur in nature.

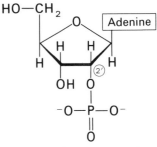

Adenosine 5'-monophosphate

Adenosine 3'-monophosphate

Adenosine 2'-monophosphate

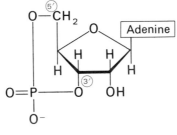

Adenosine 3',5'-cyclic monophosphate
(cyclic-AMP)

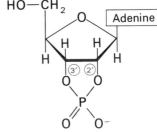

Adenosine 2',3'-cyclic monophosphate

Figure 5.7

Structures of the adenosine monophosphate derivatives: adenine mononucleotides.

Figure 5.8 summarizes the general form of these nucleoside phosphate derivatives together with their nomenclature. As you can see, the various nucleoside triphosphates containing either D-ribose or 2-deoxy D-ribose differ only in the identity of the base part.

The predominant form of free mononucleotides in cells has the phosphate group in the 5' position. This is because the major pathway for the enzymatic hydrolysis of nucleic acids yields nucleoside 5' monophosphates as initial products. In addition to monophosphates, nucleoside 5' di- and triphosphates also occur in cells and serve the following major functions:

1.   **Transfer of chemical energy.** ATP is the central carrier of chemical energy in the cell. It is produced from the phosphorylation of ADP through the energy-yielding processes of metabolism, respiration and photosynthesis. In addition to ATP, other nucleoside triphosphates such as UTP, GTP and CTP serve as carriers of chemical energy in certain specific biosynthetic reactions.

Details of the naming and structure of nucleoside monophosphates.

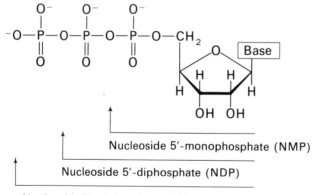

Name	Base		Name	Base
Deoxyadenosine 5'-monophosphate	Adenine		Adenosine 5'-monophosphate	Adenine
Deoxyguanosine 5'-monophosphate	Guanine		Guanosine 5'-monophosphate	Guanine
Deoxycytidine 5'-monophosphate	Cytosine		Cytidine 5'-monophosphate	Cytosine
Thymidine 5'-monophosphate	Thymine		Uridine 5'-monophosphate	Uracil

General structure and abbreviations of nucleoside phosphate derivatives.

Ribonucleoside 5'-mono-, di-, and triphosphates

Base	Abbreviations		
Adenine	AMP	ADP	ATP
Guanine	GMP	GDP	GTP
Cytosine	CMP	CDP	CTP
Uracil	UMP	UDP	UTP

Nucleoside 5'-monophosphate (NMP)

Nucleoside 5'-diphosphate (NDP)

Nucleoside 5'-triphosphate (NTP)

Deoxyribonucleoside 5'-mono-, di-, and triphosphates

Base	Abbreviations		
Adenine	dAMP	dADP	dATP
Guanine	dGMP	dGDP	dGTP
Cytosine	dCMP	dCDP	dCTP
Thymine	dTMP	dTDP	dTTP

Figure 5.8

Summary of nucleoside mono-, di-, and triphosphates.

2. **Carrier of reactive groups in biosynthesis.** Nucleoside phosphates can serve to energize and act as carriers for building block units in biosynthesis. For example, uridine diphosphate, UDP, is a specific carrier for sugar residues in polysaccharide biosynthesis. In the biosynthesis of glycogen, UDP-glucose serves as the reactive monomer unit (see Figure 5.9). Amino acid units are activated for protein biosynthesis by reaction with ATP to form an amino acyl adenylate in which AMP is linked to an amino acid group by a mixed phosphoric acid-amino acid anhydride bond (see Figure 5.9).

3. **Nucleic acid biosynthesis.** The nucleoside triphosphates are the reactive monomer species in the formation of nucleic acids.

4. **Components of some enzymes.** Adenine nucleotides are part of the molecular structure of several important coenzymes including: NAD^+, $NADP^+$, FAD, coenzyme A and coenzyme B_{12}.

5. **Metabolic regulators.** Cyclic nucleotides act as intracellular regulators of metabolism (see Chapter 12). Cyclic AMP (Figure 5.7) is the most common of these chemical messengers. ATP and ADP serve as central feedback regulators in cellular metabolism.

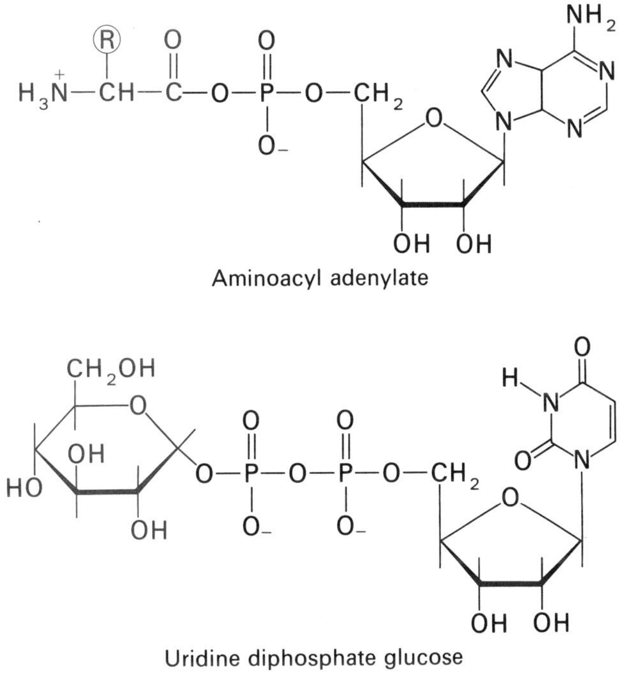

Figure 5.9

In uridine diphosphate glucose and aminoacyl adenylate, illustrated here, the nucleoside phosphate portion serves as a carrier and activator for the amino acyl group or glycosyl group. These two groups are shown in color; they are the building blocks for polysaccharides of glucose and proteins, respectively.

Aminoacyl adenylate

Uridine diphosphate glucose

5.2

Polynucleotides: The Covalent Backbone of Nucleic Acids

Nucleoside monophosphates form the repeating units of nucleic acids. The basic features of the covalent structure of a single nucleic acid side chain are shown in Figure 5.10. Both DNA and RNA are polynucleotides with a covalent backbone like the one shown in this figure. In DNA, the pentose is 2′ deoxy D-ribose and the bases are almost always A, G, C and T. In RNA, D-ribose is the pentose unit and A, G, C and U are the predominant bases.

The covalent backbones of both DNA and RNA consist of phosphodiester-linked pentose groups, with purine and pyrimidine bases acting as side-chain groups in analogy with the R-groups of amino acid residues in

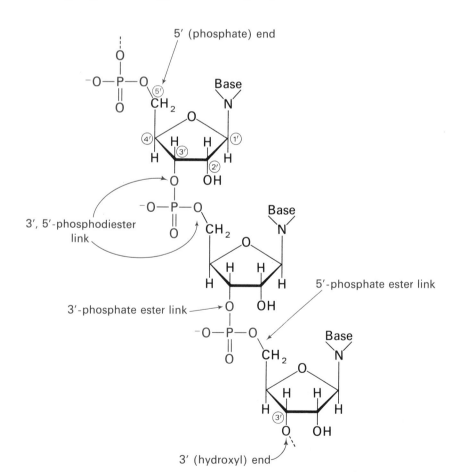

Figure 5.10

A trinucleotide fragment of a ribonucleic acid, illustrating the general covalent structure common to all nucleic acids.

proteins. The sequence of bases in DNA and RNA is crucial to the informational nature of nucleic acid molecules because it constitutes the means by which the amino acid sequences of proteins are specified.

DNA and RNA differ not only in the nature of the pentose group and certain bases but also in the overall structure of the nucleic acid molecule. DNA is ordinarily found as a double-stranded molecule in which two helical single strands are joined together by hydrogen bonds between bases on the respective strands. RNA is generally a single-stranded molecule, and may be coiled upon itself.

Figure 5.10 illustrates another important point: every nucleic acid has a 3′ end and a 5′ end. This *polarity*, or direction, of a nucleic acid chain arises from the use of nucleotides as building blocks for nucleic acids. You should be sure you understand the concept of polarity because of the important role it plays in protein biosynthesis, as we will see in Chapter 13.

Nucleic acid structures can also be represented using a shorthand notation similar to the one used for proteins and peptides. In the nucleic acid notation, pentose rings are represented by vertical lines; bases by the appropriate letter abbreviations; and the phosphodiester link by a "P" with two slanted lines, as shown in Figure 5.11. By convention, the nucleotide sequence, reading from left to right, is always 5′ → 3′.

Figure 5.11

Two examples illustrating shorthand notation for nucleic acid sequences. The sequence of bases is always written in the 5′ → 3′ direction. The figure shows shorthand notation for a short RNA fragment. If a DNA strand were to be represented, one would add the prefix "d-" before each base letter to denote that deoxyribose is present.

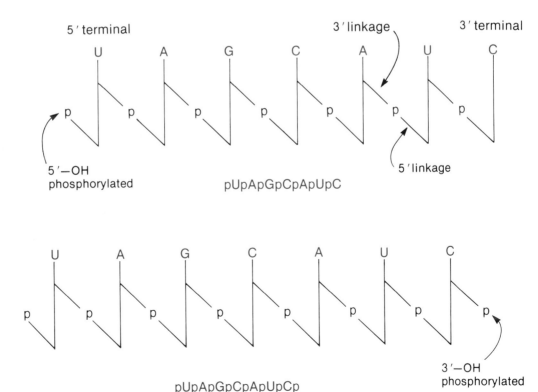

5.3

Cleavage of Nucleic Acids

The phosphodiester links of DNA and RNA are hydrolyzable and may be cleaved either chemically or enzymatically. For example, all of the hydrolyzable linkages in either DNA or RNA can be cleaved by reaction with $12M$ perchloric acid at $100°C$ for one hour. In alkaline solution, RNA is readily hydrolyzed to a mixture of 2′ and 3′ nucleoside monophosphates. Because hydrolysis in base involves a cyclic 2′—3′ monophosphate intermediate, DNA, which lacks a 2′—OH group, can be hydrolyzed in base only under extreme conditions.

Nucleic acids taken in as part of the diet are broken down in the small intestine by pancreatic nucleases to form a mixture of nucleoside monophosphates and oligonucleotides. This mixture is further reduced by enzymes in the small intestine to yield a mixture of nucleosides and orthophosphates. Nucleosides are absorbed and metabolized primarily in the liver, kidneys, spleen and bone marrow.

The enzyme-catalyzed hydrolysis of nucleic acids has proven to be a valuable tool in determining the base sequences of nucleic acids, just as the protease-catalyzed hydrolysis of proteins is helpful in finding amino acid sequences. The reason for this is that the differences in specificity properties of nucleases from various sources permit the formation of fragments of nucleic acids with overlapping sequences, as we saw in the case of proteins in Chapter 2. Enzyme-catalyzed hydrolysis has been a particularly successful technique for determining the base sequences of certain types of RNA.

Either the 3′- or the 5′-phosphate ester link can be hydrolyzed by nucleases. As shown in Table 5.1, specific nucleases may act only on DNA, RNA or both. *Exo*nucleases act only on the *end* of a nucleic acid chain, similar to the way exopeptidases act (Chapter 2). Exonucleases can also exhibit a preference for either the 5′ or the 3′ end of the nucleic acid strand. *Endo*nucleases catalyze the hydrolysis of certain phosphate ester links *within* the body of the nucleic acid chain, just as endopeptidases only act within the bodies of proteins. Many endonucleases act on only selected internucleotide links. For example, bovine pancreatic ribonuclease hydrolyzes only the 5′ ends of linkages which have the 3′ part of the link attached to a pyrimidine nucleotide (see Problem **7** at the end of this chapter). Some of the important nucleases and their specificities are listed in Table 5.1.

Nucleotides and nucleic acids have many interesting functions. In Part II of this text we will see that certain nucleoside phosphates are important in metabolism. In Part III we will learn that nucleoside phosphates serve as building blocks and controlling agents in the biosynthesis of small molecules. The most exciting function of nucleic acids is in protein and nucleic acid biosynthesis, which forms the basis of molecular genetics. If you want to learn more at this point about the biological role of nucleic acids, you should read

Table 5.1 Some nucleic acid hydrolases and their specificities.

Enzyme	Source	Action
Exonucleases:		
Phosphodiesterase	Rattlesnake venom	Hydrolyzes 3′ phosphate ester links in either DNA or RNA (single strand). It requires a free 3′-OH group to begin stepwise hydrolysis.
Phosphodiesterase	Spleen	Hydrolyzes 5′ phosphate ester links in either DNA or RNA, starting from a free 5′-OH end.
Endonucleases:		
Deoxyribo-nuclease I	Pancreas	Hydrolyzes some of the 3′ links of DNA to yield a mixture of oligonucleotides and 5′ mononucleotides
Deoxyribo-nuclease II	Spleen, bacteria	Hydrolyzes some of the 5′ links of DNA. Both strands of double-stranded DNA are cleaved at the same point simultaneously.
Ribonuclease I	Bovine pancreas	Hydrolyzes the 5′ ends of RNA linkages where the 3′ end is attached to a pyrimidine nucleotide.
Ribonuclease T_1	Aspergillus oryzae (fungus)	Hydrolyzes the 5′ ends of RNA linkages where the 3′ end is attached to a guanine nucleotide.
Restriction endonucleases	Bacteria	Cleaves DNA into sections of the double strand which have specific unique base sequences. The particular sequence depends on the species of bacterium.

all or part of Chapter 13. For now, let's briefly discuss some general facts about the major types of nucleic acids.

5.4

DNA and RNA

DNA

Friedrich Miescher first isolated DNA from cell nuclei in 1869 and called the newly discovered substance "nuclein"—a forerunner of the term *nucleic acid*. Even though DNA was studied extensively during the succeeding years, its biological role as the carrier of genetic information remained unclear until the late 1940s, when Avery and his coworkers showed that purified DNA can transfer hereditary traits from one bacterial strain to another. In the early 1950s, the X-ray crystallographic work of Watson, Crick and others established the double helix structure of DNA. In the Watson–Crick model, two single strands of opposite polarity, joined primarily by base pairing, wind around an axis to form a double helix.

DNA is found in all procaryotic and eucaryotic cells as well as in one class of viruses. In diploid eucaryotic cells, one or more DNA molecules exist in combination with polyamines and basic proteins called *histones*. This combination forms the major part of the chromosomal structure of the cell nucleus. In addition, eucaryotes have separate DNA molecules in such organelles as mitochondria and chloroplasts. Procaryotic cells have one chromosome composed of one DNA double helix combined with positively-charged polyamines. In the DNA structure of both procaryotes and eucaryotes, the positively charged molecules associated with the DNA help to counterbalance the negatively charged phosphate groups of the DNA covalent backbone. In addition, small molecules of DNA called *plasmids* or *episomes* occur in the cytoplasm of some bacteria. Because of their small size, plasmids carry only a limited amount of genetic information. In DNA viruses, the basic structure is formed from a single DNA molecule and associated protein molecules. We will learn more about DNA viruses later in this chapter.

Base Composition of DNA

The base composition of DNA can vary considerably, depending on the source. This variation was first observed in the 1950s by Chargaff and his colleagues, who determined the base compositions of DNA isolated from many different species. Table 5.2 lists some of the base compositions of DNA.

The smallest variation in base composition is between species that are genetically related. The data in Table 5.2 clarifies the important relationships originally observed by Chargaff:

$$\% \text{ [pyrimidines]} = \% \text{ [purines]}$$
$$\% \text{ [adenine]} = \% \text{ [thymine]}$$
$$\% \text{ [cytosine]} = \% \text{ [guanine]}$$

The basis for these observations is the fact that a purine on one strand of the DNA double helix specifically hydrogen-bonds (base-pairs) with a pyrimidine base on the other strand. The purine bases adenine and guanine and the pyrimidine bases cytosine and thymine are the only bases found in most types of DNA. In some species, however, minor bases do occur. For example, 5-methyl cytosine is present in the DNA of certain higher plants and animals. Wheat germ is a particularly rich source of this base. Because 5-methyl cytosine base pairs as cytosine, it can be considered equivalent to cytosine.

Table 5.2 Experimentally determined base compositions for DNA from various species. The ratio (A + T)/(G + C) remains fairly constant for cells of higher organisms, while exhibiting wide variations for bacterial cells.

Species	Mole percent				$\dfrac{(A + T)}{(G + C)}$
	A	G	C	T	
Eucaryotes:					
Man	30.9	19.9	19.8	29.4	1.52
Mouse	29.0	21.5	20.7	29.0	1.37
Wheat germ	27.3	22.7	22.8*	27.1	1.19
Bacteria:					
E. coli	24.7	26.0	25.7	23.6	0.93
Sarcina lutea	13.4	37.1	37.1	12.4	0.35
B. cereus	33.5	17.3	16.0	33.2	2.00

* Sum of cytosine + 5-methylcytosine

Molecular Weight of DNA

Chromosomal DNA in eucaryotes and procaryotes are very large molecules with a molecular weight in excess of 1×10^9. Because of their size, unbroken DNA molecules are difficult to isolate and DNA preparations from eucaryotes generally consist of a mixture of DNA fragments. However, the intact DNA of the procaryote *E. coli* has been isolated as a circular molecule having a molecular weight of 2×10^9. Mitochondrial DNA is much smaller than chromosomal DNA and has a molecular weight of approximately 1×10^7.

DNA Double Helical Structure

The DNA molecule is composed of two polynucleotide chains joined along their length and wound about an axis to yield a double helix. The basic helix shape is similar to the α-helix in proteins. This structure was first proposed in 1953 by Watson and Crick, who were awarded the 1962 Nobel Prize in medicine for their contributions. The DNA double helix is shown in Figure 5.12. The strands of the double helix are joined by hydrogen bonds between the bases: A to T and G to C. The base sequence of each strand of the double helix is *complementary* to the other, as illustrated in Figure 5.12(a). The strands of the double helix are antiparallel, or of opposite polarity: one strand runs $3' \rightarrow 5'$ with respect to the internucleotide links, and the other runs in the $5' \rightarrow 3'$ direction (see Figure 5.10). The plane of each base is perpendicular to the long axis of the double helix, which results in the stacking effect shown in Figure 5.12(b). The overlapping of planar purine and pyrimidine rings creates hydrophobic interactions between the bases, which contribute significantly to the overall stability of the double helix structure.

Figure 5.12

The DNA double helix.
(a) The pentose-
phosphodiester backbone
is represented by the two
ribbons wound around
the axis to form a
right-handed helix. The
relationship of the bases
is such that adenine on
one chain base pairs
with thymine on the
other and cytosine on
one chain hydrogen
bonds with guanine on
the other. Note that the
sequence of bases in each
strand is such that there
is normally a base pair
formed between opposite
bases at each nucleotide
unit in the double helix.
(b) A representation of
one of the strands of the
DNA double helix
structure looking down
along the long axis. The
purine and pyrimidine
bases are perpendicular to
the long axis and
partially overlap one
another.

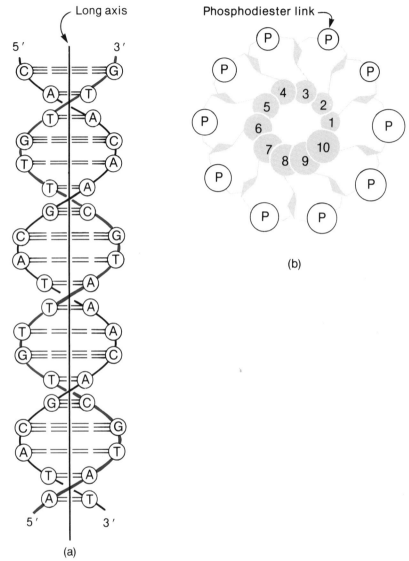

(a)

(b)

Just as proteins can become denatured (as we learned in Chapter 2), the secondary structure of DNA can also be disrupted. When a sample of DNA is heated to a certain temperature, the hydrogen bonds in the double helix are disrupted and a predominantly single-stranded random form of DNA results. The temperature at which this change takes place is termed the melting temperature (T_m), and is determined by experimentally measuring the melting curve for a DNA sample (Figure 5.13). The value of T_m is related to the base composition of the DNA sample tested. The greater the % [G + C], the greater the value of T_m. This reflects the fact that more energy is needed to disrupt the three hydrogen bonds between (G + C) than the two hydrogen bonds between (A + T).

Figure 5.13

A typical DNA melting curve, showing the sharp increase in ultraviolet light absorption by the DNA as the hydrogen bonds in the helix structure are disrupted at the melting temperature (T_m).

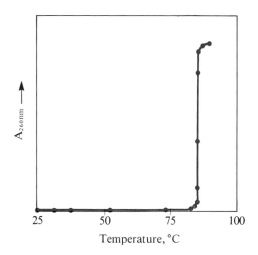

RNA

RNA constitutes 5–10% of the dry weight of a cell. There are three types of cellular RNA:

1. Messenger RNA (m-RNA), which acts as a template for the synthesis of protein chains.
2. Ribosomal RNA (r-RNA), which acts as the nucleic acid component in the structure of ribosomes where protein synthesis is carried out.
3. Transfer RNA (t-RNA), which acts as a specific amino acid carrier in the formation of a polypeptide chain.

Table 5.3 gives some of the characteristics of the three basic types of RNA for a simple bacterial cell such as *E. coli*. The RNA molecules of eucaryotic

Table 5.3 Physical Properties of Ribonucleic Acids of *E. coli*.

Type	Range of numbers of NMP units	Percent of total RNA in cell
t-RNA	75–90	16
m-RNA	75–3000*	2
r-RNA	5 S:ca. 100 ⎫ 16 S:ca. 1500 ⎬† 23 S:ca. 3100 ⎭	82

* The size of the m-RNA molecule is determined by the number of amino acid residues to be synthesized in the protein.

† The terms 5 S, 16 S, and 23 S refer to the rate at which a given molecular component of a ribosomal RNA preparation sinks, or sediments, in the high gravitational field of an ultracentrifuge. A heavier (larger) molecule sediments more rapidly and therefore has a higher *sedimentation coefficient*. The sedimentation coefficient is expressed in Svedberg (S) units after the Swedish physicist T. Svedberg who invented the ultracentrifuge in 1925.

cells are of the same basic types. Unlike DNA, RNA generally exists as a single-stranded molecule, although portions of an RNA strand can coil back upon themselves to form small helix structures. We will discuss the structure and function of these forms of RNA in more detail in Chapter 13.

In addition to the forms given above, RNA also exists as the genetic material in certain viruses. Just as the ribosome is a complex RNA-protein structure, viruses are also aggregates of nucleic acid and protein molecules, as we shall see in Section 5.5.

5.5

Nucleic Acid–Protein Complexes

Nucleic acids are often found associated with proteins in nature. In Section 5.4 we learned that in the somatic cells of plants and animals, the chromosomal DNA, or *chromatin*, is associated with proteins. Included in these DNA-associated proteins is a group called *histones*. These contain a large proportion of residues of lysine, arginine, or both, depending on the particular histone, and thus complex strongly with the negatively charged phosphodiester groups of the DNA backbone. Because the interaction between DNA and histones is not random in nature, the eucaryotic chromosome is a true nucleic acid–protein complex. Other important nucleic acid–protein complexes include ribosomes and viruses.

Ribosomes

Ribosomes are subcellular structures where protein synthesis takes place. While ribosomes of procaryotic cells differ in size and structural details from

Figure 5.14

Dissociation of the *E. coli* ribosome into r-RNA and proteins. Under the appropriate conditions, the isolated ribosomal RNA and proteins will spontaneously reassemble to form intact and functional ribosomal subunits.

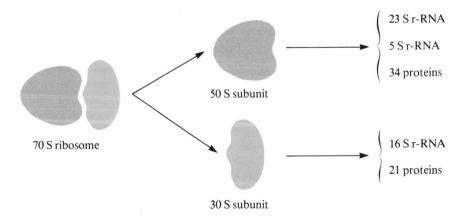

70 S ribosome

50 S subunit

{ 23 S r-RNA

5 S r-RNA

34 proteins

30 S subunit

{ 16 S r-RNA

21 proteins

those of eucaryotes, the basic features are the same for both. The *E. coli* ribosomal structure has been studied extensively. The intact 70 S ribosome (see Table 5.3) consists of two subunits, a 50 S and a 30 S subunit. These two subunits combine to form an intact ribosome having a diameter of about 200 Å and a molecular weight of about 2.5×10^6. The nucleoprotein structure of the ribosome can be dissociated by chemical treatment into its parent components, as shown in Figure 5.14. Studies on the reconstitution of the 30 S and 50 S ribosomal subunits have shown that ribosomal proteins are essential to the structure and function of the ribosome.

Viruses

Viruses are infectious inanimate particles consisting of a nucleic acid molecule surrounded by a protective protein coat. The protein coat protects the viral nucleic acid from the action of nucleases. Viral proteins can also serve structural functions, as in the case of the bacterial virus bacteriophage T4 (see Figure 5.15). Viruses cannot carry out energy metabolism and can perpetuate themselves only by infecting a host cell. When a virus infects a host cell, the viral genetic material (DNA or RNA) is injected into the cell. The cell's protein and nucleic acid biosynthetic apparatus then produces new viral nucleic acid and proteins using the genetic information carried by the injected DNA or RNA. Spontaneous assembly of the viral proteins and nucleic acid leads to the formation of new viruses. Many viruses ultimately destroy the host cell and are thus said to be *pathogenic*, or disease-producing. Some viruses cause the host cell to develop abnormal growth patterns and cell surfaces. These are termed *oncogenic*, or tumor-producing, viruses.

Figure 5.15

Molecular structure of two virus particles.
(a) The structure of the tobacco mosaic virus (TMV). The figure shows a simplified view of how the identical protein subunits associate with the RNA to form the helical virus particle structure.
(b) The structure of the bacteriophage T2. While many viruses are simply nucleic acid in an icosahedral protein capsule, the T-even bacteriophage have in addition a tubular structure and protein fibers to facilitate the injection of its DNA into a host cell. The action of the virus can be compared to that of an eye-dropper. The diameter of the tube is just sufficiently large enough to allow the passage of the DNA molecule into the host cell. The sheath surrounding the tube contracts during the infecting process thus causing the inner tube to be injected into the host cell.

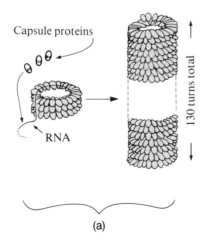

Capsule proteins

RNA

130 turns total

(a)

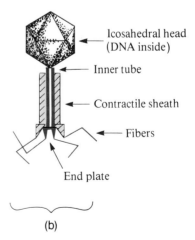

Icosahedral head (DNA inside)

Inner tube

Contractile sheath

Fibers

End plate

(b)

Table 5.4 The composition and size of selected viruses.

	Virus particle weight × 10^6 Daltons*	Nucleic acid	% Nucleic acid	Number of strands
Plant viruses				
Tobacco mosaic	40	RNA	6	1
Bushy stunt (tomato)	10.6	RNA	15	1
Animal viruses				
Encephalitis (equine)	50	RNA	4.4	1
Herpes (mammals)	700	DNA	9.5	2
Influenza Type A (human)	280	RNA	0.08	1
Mumps (human)	70	RNA	3.5	1
Poliomyelitis (human)	6.7	RNA	30	1
Smallpox (human)	3200	DNA	5.6	2
Bacterial viruses (bacteriophage)				
T2, T4, T6 (*E. coli*)	220	DNA	61	2
MS2, R17 (*E. coli*)	3.6	RNA	32	1

* With very large molecules such as DNA, RNA or supra-molecular nucleic acid–protein aggregates, it is more convenient to refer to the mass of one molecule or particle rather than a mole (6.02×10^{23}) of particles. We do this in units called *Daltons* where 1 Dalton equals the mass of 1 hydrogen atom.

The nucleic acid–protein structure of many viruses is well understood. Some viruses are rudimentary, having a small nucleic acid with only 3 genes. Other viruses have more complex structures and therefore more genes; in some cases as many as 250 or more. Table 5.4 gives the composition and size of selected viruses.

The first virus to be crystallized and identified as a ribonucleoprotein was the tobacco mosaic virus (TMV) in 1935. Its structure is representative of a general class of viruses having a helical rod-like form as shown in Figure 5.15(a). TMV illustrates a common type of viral structure in which the nucleic acid is surrounded by a structural coat composed of many identical protein molecules or many of a few kinds of proteins. The TMV structure consists of a single RNA helix closely surrounded by 2130 identical protein subunits, yielding a virus particle with a total length of about 3000 Å and a diameter of about 180 Å. Under appropriate conditions, the protein subunits can be dissociated from the RNA and then reassembled to produce an infective virus again. This is an example of the self-assembly of supramolecular complexes often observed in biochemistry and molecular biology. A viral structure can also be complex, as in the case of *E. coli* bacteriophage T2 shown in Figure 5.15(b). The DNA of the T2 bacteriophage is encapsulated by a protein coat having an icosahedral shape. This is an arrangement common to many types of viruses.

Suggestions for Further Reading

Lehninger, A. L. *Biochemistry*. 2d ed. New York: Worth Publishers, 1975. A readable standard text having an excellent discussion of nucleotides and nucleic acids in Chapter 12.

Watson, J. D. *The Molecular Biology of the Gene*. 3d ed. Menlo Park, Cal.: W. A. Benjamin, Inc., 1976. This is a classic in its area and is a good source for students wishing to go beyond the level of this text.

Scientific American
Butler, P. J. G., and Klug, A. "The Assembly of a Virus." November 1978.
Holley, R. W. "The Nucleotide Sequence of a Nucleic Acid." February 1966.
Nomura, M. "Ribosomes." October 1969.

Problems

1. For each of the terms below, provide a short definition or explanation.

deoxyribonucleic acid	nucleic acid	purine
endonuclease	nucleoside	pyrimidine
exonuclease	phosphodiester link	ribonucleic acid
mononucleotide	polarity	

2. We discussed a number of metabolites in this chapter that are important to the structure and function of nucleic acids. Give the structure of each metabolite listed.

adenine	D-ribose
cytosine	thymine
2-deoxy D-ribose	uracil
guanine	

3. What is the predominant form of nucleoside phosphate esters in the cell? Draw the structure of one such nucleoside phosphate in which the base is guanine.

4. What are the major functions of nucleoside phosphates in the cell?

5. Draw the structures of all possible oligo-deoxyribonucleotides containing 4 mononucleotide units and a base composition of one mole each of A, G, C and T. Draw out one of the possible structures completely, showing all functional groups. For the remaining structures, use the shorthand notation shown in Figure 5.11.

6. Draw the structure of the tetradeoxyribonucleotide written in shorthand notation as:

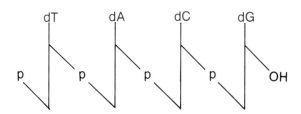

7. Draw the structures of the products obtained from the hydrolysis of the given pentaribonucleotide by the action of each of the following enzymes:

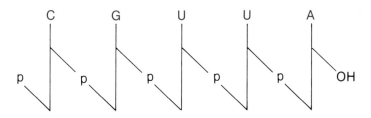

(a) bovine pancreatic ribonuclease I
(b) ribonuclease T_1

8. Refer to Table 5.2. From which source given in this table would you expect to obtain DNA having the highest T_m value?

9. There are 10 bases per turn on each strand of the DNA double helix, and each turn has a height of 34 Å (34×10^{-8} cm). Given that the average molecular weight of a nucleotide residue in DNA is about 325, estimate the length of the *E. coli* DNA (MW $= 2 \times 10^9$) assuming a perfectly linear molecule.

CHAPTER 6

Lipids

Many of life's crucial processes occur in nonaqueous surroundings. We have seen how the interior of a protein molecule exhibits essentially apolar (hydrophobic) properties. The active centers of many enzymes are also predominantly hydrophobic. However, many other enzymes and proteins do not exist in free solution in a cell, but instead form part of the structure of a lipid membrane. Membrane structure is essential to the operation of many vital processes in living organisms, including photosynthesis, respiration, and active transport. In this chapter we will consider the nature of biological membranes. But before we do, we must look at some of the basic aspects of lipid biochemistry.

6.1

Characteristics of Lipids

The term **lipid** refers to substances that can be extracted from living matter using hydrocarbon solvents such as ligroin, benzene, ethyl ether, or chloroform. Proteins, carbohydrates, and nucleic acids are essentially insoluble in these nonpolar solvents. The conclusion that lipids are fat-soluble is perhaps the only generalization about lipids we can make, since they exhibit a large range of both functional and structural diversity. Functions of lipids include

1. Energy storage and transport;

2. Membrane structure;

3. Protective coat, cell-wall component;

4. Chemical messengers.

In the past, the study of lipids was hampered by a lack of good techniques for separating complex mixtures of these compounds. Today the development of such techniques as gas chromatography and other chromatographic separation methods has enabled the field of lipid biochemistry to develop into one of the exciting areas of current research. Table 6.1 summarizes some of the major types of lipids and illustrates the many different classes of compounds that fall into the category of lipids.

Table 6.1   Major Types of Lipids.

Category	Description
I.	Fatty acids: long-chain aliphatic carboxylic acids
II.	Fatty alcohols: long-chain aliphatic alcohols
III.	Neutral
	A. mono, di- and tri-acyl glycerols (esters with glycerol)
	B. glycerol ethers
	C. waxes: esters of fatty acids with any alcohol other than glycerol
IV.	Phosphoglycerides: phosphatidic acid derivatives (many are associated with membranes)
V.	Sphingolipids: generally associated with nervous-system tissue
VI.	Terpenes: includes a variety of unsaturated compounds such as essential oils and flavorings, vitamin A, visual pigments of the retina, and chlorophyll
VII.	Steroids: fused-ring alicyclic compounds including cholesterol and steroid hormones
VIII.	Conjugated lipids:
	A. lipoproteins (soluble in water)
	B. proteolipids (insoluble in water—soluble in fat solvents)
	C. lipopolysaccharides
IX.	Prostaglandins: lipids produced from polyunsaturated fatty acids that have intense biological activity
X.	Hydrocarbons: saturated and unsaturated hydrocarbons are ubiquitous in nature

Many of the lipids summarized in Table 6.1 have the physical property of being **amphipathic**. The term "amphipathic," originally used by Hartley in 1936, describes ionic or polar derivatives of hydrocarbons that have one (polar) part "in sympathy" with an aqueous environment and a hydrocarbon (hydrophobic) part which is not. Consider the substance stearic acid, a linear carboxylic acid with 18 carbon atoms:

$$CH_3-CH_2-CH_2-CH_2-CH_2-CH_2-CH_2-CH_2-CH_2-CH_2-CH_2-CH_2-CH_2-CH_2-CH_2-CH_2-CH_2-C\overset{\displaystyle O}{\underset{\displaystyle O^-}{<}}$$

Stearate anion

Since most fatty acids have pK_a values of less than 5, the carboxyl function becomes ionized at pH 7. This is clearly an amphipathic molecule since it contains a *hydrophilic* (water-loving) ionic carboxylate group at one end and a *hydrophobic* (water-hating) hydrocarbon chain. In an aqueous environment, the stearate molecules arrange themselves spontaneously in such a way as to minimize contact between the hydrophobic groups and water. This can produce a number of different structures, as shown in Figure 6.1. In all cases shown in

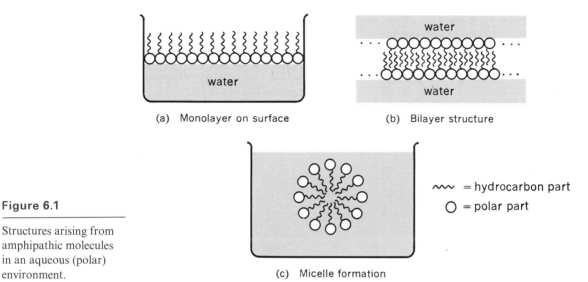

Figure 6.1

Structures arising from amphipathic molecules in an aqueous (polar) environment.

the figure, the structures are arranged to minimize contact between the apolar hydrocarbon part of the stearate ion and the water. It is interesting to note that biological membranes conform quite well to the simple bilayer structure of Figure 6.1(b). Amphipathic molecules such as phospholipids are essential components of these natural membranes. Micelle formation is also common. For example, soap solutions appear opalescent due to the light-scattering properties of micelles, formed from the fatty-acid anions (such as stearate ion) which constitute the soap.

6.2

Fatty Acids

Fatty acids are rarely found free in nature, but occur as esters in combination with an alcohol function. We can make some generalizations regarding fatty acids, even though there are exceptions, as we will see.

1.   Fatty acids are generally straight-chain monocarboxylic acids.

2.   Fatty acids generally have an even number of carbon atoms. [This is significant because they are synthesized primarily two carbons at a time (see Chapter 12).]

3.   Fatty acids can be saturated or they can have one or more double bonds.

Table 6.2 Some Naturally Occurring Fatty Acids.

Acid	Number of carbons	Structure	Sources
Saturated fatty acids			
Caprylic	8	$CH_3(CH_2)_6COOH$	Butter fat, coconut oil
Capric	10	$CH_3(CH_2)_8COOH$	Coconut oil
Lauric	12	$CH_3(CH_2)_{10}COOH$	Coconut oil
Myristic	14	$CH_3(CH_2)_{12}COOH$	Vegetable seed oil
Palmitic	16	$CH_3(CH_2)_{14}COOH$	Animal, vegetable fats and oils
Stearic	18	$CH_3(CH_2)_{16}COOH$	Animal fats, vegetable fats and oils
Arachidic	20	$CH_3(CH_2)_{18}COOH$	Peanut oil
Unsaturated fatty acids			
Palmitoleic	16	$CH_3(CH_2)_5CH{=}CH(CH_2)_7COOH$	Butter fat
Oleic	18	$CH_3(CH_2)_7CH{=}CH(CH_2)_7COOH$	All fats and oils
Linoleic	18	$CH_3(CH_2)_4CH{=}CH{-}CH_2{-}CH{=}CH(CH_2)_7COOH$	Vegetable oils
Linolenic	18	$CH_3CH_2CH{=}CHCH_2CH{=}CHCH_2CH{=}CH(CH_2)_7COOH$	Vegetable oils
Arachidonic	20	$CH_3(CH_2)_4(CH{=}CHCH_2)_3CH{=}CH(CH_2)_3COOH$	Fish oils

Since many fatty acids were initially isolated from natural oils or fats, their nomenclature reflects these natural origins. Table 6.2 summarizes the names, structures, and sources of several fatty acids. The table is highly abbreviated, as many naturally occurring fatty acids have been identified. Chemical treatment of a pure sample of a fat or oil can yield at least four and as many as twelve different fatty acids.

While linear-chain fatty acids predominate in nature, many other types are known. For example, wool fat and bacterial sources yield branched-chain fatty acids. Cyclic fatty acids also exist. For example, the unsaturated cyclic fatty acid chaulmoogric acid is an important agent for the treatment of leprosy:

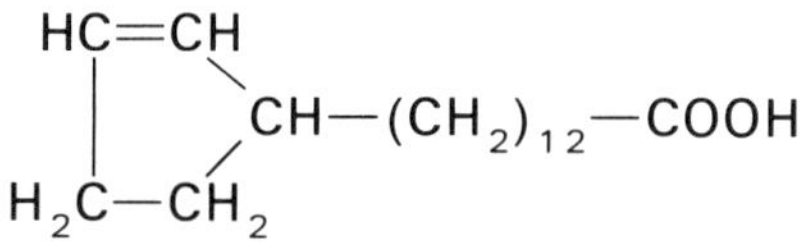

$$HC{=}CH \diagdown$$
$$\quad CH{-}(CH_2)_{12}{-}COOH$$
$$H_2C{-}CH_2 \diagup$$

Chaulmoogric acid

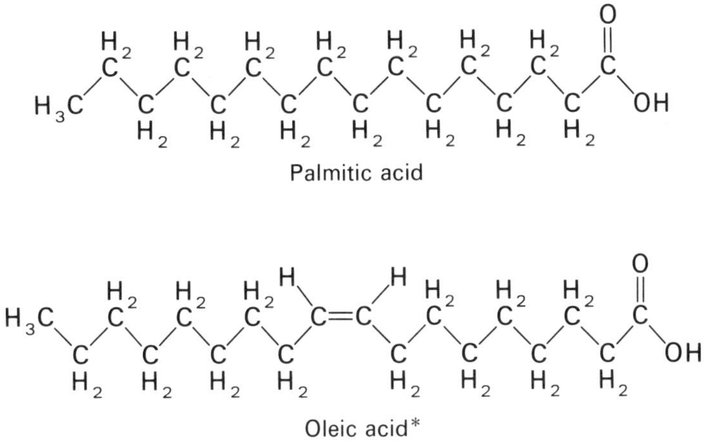

Figure 6.2

Structures of two fatty acids. The saturated chains are shown in an extended form.

Palmitic acid

Oleic acid*

*Note *cis*-conformation

The actual shape of a fatty acid develops from the shape of the parent hydrocarbon, as shown in Figure 6.2. It is important to note that the **double-bond configuration of naturally occurring fatty acids is generally** *cis*:

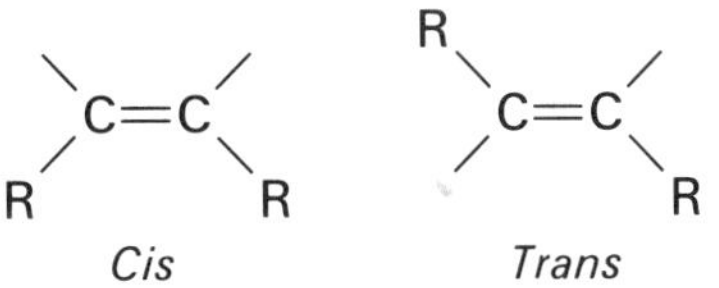

Cis   *Trans*

The fact that nature prefers the more bulky *cis*-unsaturated fatty acids may relate to the importance of these compounds in the structure of biological membranes.

Higher animals can biosynthesize saturated and monounsaturated fatty acids from other sources such as carbohydrates. Linoleic and linolenic acids and higher polyunsaturated fatty acids cannot be produced in higher animals and are therefore termed *essential fatty acids*. Arachidonic acid can be biosynthesized from linoleic acid and therefore is not an essential part of the diet. However, mammals cannot survive on a diet that does not contain linoleic and linolenic acids. Fortunately, plant sources, particularly vegetable oils, are rich in these essential fatty acids.

A salt of a fatty acid is generally called a *soap*. The cleaning action of soap relies on the amphipathic character of the soap molecules, as shown

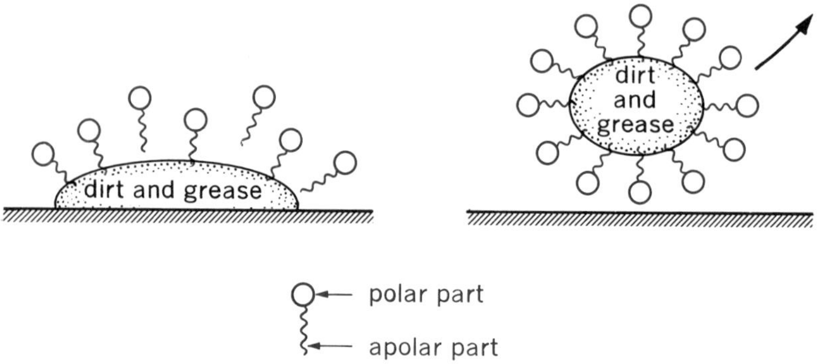

Figure 6.3

Cleaning action of soap. The soap molecules disrupt the greasy materials holding dirt to the surface by attaching themselves to the grease molecules. The polar parts of the associated soap molecules make the dirt and grease particles stable in aqueous solution, so they can be washed away in water.

polar part

apolar part

schematically in Figure 6.3. Synthetic detergents are based on this type of structure and work in much the same way, as we can see from some examples:

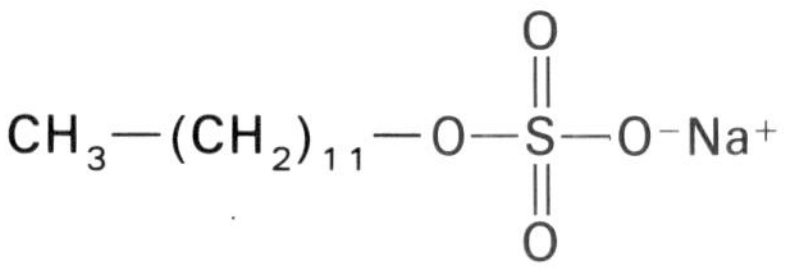

$$CH_3-(CH_2)_{11}-O-\overset{\overset{\displaystyle O}{\|}}{\underset{\underset{\displaystyle O}{\|}}{S}}-O^-Na^+$$

Sodium dodecyl sulfate

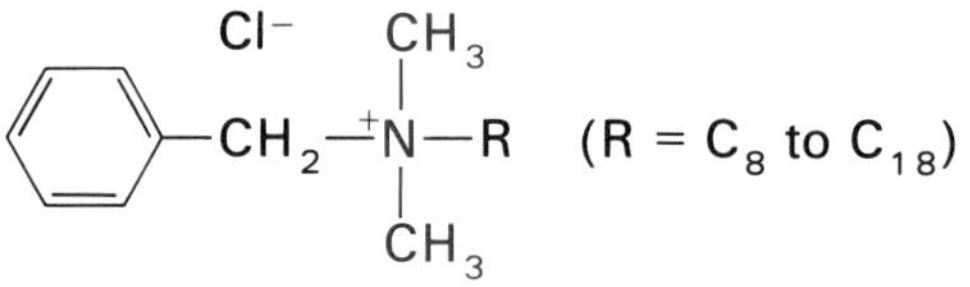

$$Cl^-\quad CH_3$$
$$\text{(benzene ring)}-CH_2-\overset{+}{N}-R \quad (R = C_8 \text{ to } C_{18})$$
$$CH_3$$

Quaternary ammonium detergent

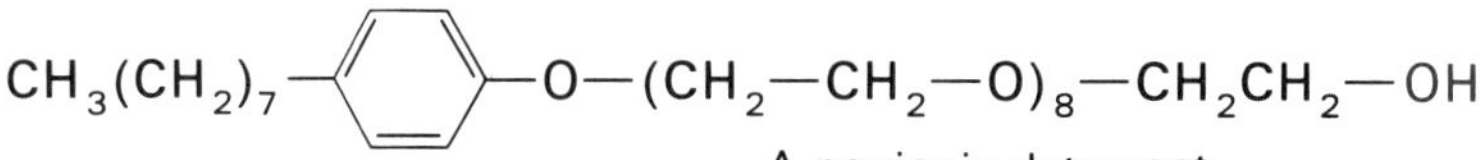

$$CH_3(CH_2)_7-\text{(benzene ring)}-O-(CH_2-CH_2-O)_8-CH_2CH_2-OH$$

A nonionic detergent

6.3

Neutral Lipids

Triacylglycerols

Perhaps the most important chemical feature of fatty acids is the formation of **triacylglycerols**—esters of fatty acids with each of the three hydroxyl groups of glycerol:

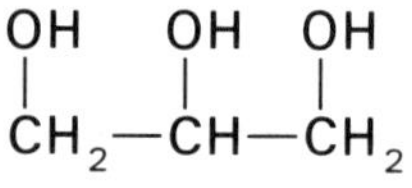

$$\overset{\displaystyle OH}{\underset{\displaystyle CH_2}{|}}\quad\overset{\displaystyle OH}{\underset{\displaystyle CH}{|}}\quad\overset{\displaystyle OH}{\underset{\displaystyle CH_2}{|}}$$
$$CH_2-CH-CH_2$$

Glycerol

The general form of a triacylglycerol (or triglyceride) is shown in Figure 6.4.

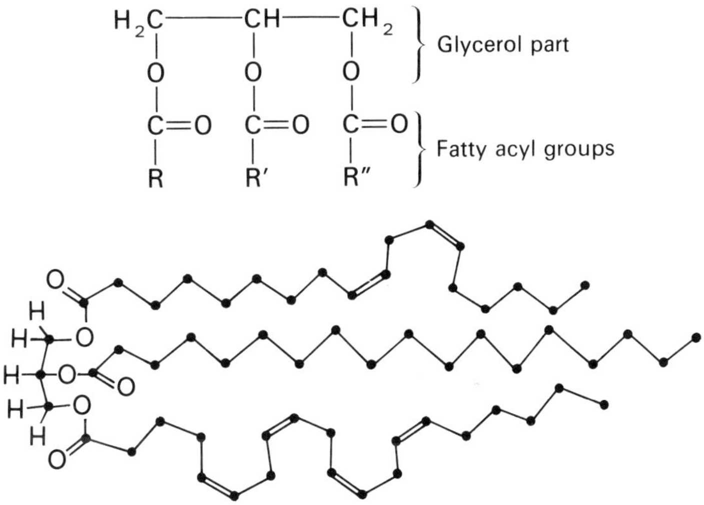

Figure 6.4

General structure of a triacylglycerol molecule. The specific example shows the glycerol part of the molecule at the left of the figure. Attached to it are a saturated acid (stearic acid) and two polyunsaturated fatty acyl groups (linolenic acid at top and arachidonic acid at bottom of figure).

The number of naturally occurring triacylglycerol molecules in which all three fatty acid groups are the same is relatively small. **Mixed triacylglycerols tend to be much more common.** The nomenclature of these compounds is illustrated by the examples in Table 6.3. In the case of the second example given

Table 6.3   Trivial Names of Some Triacylglycerols (the systematic name is given in parentheses).

Structure		Name
$H_2C-O-\overset{O}{\overset{\|\|}{C}}(CH_2)_{16}-CH_3$	(Stearic)	Tristearin (Tristearoylglycerol)
$HC-O-\underset{O}{\overset{\|\|}{C}}(CH_2)_{16}-CH_3$	(Stearic)	
$H_2C-O-\underset{O}{\overset{\|\|}{C}}(CH_2)_{16}-CH_3$	(Stearic)	
$H_2C-O-\overset{O}{\overset{\|\|}{C}}-(CH_2)_{10}-CH_3$	(Lauric)	1-lauro 2-oleo 3-palmitin (1-lauroyl 2-oleoyl 3-palmitoylglycerol)
$HC-O-\overset{O}{\overset{\|\|}{C}}-(CH_2)_7-CH=CH(CH_2)_7CH_3$	(Oleic)	
$H_2C-O-\overset{O}{\overset{\|\|}{C}}-(CH_2)_{14}-CH_3$	(Palmitic)	

in the table, three different structures are possible depending on the position of the fatty acyl groups in the triacylglycerol (see Problem **2** at the end of this chapter). Since fats are *not* informational molecules having a predetermined structure and sequence, this variability of structure in fats is not particularly important, as long as a given fatty acid composition is maintained overall.

The physical properties of neutral fats reflect the fatty acid makeup of the fat. As a general rule, the melting point of a fatty acid decreases with increasing unsaturation and decreasing molecular weight. The lower melting point of the unsaturated fatty acids may be due to the increased bulkiness imposed upon the molecular structure by the carbon–carbon double bonds in the *cis* configuration in natural fatty acids. The melting point of a triacylglycerol is a function of the melting points of the constituent fatty acids. Table 6.4 presents the melting points of a series of fatty acids as well as some representative triacylglycerols.

Table 6.4 Melting Points of Some Fatty Acids and Triacylglycerols.

Fatty acids (all C_{18})	Number of double bonds	Melting point (°C)
Stearic	0	70
Oleic	1	14
Linoleic	2	5
Linolenic	3	−11
Arachidonic	4	−50

Triacylglycerols	Percent saturated fatty acids	Percent unsaturated fatty acids	Melting point (°C)
Beef tallow	49–50 (mostly palmitic)	50–51 (mostly oleic)	37–47
Olive oil*	10	90	17–21

* Fats are solid at room temperature; oils are liquid at room temperature.

Oils generally have a high percentage of unsaturated fatty acyl groups; fats tend to have a high percentage of saturated fatty acyl groups. However, we can observe considerable variation in the fatty acid composition of similar triacylglycerols from species to species and even from organ to organ. Fatty acid composition also depends on climate. For example, a sample of linseed oil from flax grown in the cool climate of Switzerland contains over *twice* as many unsaturated fatty acids as linseed oil from flax grown in a Berlin greenhouse, maintained at a temperature between 20 and 25°C. Thus, for the same species, the triacylglycerol from plants in a colder environment has a lower melting point. Similarly, the outer layers of adipose tissue in mammals tend to be more unsaturated than the inner layers.

The chemistry of triacylglycerols tends to follow that of the functional groups present: ester, alkene, and poly-alkene. Since animal and vegetable fats serve as important articles of commerce, several features of the chemistry of these compounds are worth mentioning.

Saponification is a term derived from the Latin for "soap making" and refers to the base-catalyzed hydrolysis of ester linkages in triacylglycerols. Animal fat hydrolyzes in the presence of hot concentrated sodium or potassium hydroxide to produce the alkali metal salt of the fatty acids (soap) plus glycerol:

$$
\begin{array}{l}
\text{H}_2\text{C} - \text{O} - \overset{\displaystyle O}{\overset{\|}{\text{C}}} - \text{R} \\[4pt]
\qquad\qquad\overset{\displaystyle O}{\overset{\|}{}} \\
\text{HC} - \text{O} - \overset{\|}{\text{C}} - \text{R}' + 3\,\text{NaOH} \xrightarrow[\Delta]{\text{H}_2\text{O}} \\[4pt]
\qquad\qquad\overset{\displaystyle O}{\overset{\|}{}} \\
\text{H}_2\text{C} - \text{O} - \overset{\|}{\text{C}} - \text{R}''
\end{array}
$$

(1)

$$
\begin{array}{ll}
\text{H}_2\text{COH} & \text{R} - \overset{\displaystyle O}{\overset{\|}{\text{C}}} - \text{O}^-\text{Na}^+ \\[4pt]
\text{HC} - \text{OH} & \text{R}' - \overset{\displaystyle O}{\overset{\|}{\text{C}}} - \text{O}^-\text{Na}^+ \\[4pt]
\text{H}_2\text{COH} & \text{R}'' - \overset{\displaystyle O}{\overset{\|}{\text{C}}} - \text{O}^-\text{Na}^+
\end{array}
$$

Glycerol can form esters with inorganic oxy-acids as well as with carboxylic acids. For example, L-glycerol 3-phosphate is an important metabolic intermediate in glucose metabolism (see Chapter 8). Also familiar is the ester of glycerol formed with three equivalents of nitric acid; this reaction produces glyceryltrinitrate (nitroglycerine), an important explosive and a medicine for lowering blood pressure.

$$
\begin{array}{ll}
\text{H}_2\text{C} - \text{OH} & \text{H}_2\text{C} - \text{O} - \text{NO}_2 \\[4pt]
\text{HO} - \text{C} - \text{H} \quad \overset{\displaystyle O}{} & \text{HC} - \text{O} - \text{NO}_2 \\[4pt]
\text{H}_2\text{C} - \text{O} - \overset{\|}{\text{P}} - \text{O}^- & \text{H}_2\text{C} - \text{O} - \text{NO}_2 \\[4pt]
\qquad\qquad\quad \text{O}^- &
\end{array}
$$

L-Glycerol 3-phosphate   Glyceryltrinitrate (nitroglycerine)

Triacylglycerols undergo reactions characteristic of the carbon–carbon double bond if unsaturated fatty acyl groups are present. A typical reaction of the $\text{C}{=}\text{C}$ functional group is the addition of various reagents across the double bond. Two such reactions important in fat chemistry are shown in Eq. (2) on the next page.

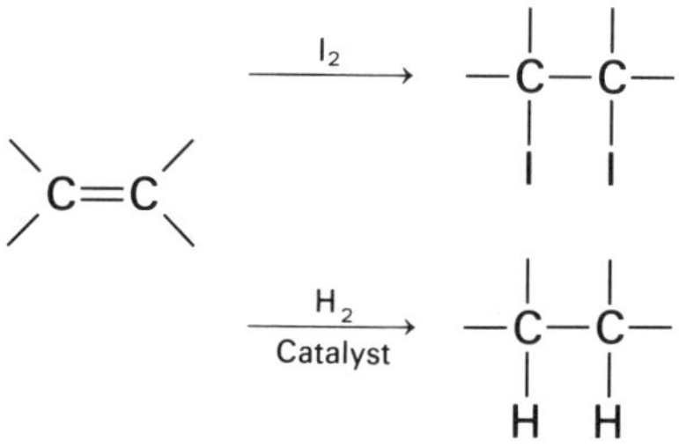

$$\text{C=C} \xrightarrow{I_2} \quad -\overset{|}{\underset{|}{C}}-\overset{|}{\underset{|}{C}}- \tag{2}$$

$$\xrightarrow[\text{Catalyst}]{H_2} \quad -\overset{|}{\underset{H}{C}}-\overset{|}{\underset{H}{C}}-$$

The reaction of a fat or oil with iodine yields the *iodine number* for that sample, defined as the number of grams of I_2 that combine with 100 grams of fat. Thus, the iodine number measures the degree of unsaturation in a given sample of a triacylglycerol. The reaction with hydrogen (hydrogenation) is an important commercial process. Industrial chemists can hydrogenate relatively inexpensive vegetable oils to produce more valuable fat products such as shortening and margarine. Oils possessing a high degree of unsaturation, such as linseed oil, can be reacted with molecular oxygen by a complex polymerization reaction, producing a durable film. Such oils are called *drying oils*, and form the basis for most oil-based paints, linoleum, and oilcloth.

Other Neutral Lipids

Glycerol ethers are neutral lipids in which a hydroxy group of glycerol is combined with the hydroxy group of a long-chain saturated or unsaturated alcohol to form an ether linkage. The remaining —OH groups of glycerol ethers may be esterified with fatty acids or may be present as free —OH groups, depending on the compound. For example, batyl alcohol is a glycerol ether found in shark and whale oils.

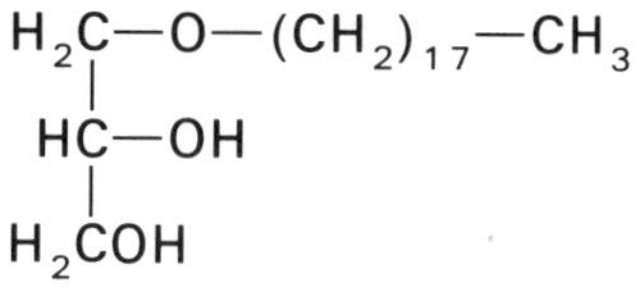

$$H_2C-O-(CH_2)_{17}-CH_3$$
$$HC-OH$$
$$H_2COH$$

Batyl alcohol

Waxes are esters of fatty acids with alcohols other than glycerol. Plant surface lipids contain waxes composed of esters of $C_{10}-C_{30}$ fatty acids with $C_{10}-C_{30}$ aliphatic alcohols. Beeswax contains a large proportion of an ester of palmitic acid with a C_{30} alcohol as shown below.

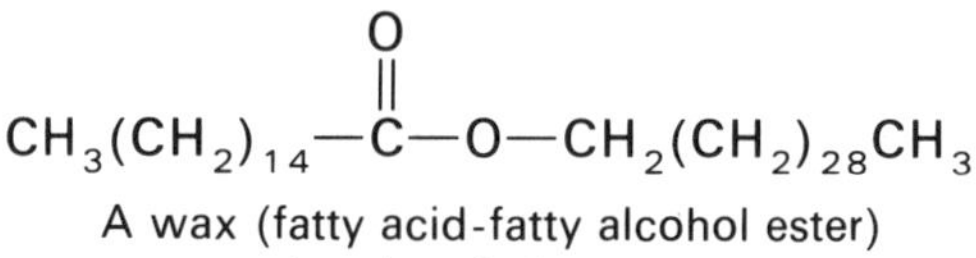

$$CH_3(CH_2)_{14}-\overset{O}{\overset{||}{C}}-O-CH_2(CH_2)_{28}CH_3$$

A wax (fatty acid–fatty alcohol ester)
abundant in beeswax

Fat Digestion and Absorption

Most of the lipids in the diet are in the form of triacylglycerols. Digestion of fats takes place in the small intestine. Here dietary lipids are acted upon by hydrolases called *lipases*. Because the lipases are water soluble and their substrates are not, the lipid material is converted into small globules (emulsified) by bile salts. Bile is produced in the liver and stored in the gall bladder, which releases it into the small intestine. The principal emulsifying agents in bile are the sodium salts of taurocholic acid and glycocholic acid, both steroid derivatives (see Figure 6.7 on page 196). Pancreatic lipase hydrolyzes two of the three ester links in triacylglycerols as shown below:

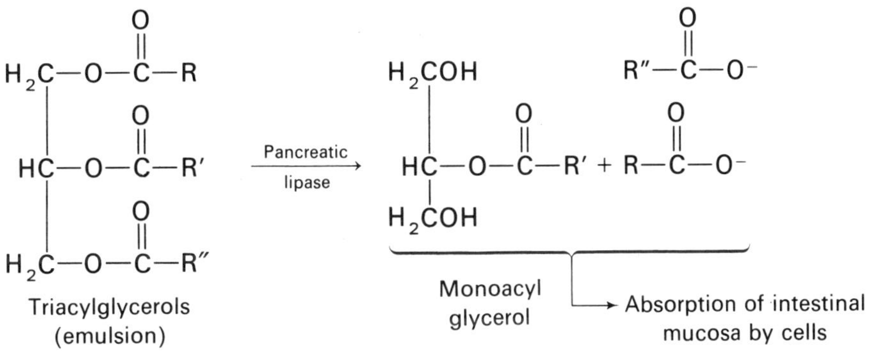

$$
\begin{array}{l}
H_2C-O-\overset{\displaystyle O}{\overset{\|}{C}}-R \\[2mm]
HC-O-\overset{\displaystyle O}{\overset{\|}{C}}-R' \quad \xrightarrow[\text{lipase}]{\text{Pancreatic}} \\[2mm]
H_2C-O-\overset{\displaystyle O}{\overset{\|}{C}}-R''
\end{array}
\qquad
\begin{array}{l}
H_2COH \\[2mm]
HC-O-\overset{\displaystyle O}{\overset{\|}{C}}-R' + R-\overset{\displaystyle O}{\overset{\|}{C}}-O^- \\[2mm]
H_2COH
\end{array}
\qquad
R''-\overset{\displaystyle O}{\overset{\|}{C}}-O^-
$$

Triacylglycerols
(emulsion)

Monoacyl glycerol → Absorption of intestinal mucosa by cells

Monoalkyl glycerols and fatty acids are absorbed through the intestinal lining and reconverted into triacylglycerols which then enter the blood via the lymph system.

Lipids are transported in the blood in several forms:

1. **Chylomicrons** are large micelles consisting mostly of triacylglycerols and about 1% protein.

2. **Very low density lipoproteins (VLDL)** contain about 10% protein by weight and transport primarily triacylglycerols.

3. **Low density lipoproteins (LDL)** contain about 20% protein by weight and are important in cholesterol transport. In the clinical laboratory, testing for blood levels of these is important in diagnosing early stages of *atherosclerosis* in which cholesterol is deposited on the inside walls of the arteries, thus impeding the flow of blood (see Fig. 6.5, p. 194).

4. **High density lipoproteins (HDL)** contain about 40% protein and are important in phospholipid transport.

5. **Serum albumin** binds and transports free fatty acids in the blood.

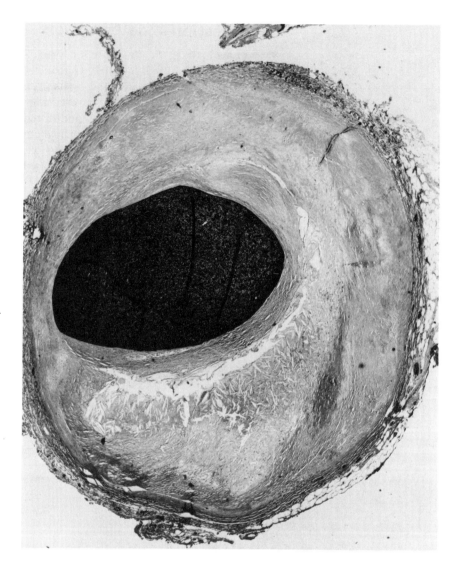

Figure 6.5

This cross-sectional photo of an atheriosclerotic artery taken from near the heart shows how material has built up inside the vessel wall to reduce the area through which blood can flow. The dark area is the passage through the artery; the white streaks are cholesterol. (Courtesy of Dr. John T. Fallon, Massachusetts General Hospital.)

6.4

Phosphoglycerides

Phosphoglycerides, an important class of amphipathic lipids, are abundant and essential components of biological membranes. The parent compound is *phosphatidic acid*, which is not present in free form in cells except as an intermediate in the biosynthesis of other phosphoglycerides. Figure 6.6 shows the structure of phosphatidic acid, as well as the structures of some common

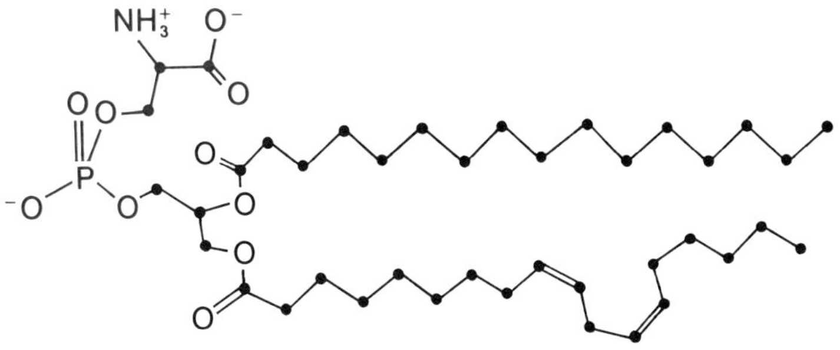

Phosphatidyl serine

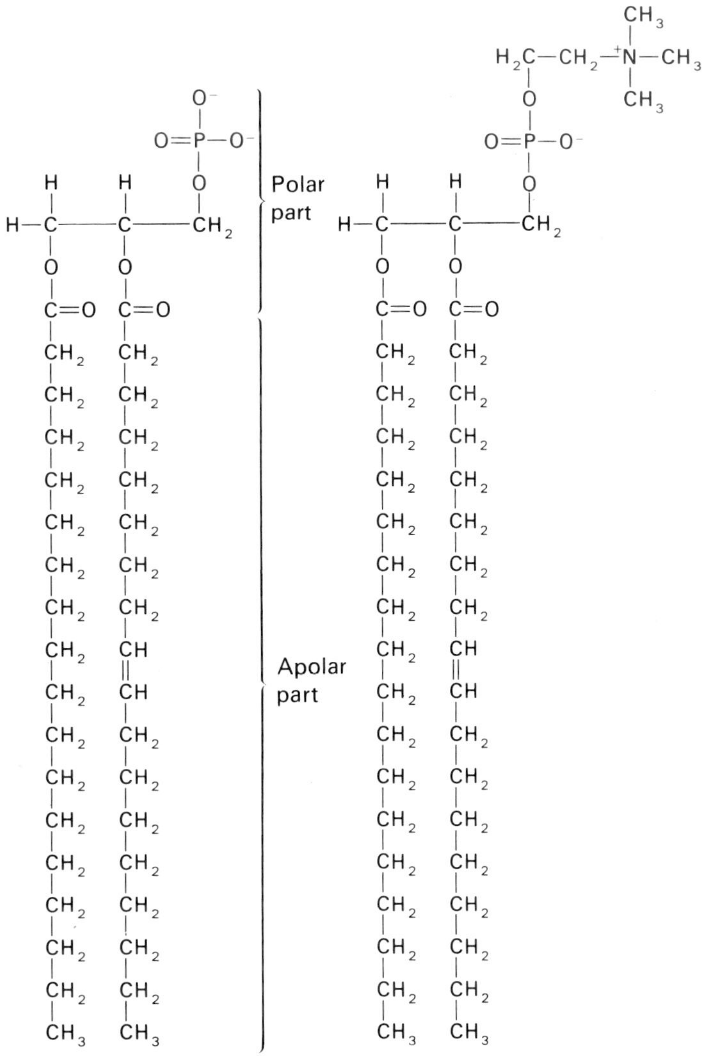

Phosphatidic acid

Phosphatidyl choline
(lecithin)

Figure 6.6

Some phosphoglyceride structures. The fatty acyl groups are representative of the average composition of the natural substances.

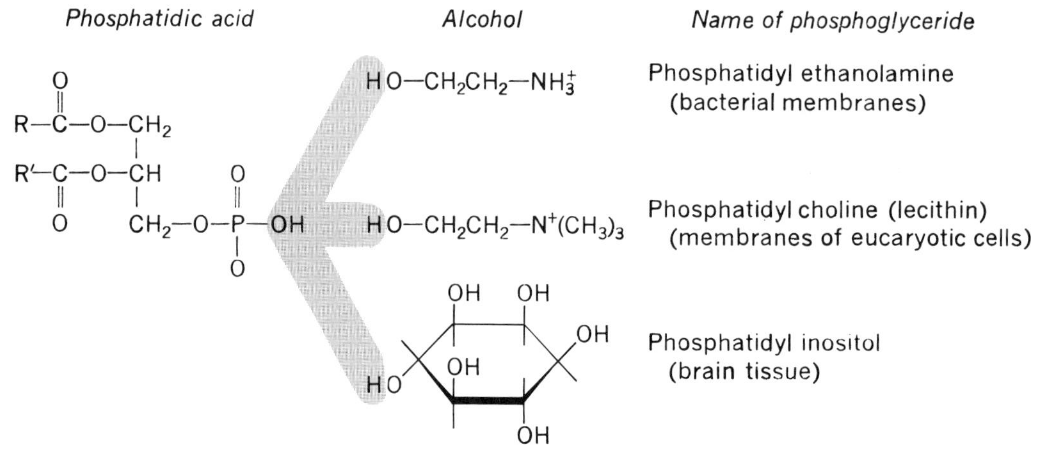

Phosphatidic acid · Alcohol · Name of phosphoglyceride

HO—CH₂CH₂—NH₃⁺ → Phosphatidyl ethanolamine (bacterial membranes)

HO—CH₂CH₂—N⁺(CH₃)₃ → Phosphatidyl choline (lecithin) (membranes of eucaryotic cells)

Phosphatidyl inositol (brain tissue)

Figure 6.7

Polar groups esterified to phosphatidic acid form various phosphoglycerides.

phosphoglycerides derived from it. The nature of the polar group which is ester-linked to the phosphate gives the phosphoglyceride its particular identity. Figure 6.7 gives three such polar groups and shows how they fit into the general phosphoglyceride structure.

6.5

Sphingolipids

Sphingolipids are amphipathic nonglycerol-containing lipids, particularly abundant in brain and nerve tissue. These lipids are derived from *sphingosine*:

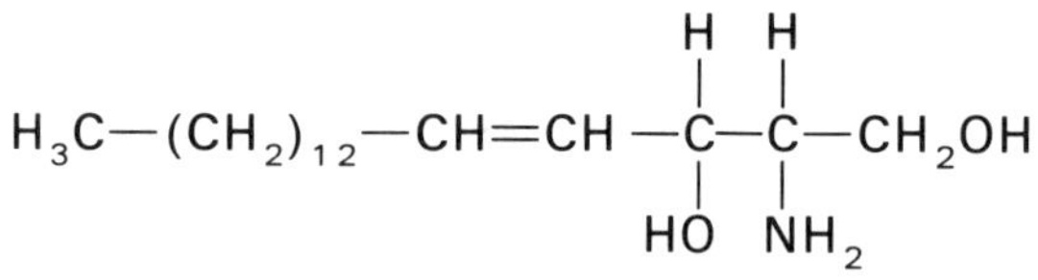

$$H_3C—(CH_2)_{12}—CH=CH—\overset{\overset{\displaystyle H}{|}}{C}—\overset{\overset{\displaystyle H}{|}}{C}—CH_2OH$$

with HO and NH₂ below the two carbons

Sphingosine

Ceramides are N-acyl fatty acid derivatives of sphingosines. While ceramides are widely distributed in plant and animal tissues, they generally are not major lipid constituents. Certain ceramide derivatives are plentiful. For example, *cerebrosides* are found in copious amounts in the myelin sheath of nervous tissue. As the structure below shows, cerebrosides are glycolipids because a sugar moiety is linked to the lipid part by a glycoside link.

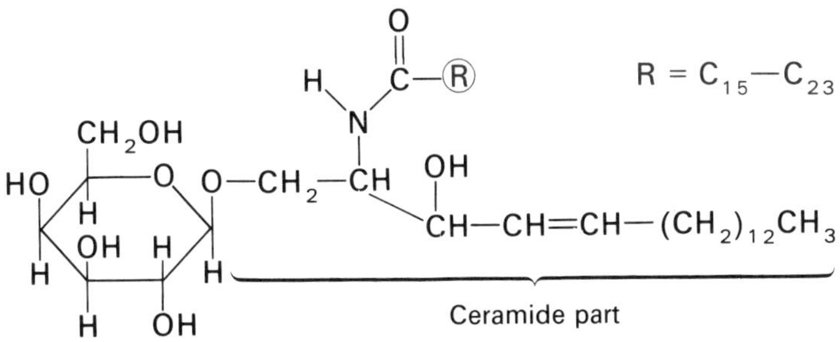

$R = C_{15} - C_{23}$

A cerebroside

The most abundant of the sphingolipids is sphingomyelin, found in brain and nerve tissue and in the lipid portion of blood. The structure of a sphingomyelin is shown below and is representative of the amphipathic nature of these membrane lipids.

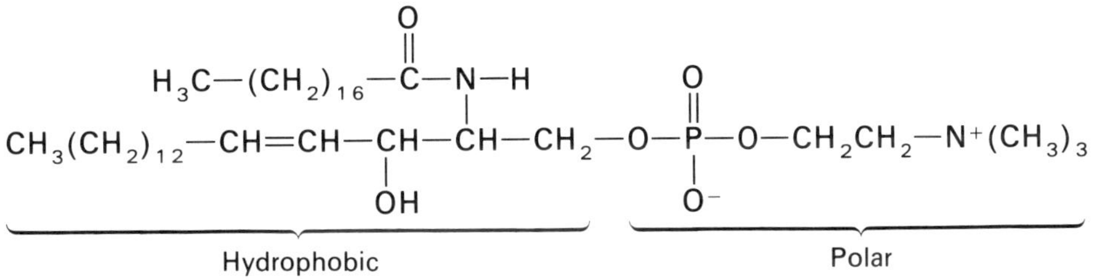

Sphingomyelin: The fatty acyl group can vary from C_{16} to C_{24}

6.6

Terpenes

We can consider terpenes as lipids that contain the structural framework of the isoprene unit:

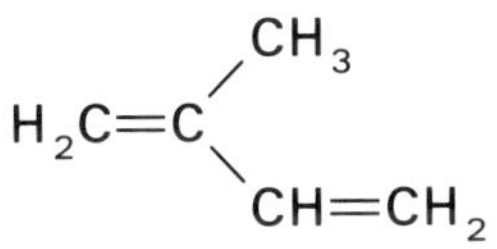

Isoprene

Many unusual natural products are terpenes, and the organic chemistry of these substances is both challenging and interesting (see Suggestions for Further Reading at the end of this chapter). Biologically, terpenes serve in many diverse roles.

Squalene, for example, is an important natural hydrocarbon present as a major component of sebum (skin oil), in shark liver oil, and as a biosynthetic precursor to cholesterol and, hence, the other steroids.

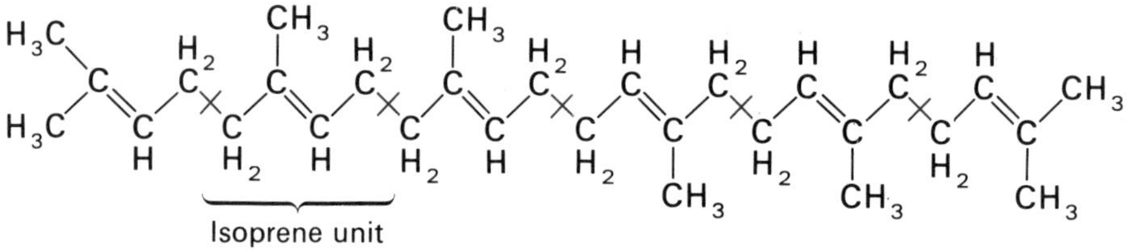

Squalene
(Isoprene units are separated by × s)

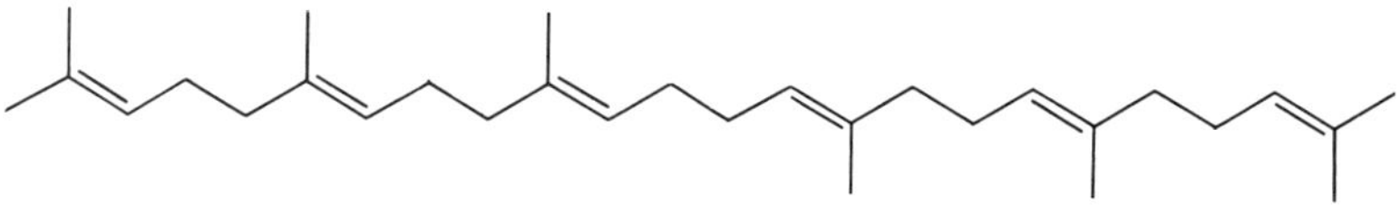

Squalene written in shorthand notation

The relationship between squalene and steroids can be seen more easily if we rewrite the squalene structure in a slightly different format as shown below.

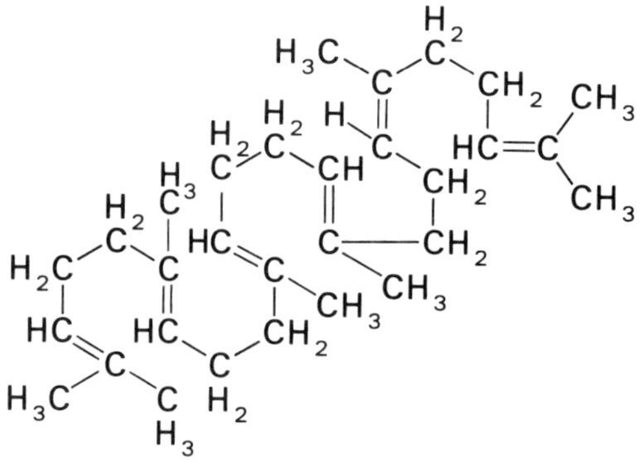

Squalene
(Compare with Figure 6.7)

Examples of other substances that are terpenes are given throughout the text. For example, Figure 10.4 (Sec. 10.2) shows the phytol side chain of chlorophyll and β-carotene, both important in photosynthesis. As we saw in Chapter 3, the fat-soluble vitamins A, E and K are terpene derivatives. Many essential oils and fragrances are terpenes, including *geraniol*, a major constituent of oil of roses.

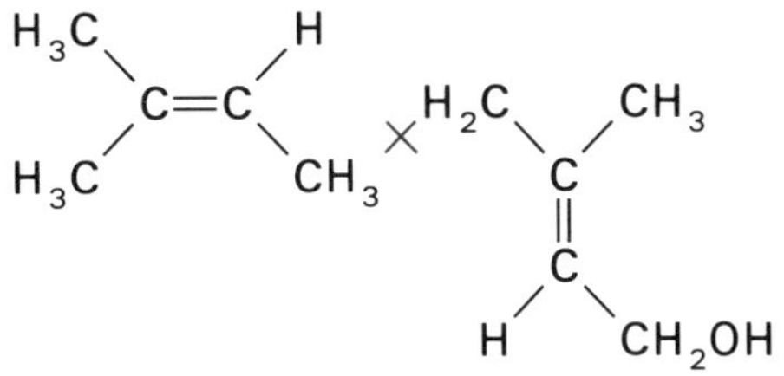

Geraniol
(two isoprene units)

The biochemistry of vision requires the terpene *retinal* which is derived from vitamin A. In the retina of the eye, visual cells contain a pigment called *rhodopsin*, or visual purple. This conjugated protein consists of two parts: opsin, the apoprotein, and the prosthetic group 11-*cis*-retinal. When light strikes the rhodopsin molecule, the 11-*cis*-retinal is photoisomerized to all-*trans*-retinal and dissociates from the protein. This isomerization and dissociation is ultimately transformed into a nerve impulse by a mechanism that still is not clearly understood. The visual cycle is completed when the all-*trans*-retinal is photoisomerized back into 11-*cis* form and spontaneously reassociates with opsin to form rhodopsin.

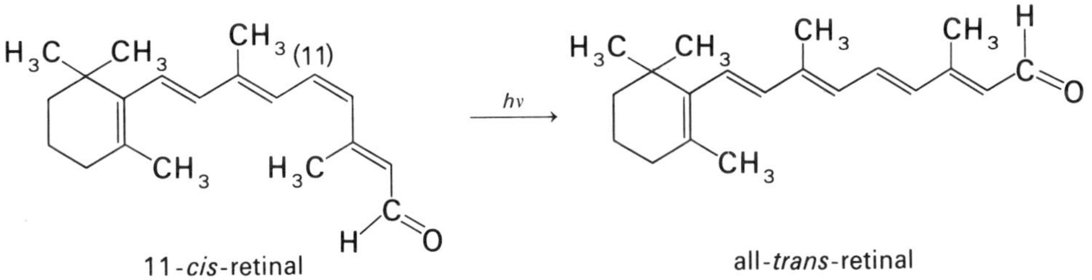

11-*cis*-retinal

all-*trans*-retinal

6.7

Steroids

Both terpenes and steroids are nonsaponifiable lipids, meaning that alkaline hydrolysis does not produce a soap. The general structure common to all steroids is the cyclopentanoperhydrophenanthrene skeleton:

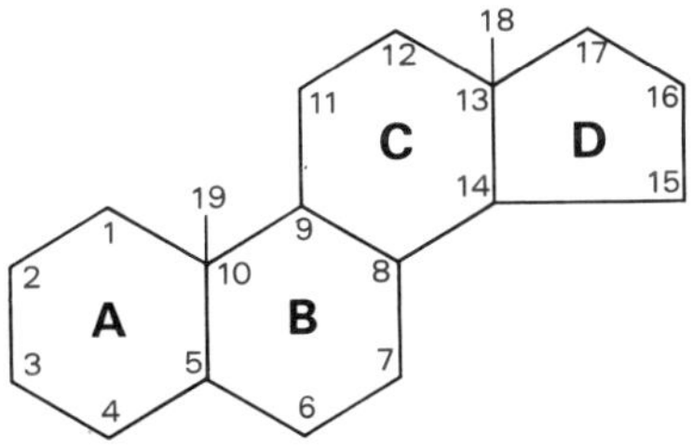

The identity of any given steroid depends on the substituents at each position, the location of any double bonds, and the stereochemistry of any asymmetric positions. In higher animals, steroids fall into three main groups: bile salts, steroid hormones, and steroid membrane components.

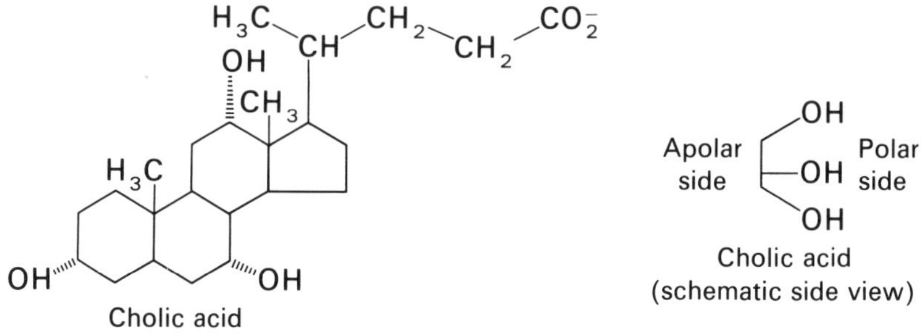

Figure 6.8

Structure of cholic acid, showing the origin of its amphipathic nature.

Bile salts act as emulsifiers, aiding in the digestion of fats in the small intestine. They are derivatives of cholic acid, shown in Figure 6.8. Cholic acid is amphipathic since it has both polar and apolar faces. This example illustrates the importance of stereochemistry in the biochemistry of steroids.

Cholesterol and Steroid Hormones

Cholesterol is an important steroid, not only because it is a membrane component, but also because it is the general biosynthetic precursor for other steroids including steroid hormones and bile salts. Cholesterol is abundant in brain and other nervous tissues, reflecting the importance of membrane function in these tissues. As a membrane lipid, cholesterol is found in the cell membranes of higher organisms, but not in bacterial or mitochondrial membranes. The structure of cholesterol is:

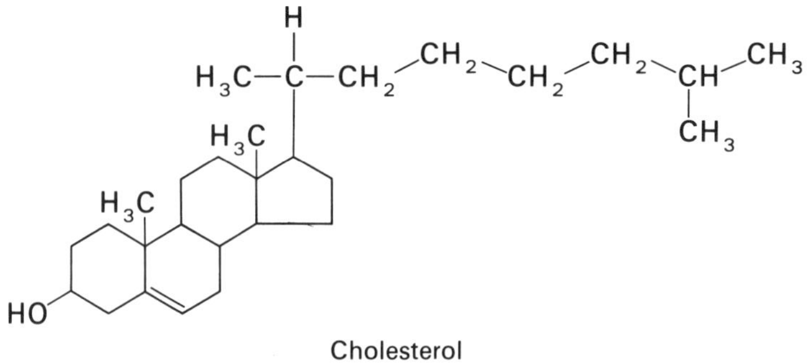

Cholesterol

In humans, cholesterol is obtained directly from the diet and also biosynthesized from acetate via squalene in the liver. The total amount of cholesterol in the blood depends to a large extent on diet, age and sex. A normal level is about 1.7 grams per liter of blood, but in older people it can increase to 2.5 g/L or higher. Certain cardiovascular diseases such as atherosclerosis, or "hardening of the arteries," have been partially attributed to high cholesterol levels.

Steroid hormones are a class of steroids that elicit intense physiological activity in minute amounts. They fall into three important groups: the *progestins*, the *sex hormones* and the *adrenal cortical hormones*. All of these steroid hormones act by controlling the expression of specific genes in chromosomes; that is, they initiate or inhibit the biosynthesis of specific proteins.

The principal progestin is *progesterone* (Figure 6.9) which is essential in the maintenance of pregnancy and in the regulation of the menstrual cycle in females. Many oral contraceptives are synthetic analogs of progesterone. They are effective in preventing unwanted pregnancy because elevated levels of progestins early in the menstrual cycle inhibit ovulation. Progesterone is also a biosynthetic precursor for the adrenal cortical hormones.

The principal male sex hormone, or *androgen*, is *testosterone*. It affects the reproductive system and promotes facial hair growth in both males and females. Some individuals have the inherited tendency of testosterone causing their head hair follicles to die, resulting in baldness. This is one reason why hair loss in males tends to run in families.

Estrogens are a group of female sex hormones in which the steroid *A* ring is aromatic. Estrogens are responsible for maintaining female characteristics. *Estradiol*, the major estrogen, acts in concert with progesterone to regulate the menstrual cycle. Synthetic estrogens form the basis for some oral contraceptives. *Diethyl stilbestrol*, another synthetic estrogen, had been used as a growth

Figure 6.9

The structures of some steroid hormones and other steroid derivatives which stimulate intense physiological activity. Note the similarities and differences between the structures.

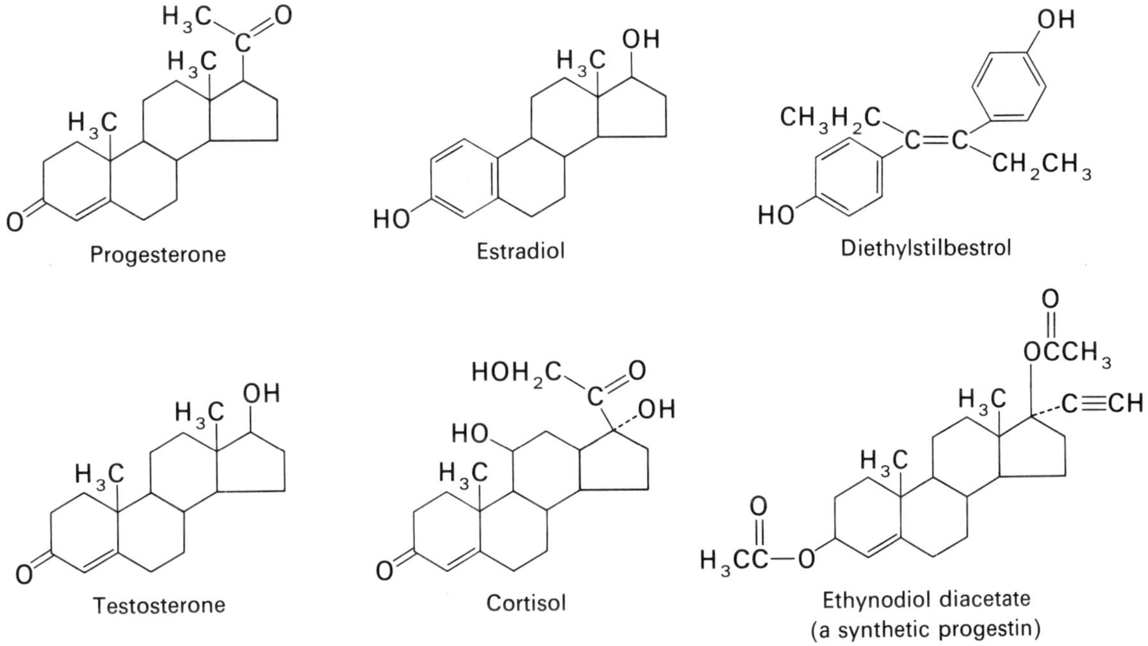

Progesterone

Estradiol

Diethylstilbestrol

Testosterone

Cortisol

Ethynodiol diacetate
(a synthetic progestin)

promoter for livestock. Its use has been discontinued due to studies implicating diethyl stilbestrol with cancer in rats. Figure 6.9 shows the structures of some of the substances we have been discussing.

The adrenal cortical hormones are produced in the outer layer, or cortex, of the adrenal glands. One of these hormones, *cortisol* (Figure 6.9), regulates carbohydrate and fat metabolism. Because of its central role as a regulator, cortisol is essential to life. Cortisol and a closely related compound, *cortisone*, are potent anti-inflammatory drugs used in alleviating severe arthritis and acute skin inflammations. Because these substances inhibit calcium absorption through the intestines, prolonged use can cause serious side effects such as muscle and bone attrition similar to that caused by vitamin D deficiency.

Steroids and their derivatives are widely distributed in nature and have interesting functions and properties (see Suggestions for Further Reading). For example, *digitalis* is a mixture of steroid glycoside derivatives from the garden plant purple foxglove (*Digitalis purpurea*). In anything larger than minute amounts, digitalis is a potent poison. In amounts of 0.1 mg it is an important drug in the treatment of congestive heart disease because it increases the tone of heart muscle tissue.

6.8

Lipid Membranes

Lipid membranes are vital for all living systems. The cell membrane acts as a permeability barrier between the living cell and its environment. Cell membranes are generally impermeable to polar molecules and ions, so mechanisms exist that permit the controlled entry of essential polar metabolites into the cell. This comes under the heading of *active transport*, discussed in Chapter 7. Many essential processes in living systems occur entirely within a membrane structure. We will see in Chapters 9 and 10 how the mitochondrial and chloroplast membranes play an intimate role in respiration and photosynthesis, respectively.

Earlier, in Section 6.1, we showed how amphipathic molecules spontaneously form micelles and bilayer structures in an aqueous (polar) environment. The structure of biological membranes is maintained almost entirely by such noncovalent apolar interactions. We can readily demonstrate this by the ease with which membranes are disrupted by treatment with detergent solutions or fat solvents such as ethanol or acetone.

Most biological membranes contain about 40% lipid by weight. In addition, membranes contain various proteins which either assist in maintaining the stability of the membrane, or act as enzymes in such membrane functions as respiration or transport of metabolites. In many cases, the membrane-bound enzymes seem to require a lipid environment in order to be active. Table 6.5

Table 6.5 Some Major Constituents of Membranes.

Component	Types	Function
Lipids	Phosphoglycerides	Form the basic bilayer structure.
	Cholesterol	Plasma membrane component of algae, fungi, and higher organisms.
Proteins	Enzymes	Catalyze reactions associated with the membrane function.
	Transport proteins	Mediate the transfer of material across the membrane barrier.
	Structural protein	A specific protein studied primarily in mitochondrial membranes; it complexes with phosphoglycerides.

Figure 6.10

A portion of the cell envelope of *B. subtilis*, enlarged to show the trilaminar cell membrane structure. Electron micrograph at 87,000×. (Courtesy Roger M. Cole, PhD., M.D., Chief, Laboratory of Streptococcal Diseases, National Institutes of Health, Bethesda, MD.)

summarizes some of the types of substances that make up membrane structure. As the table implies, membranes are not just simple lipid films that separate cells from their environment. Instead they are quite complicated in structure and function, particularly with respect to the role of the membrane-bound proteins.

There are a number of models for the structure of biological membranes. All models must be consistent with the observation that high-magnification electron micrographs of transverse sections showing a cell membrane generally reveal a tri-laminar structure, as shown in Figure 6.10. The simplest membrane model is the *unit membrane* model proposed by J. D. Robertson and shown in

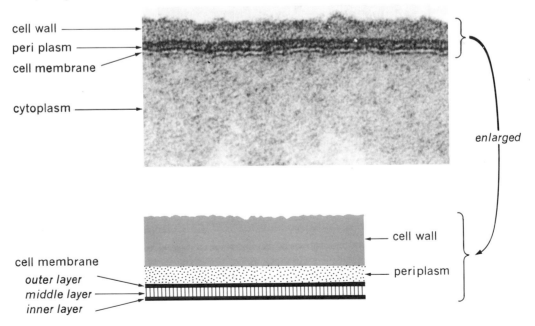

Figure 6.11

A unit membrane model. This model is certainly an oversimplification. Actual biological membranes are better represented by the fluid-mosaic model shown in Figure 6.12.

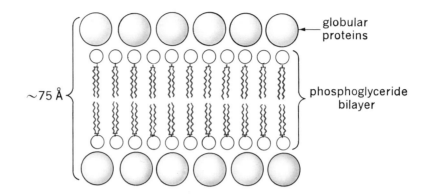

Figure 6.11. The unit membrane is certainly an oversimplification, but it is a useful basis for understanding membrane structure. Since the proposal of the unit membrane model, researchers have established that the participation of proteins in the lipid part of the membrane can be much more extensive than is implied by the simple unit membrane hypothesis. Perhaps a more realistic structure is the *fluid mosaic* model recently proposed by S. Singer. Here, we can think of the membrane as proteins floating about in a mobile sea of lipid, pictured in Figure 6.12. The fluid mosaic model is consistent with the fact that some membrane proteins are easily removed from the membrane, while others are so deeply embedded that it is extremely difficult to separate them from the membrane intact. As we will see in Chapter 9, some of the cytochromes of the mitochondrial respiratory chain of electron-transferring heme proteins are so deeply embedded in the inner mitochondrial membrane that it has been difficult to study the behavior of these important proteins outside of the mitochondrion.

6.9

Specialized Functions of Lipids

Lipids can serve many diverse purposes in nature beyond those already discussed. The following cases will illustrate the wide range of lipid function and the central role of lipids in biological systems.

Plant Surface Lipids

Plant surface lipids are vital to green plants living in arid environments in which water must be conserved. A coating of wax and hydrocarbon material on the plant surfaces helps to guard against evaporation of water. This is particularly important in tropical plants. For example, mangrove leaves contain aliphatic hydrocarbons related to paraffin wax, in addition to fatty acid–fatty alcohol esters. In this case lipids are serving in a protective role.

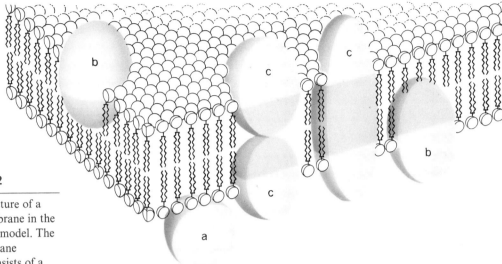

Figure 6.12

Overall structure of a
typical membrane in the
fluid mosaic model. The
basic membrane
structure consists of a
phospholipid bilayer
about 45 Å thick. The
figure includes various
types of membrane
proteins. Some
membrane proteins are
bound to the membrane
surface (a); others are
partially submerged in
the membrane structure
(b). Still other membrane
proteins extend through
the membrane either as a
single molecule or as a
pair (or possibly more)
of protein molecules (c).
(From R. A. Capaldi,
"A Dynamic Model of
Cell Membranes,"
Copyright © 1974 by
Scientific American, Inc.
All rights reserved.)

Pheromones

Pheromones are substances used by insects and other species for communication. Most airborne pheromones appear to be lipids, while waterborne pheromones for species such as barnacles, lobsters and snails have been found to be either proteins or lipids.

Pheromone substances in insects have been studied extensively. For example, the sex attractant (pheromone) used by female European corn borers to attract males is *cis*-11-tetradecenylacetate (see Figure 6.13). Only minute amounts of sex pheromones are necessary to cause a response because the sensitivity of male insects to these substances is acute. For instance, the male silkworm moth (*Bombyx mori*) can detect its specific sex attractant in concentrations as low as 100 molecules/cubic centimeter of air. At standard temperature and pressure, this works out to a molar concentration of approximately 3.7×10^{-18} moles/liter! Individual species seem to have unique pheromones or mixtures of pheromones. Figure 6.13 gives the sources and structures for some of the many known insect sex pheromones.

In addition to sex pheromones, insects use other chemical substances to communicate. Alarm pheromones, aggregating pheromones and trail-marking pheromones have been identified and studied. These substances are particularly evident in the case of social insects such as ants and bees. For example, a queen bee secretes 9-hydroxydecenoic acid to cause the clustering of worker bee swarms. Another example is the alarm pheromones of ants. In one species of ant, a disturbed worker will release a mixture of substances which alarms and

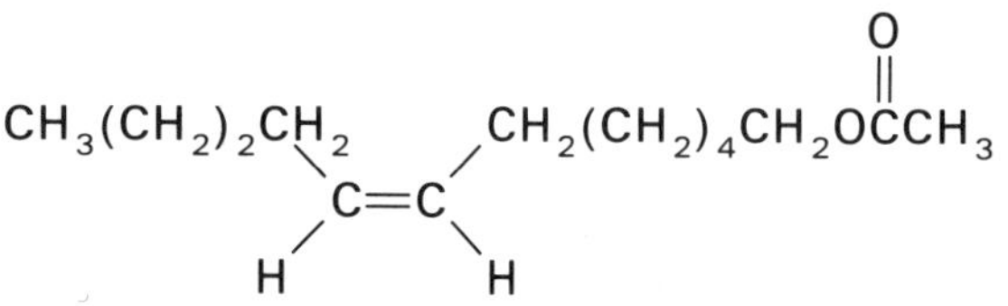

$$CH_3(CH_2)_2CH_2 \quad CH_2(CH_2)_4CH_2O\overset{\overset{\displaystyle O}{\|}}{C}CH_3$$

$$C=C$$
$$H \quad H$$

Cabbage looper moth (cabbage family pest)

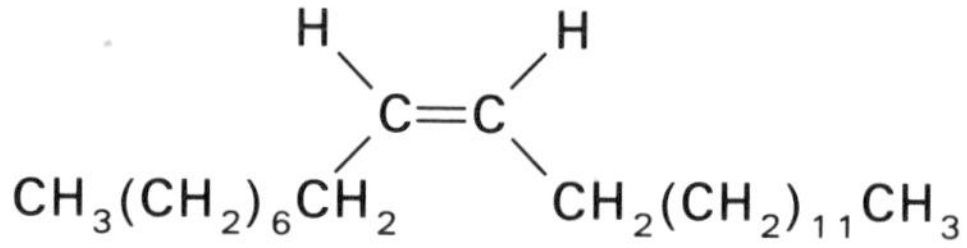

$$H \quad H$$
$$C=C$$
$$CH_3(CH_2)_6CH_2 \quad CH_2(CH_2)_{11}CH_3$$

Common house fly (general pest)

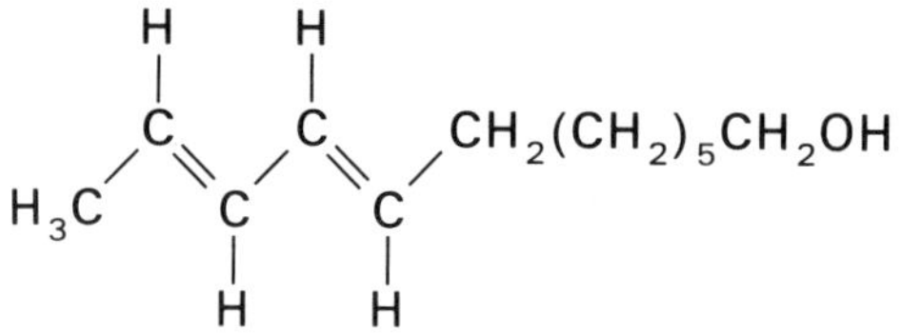

Figure 6.13

The sources and structures of some insect sex pheromones. Geometrical isomerism is important in these substances; changing from a *cis* form to a *trans* form may mean a complete loss of activity.

$$H \quad H$$
$$C \quad C$$
$$H_3C \quad C \quad C \quad CH_2(CH_2)_5CH_2OH$$
$$H \quad H$$

Codling moth (major apple pest)

attracts other workers of the same species—a sort of olfactory civil defense system. Alarm pheromones isolated from ants are shown below.

$$CH_3(CH_2)_9CH_3 \qquad CH_3(CH_2)_{11}CH_3$$
Undecane \qquad Tridecane

$$CH_3(CH_2)_9CH_2\overset{\overset{\displaystyle O}{\|}}{C}CH_3$$
2-Tridecanone

Clearly there is a fascinating overlap between the chemistry of natural products and insect behavior in the study of pheromones as chemical messengers.

Synthetic sex pheromones are currently being used to disrupt the breeding cycles of certain insect pests. Paper strips impregnated with a pheromone in small amounts attract males into a trap, and larger amounts disorient the males, thus preventing them from finding female breeding partners. Because pheromones are not toxic and are also highly specific for a given insect species, they provide a useful alternative to conventional chemical pesticides.

It has recently become apparent that hydrocarbons also serve as chemical messengers for certain marine animals. For example, it is known that the feeding rate of lobsters is stimulated by certain hydrocarbons. The disruption of chemical communications in the sea may be one of the more serious effects of an oil spill.

Prostaglandins

Prostaglandins are lipid chemical messengers in the human body whose chemical basis of action is not yet fully understood. Since their discovery in the 1930s, many prostaglandins have been isolated and identified. Prostaglandins are fatty acids with 20 carbons, including a 5-carbon ring. They fall into four main classes, denoted PGA, PGB, PGE and PGF. Accompanying the shorthand abbreviation for the classes is a subscript denoting the number of double bonds outside the ring.

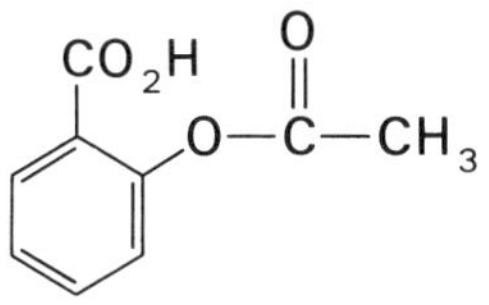

Prostaglandins exert a physiological response in minute concentrations. They are of considerable biomedical interest because of their role in the treatment of inflammation and allergies. For example, *rheumatoid arthritis* affects over 5 million Americans and causes severe inflammation of the joints. Aspirin, a common household drug, is still one of the best anti-inflammatory drugs for treating this disease. There is experimental evidence that aspirin may inhibit prostaglandin biosynthesis from polyunsaturated fatty acids and therefore prevent the physiological responses normally caused by prostaglandins. Figure 6.14 shows the structures of some well-known prostaglandins.

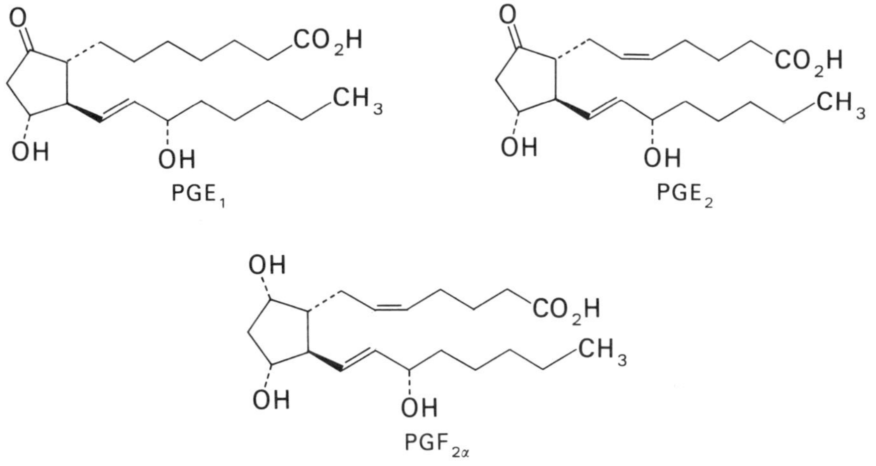

Acetyl salicylic acid
(aspirin)

PGE$_1$

PGE$_2$

PGF$_{2\alpha}$

Figure 6.14

The structures of some prostaglandins. Note the difference in the functional groups present in the E and F prostaglandins.

There is considerable variation in the physiological effects of various prostaglandins even though the structural differences may appear to be minor. PGE$_2$ tends to suppress inflammation while PGE$_1$ has the opposite effect. PGE$_1$ is also a strong fever-inducing agent, and it is thought that the fever-reducing effect of aspirin has its basis in prostaglandin suppression. Prostaglandins of the F family induce allergic responses such as *asthma*, while E prostaglandins, when inhaled in minute amounts, reverse the effect. These remarkable molecules are short-lived in the body and, once formed, are rapidly oxidized and excreted after they enter the blood stream. Because of their clinical applications, there is considerable current research on prostaglandins.

Suggestions for Further Reading

Fieser, L. F. and Fieser, M. *Organic Chemistry*. Lexington, Mass.: D. C. Heath, 1956. This text and its later editions offer a first-rate review of the organic chemistry of lipids, including many interesting examples; in particular, see Chapters 15, 37, and 38.

Lehninger, A. L. *Biochemistry*. 2d ed. New York: Worth Publishers, 1975. Chapters 11 and 24 provide a detailed coverage of lipids in a lucid manner.

Sondheimer, E. and Simeone, J. B. *Chemical Ecology*. New York: Academic Press, 1970. This exciting book contains 300 pages on the ways in which plants and animals use chemicals to communicate and regulate.

White, A.; Handler, P.; Smith, E.; Hill, R.; and Lehman, I. *Principles of Biochemistry*. New York: McGraw-Hill, 1978. Chapters 41–47 of this standard text provide an excellent discussion of steroid hormones and their physiological role.

Scientific American

Benson, A. A. and Lee, R. F. "The Role of Wax in Oceanic Food Resources." March 1975.

Capaldi, R. A. "A Dynamic Model of Cell Membranes." March 1974.

Fieser, L. F. "Steroids." January 1955.

Fox, C. Fred. "The Structure of Cell Membranes." February 1972.

Hokin, L. E. and Hokin, M. R. "The Chemistry of Cell Membranes." October 1965.

Pike, J. E. "Prostaglandins." November 1971.

Williams, C. M. "Third Generation Pesticides." July 1967.

Problems

1. For each of the terms given below, provide a brief definition or explanation.

Amphipathic	Micelle	Soap
Fatty acid	Phospholipid	Steroid
Lipid	Pheromone	Terpene
Membrane	Prostaglandin	Triacylglycerol

2. A sample of fat from a natural source is saponified and the fatty acids present identified as stearic and palmitic acids. Draw the structures for all possible triacylglycerol species present in the original sample of fat.

3. Think back to your organic chemistry course and give the *systematic* name for each of the fatty acids in Table 6.2.

4. Is batyl alcohol an amphipathic substance? Give your reason.

5. Draw the structures of the products of the hydroxide-ion-catalyzed complete hydrolysis of:
   (a) phosphatidyl choline
   (b) 1-myristoyl-2,3-dioleoylglycerol
   (c) a cerebroside having palmitic acid as the fatty acyl groups.

6. Fatty acids form strong complexes with such metal ions as Mg^{2+} and Ca^{2+}, where the carboxyl function binds to the metal ion. Predict the result of adding a solution of Ca^{2+} to a soap solution in a beaker.

7. Cortisone is a widely used anti-inflammatory drug formed by the oxidation of the secondary alcohol functional group of cortisol (Figure 6.8) at the number 11 position. Draw the structure of cortisone. (See also Appendix B.)

8. Give the systematic names for the sex pheromones shown in Figure 6.12.

Generating Energy: Catabolic Sequences

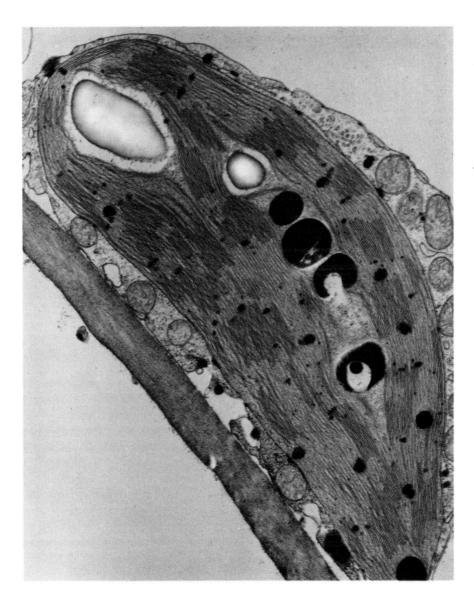

CHAPTER 7

Introduction to Metabolism

Living cells are in a state of continuous biochemical activity as a result of growth and maintenance of cell structure. For example, in a growing culture of a procaryote such as *E. coli*, reproduction by cell division occurs about once every 20 minutes. A culture medium need only contain glucose for a carbon source and inorganic salts to provide nitrogen, phosphorus, and other required simple substances. The fact that one *E. coli* cell is able to become two cells every 20 minutes means that it has within its cell membrane all the chemistry needed to synthesize, using just a few simple precursors, all the proteins, nucleic acids, carbohydrates, lipids, coenzymes, and other biomolecules that a living cell requires.

In this chapter, and those which follow, we will learn how cells gain energy from foodstuffs and how cells use energy and foodstuffs to produce the substances they need. Before we discuss the chemistry of these processes in detail, we must first go over some basic concepts concerning cells and their functions.

7.1

Metabolic Pathways and Cell Function

A cell must have biochemical pathways available for the stepwise degradation of foodstuffs to produce energy, as well as for the stepwise biosynthesis of required substances. We must begin our discussion of these processes by defining some important terms. The chemical changes that constitute the pathways for food molecule degradation and biosynthesis collectively define the *metabolism* of a cell. As we saw in Chapter 1, processes leading to the breakdown of molecules and the generation of energy are *catabolic* processes and define the *catabolism* of the organism. The processes of biosynthesis which consume both materials and energy are *anabolic* processes.

All processes of metabolism are *intermediary* in nature; that is, the transformation of one substance into another generally involves a sequence of steps in which discrete and identifiable chemical substances, or *intermediates*, are the reactants and products. The intermediary nature of metabolism makes it possible for a few simple substances to act as precursors for all the chemical needs of a cell, as shown schematically in Figure 7.1. Each step in the metabolic reaction sequence, or *pathway*, shown in Figure 7.1(b) is catalyzed by a given enzyme. Any particular cell contains thousands of different enzymes; therefore, a metabolic "map" showing all the reactions occurring in even a simple cell would be highly complicated. Such a complete map does not yet exist since only

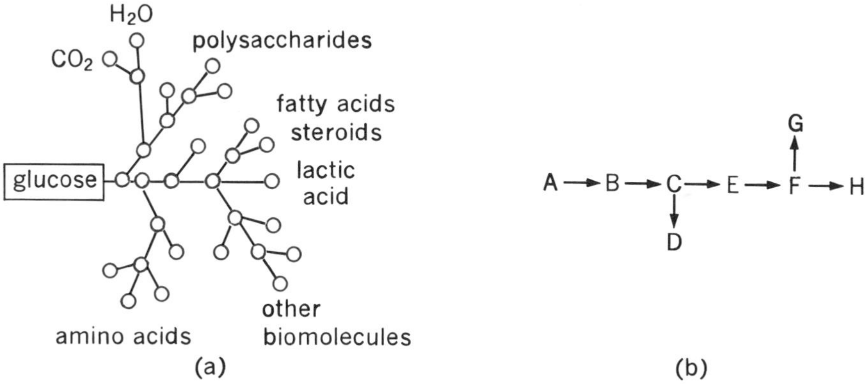

Figure 7.1

Schematic representations of intermediary metabolism. Not only are the end products of each branch available to the cell (a), but all the intermediates produced in the various steps of the pathways are also available. Moreover, most reactions of metabolism are reversible, permitting a close relationship between the metabolites involved.

(a)

(b)

a fraction of the metabolic processes in even a simple bacterial cell are currently understood. The situation in the cells of humans is vastly more complex. As Figure 7.1(a) shows, the pathways of cellular metabolism are interrelated, which adds to the complexity.

In a viable cell, all metabolic reactions are subject to **regulation by control mechanisms**. In Chapter 3 we saw how the activity of certain enzymes, which serve as "pacemakers" in metabolic pathways, can be controlled either by *feedback* (allosteric) inhibition or by *genetic* control. It is important to recognize that a metabolic pathway is *not a static situation*. The reactions that constitute a metabolic pathway are enzyme-catalyzed reactions in which one intermediate normally transforms into another on a continual basis. Therefore, the concentration of a **given intermediate**, or *metabolite*, in such a pathway is *rate controlled*. This means that the concentration of a given metabolite in a cell is governed by the relationship between the rate at which it forms and the rate at which it transforms into other substances. A cell metabolite whose concentration is constant as the result of a balance between consumption and formation is said to be in a *steady state*. For example, heavy muscular activity brings about an increased rate of formation of the metabolite lactic acid; its steady-state concentration in the muscle tissues increases, as does physical discomfort associated with muscle fatigue. Upon return to normal levels of muscular activity, the consumption of lactic acid is able to catch up with the excess amount and the metabolite falls back to its normal physiological level.

In Part II of this book we will learn about energy-yielding metabolic processes; in Part III, we will study some of the major energy-requiring processes of biosynthesis. Since a living cell is a highly coordinated biochemical entity, it is important to recognize that the processes of catabolism and anabolism do not proceed independently of one another. Furthermore, within each "side" of intermediary metabolism there are many close interrelationships, so that it is fair to say that there are few, if any, individual reactions in a living cell that occur completely independently of all others. This is confirmed by the effect of inborn errors of metabolism in which a single defective enzyme can completely disrupt seemingly unrelated metabolic pathways. Throughout our discussion of metabolism, we'll see examples of these hereditary defects in metabolism. At the back of this text there is a metabolic chart showing some of the principal processes of intermediary metabolism. You should refer to this chart as you read to help you integrate the individual parts of metabolism with the overall picture.

In our discussion of the major features of intermediary metabolism in the following chapters we will consider not only the chemistry involved, but also the enzymology, regulation and *energetics* of some of these processes, particularly those which yield energy to the cell.

7.2

Energy Flow in the Biosphere

In Chapter 1 we noted that cells fall into two classes: heterotrophs, which obtain food from other cells, and autotrophs, which create their own food from simple inorganic substances and sunlight. These two types of cells are closely related by the overall processes of the *carbon cycle*, shown in Figure 7.2.

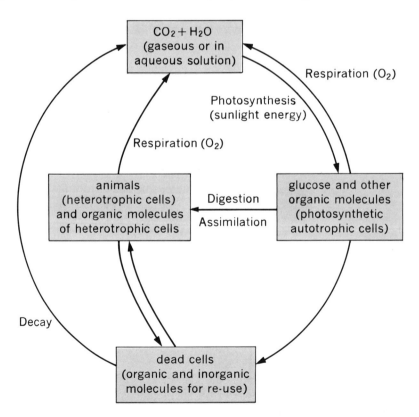

Figure 7.2

The flow of carbon in the biosphere. Neither autotrophic nor heterotrophic cells can live without the other.

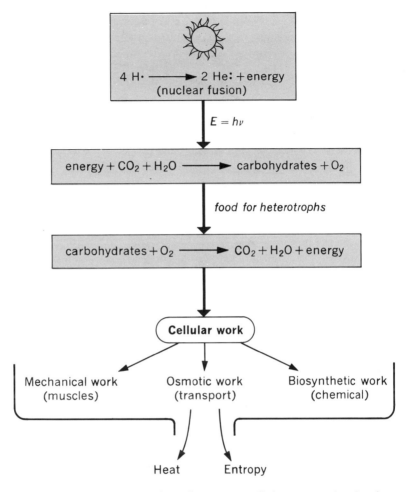

Figure 7.3

The flow of energy in the biosphere.

Photosynthetic autotropic cells *conserve* light energy in the form of carbohydrates, which are then utilized by nonphotosynthetic hererotrophic cells as a source of not only energy, but also of raw materials for biosynthesis. The "flow" of energy from the sun through various stages of utilization is shown schematically in Figure 7.3.

We can note many similarities between metabolic processes in simple procaryotic cells such as bacteria and those in more complicated eucaryotic species such as the cells of higher animals. In Chapter 3 we saw that the primary structure of the protein cytochrome c in such evolutionary diverse sources as yeast and man is highly homologous. This similarity at the molecular level is also apparent at the level of cellular metabolism. Because of the universality of major metabolic processes, scientists were able to use simple procaryotes to study intermediary metabolism, and could extend many of the conclusions gained from such studies to the corresponding processes in the cells of higher animals.

While the basic features of intermediary metabolism are surprisingly similar for all organisms, there is considerable variation in the details from one organism to another. We can classify organisms with respect to how they gain energy. Let's examine these classifications and discuss how the energy metabolism of each category differs from the others.

Autotrophic Cells (Self-Feeding)

Chemoautotrophs gain energy by oxidizing reduced inorganic molecules such as H_2S, S, NH_3, NO_2, H_2, and CO, thus producing the energy needed to reduce CO_2 (used as a carbon source for carbohydrates) and providing a driving force for energy-requiring processes. These nonphotosynthetic organisms include certain specialized families of bacteria which can satisfy all their material and energy needs from the uptake of simple inorganic molecules.

Photoautotrophs gain energy by converting sunlight into chemical energy (see Chapter 10). These cells include all photosynthetic plant cells, algae, the procaryotic blue-green algae, and photosynthetic sulfur bacteria.

Heterotrophic Cells (Feeding on Others)

Photoheterotrophs require organic compounds other than CO_2 as a carbon source. Since organic compounds in nature come almost exclusively from the degradation of once-living material, these cells are heterotrophs. Such organisms include the purple nonsulfur bacteria.

Chemoheterotrophs require organic compounds both as sources of carbon and as sources of energy. This class covers a vast range of organisms, including higher animals, fungi, protozoa, and most bacteria.

These four main categories are somewhat arbitrary in that there can be overlapping between classes. For example, in the absence of light, many photoautotrophic algae can grow as chemoheterotrophs. We can also make a further distinction among the chemoheterotrophs, between *aerobic* and *anaerobic* organisms.

Aerobic organisms couple the oxidation of organic molecules with the reduction of O_2, producing water. Oxygen is said to be the *terminal electron acceptor* in this case. Obligate aerobes cannot exist in the absence of oxygen.

Anaerobic organisms are far more common. They fall into two types: *strict anaerobes* and *facultative anaerobes*. Strict anaerobes use organic molecules as terminal electron acceptors in oxidative metabolism and are poisoned by the presence of oxygen. *Clostridium botulinus*, a strict anaerobe which is sometimes present in improperly canned low-acid food, produces a deadly bacterial toxin that causes botulism, or food poisoning. Facultative anaerobes can utilize either O_2 or an organic molecule as a terminal electron acceptor. Humans are facultative anaerobes, since many of the cells in our body function anaerobically all the time, while others may function aerobically or anaerobically depending on circumstances. Yeasts are also facultative anaerobes because they metabolize glucose to CO_2 and H_2O in the presence of O_2. In the absence of oxygen, they metabolize glucose to produce CO_2 and a reduced organic molecule such as ethanol. Therefore, if you want to make beer in your closet, do not let air get into your fermentation flask!

$$\text{Aerobic: } C_6H_{12}O_6 + 6\,O_2 \xrightarrow{\text{Yeast}} 6\,CO_2 + 6\,H_2O$$

$$\text{Anaerobic: } \quad C_6H_{12}O_6 \xrightarrow{\text{Yeast}} 2\,CO_2 + 2\,CH_3CH_2OH$$

7.3

Energy Conservation in Living Systems

We tend to think of energy conservation in the context of thrift, where we turn thermostats to lower settings and do not leave lights on needlessly. In a chemical context, the term "energy conservation" has a somewhat different meaning. In a heterotrophic cell, a reduced organic molecule such as glucose is oxidized to form CO_2 and H_2O:

$$C_6H_{12}O_6 + 6\,O_2 = 6\,CO_2 + 6\,H_2O + \text{energy}$$

This process liberates a substantial amount of energy, which is then available to the cell for performing useful work. This energy is *conserved*, or transformed from a "storage" form of energy to a "biosynthetically useful" form of energy utilizable by the cell. Since cells function *isothermally* (at a constant temperature), this useful form of cellular energy cannot be heat because work can only be gained from heat energy when heat is transferred from a hot body to a cooler body. Since all cellular processes are essentially chemical in nature, **energy from the catabolism of foodstuffs is conserved in chemical form, as the chemical energy in phosphorus-oxygen bonds in adenosine triphosphate (ATP).**

Before we can discuss the role of ATP in the transfer of energy in the cell, we must first review some of the pertinent parts of thermodynamics and chemical equilibrium. This will give us the tools for understanding the fundamentals of energy changes in living systems.

7.4

Thermodynamics and Chemical Equilibrium

Energy is the capacity to perform *work*, where work can be expressed in several different ways. We can visualize the mechanical work of raising an object against the force of gravity, or the pressure-volume work of a piston compressing a gas in a cylinder. In the context of a chemical reaction, we can speak meaningfully of *chemical work*. Energy can change from one form to another. Therefore, the chemical energy gained from the breakdown of food in an animal is expressed as the mechanical work of muscle contraction.

We study energy changes with respect to a given *system*, which is that part of the universe we have isolated for study. Our system can be a chemical reaction mixture in a flask, or a cell suspension in a beaker. The *surroundings* are the rest of the universe outside the system. A system can gain or lose energy to the surroundings as a result of chemical or physical processes. Dealing with such changes is what thermodynamics is all about.

The energy of a system can change in two different ways:

1. The flow of *Heat* $\equiv q$ (energy transferred due to a temperature difference).

2. The performance of *Work* $\equiv w$ (all other forms of energy).

For all energy changes, the **First Law of Thermodynamics** states that **for any process, the total energy change in the system is equal but opposite in sign to the total energy change of the surroundings**. This is a conservation-of-energy law, where the total energy of the system and surroundings (i.e., the universe) is constant, no matter what interconversion of forms of energy takes place.

We can use this law to deal with energy changes associated with chemical processes, where the internal energy change ΔE is equal to the total energy of the products minus the total energy of the reactants. This quantity does not take into account any external pressure-volume work done by the system if it expands or contracts during the course of the reaction. However, most processes of interest to biologists and biochemists occur at constant pressure (i.e., open to the atmosphere). For reactions at constant pressure we use an expression called the **enthalpy change**, or *heat of reaction*, given by the symbol ΔH:

$$\Delta H = \Delta E + \text{(pressure-volume work)}$$

For a reaction at constant pressure, the heat gained or lost by the system is equal to ΔH. If reactants and products are in the *standard state* (1 atmosphere pressure for gases, unit concentration for solutes, and 25°C) we use the symbol ΔH^0 to denote the *standard* enthalpy change. The superscript zero denotes standard conditions for any other thermodynamic quantity as well.

Consider the process

$$CH_4 + 2O_2 = CO_2 + 2H_2O$$
$$\Delta H^0 = -213,000 \text{ calories/mole}$$

(1)

When one mole of methane is burned it releases 213,000 calories. The negative sign of ΔH^0 thus signifies that the system is *losing* heat to the surroundings. A positive sign means that energy is *absorbed* from the surroundings. The calorie is a unit of energy, defined as the quantity of heat needed to raise the temperature of one gram of water from 14.5 to 15.5°C. If we consider the reaction in Eq. (1) going in the reverse direction, the result is Eq. (2):

$$CO_2 + 2H_2O = CH_4 + 2O_2$$
$$\Delta H^0 = +213,000 \text{ calories/mole}$$

(2)

The First Law of Thermodynamics does not allow us to determine whether the reaction in Eq. (1) or Eq. (2) proceeds spontaneously. Experience suggests that the process given in Eq. (1) is the one that proceeds spontaneously, yet there is nothing in the First Law that tells us this. To predict in which direction a process will proceed spontaneously, we need a new quantity, the **entropy**, S.

Entropy is a measure of disorder in a system. Body heat, for example, represents in part the creation of entropy as a result of our life processes. Body heat causes nearby gas molecules to gain kinetic energy, thus becoming more disordered. The creation of entropy is the price that must be paid for a real process to occur with the production of useful work. Therefore, we can view entropy created as a result of a real process as work which is forever lost.

The Second Law of Thermodynamics says that spontaneous processes proceed with a net increase in the entropy of the universe. Using the Second Law, we can tell whether the processes in Eqs. (1) and (2) proceed spontaneously. To do so we have to measure the entropy changes (ΔS) of the system and the surroundings for each process. If the sum of the two entropy changes is greater than zero, then the universe gains entropy and, by the Second Law, the process is spontaneous. It is not difficult to measure ΔS of the system, but measuring ΔS of the surroundings is hard to do experimentally. Because of the difficulty in measuring *all* the effects of a process on the surroundings, it is much more convenient to define spontaneity in terms of a quantity related only to the system. We call this quantity **free energy** and give it the symbol G.

The free-energy change ΔG of a reaction is a measure of the maximum work obtainable from a given process; therefore, ΔG is a measure of the *driving force* behind a chemical reaction. The free-energy change is defined by the important relation

$$\Delta G = \Delta H - T\Delta S$$

(3)

where T is the temperature of the reaction. Eq. (3) defines the two factors that determine the net driving force for a given reaction:

1. ΔH: the change in internal energy at constant pressure.

2. $T\Delta S$: the contribution due to the change in entropy.

A spontaneous process is characterized by a *loss* of free energy; that is, ΔG is negative. A process at equilibrium has no net driving force for change, thus ΔG is zero.

When reactants and products are in the standard state, the free-energy change is termed ΔG^0, in accordance with the convention stated earlier. However, in dealing with biochemical processes, it is convenient to modify our definition of standard state somewhat. Ordinarily, the standard state for all dissolved substances is unit concentration: 1 mole per liter for solutes which do not physically interact with other substances in the solution. Since the concentration of H_3O^+ under physiological conditions is about 1×10^{-7} mole per liter, it is convenient to define a "physiological standard state" in which the concentration of H_3O^+ is $10^{-7} M$, all other solutes are at unit concentration, all gases are at 1 atmosphere pressure, and temperature is 25°C, as before. In this case we designate the (physiological) standard free-energy change as $\Delta G^{0\prime}$.

Free energy and chemical equilibrium are closely related. Consider the process

$$\text{L-aspartate} \rightleftharpoons \text{fumarate} + \text{NH}_4^+ \tag{4}$$
$$\Delta G^0 = +3580 \text{ calories/mole}$$

As written, this reaction is not spontaneous if one mole of each of the three reactants and products are mixed together in 1 liter of water at 25°C. In fact, under these conditions fumarate ion would react spontaneously with ammonium ion to produce L-aspartate. The equilibrium constant expression for this reaction is

$$K_{eq} = \frac{[\text{fumarate}] [\text{NH}_4^+]}{[\text{L-aspartate}]}$$

If the ratio of the concentrations of products to reactants is *less* than the value of K_{eq}, the system proceeds to form products until sufficient reactant has been consumed and products have been formed for the concentration ratio to reach the value of K_{eq}. When chemical change ceases to occur, the system is said to be at equilibrium. The important relationship between ΔG^0 and K_{eq} is

$$\Delta G^0 = -2.3 \, RT \log(K_{eq}) \tag{5}$$

where $R =$ the gas constant $= 1.99$ cal/(mole-deg Kelvin) and the temperature is given in degrees Kelvin. Using this relationship for the reaction in Eq. (4), we obtain a value of $K_{eq} = 2.5 \times 10^{-3} M$. Since the equilibrium constant is less than 1, the equilibrium in Eq. (4) favors reactants—the reaction is not spontaneous. The relation between ΔG and equilibrium is summarized below:

Equilibrium and the Relationship between K_{eq} *and* ΔG^0			
	ΔG^0	K_{eq}	*Position of equilibrium*
Exergonic	negative	>1	Products favored at equilibrium
Endergonic	positive	<1	Reactants favored at equilibrium

The value of ΔG is strongly influenced by factors such as the concentrations of reactants and of products. The concentration dependence of ΔG for a chemical process is given by Eq. (6). For the generalized reaction

$$a\text{A} + b\text{B} \rightleftharpoons c\text{C} + d\text{D},$$

$$\Delta G = \Delta G^0 + 2.3 \, RT \log\left(\frac{[\text{C}]^c[\text{D}]^d}{[\text{A}]^a[\text{B}]^b}\right) \tag{6}$$

A quick examination of Eq. (6) will be instructive. When the system is in the standard state, the concentrations of reactants and products are $1M$, and Eq. (6) simply becomes $\Delta G = \Delta G^0$. When the ratio of the concentrations of reactants and products is equal to the value of K_{eq}, the system is in a state of rest, or equilibrium; there is *no driving force for change* under the prevailing conditions of T and P. At equilibrium, $\Delta G = 0$ and Eq. (6) becomes identical to Eq. (5). If we add additional reactant or product to the system at equilibrium, *a change must occur*, as predicted by Eq. (6). This observation is the basis for an important concept known as **Le Châtelier's Principle**, which states that **if a system at equilibrium is perturbed, then the system will react or change to seek a new equilibrium position that minimizes the effect of the perturbation**.

Figure 7.4 illustrates Le Châtelier's Principle. The perturbation shown in the figure is analogous to removing one of the products of a chemical system at equilibrium. For example, if we remove fumarate ion from the system given in Eq. (4) (perhaps by reacting it with something else), then additional reactant *must react* to form more products until the system is again at equilibrium. We can therefore say that removal of products "drives the equilibrium to the right." As we will observe in later chapters, this application of Le Châtelier's Principle is a common occurrence in the reactions of intermediary metabolism.

A final important point in our review of thermodynamics and equilibrium is that free-energy changes are additive in coupled chemical reactions where the product of one reaction is the reactant of another. A good example is the

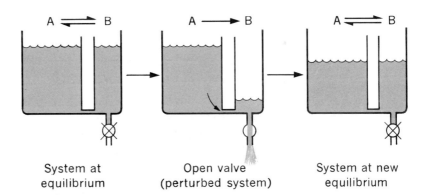

Figure 7.4

Illustration of
Le Châtelier's Principle.

System at
equilibrium

Open valve
(perturbed system)

System at new
equilibrium

set of reactions of the Krebs Tricarboxylic Acid Cycle (see Chapter 9), which is catalyzed by the enzyme *aconitase*:

Reaction	$\Delta G^{0\prime}$
citrate $\rightleftharpoons$ *cis*-aconitate + H_2O	+2.05 kcal/mole
cis-aconitate + H_2O $\rightleftharpoons$ isocitrate	−0.50 kcal/mole
overall: citrate $\rightleftharpoons$ isocitrate	+1.55 kcal/mole

For a metabolic pathway like the one shown below involving a series of sequential chemical reactions,

$$A \overset{1}{\rightleftharpoons} B \overset{2}{\rightleftharpoons} C \overset{3}{\rightleftharpoons} D \overset{4}{\rightleftharpoons} E \overset{5}{\rightleftharpoons} \text{Product}$$

Overall reaction

the overall free-energy change is equal to the algebraic sum of the $\Delta G^{0\prime}$ values of the separate steps:

$$\Delta G^{0\prime}_{\text{Overall}} = \Delta G^{0\prime}_1 + \Delta G^{0\prime}_2 + \Delta G^{0\prime}_3 + \Delta G^{0\prime}_4 + \Delta G^{0\prime}_5$$

The overall equilibrium $A \rightleftharpoons$ Product will favor product if the algebraic sum of the standard free-energy values for the separate steps is negative ($\Delta G^{0\prime}_{\text{overall}} < 0$), even though some of the individual steps might be endergonic ($\Delta G^{0\prime} > 0$).

This completes our brief review of thermodynamics. We will find many applications of these basic concepts concerning free energy and equilibrium. We are now ready to discuss some of the basic features of cellular energetics.

7.5

Adenosine Triphosphate

Cellular energy transfers other than oxidation-reduction reactions utilize the energy stored in the P—O bonds of phosphoric acid anhydride and phosphate ester derivatives. The central biomolecule in this energy transfer system is adenosine triphosphate (ATP). The structures of ATP and related substances are shown in Figure 7.5.

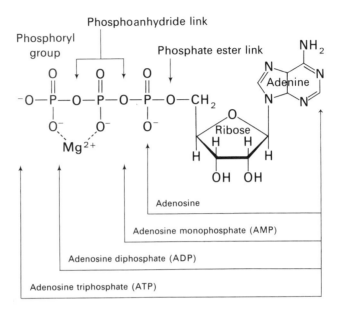

Figure 7.5

Structures of adenosine triphosphate and related substances. The ionic form shown is the predominant form at pH 7. ATP acts to donate a phosphoryl group to a phosphoryl group acceptor. In the cell, ATP and ADP are not present in the free form, but rather as a complex with Mg^{2+} ion, as represented here.

ATP and Cellular Energetics

ATP serves as a central "medium of exchange" that links energy-yielding biochemical reactions with energy-requiring processes in the cell. The action of ATP centers on its ability to become a high-energy *phosphoryl-group donor* to a phosphoryl-group acceptor such as an R—OH type of compound. The hydrolysis reaction between ATP and HOH is representative of a phosphoryl-group donor-acceptor reaction:

$$ATP + HOH \rightleftharpoons ADP + \begin{Bmatrix} H_3PO_4 \\ H_2PO_4^- \\ HPO_4^{2-} \\ PO_4^{3-} \end{Bmatrix} \equiv P_i \qquad (7)$$

$\Delta G^{0'} = -7.3$ kcal/mole

The reaction shown in Eq. (7) produces adenosine diphosphate (ADP) and an equilibrium mixture of phosphate ion species (the predominant forms at pH 7 are shown in color). Rather than specify the relative amounts of the four phosphate species, we will use the symbol P_i to designate the equilibrium mixture of free phosphate-ion forms. The free-energy change for this reaction is negative in the standard state, indicating a large driving force behind this reaction under standard conditions.

The relationship among ATP, ADP, and energy-generating and energy-consuming processes is summarized in Figure 7.6. In all cases, energy-requiring

Figure 7.6

The relationship among ATP, ADP, and metabolism. ATP acts as the medium of exchange for chemical energy in the cell. Catabolic processes replenish the supply of ATP in the cell by driving the rephosphorylation of ADP.

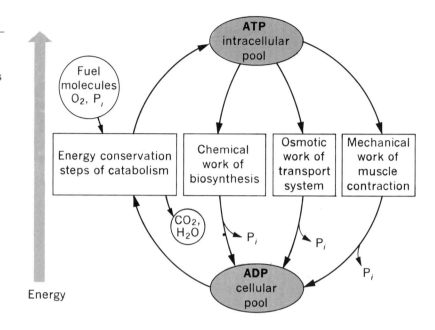

processes deplete the cellular pool of ATP while energy-generating processes replenish the pool. Replenishment always occurs by the rephosphorylation of ADP:

$$ADP + P_i \rightleftharpoons ATP + H_2O \qquad \Delta G^{0\prime} = +7.3 \text{ kcal/mole}$$

Note that this reaction is simply the reverse of Eq. (7); therefore, the free-energy change is equal but opposite in sign. When AMP is produced, it is not rephosphorylated directly to ATP, but instead takes two steps:

$$AMP + ATP \rightleftharpoons 2\,ADP$$

$$2\,ADP + 2\,(P_i \text{ donor}) \rightleftharpoons 2\,ATP$$

The actual concentrations of ATP and ADP within a cell vary depending on the organelle or location within the cell's internal structure. To get a rough idea of the cellular concentrations of ATP and ADP, we can cite the concentrations of ATP, ADP, and P_i in the human red blood cell:

$$[ATP] = 1.4 \times 10^{-3} M;$$

$$[ADP] = 0.15 \times 10^{-3} M;$$

$$[P_i] = 1.0 \times 10^{-3} M$$

In some cases, a molecule becomes activated for further reaction by *adenylation* rather than by phosphorylation. In protein biosynthesis, amino acids are activated as a first step in the formation of a peptide link by the adenylation reaction:

$$\text{amino acid} + \text{ATP} \rightleftharpoons (\text{AMP-amino acid}) + \;^-\text{O}-\overset{\overset{\displaystyle O}{\|}}{\underset{\underset{\displaystyle \;^-\text{O}}{|}}{\text{P}}}-\text{O}-\overset{\overset{\displaystyle O}{\|}}{\underset{\underset{\displaystyle \;^-\text{O}}{|}}{\text{P}}}-\text{O}^- \equiv \text{PP}_i$$

Pyrophosphate
anion

We will discuss this process in more detail in Chapter 13.

As Figure 7.6 illustrates, ATP is a central species in cellular energetics. Energy is conserved in the enzymatic *coupling* of an exergonic step in catabolism with the phosphorylation of ADP. An example of this is given in Eq. (8), which shows the basic features of an energy-conservation step in glycolysis (see Chapter 8):

$$\Delta G^{0\prime} \text{ (kcal/mole)}$$

I. $\text{CH}_2{=}\underset{\underset{\displaystyle \text{OPO}_3^{2-}}{|}}{\text{C}}{-}\text{CO}_2^- + \text{H}_2\text{O} \rightleftharpoons \text{CH}_3{-}\underset{\underset{\displaystyle O}{\|}}{\text{C}}{-}\text{CO}_2^- + \text{P}_i$ $\qquad -14.8$

Phosphoenol pyruvate $\qquad\qquad$ Pyruvate

II. $\text{ADP} + \text{P}_i \rightleftharpoons \text{ATP} + \text{H}_2\text{O}$ $\qquad +7.3$

Net: phosphoenol pyruvate + ADP $\rightleftharpoons$ pyruvate + ATP $\qquad -7.5$

$$(8)$$

The overall driving force for the transfer of a phosphoryl group from phosphoenol pyruvate (phosphoryl-group donor) to ADP (phosphoryl-group acceptor) is strongly exergonic. The excess free energy is liberated as heat. In a cell, the overall reaction of Eq. (8) is catalyzed by the enzyme *pyruvate kinase*. (Remember from Chapter 3 that a kinase is an enzyme that catalyzes the transfer of a phosphoryl group from a phosphoryl-group donor to a phosphoryl-group acceptor.) The cell can draw upon the ATP produced by reactions such as Eq. (8) to provide a source of chemical work.

ATP as a Medium of Exchange for Phosphate-Bond Energy

We can compare the energy content of various phosphate compounds occurring in living systems on the basis of their free energies of hydrolysis [see

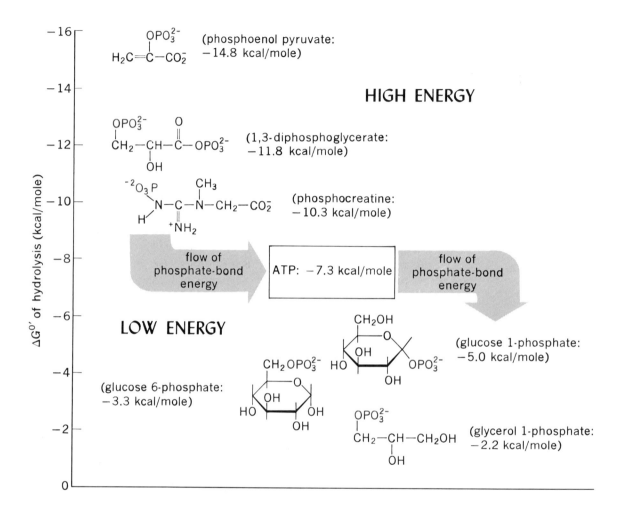

Figure 7.7

The relationship of ATP and some other phosphate compounds of biological importance. The intermediate value for the free energy of hydrolysis of ATP is consistent with the fact that phosphate-bond energy transfers from high-energy to low-energy phosphate compounds through ATP.

Eq. (7)]. Figure 7.7 compares some important phosphate compounds with ATP. Note that the free energy of hydrolysis of ATP is roughly midway between those of the high-energy and low-energy phosphate compounds. In addition to the molecules shown in Figure 7.7, Table 7.1 lists some biologically important phosphate compounds and their free energies of hydrolysis.

The direct phosphorylation of a low-energy phosphoryl-group acceptor such as glucose by a high-energy phosphoryl-group donor such as phosphoenol pyruvate has not been observed in living systems. For such a process to occur, two steps must take place:

phosphoenol pyruvate + ADP $\longrightarrow$ pyruvate + ATP

ATP + glucose $\longrightarrow$ glucose 1-phosphate + ADP

Table 7.1   Free Energy of Hydrolysis for Some Important Phosphate Compounds. Although the transfer of a phosphoryl group to HOH is but one of many possible types of phosphoryl-group transfers encountered in metabolism, use of the standard free energy of hydrolysis lets us compare the energy content of different phosphoryl donors using a common point of reference. (See also Problem **5** at the end of this chapter.)

Compound and reaction	$\Delta G^{\circ\prime}_{\text{hydrolysis}}$ (kcal/mole)
Phosphoenol pyruvate + HOH = pyruvate + P_i	−14.8
1,3-diphosphoglycerate + HOH = 3-phosphoglycerate + P_i	−11.8
Phosphocreatine + HOH = creatine + P_i	−10.3
Acetyl phosphate + HOH = acetate + P_i	−10.1
PP_i (pyrophosphate) + HOH = $2P_i$	−8.0
ATP + HOH = AMP + PP_i	−7.6
ATP + HOH = ADP + P_i	**−7.3**
Glucose 1-phosphate + HOH = glucose + P_i	−5.0
Fructose 6-phosphate + HOH = fructose + P_i	−3.8
Glucose 6-phosphate + HOH = glucose + P_i	−3.3
Glycerol 1-phosphate + HOH = glycerol + P_i	−2.2

This may seem like an indirect process, but we must remember that phosphoryl-group transfers, like other chemical reactions in living systems, are enzyme-catalyzed. Since a large part of the chemical work done by a cell is for protein biosynthesis, living cells have evolved a system that minimizes the number of different enzymes required for efficient metabolism. Figure 7.8 shows how the use of ATP as a medium of exchange for phosphate-bond energy is consistent with this overall efficiency.

Figure 7.8

An illustration of the efficiency gained by the use of ATP as an intermediary in phosphoryl-group transfers from high-energy donors to low-energy acceptors. Each arrow represents a phosphoryl-group transfer catalyzed by a unique enzyme. The hypothetical case of direct transfer would require many more enzymes than in the actual case, where ATP acts as an intermediary.

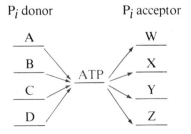

Hypothetical case

Direct phosphorylation of a low-energy acceptor by a high-energy donor

P_i donor P_i acceptor

A   W
B   X
C   Y
D   Z

16 different enzymes

Actual case

Phosphoryl-group transfer via ATP

P_i donor P_i acceptor

A   W
B   X
      ATP
C   Y
D   Z

8 different enzymes

Driving Force for Biochemical Processes

In Chapter 1 we said that nature relies heavily on polymeric molecules with recurring structural linkages, such as the peptide link in proteins and the glycoside link in polysaccharides. Both the peptide bond and the glycoside linkage are *thermodynamically unstable* in H_2O:

$$\Delta G^{0'} = -0.5 \text{ kcal/mole}$$

$$\Delta G^{0'} = -2.0 \text{ kcal/mole}$$

The negative free energy of hydrolysis for these processes means that the *formation* of these linkages in an aqueous environment is thermodynamically unfavorable.

To illustrate how ATP can drive an endergonic process, let's consider the formation of maltose from two glucose molecules:

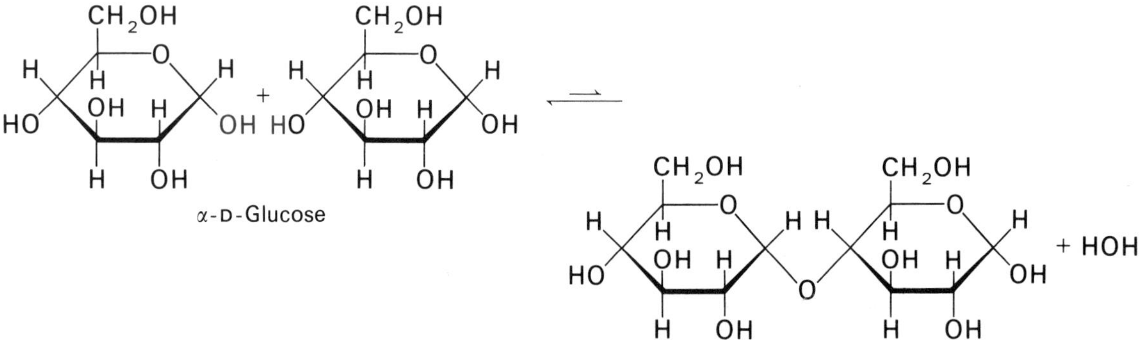

Maltose

$$\Delta G^{0'} = +2.0 \text{ kcal/mole}$$

The standard free-energy change corresponds to a value of $K_{eq} = 3.5 \times 10^{-2}$. This means that at equilibrium, about 5 moles of glucose are present for every mole of maltose—not an efficient conversion. However, we can make the yield much better by coupling the formation of maltose to the hydrolysis of ATP, as shown at the top of the next page. The energetics are vastly improved and the value of $K_{eq} = 1.0 \times 10^4$ for the coupled reaction means that the overall formation of maltose proceeds with better than 99.99% yield.

In any set of chemical processes *coupled by a common intermediate*, we can sum up the free-energy changes of each step in the overall process, thus giving us the overall free-energy change for the coupled reaction. Therefore, an

α-D-Glucose + ATP ⇌ α-D-Glucose 1-phosphate + ADP $\Delta G^{0\prime}$ = −2.5 kcal/mole

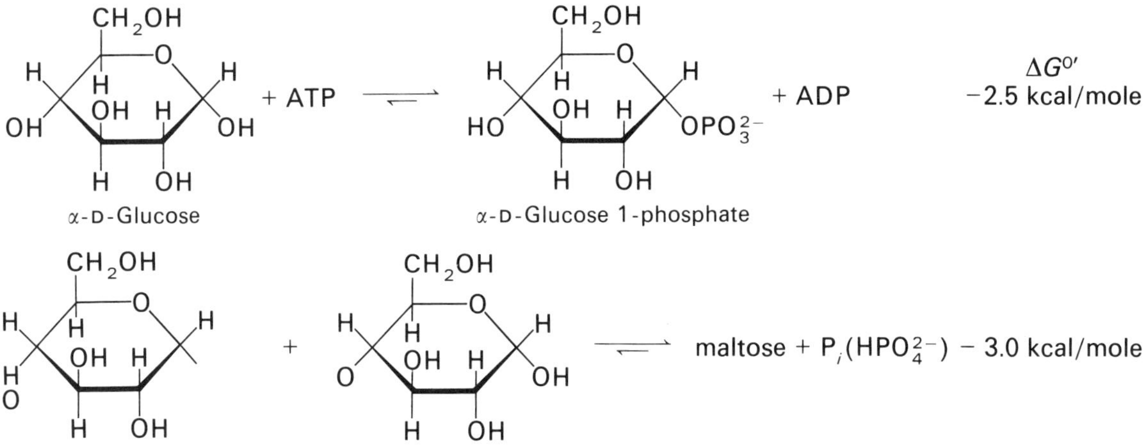

⇌ maltose + P_i (HPO_4^{2-}) − 3.0 kcal/mole

net: 2 glucose + ATP ⇌ maltose + P_i + ADP   −5.5 kcal/mole

exergonic reaction can drive an endergonic reaction if the two processes are coupled through a common intermediate. In the above example, the common intermediate is glucose 1-phosphate.

7.6

Overview of Catabolism

In the following chapters we will discuss the sequences of enzyme-catalyzed reactions that replenish the intracellular pool of ATP. To accomplish this replenishment, these processes must produce high-energy phosphate compounds, permitting the phosphorylation of ADP. We will consider catabolism in a general way, without specific reference to any individual organism. We can do this because the general form of intermediary metabolism is common to almost all organisms. Because of its importance, we will consider numerous applications of intermediary metabolism to human biochemistry and physiology throughout the text. While we will deal mainly with the metabolism of heterotrophic cells, in Chapter 10 we will consider photosynthesis, the primary energy source for most autotrophs.

Figure 7.9 provides an overview of the processes we will discuss in Chapters 8–11. You should keep in mind that the extent to which the cell uses each of the primary food sources (lipid, carbohydrate, and protein) remains under close control.

The processes shown in Figure 7.9 are closely interrelated. It is important to remember the interplay of one metabolic pathway with other pathways as you learn about each separately. The chart of intermediary metabolism at the

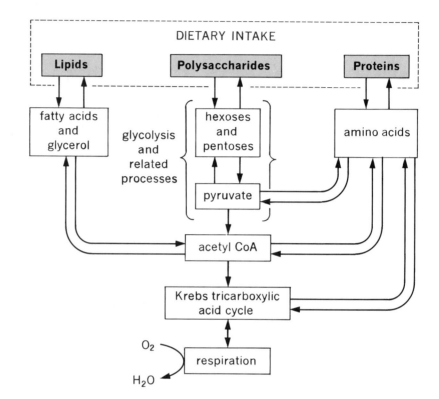

Figure 7.9

A general overview of metabolism. The arrows indicate relationships by chemical-reaction sequences. As the cell breaks down the original food molecules, some of the energy produced is conserved by the formation of ATP. The remainder of the energy is liberated to the environment as heat.

back of the text will help you appreciate these interrelationships. You should refer to the chart regularly as you study the following chapters so that your view of metabolism does not become a meaningless jumble of individual steps.

Before considering what happens to metabolites in a cell, it is worth asking how the metabolite came to be in the cell in the first place. We will do this on two levels. First, we will consider carbohydrate, lipid and amino acid digestion and absorption in humans in Section 7.7. Then we will deal with the more general question of how cells absorb metabolites on the molecular level by discussing transport processes in Section 7.8.

7.7

Digestion

Digestion in higher animals is basically a hydrolytic process. The large molecules such as polysaccharides, proteins, nucleic acids and lipids which constitute the bulk of the diet must be broken down into small molecules in order to permit their absorption across the lining of the small intestine.

Carbohydrate Digestion and Absorption

The breakdown of dietary polysaccharides in humans begins in the mouth, where salivary α-amylase hydrolyzes some of the α-1 → 4 glycoside links. Little further breakdown of carbohydrates occurs until the food reaches the small intestine, where the major part of starch digestion takes place. Here pancreatic α-amylase catalyzes the hydrolysis of α-1 → 4 linked polysaccharides into primarily glucose and maltose (see Figure 4.7, page 153). Where branching occurs as in amylopectin or glycogen, isomaltose, an α-1 → 6 linked disaccharide, is also produced. In addition to the disaccharides created by the action of α-amylase, other dietary disaccharides include lactose (from milk) and sucrose. Disaccharides are broken down by specific enzymes located on the large enfolded surface formed by the cells lining the small intestine. This surface has been termed the "brush border" of the intestinal mucosa and represents an enormous area for efficient absorption of digested food. The enzymes associated with carbohydrate digestion are given in Table 7.2.

Table 7.2 Enzymes Required for the Complete Digestion of Carbohydrates.

Enzyme	Substrate	Major products
α-Amylase	α-1 → 4 linked polysaccharides	Glucose, maltose and isomaltose
Maltase I and II	Maltose	Glucose
Maltase III and IV	Maltose and sucrose	Glucose and fructose
Maltase V	α-1 → 6 linked disaccharides	Glucose (from isomaltose)
		Glucose and fructose (from palatinose)
β-Galactosidases	β-linked oligosaccharides	A wide variety, depending on the substrate
Lactase	Lactose	Galactose and glucose

Lactose digestion is unusual because lactose is found only in milk. In most mammals, the digestive enzyme *lactase* present in infants disappears after weaning. A hereditary lack of this enzyme in infants can lead to serious gastroenteritis because the undigested and unabsorbed lactose promotes the growth of undesirable intestinal bacteria. Such infants can be placed on a milk-free diet and will eventually outgrow the effects of this deficiency.

Galactose, glucose and fructose are absorbed through the intestinal lining and carried via portal circulation to the liver. Figure 7.10 summarizes the major transformations of these sugars once they are in the liver. The liver is a key organ for regulating the amount of glucose in the blood. After a meal, glucose is absorbed and blood sugar level rises. This increase is countered by an increase in the rate of the processes that remove glucose from the blood: storage, as

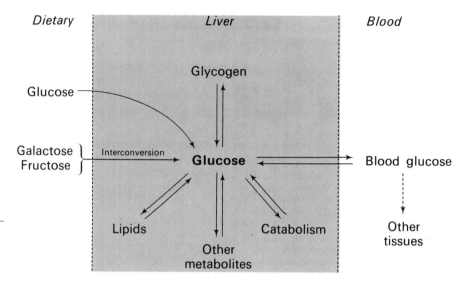

Figure 7.10

Major transformations
of dietary
monosaccharides.

glycogen or lipids, and catabolism, which generates ATP. Therefore, control
of the concentration of glucose in the blood depends on the balance among
intake, storage and catabolism. Any defects in this regulation can be detected
clinically by administering glucose and then measuring the blood sugar level
over a period of several hours. This produces a graph called a *glucose tolerance
curve*. Figure 7.11 shows glucose tolerance curves for both a normal person and
one with *diabetes mellitus*. In the case of a diabetic, blood sugar level can
exceed the renal threshold and cause glucose to be excreted directly into the
urine. Because the presence of glucose can be detected easily (see Section 3.10),
urinalysis for blood sugar has become a routine part of most physical exams.

Figure 7.11

Glucose tolerance curves
for a normal and a
diabetic individual.
These are idealized
curves because the actual
curves depend on factors
such as age and weight.
Ordinarily, data is
obtained using the
following test: 100 grams
of glucose in the form
of a fruit-flavored drink
are ingested and samples
of blood are taken and
analyzed for glucose as a
function of time. The
renal threshold for any
metabolite is the
concentration limit
above which it is passed
directly into the urine.

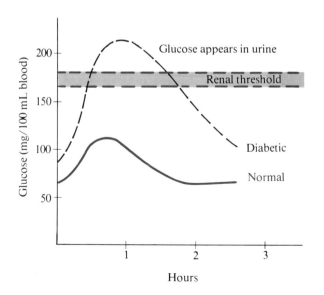

The large overshoot in blood glucose level in a diabetic is caused by an impairment of the ability to absorb glucose from the blood. The protein hormone insulin is required for normal absorption of glucose from the blood by body tissues and is thus a key part of the regulatory apparatus for blood glucose. In diabetes mellitus, production of insulin is generally impaired. Without the hormone, elevated glucose levels occur. As we shall see in Chapter 11, this causes not only an imbalance in carbohydrate metabolism but an imbalance in lipid metabolism as well.

Lipid Digestion and Absorption

We have already discussed lipid digestion and absorption in some detail in Chapter 6. There we learned how certain blood plasma lipoproteins are required for lipid transport in the blood (p. 193). The important plasma lipoproteins are summarized in Table 7.3.

Table 7.3 Plasma Lipoproteins.

Plasma lipoprotein	Transports	% Protein
Chylomicrons	Triglycerides	1
Very Low Density (VLDL)	Triglycerides	10
Low Density (LDL)	Cholesterol esters	25
High Density (HDL)	Phospholipids, cholesterol esters	50

In a normal person, ingestion of triglycerides causes an increase in blood triglyceride levels followed by a decrease to a fasting level as the excess is either catabolized or made into adipose tissue. This is shown in Figure 7.12. The figure shows the curve for both a normal and an abnormal case. Lipid storage and transport abnormalities are not uncommon. *Hyperlipidemia*, or excessive

Figure 7.12

Variation of blood triglyceride level as a function of time, following the ingestion of dietary triglycerides. The abnormal curve corresponds to a condition of hyperlipidemia.

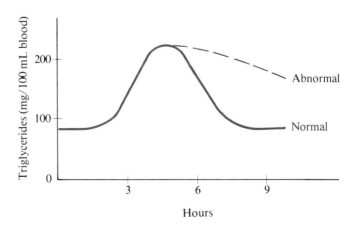

blood lipids, often arises from increased concentrations of one or more of the plasma lipoproteins. Elevated levels of low-density lipoproteins can lead to large amounts of cholesterol in the blood. This can be due to either hereditary or dietary factors, or both. The excess cholesterol tends to be deposited on the walls of arteries. Over a period of time these lipid deposits combine with collagen to form a hard plaque. Such deposits can lead to blood clot formation and interference with normal circulation. In severe cases, a total blockage or thrombosis at a given point can occur, shutting off blood supplied to tissues downstream from the blockage. The tissues, deprived of oxygen and nutrients, degenerate and die.

Protein Digestion and Absorption

Figure 7.13 summarizes the transformations of dietary proteins in the body. Protein digestion begins in the stomach, where the proteolytic enzyme *pepsin* catalyzes the hydrolysis of peptide links in proteins to yield a mixture of polypeptides. The acidic contents of the stomach are neutralized in the alkaline medium of the small intestine. Here, polypeptides are acted upon by trypsin, chymotrypsin, carboxypeptidases and elastase. These enzymes break down the dietary proteins and polypeptides into small oligopeptide fragments and free amino acids. Proteolytic enzymes in the cells of the intestinal lining complete the hydrolysis of the short oligopeptides into amino acids.

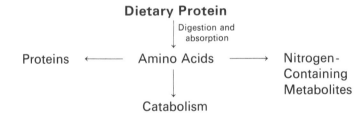

Figure 7.13

Major transformations of dietary protein.

Once absorbed, amino acids are carried principally to the liver and, to a lesser extent, to the kidneys. The liver is the major location for amino acid metabolism (see Chapters 11 and 12) and is responsible for:

1. synthesis of purines, pyrimidines, uric acid, urea and other nitrogen-containing compounds;

2. regulation of the concentrations of amino acids in the blood which are used by other tissues;

3. amino acid biosynthesis and catabolism.

In normal adults, the intake of nitrogen compounds is balanced by their excretion. This is a dynamic equilibrium in which proteins in the body are constantly being synthesized and broken down. This "turnover" of protein in a normal adult male is about 400 grams per day. Infants and children must produce new tissue and therefore take in more nitrogen (as protein) than is excreted. This is the state of *positive nitrogen balance*. In extreme old age and cases of starvation a breakdown of body protein to supply energy results in a *negative nitrogen balance* in which more nitrogen compounds are excreted than taken in.

7.8

Entry of Metabolites into the Cell: Transport

The movement of metabolites across the membranes of the cell is a vital process in all living cells. Since lipid membranes are impermeable to polar molecules and ions, mechanisms must exist to provide a means for the *translocation* or movement of specific metabolites across a membrane barrier.

We can define two basic types of transport processes: **active transport** and **passive transport**, as shown in Figure 7.14. The main feature distinguishing active from passive transport is that passive transport requires no energy, while active transport needs energy to drive the endergonic process. In either case, we can observe a high degree of specificity for transport processes in living systems.

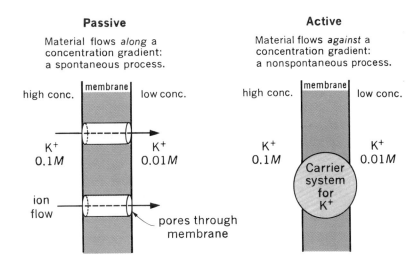

Figure 7.14

Characteristics of active and passive transport across a cell membrane.

This means that, like enzyme-substrate interactions, active and passive transport in living cells involve a specific interaction between a carrier substance in the cell membrane and the metabolite transported across the membrane. This concept is presented schematically in Figure 7.15, which illustrates the essential features of any transport process.

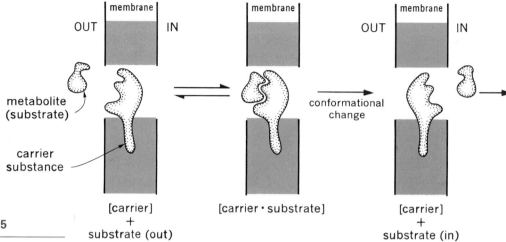

Figure 7.15

Carrier-mediated
transport. The carrier
substance can change
shape upon the binding
of substrate, proceeding
either spontaneously
(passive transport) or by
a process coupled to an
energy source (active
transport).

The driving force across the membrane can be either a negative concentration gradient in the case of passive transport, or an energy-requiring mechanism in the case of active transport. As an example of an energy-requiring mechanism, a change in the shape of the carrier substance may permit passage of the metabolite across the membrane. In either case, binding forms a crucial part of the transport process in cells, since it is here that a carrier substance's specificity for a given metabolite originates (see Figure 7.15). *Saturation* is a characteristic of *carrier-mediated* transport processes, just as it is of enzyme reactions. This state is indicated by the leveling off of the curve in Figure 7.16.

We can summarize the essential features of a biological transport process in three steps:

1. **Binding** The binding of the metabolite molecule being transported across the cell or mitochondrial membrane involves a specific interaction between the metabolite and a *receptor site* of the carrier substance.

Figure 7.16

Kinetics of
carrier-mediated
transport. The similarity
between this curve and
the velocity/substrate
curve for an
enzyme-catalyzed
reaction (Figure 3.13,
p. 110) is meaningful
because substrate
binding is the crucial
step in both systems.

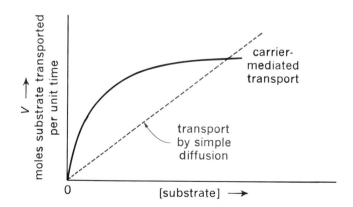

Researchers have identified several metabolite *binding proteins* in various cell membranes, and believe them to be part of the specific transport system they are associated with.

2. **Translocation** Somehow, a conformational change occurs which permits the transfer of *bound* metabolite across the membrane barrier. Biological transport systems are *vectorial* in nature, meaning that the flow of metabolite via a given transport system is unidirectional. Therefore, the process of translocation must also cause a great decrease in the affinity of the receptor site for the metabolite, readily releasing the metabolite inside the membrane.

3. **Energy coupling** There are two general processes that drive transport systems. The first is simply the transfer of solute from a region of high concentration to a region of lower concentration. This is termed passive transport, or facilitated diffusion. The second method couples a metabolic energy source to an active transport process, carrying a solute from a region of low concentration to one of higher concentration inside the cell. The means by which energy coupling in active transport occurs are not well understood.

Transport is an important property in enabling a cell to precisely control its internal environment, irrespective of changes which may occur outside. For example, in mammals, the ionic composition of intracellular fluid differs considerably from that of extracellular fluid. This difference is maintained by many active transport systems functioning to keep the proper solute balance. The principal ions involved are shown in Figure 7.17. The most important thing to note in this figure is the large difference in concentration of ionic solutes inside and outside the cell. Since Na^+ ion can "leak" into the cell or be carried in along with a neutral metabolite in an "ion-coupled" transport process, the cell must continuously "pump" Na^+ out of the cell. Similarly, K^+ ion is constantly "pumped" into the cell.

In bacteria, active transport is essential to the existence of the organism. Inside the bacterial cell, the concentration of metabolites is tightly controlled, but the bacterium must exist in a highly variable and usually dilute solution of nutrients over which it has no control. Therefore, active transport processes enable the bacterial cell to accumulate needed substances from the environment. The bacterial cell devotes a considerable fraction of its metabolic energy resources to do this, up to 30% of its total energy output.

We will not discuss in any detail what researchers have learned about various transport systems. However, you may wish to consult the suggested references at the end of the chapter for further information. What is important to remember is that the entry of a metabolite into a cell or a subcellular organelle takes place in a highly controlled and specific manner by one of the general mechanisms discussed.

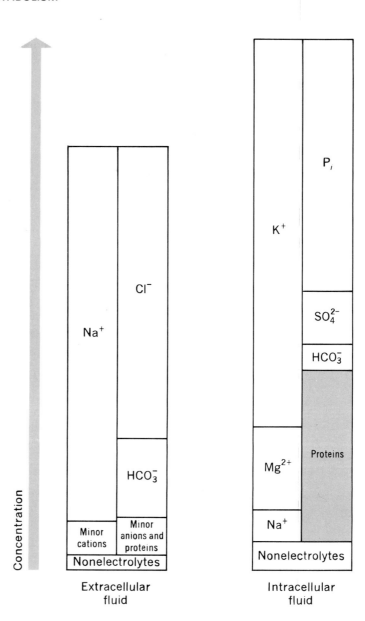

Figure 7.17

Relative solute
concentrations of
intracellular and
extracellular fluids in
human skeletal muscle.

7.9

Blood

About 8% of your body weight is blood. It is composed of a liquid portion
(plasma) and cellular components: erythrocytes (red blood cells), leukocytes
(white blood cells) and platelets (thrombocytes). Its primary function is to trans-

Table 7.4 Major components of blood plasma, along with their approximate concentrations. The actual amount of a given substance depends on such factors as age, sex, and physiological state.

Component	Normal concentration range (mg/100 mL)	Component	Normal concentration range (mg/100 mL)
Nitrogen compounds:		*Lipids:*	
Urea	20–30	total of all lipids	285–675
Amino acids	35–65	Triglycerides	80–240
Bilirubin	0.2–1.4	total cholesterol	130–260
Creatine	0.2–0.9		
Creatinine	1–2	*Inorganic components:**	
Uric acid	2–6	*Anions*	
Plasma proteins	5700–8000	total	142–150
Carbohydrates:		HCO_3^-	24–30
		Cl^-	100–110
Glucose	65–90	Phosphate	1.6–2.7
Fructose	6–8		
Pentoses	2–4	*Cations*	
Polysaccharides	70–105	total	142–158
Organic acids:		Na^+	132–150
		Ca^{2+}	4.5–5.6
Lactic acid	8–17	Mg^{2+}	1.6–2.2
Citric acid	1.4–3.0	K^+	3.8–5.4
Pyruvic acid	0.4–2.0		
Others	2–5		

* Concentrations of inorganic components are given in milliequivalents/liter, normality × 1/1000.

port oxygen and metabolites to cells and to transport CO_2 and waste products away. Blood is a complex mixture of many substances. Table 7.4 summarizes some of the major components of blood plasma and their approximate concentrations. As Table 7.4 shows, proteins are a major component of the dissolved material in blood plasma. Because of their importance, we will consider some of these in more detail.

Plasma Proteins

The proteins found in the plasma portion of blood have a variety of interesting properties and functions. Table 7.5 (p. 238) lists the major types of plasma proteins and their functions.

The most abundant protein in blood plasma is *serum albumin*. We have already seen in Chapter 6 how it serves as a transport protein for free fatty acids. Its other crucial function is to maintain the osmotic pressure of the blood relative to that of the tissues.

The γ globulins are the second most abundant of the serum proteins. These proteins act as **antibodies** which form a system of defense against foreign

Table 7.5 Major Types of Blood Plasma Proteins.

Protein	Function
Albumin	Osmotic regulation; fatty acid transport
α_1 Globulins	Family includes high density lipoproteins
α_2 Globulins	Family includes prothrombin (blood clotting) and ceruloplasmin (copper transport)
β Globulins	Family includes very low density lipoproteins and low density lipoproteins and transferrin (iron transport)
Fibrinogen	Blood clotting
γ Globulins	Antibodies

proteins and other **antigens**. Specific γ globulins are formed by the immune system in response to a specific antigen. This is the basis for immunization against tetanus, polio, diphtheria and many other diseases. The success of multiple immunizations proves that there are many different antibodies in the γ globulin fraction of normal blood. Antibodies react with antigens to form an insoluble precipitin complex as shown in Figure 7.18. This deactivates the antigen and permits its removal and breakdown by the white blood cells.

Blood clotting requires *fibrinogen*, another major protein of blood plasma. Fibrinogen is a soluble protein which is converted into an insoluble fibrous polymer called *fibrin* as a result of trauma, in order to prevent blood loss. The

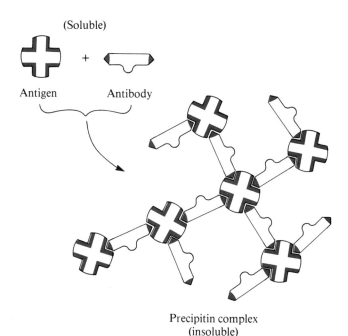

Figure 7.18

The action of an antibody and its specific antigen to form an inactive precipitin complex. The precipitated antigen-antibody complex is ingested and broken down by white blood cells.

major steps of the blood clotting process are given below:

$$\text{Prothrombin (inactive)} \xrightarrow{\text{Thromboplastin, Ca}^{2+}} \text{Thrombin (active)}$$

$$\text{Fibrinogen} \xrightarrow{\text{Thrombin}} \text{Fibrin monomer}$$

$$\text{Fibrin monomer} \longrightarrow \text{Fibrin polymer (soft clot)}$$

$$\text{Fibrin polymer} \xrightarrow{\text{Fibrinase, Ca}^{2+}} \text{Fibrin (hard clot)}$$

Because uncontrolled blood clotting is seriously detrimental, there are at least seven other proteins that are essential to the normal clotting process. These act as controlling factors, primarily in the activation of prothrombin. Ordinarily, blood takes five to eight minutes to form a clot at body temperature. A number of hereditary diseases prevent blood from clotting or lengthen the clotting time. *Hemophilia* is a hereditary disease in which one of three protein factors (depending on the type of hemophilia) involved in the activation of prothrombin is lacking or inactive. Persons with hemophilia hemorrhage as a result of even a minor injury.

Because the liver is the site of prothrombin and fibrinogen biosynthesis, conditions affecting the liver such as hepatitis and chronic alcoholism can interfere with the clotting process. Vitamin K (Section 3.4) is required for prothrombin biosynthesis, and a deficiency of this fat-soluble vitamin can increase clotting time. Calcium ion is also essential to the clotting process and blood will not clot in its absence. On the other hand, blood stored for later use contains an anticoagulant such as citrate ion. Citrate is a non-toxic ion that forms a strong calcium citrate complex, thus preventing free Ca^{2+} from being involved in the clotting process.

7.10

Human Nutrition

In order to maintain a constant weight and good health, an "average" adult male consumes about 1000 pounds of food per year. This food must be a balanced diet consisting of the six primary dietary components:

proteins	vitamins
carbohydrates	minerals
fats	water

The relative amounts of proteins, carbohydrates and fats required for a good diet depend on age, sex, work load and physiological state. For the hypothetical average American, it is recommended that 15% of the calories come from proteins, 40% from fats and 45% from carbohydrates.

The body is a marvelously efficient mechanism for converting food energy into useful cellular and muscular work. As we shall see in Chapter 9, the efficiency for the conversion of food energy to ATP is about 40%, which is exceptionally high compared with non-biological energy production. An average 70 kilogram adult man requires an intake of approximately 1700 kcal/day to maintain life processes and body weight, without considering added activity such as muscular work. This corresponds to a basal metabolic rate (BMR) of 24 kcal/day per kilogram of body weight. Any exercise or other activity requires added food intake. Table 7.6 gives some activities and the approximate additional kcal needed by the average 70 kilogram man.

Table 7.6 Kilocalories of Food Energy Needed in Addition to BMR by an "Average" 70 Kilogram Man Performing Various Activities.

Activity	Added kcal/hr needed
Swimming	550
Running	490
Walking (5 miles/hour)	240
Driving a car	60
Writing (sitting)	30

The caloric content of foods depends on the amount of carbohydrates, fats and proteins they contain. The available calories for these three food components are approximately:

Carbohydrates	4 kcal/gram
Fats	9 kcal/gram
Proteins	4 kcal/gram

For example, one pound of adipose tissue (85% lipid, 15% water) can be catabolized to yield 3500 kcal of energy. Therefore, one must swim for 3500/550 = 6.4 hours to consume the energy in one pound of body fat. This is something to think about when ordering a hot fudge sundae. (See Problem **8** at the end of this chapter.)

In Chapter 3 we learned about the nutritional roles of vitamins and minerals. In Chapters 4 and 6 we discussed the nutritional importance of carbohydrates and lipids, respectively. We will learn more about nutrition in Chapter 11, which describes some of the metabolic consequences of improper diet and the reasons why some amino acids are essential in the diet.

Suggestions for Further Reading

Lehninger, A. L. *Biochemistry*. 2d ed. New York: Worth Publishers, 1975. This standard comprehensive text does an excellent job discussing bioenergetics in Chapter 15.

White, A.; Handler, P.; Smith, E.; Hill, R. and Lehman, I. *Principles of Biochemistry*. 6th ed. New York: McGraw-Hill, 1978. Chapter 11 of this standard text contains a clear and thorough introduction to metabolism.

Scientific American
Green, D. E. "Enzymes in Teams." September 1949.
Holter, H. "How Things Get into Cells." September 1961.
Lehninger, A. L. "Energy Transformations in the Cell." May 1960.
———. "How Cells Transform Energy." September 1961.
Margaria, R. "Sources of Muscular Energy." March 1972.
Solomon, A. K. "Pumps in the Living Cell." August 1962.
Stumpf, P. K. "ATP." April 1953.

Problems

1. For each of the terms given below, provide a brief definition or explanation.

Active transport	Energy conservation	Le Châtelier's Principle
Antibody	Entropy	Phosphoryl-group donor
Basal Metabolic Rate	Equilibrium constant	Spontaneous process
Carrier system	Exergonic	Steady state
Endergonic	Free energy	

2. Draw out the structure of each of the following molecules:

   (a) ATP   (b) ADP   (c) AMP

3. Using Eq. (6) of this chapter (page 219) and the following data, calculate the actual free-energy change $\Delta G'$ for the hydrolysis of ATP in a human erythrocyte at 27°C.

   $$ATP = 1.4 \times 10^{-3}M; \quad ADP = 0.14 \times 10^{-3}M; \quad P_i = 1.0 \times 10^{-3}M$$

   (In aqueous systems, the concentration of H_2O is taken as unity: $1M$.)

4. The reaction of glycolysis, catalyzed by the enzyme aldolase, is

   D-fructose 1,6-diphosphate (FDP) $\rightleftharpoons$

   dihydroxyacetone phosphate (DHP) + D-glyceraldehyde 3-phosphate (G3P)

   $$\Delta G^{0\prime} = +5.8 \text{ kcal/mole at 27°C.}$$

   (a) Calculate K_{eq} for this process. [Use Eq. (5), page 218.]
   (b) Show that the bioenergetic picture for this important reaction is not as grim as the standard free energy implies, given that the steady-state concentrations of the intermediates involved (human erythrocytes) are [FDP] $= 3.1 \times 10^{-5}M$; [DHP] $= 1.4 \times 10^{-4}M$; [G3P] $= 1.9 \times 10^{-5}M$. [Hint: See Eq. (6).]

5. Table 7.1 (p. 225) gives the names of a number of important phosphoryl group donors. Many of them have structural similarities and can be grouped according to the organic or inorganic functional group that forms the energy-rich part of the

molecule. For each functional group given below, list the name(s) and structure(s) of the compounds in Table 7.1 that contain it.

(a) phosphoric acid anhydride
(b) guanidinium function (see Chapter 2)
(c) mixed carboxylic acid-phosphoric acid anhydride
(d) phosphate ester

6. Using appropriate equations, show how ATP mediates the transfer of a phosphoryl group from phosphoenol pyruvate to glycerol 1-phosphate.

$$
\underset{\text{Phosphoenol pyruvate}}{CH_2=CH-CO_2^-} \qquad \underset{\text{Glycerol 1-phosphate}}{CH_2-CH-CH_2-OPO_3^{2-}}
$$

with OPO_3^{2-} on phosphoenol pyruvate and OH, OH on glycerol 1-phosphate.

7. Creatine phosphate of muscle cells acts as an energy reservoir when the intracellular concentration of ATP gets too high; the ATP reversibly transfers a phosphoryl group to creatine, generating creatine phosphate and ADP. When the cellular concentration of ATP drops rapidly as a result of maximum muscular activity, creatine phosphate can phosphorylate ADP, providing a rapidly available supply of phosphate bond energy. Would you expect the free energy of hydrolysis of creatine phosphate to be greater than, equal to, or less than that of ATP? Why?

$$
\underset{\text{Creatine}}{H_2N-\overset{NH}{\underset{\underset{\text{NH}}{}}{C}}-N-CH_2-CO_2^-} \qquad \underset{\text{Creatine phosphate}}{^{2-}O_3P-\overset{NH}{\underset{\underset{H}{}}{C}}-N-CH_2-CO_2^-}
$$

(Creatine has CH_3 on the central N; Creatine phosphate has NH double-bonded to C, and CH_3 on N.)

8. Calculate the calories in the hot fudge sundae referred to in Section 7.10 using the data given below. Assume the sundae is made from 100 g of ice cream, 30 g of fudge and 30 g of whipped cream. Using the data in Table 7.6, page 240, calculate how many miles you would have to walk at 5 miles/hour in order to use up the calories from the sundae. (Assume the data in Table 7.6 applies to both men and women.)

Food	% H_2O*	% protein*	% fat*	% carbohydrate*
ice cream	62	4	12	21
fudge topping	24	2	11	63
whipped cream	73	3	20	4

* % by weight (g/100 g)

9. Calculate your basal metabolic rate using the data in Section 7.10 (remember that 2.2 pounds = 1 kilogram). Assume the data given applies to both men and women. Estimate whether your usual diet has the appropriate amount of calories (kcal) and correct relative amounts of proteins, lipids and carbohydrates. To do this you should estimate your daily needs for your level of activity, using the data in Table 7.6 as a guide.

CHAPTER 8

Carbohydrate Catabolism: Fermentation and Glycolysis

Glucose is a key food molecule for most living organisms. One of the unifying features of cellular metabolism is that virtually all cells initially catabolize glucose using the same basic pathway. The universality of this glucose catabolism pathway suggests that it became a part of cellular metabolism during an early period in the evolution of living organisms.

In this chapter we will learn how cells break down glucose into smaller organic molecules, obtaining at the same time a modest return of conserved energy. We start by discussing fermentation—a process of glucose catabolism that can proceed in the absence of oxygen.

8.1

Fermentation

Anaerobic catabolism, or fermentation, of carbohydrates or other fuel molecules provides the simplest and most rudimentary means for degrading molecules to obtain energy. While fermentation is the principal means for conserving energy in anaerobic cells, it serves an essential function in the metabolism of most aerobic organisms as well. In both aerobes and facultative anaerobes, the end products of the enzymatic steps of anaerobic fermentation provide the necessary fuel for the oxidative processes of respiration. Figure 8.1 shows the role of fermentation in aerobic and anaerobic cells. One reason for discussing fermentation is that glucose (as starch or other polysaccharides) serves as a primary food source for most living organisms, including humans.

Figure 8.1

The role of fermentation in aerobic and anaerobic cells. Although glucose is certainly the most widespread fuel molecule, other monosaccharides and other types of reduced compounds can also serve as fuel sources.

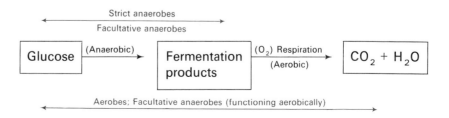

We are familiar with two closely related types of glucose fermentation: **homolactic fermentation,** or **glycolysis,** and **alcoholic fermentation.** Glycolysis is found in many microorganisms (including *lactobaccillus,* which sours milk), as well as in the cells of most higher animals, including mammals. The overall reaction for glycolysis is

$$C_6H_{12}O_6 \xrightarrow{\text{Glycolysis}} 2\,CH_3\!-\!\overset{\displaystyle OH}{\underset{\displaystyle |}{CH}}\!-\!COOH$$

D-Glucose   L-Lactic acid

The results of alcoholic fermentation are familiar to all of us. This pathway is most commonly associated with yeasts and other microorganisms. The steps are virtually identical to glycolysis except in the last two enzymatic reactions, which produce ethanol and CO_2 as the final products:

$$C_6H_{12}O_6 \xrightarrow[\text{fermentation}]{\text{Alcoholic}} 2\,CH_3CH_2OH + 2\,CO_2$$

D-Glucose   Ethanol

Although ethanol is shown here as a product of alcoholic fermentation, the end product can be other similar substances, depending on the organism.

We will describe glycolysis in detail because of its universal nature and its direct applicability to mammalian metabolism. Since it was the first metabolic pathway studied in detail, it is the most completely understood of all metabolic pathways. In Section 7.3 we mentioned how Buchner studied glycolysis in cell-free yeast extracts in the 1890s. Although many workers contributed to our present understanding of glycolysis, the work of Embden and Meyerhof was particularly important; in recognition of their contributions, glycolysis is also referred to as the Embden–Meyerhof Pathway.

8.2

Glycolysis

The overall reaction and energetics of glycolysis are shown in Figure 8.2. The conversion of glucose to two molecules of lactate involves chemical changes of structure that occur in a multi-step process. Part of the total free-energy change in going from D-glucose to L-lactate is conserved by coupling some of the exergonic steps in this sequence with the phosphorylation of ADP. As the energetics imply, glycolysis is not a particularly efficient process—only a third of the total energy available is conserved as ATP. Furthermore, comparatively little of the available energy in the glucose molecule is utilized. The total free-

Exergonic:

D-glucose $\longrightarrow$ 2 lactate $\qquad \Delta G^{0\prime} = -47.0$ kcal/mole

Endergonic:

$2\,ADP + 2\,P_i \longrightarrow 2\,ATP + 2\,H_2O \qquad \Delta G^{0\prime} = +14.6$ kcal/mole

---

Overall:

D-glucose $+ 2\,ADP + 2\,P_i \longrightarrow$ 2 lactate $+ 2\,ATP + 2\,H_2O \qquad \Delta G^{0\prime} = -32.4$ kcal/mole

Figure 8.2

Overall reaction and
energetics of anaerobic
glycolysis.

| Free-energy profit: can be used for cellular work | Excess free energy: liberated as heat, increasing the entropy of the environment |

energy change for the complete oxidation of glucose to CO_2 and H_2O is

$$C_6H_{12}O_6 + 6\,O_2 \longrightarrow 6\,CO_2 + 6\,H_2O \qquad \Delta G^{0\prime} = -686.0 \text{ kcal/mole}$$

This is much greater than the standard free-energy change of -47 kcal/mole in glycolysis. Obviously, then, the complete oxidative catabolism of glucose, starting with glycolysis, makes much more efficient use of glucose as a fuel molecule. Because of their more modest energy-generating metabolism, strict anaerobes are not capable of supporting the metabolic demands of more advanced organisms, so they are much more primitive than aerobes and facultative anaerobes.

In anaerobic glycolysis we find no net change in the oxidation state of the system, as shown by the following half-reactions:

Oxidation: $\quad C_6H_{12}O_6 \longrightarrow 2\,CH_3\overset{\displaystyle O}{\overset{\|}{C}}{-}COOH + 4\,H^+ + 4\,e^-$

Reduction: $\quad 2\,CH_3\overset{\displaystyle O}{\overset{\|}{C}}{-}COOH + 4\,H^+ + 4\,e^- \longrightarrow 2\,CH_3{-}\overset{\displaystyle OH}{\overset{|}{C}H}{-}COOH$

Net: $\quad C_6H_{12}O_6 \longrightarrow 2\,CH_3{-}\overset{\displaystyle OH}{\overset{|}{C}H}{-}COOH$

We can understand how this happens by considering the relationship between the structures of glucose and lactic acid shown on the next page.

There is no *net* change in the oxidation state of the overall system in going from glucose to lactic acid, yet there is a profound change in the oxidation state of the carbon atoms within the molecule. The oxidation states of carbons 2 and 5 in

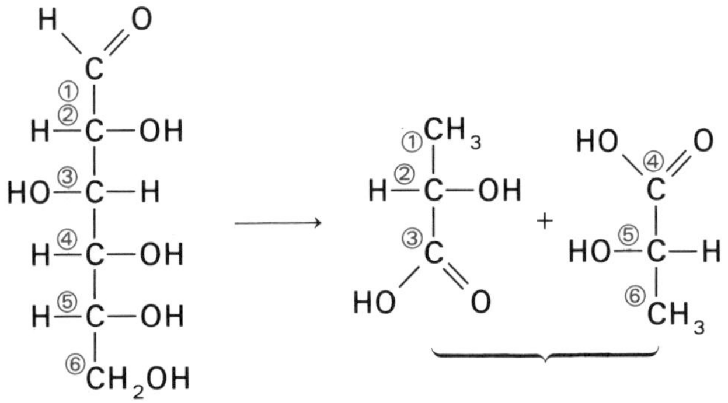

$$\text{D-Glucose (open chain)} \longrightarrow \text{2 L-Lactic acid} \tag{1}$$

D-Glucose (open chain)   2 L-Lactic acid

the original glucose molecule remain unchanged at the level of an alcohol. Carbons 3 and 4 become more oxidized on going to two lactic acids and carbons 1 and 6 become more reduced. Clearly, glycolysis must involve oxidation-reduction reactions at one or more steps. An important point illustrated here is that **cells can obtain energy only by moving electrons**. An energy-producing reaction sequence must involve a redox step somewhere in the course of the sequence. The oxidation-reduction step serves to create a high-energy chemical substance which is transformed in further reaction steps into compounds of lower energy. The cell conserves some of this redox energy as the chemical energy of the phosphorus-oxygen bonds in ATP. (The background organic chemistry of oxidation-reduction is reviewed in Appendix B.)

Note on Interpreting Metabolic Charts   Figure 8.3 is the first of many charts that show pathways in intermediary metabolism. Because of the number of metabolic steps represented, a somewhat simplified notation is used to represent the reactions. Processes that are essentially irreversible are represented by a single arrow. A double-headed arrow represents a step that is reversible. The direction of flow of material in catabolism is indicated by the solid-headed arrows. The following examples show how the notation used in Figure 8.3 relates to more familiar notation:

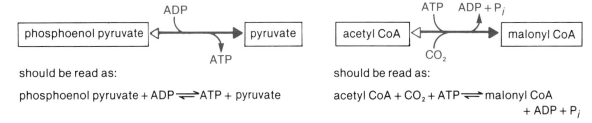

should be read as: should be read as:

phosphoenol pyruvate + ADP $\rightleftharpoons$ ATP + pyruvate   acetyl CoA + CO_2 + ATP $\rightleftharpoons$ malonyl CoA
                                                                                                                              + ADP + P_i

We will use this way of representing reactions in a metabolic pathway whenever the number of steps and intermediates involved make conventional notation unwieldy.

Figure 8.3

Entry of Dietary Sugars into Glycolysis

Entry of glucose,
galactose and fructose
into glycolysis. The step
blocked in galactosemia
is marked with a single
asterisk (*). The location
of the defect in
"fructose intolerance"
is marked with a double
asterisk (**).

In Chapter 7 we discussed the major features of carbohydrate digestion up to the point where the monosaccharides glucose, fructose and galactose are made available for catabolism. Figure 8.3 shows how these dietary monosaccharides are transformed into glycolytic intermediates. Both glucose and fructose are mainline metabolites and hereditary defects in metabolism that block their utilization are usually lethal.

A well-known inborn defect of galactose metabolism is *galactosemia*. This condition is a result of a lack of D-galactose 1-phosphate uridyl transferase

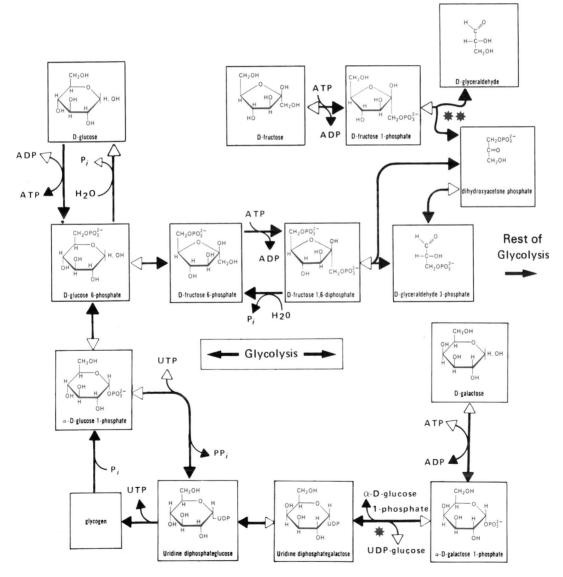

activity. The normal transformation of galactose is inhibited because this enzyme catalyzes a key step in the conversion of galactose (from lactose and other sources) to glucose 6-phosphate, as shown in Figure 8.3. The resulting inability to metabolize ingested galactose causes its accumulation in the blood. Unless this defect is recognized soon after birth, the elevated level of galactose in the blood can have a severely detrimental effect. Galactosemic infants exhibit a number of symptoms, including mental retardation and clouding of the eye lenses. Fortunately, if the condition is diagnosed in early infancy, the effects can be eliminated by removing milk and all other sources of galactose from the diet. Since an alternate pathway for galactose metabolism normally develops in all individuals during childhood, the problem of galactosemia can be out-grown.

Fructose is an important dietary monosaccharide which enters glycolysis by a means different from those used by either glucose or galactose. While fructose can be phosphorylated in the number 6 position by group-specific kinases and thus enter glycolysis at the fructose 6-phosphate level, this method of entry accounts for only a small fraction of the total amount of fructose taken in. As Figure 8.3 shows, the major route is via the much more rapid fructo-kinase-catalyzed reaction in which fructose is phosphorylated in the number 1 position. The fructose 1-phosphate produced is then broken down into two triose units which can enter the mainstream of glycolysis. Some individuals lack

Figure 8.4

Glycolysis. You can best appreciate the central role of glycolysis in metabolism by comparing this figure with the chart at the back of the book. The reaction numbers shown correspond to the step numbers in the discussion of glycolysis in the text.

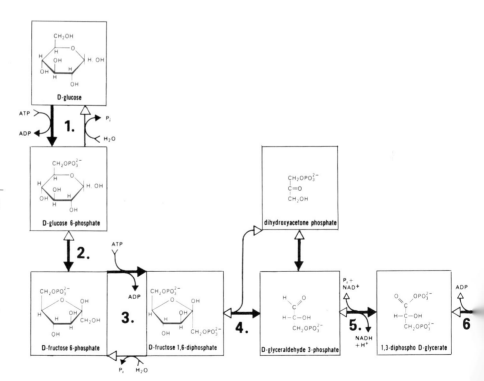

fructokinase and excrete the unmetabolized fructose in the urine. In others, the enzyme that breaks down fructose 1-phosphate is absent. The latter condition is called "fructose intolerance" and is serious because it causes an accumulation of fructose 1-phosphate which interferes with the activity of other enzyme systems.

Figure 8.3 also shows how glycogen (muscle sugar) is broken down into glucose 1-phosphate. Glycogen is important in the fine control of carbohydrate metabolism because of its rapid formation and breakdown. Some excess blood sugar can be converted to glycogen and, during periods of energy demand, glycogen can be quickly broken down to increase the blood sugar level. We will discuss glycogen formation and breakdown in detail in Chapter 11. Glycogen's role in the control of carbohydrate metabolism will be examined further in Section 8.4.

Overall Plan of Glycolysis

Glycolysis consists of ten sequential, enzyme-catalyzed reactions which take place in the cytoplasm. These steps, shown in Figure 8.4, can be divided into two broad sequences:

Sequence I (requires ATP) Glucose is phosphorylated and cleaved to form two 3-carbon triose phosphate units. This corresponds to steps 1–4 in the following discussion.

Sequence II (generates ATP) Glyceraldehyde 3-phosphate (triose phosphate) is transformed into lactate. Oxidation-reduction occurs in this series of steps. The system conserves energy by coupling exergonic steps of this sequence with the phosphorylation of ADP. This corresponds to steps 5–10 in the following discussion.

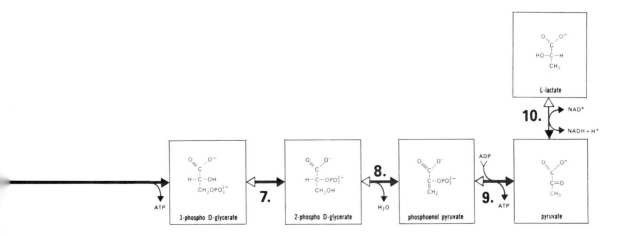

Steps of Glycolysis

Step 1   Activation of D-glucose
There is very little free D-glucose inside the cell; most intracellular glucose is in the form of phosphate esters. Upon entry into the cell, D-glucose is phosphorylated by ATP to produce D-glucose 6-phosphate:

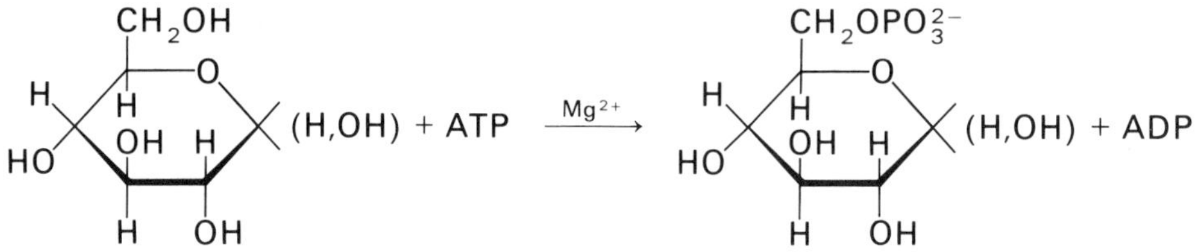

$$\Delta G^{0'} = -4.0 \text{ kcal/mole}$$

In most cells, this reaction is catalyzed by the enzyme *hexokinase*, which has a broad sugar-substrate specificity for hexoses and hexosamines. In liver tissue, we find a second type of enzyme called *glucokinase*, which is specific for D-glucose and has a much lower affinity for this substrate than does hexokinase. Glucokinase, whose synthesis is induced by insulin, has little significant effect unless the blood glucose level becomes extremely high. It is therefore an important element in the insulin-directed control of the concentration of glucose in the blood. Both glucokinase and hexokinase require either Mg^{2+} or Mn^{2+} in order to function. The divalent cation forms a complex with ATP, either $MnATP^{2-}$ or $MgATP^{2-}$, which is the form of ATP used as a substrate by most kinases. The large free-energy change corresponds to an equilibrium constant $K_{eq} = 1 \times 10^3$, which explains why so little free D-glucose occurs in the cell. The formation of glucose 6-phosphate is the first of two ATP-requiring activation reactions in glycolysis, and thus represents an initial investment of energy which will ultimately yield more in return.

Step 2   Formation of fructose 6-phosphate
Dietary fructose enters glycolysis at this step, where D-fructose is phosphorylated by ATP to yield D-fructose 6-phosphate in a reaction catalyzed by hexokinase. Fructose 6-phosphate forms primarily from glucose 6-phosphate by the following isomerization reaction catalyzed by *phosphoglucoisomerase*:

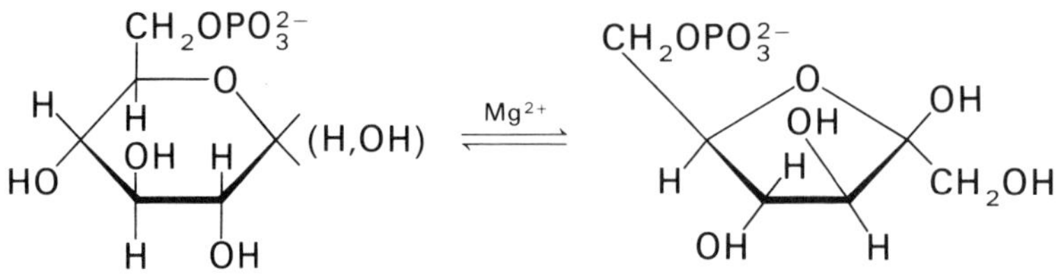

$$\Delta G^{0'} = +0.4 \text{ kcal/mole}$$

This reaction yields an equilibrium mixture containing approximately two parts glucose 6-phosphate to one part fructose 6-phosphate. In the cell this reaction proceeds to the right because the subsequent exergonic steps use up fructose 6-phosphate as fast as it forms.

Step 3   The second activation step: formation of fructose 1,6-diphosphate

This step is important in controlling the overall rate of glycolysis and is catalyzed by the regulatory enzyme *phosphofructokinase* (PFK), which actively catalyzes only the forward reaction:

$$\Delta G^{0'} = -3.4 \text{ kcal/mole}$$

Phosphofructokinase exhibits a substrate dependence which is shown in Figure 8.5. The regulatory behavior of PFK is similar to what we discussed in Chapter 3 for regulatory enzymes in general. The rate of glycolysis is regulated to maintain a fairly constant overall ratio of [ATP]/[ADP] in the cytoplasm. As we will see throughout our discussion of metabolism, the controlling factors all operate to maintain a constant [ATP]/[ADP] ratio, and therefore, these two metabolites themselves function as central controlling agents in metabolism.

Figure 8.5

Allosteric behavior of phosphofructokinase (PFK). ATP is a negative modulator of this enzyme, while both ADP and AMP act as positive modulators (see Chapter 3).

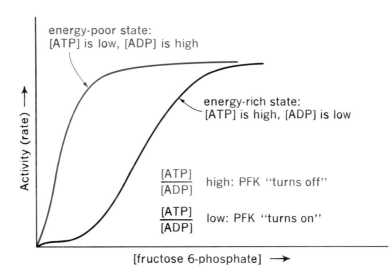

energy-poor state:
[ATP] is low, [ADP] is high

energy-rich state:
[ATP] is high, [ADP] is low

Activity (rate) →

$\dfrac{[ATP]}{[ADP]}$ high: PFK "turns off"

$\dfrac{[ATP]}{[ADP]}$ low: PFK "turns on"

[fructose 6-phosphate] →

Significantly, the reverse reaction shown below is catalyzed by a different regulatory enzyme, *diphosphofructose phosphatase*:

fructose 1,6-diphosphate + HOH $\xrightarrow{\text{Mg}^{2+}}$ fructose 6-phosphate + P$_i$

$$\Delta G^{0\prime} = -4.0 \text{ kcal/mole}$$

This enzyme controls the reverse sequence of glycolysis, producing glucose from excess pyruvate. This reaction is desirable under conditions of high [ATP]/[ADP] because the excess glucose (actually glucose 1-phosphate) can be transformed into glycogen. Diphosphofructose phosphatase is inhibited by AMP and activated by ATP. Due to the large free-energy changes and the properties of these two important regulatory enzymes, the reaction is essentially irreversible in either direction.

Step 4   The formation of two triose phosphate units

The completion of the first sequence in glycolysis requires that the hexose diphosphate be cleaved to form two triose monophosphates. This cleavage is catalyzed by the enzyme *aldolase*, whose name indicates the reaction catalyzed—namely, aldol condensation. The essentials of the aldol condensation are:

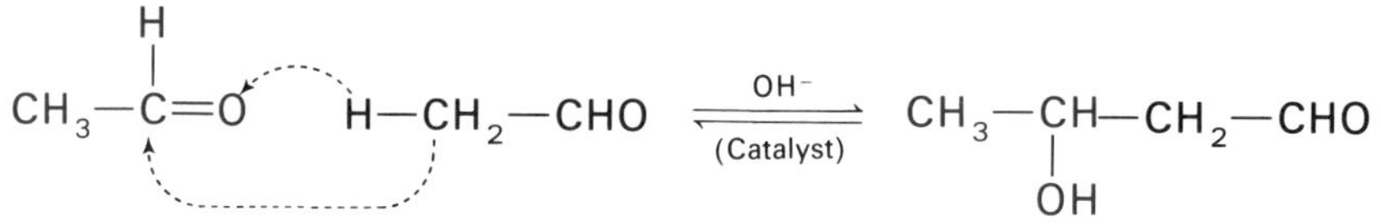

Acetaldehyde   "Aldol"   (2)

This reversible reaction is characteristic of these functional groups. With Eq. (2) in mind, let's look at the reaction catalyzed by aldolase, which may best be visualized as fructose 1,6-diphosphate in its open-chain form:

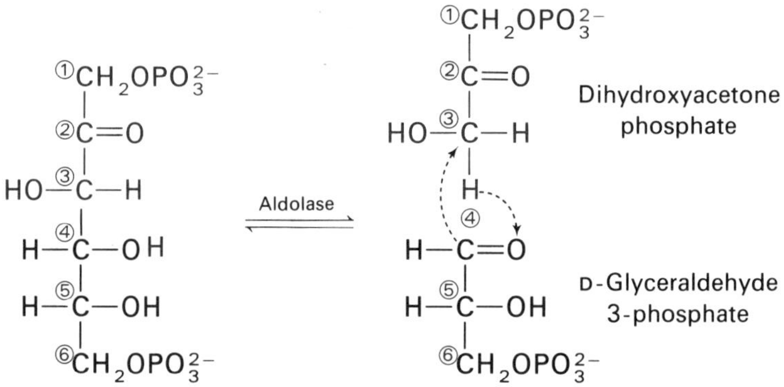

Dihydroxyacetone phosphate

D-Glyceraldehyde 3-phosphate

$$\Delta G^{0\prime} = +5.7 \text{ kcal/mole}; \ K_{eq} \cong 10^{-4} M$$

The aldolase reaction is completely analogous to the aldol condensation shown in Eq. (2). The energetics of this reaction are poor in the standard state (see Problem **3**, Chapter 7), and at first glance you might think this reaction a poor choice for a step in such a vital pathway as glycolysis. Things are not really so grim *in vivo* because the system is not in the standard state in a living cell. Instead of being at 1.0 molar concentration, the fructose 1,6-diphosphate concentration is about $10^{-4} M$, which means that in the equilibrium mixture in the cell, the concentrations of each of the products is about $10^{-4} M$—a 50% conversion!

The subsequent steps of glycolysis further degrade only glyceraldehyde 3-phosphate. The products of the aldolase reaction are in rapid equilibrium through a reaction catalyzed by the enzyme *triose phosphate isomerase*:

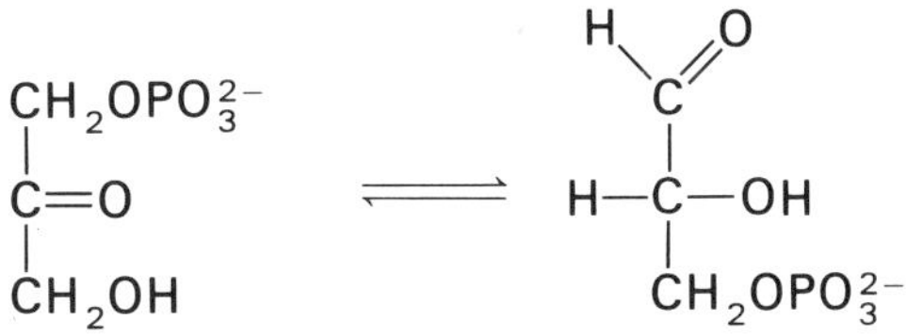

$$
\begin{array}{ccc}
\begin{array}{c}
CH_2OPO_3^{2-} \\
| \\
C=O \\
| \\
CH_2OH
\end{array}
&
\rightleftharpoons
&
\begin{array}{c}
\underset{H}{\overset{O}{\diagup}}C \\
| \\
H-C-OH \\
| \\
CH_2OPO_3^{2-}
\end{array}
\qquad \Delta G^{0\prime} = +1.8 \text{ kcal/mole}
\end{array}
$$

Dihydroxyacetone phosphate   D-Glyceraldehyde 3-phosphate

The overall result of the first sequence of glycolysis is that one glucose 6-phosphate molecule transforms into two glyceraldehyde 3-phosphate molecules, which are then available for further reaction in the second sequence of glycolysis.

Step 5   Glyceraldehyde 3-phosphate dehydrogenase   In this first step of the second sequence of glycolysis, a highly exergonic oxidation of an aldehyde functional group is coupled with the formation of a high-energy phosphate bond. The overall reaction is

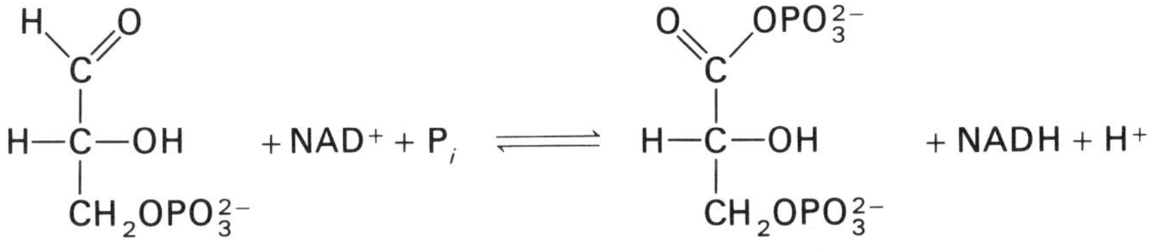

$$
\begin{array}{c}
\underset{H}{\overset{O}{\diagup}}C \\
| \\
H-C-OH \\
| \\
CH_2OPO_3^{2-}
\end{array}
+ NAD^+ + P_i
\rightleftharpoons
\begin{array}{c}
\overset{O}{\underset{}{\diagdown}}C{\overset{OPO_3^{2-}}{\diagup}} \\
| \\
H-C-OH \\
| \\
CH_2OPO_3^{2-}
\end{array}
+ NADH + H^+ \quad \Delta G^{0\prime} = +1.5 \text{ kcal/mole}
$$

D-Glyceraldehyde 3-phosphate   1,3-Diphosphoglycerate

The product of this step, an *acyl phosphate*, is a high-energy phosphoryl-group donor.

To understand how a phosphorylation couples to an oxidation, we must consider in more detail the enzyme which catalyzes the reaction shown above. Glyceraldehyde 3-phosphate dehydrogenase is a tetrameric enzyme containing four identical subunits, each with a tightly bound NAD^+ prosthetic group. Also present at the active center of each protomeric subunit is an essential —SH group. The action of the enzyme can be understood from the following three reactions:

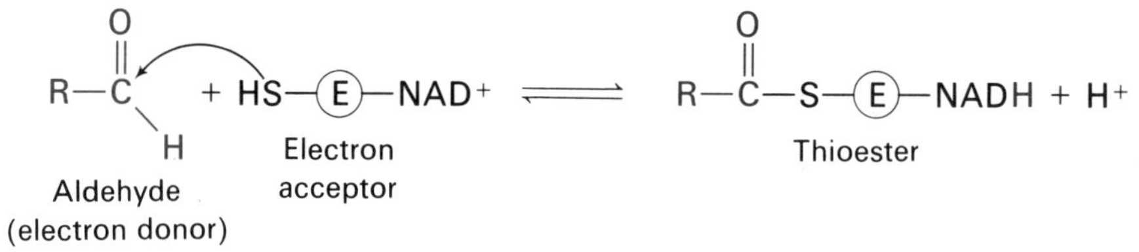

$$R-\overset{\overset{\displaystyle O}{\|}}{C}_{\diagdown H} + HS\!-\!\textcircled{E}\!-\!NAD^+ \rightleftharpoons R-\overset{\overset{\displaystyle O}{\|}}{C}-S\!-\!\textcircled{E}\!-\!NADH + H^+$$

Aldehyde (electron donor)   Electron acceptor   Thioester

The aldehyde function is oxidized to the level of a carboxylic acid, forming a thioester.

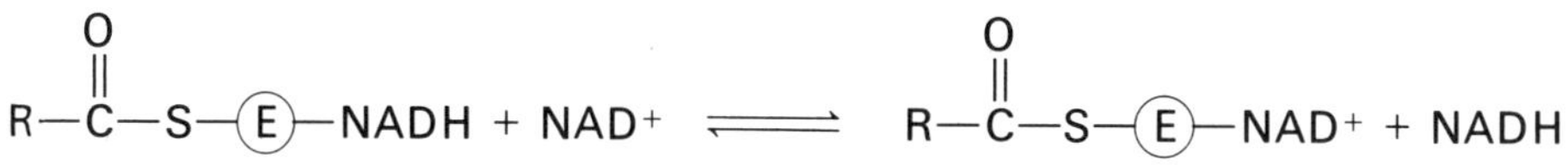

$$R-\overset{\overset{\displaystyle O}{\|}}{C}-S\!-\!\textcircled{E}\!-\!NADH + NAD^+ \rightleftharpoons R-\overset{\overset{\displaystyle O}{\|}}{C}-S\!-\!\textcircled{E}\!-\!NAD^+ + NADH$$

NAD^+ from the intracellular pool of this coenzyme reoxidizes the enzyme-bound prosthetic group.

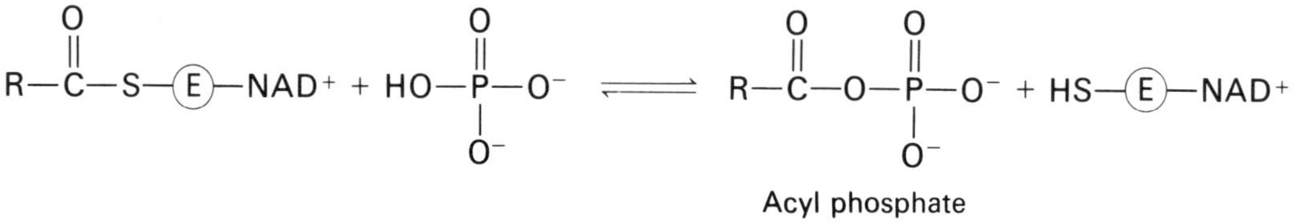

$$R-\overset{\overset{\displaystyle O}{\|}}{C}-S\!-\!\textcircled{E}\!-\!NAD^+ + HO-\overset{\overset{\displaystyle O}{\|}}{\underset{\underset{\displaystyle O^-}{|}}{P}}-O^- \rightleftharpoons R-\overset{\overset{\displaystyle O}{\|}}{C}-O-\overset{\overset{\displaystyle O}{\|}}{\underset{\underset{\displaystyle O^-}{|}}{P}}-O^- + HS\!-\!\textcircled{E}\!-\!NAD^+$$

Acyl phosphate

The acyl group is then transferred from the enzyme to an inorganic phosphate ion.

Through the steps outlined above, the large free-energy change associated with oxidation of the aldehyde group of glyceraldehyde 3-phosphate is conserved in the high-energy phosphate bond of 1,3-diphosphoglycerate. This oxidation-reduction reaction proceeds with the formation of NADH in the cytoplasm. Ultimately, NAD^+ must be regenerated, otherwise successive molecules of glyceraldehyde 3-phosphate could not be oxidized and glycolysis would cease. The cell must contain, then, a terminal electron acceptor or oxidant which regenerates the NAD^+ consumed in this step of glycolysis. In Chapter 7 we stated that in anaerobic metabolism the terminal electron acceptor is usually an organic molecule. As we will see shortly, the terminal electron acceptor in anaerobic glycolysis is pyruvate.

Glycerol, produced primarily by the hydrolysis of triglycerides in the small intestine, enters glycolysis at this level, as the following reactions show:

$$\text{Glycerol} + \text{ATP} \longrightarrow \text{Glycerol 3-phosphate} + \text{ADP} \overset{\overset{\text{NAD}^+ \quad \text{NADH} + \text{H}^+}{\curvearrowright}}{\longrightarrow} \text{Glyceraldehyde 3-phosphate}$$

Here, a phosphoryl ester is produced and the remaining primary alcohol functional group is oxidized to an aldehyde (see Appendix B).

Step 6   First energy-conservation step   In the preceding reaction, a high-energy phosphate compound was produced possessing sufficient energy to phosphorylate ADP. This phosphoryl-group transfer reaction is catalyzed by *phosphoglycerate kinase* and is an example of *substrate level phosphorylation*.

$$\begin{array}{c}
\overset{O}{\underset{\displaystyle \|}{C}}\diagdown \text{OPO}_3^{2-} \\
| \\
H\!-\!\overset{\displaystyle |}{\underset{\displaystyle |}{C}}\!-\!OH \\
CH_2OPO_3^{2-}
\end{array}
\ + \ ADP \ \underset{}{\overset{Mg^{2+}}{\rightleftharpoons}} \ \begin{array}{c}
\overset{O}{\underset{\displaystyle \|}{C}}\diagdown O^- \\
| \\
H\!-\!\overset{\displaystyle |}{\underset{\displaystyle |}{C}}\!-\!OH \\
CH_2OPO_3^{2-}
\end{array}
\ + \ ATP \qquad \Delta G^{0\prime} = -4.5 \text{ kcal/mole}$$

1,3-Diphosphoglycerate   3-Phosphoglycerate

This strongly exergonic reaction helps "pull" the preceding reaction to completion. Since two moles of 1,3-diphosphoglycerate are produced for each mole of glucose entering glycolysis initially, we have recouped all our "invested" ATP—any additional ATP generated will represent a net gain.

Step 7   Isomerization of 3-phosphoglycerate   We can compare this step and the one following to pushing a weight across a plateau to the edge of a cliff, whereupon a large decrease in potential energy takes place. These two steps rearrange the 3-phosphoglycerate molecule into the high-energy phosphate compound phosphoenol pyruvate. The isomerization of 3-phosphoglycerate is catalyzed by *phosphoglyceromutase*, which moves the phosphoryl group from the 3- to the 2-position of the glycerate anion:

$$\begin{array}{c}
CO_2^- \\
| \\
H\!-\!\overset{\displaystyle |}{\underset{\displaystyle |}{C}}\!-\!OH \\
CH_2OPO_3^{2-}
\end{array}
\ \underset{}{\overset{Mg^{2+}}{\rightleftharpoons}} \ \begin{array}{c}
CO_2^- \\
| \\
H\!-\!\overset{\displaystyle |}{\underset{\displaystyle |}{C}}\!-\!OPO_3^{2-} \\
CH_2OH
\end{array}
\qquad \Delta G^{0\prime} = +1.0 \text{ kcal/mole}$$

3-Phosphoglycerate   2-Phosphoglycerate
(80%)   (20%)

Step 8   Formation of phosphoenol pyruvate   This is the second of two reactions in glycolysis that produce a high-energy phosphate compound. The conversion of 2-phosphoglycerate to phosphoenol pyruvate is catalyzed by the enzyme *enolase*:

$$H-\underset{\underset{CH_2OH}{|}}{\overset{\overset{CO_2^-}{|}}{C}}-OPO_3^{2-} \quad \underset{}{\overset{Mg^{2+}}{\rightleftharpoons}} \quad \underset{\underset{CH_2}{||}}{\overset{\overset{CO_2^-}{|}}{C}}-OPO_3^{2-} + H_2O \qquad \Delta G^{0\prime} = +0.5 \text{ kcal/mole}$$

2-Phosphoglycerate   Phosphoenol pyruvate

This simple dehydration reaction converts a low-energy phosphate compound to a high-energy phosphate compound. The change in the distribution of energy within the three-carbon framework is more profound than might appear at first glance. The organic chemistry involved in this step and the following step is the conversion of a *diol* to a *ketone*:

$$\underset{\text{Diol}}{\overset{\overset{OH \quad OH}{|\quad\;\; |}}{CH_2-CH_2-R}} \quad \underset{\text{(Step 8)}}{\overset{-H_2O}{\rightleftharpoons}} \quad \underset{\text{Enol}}{\overset{\overset{OH}{|}}{CH_2=C-R}} \quad \overset{\text{more reduced}}{\underset{\text{(Step 9)}}{\rightleftharpoons}} \quad \underset{\text{Ketone}}{\overset{\overset{O}{||}}{CH_3-C-R}} \overset{\text{more oxidized}}{}$$

Step 9   Second energy-conservation step

Phosphoenol pyruvate (PEP) acts as a high-energy phosphoryl-group donor because the product pyruvate is much more stable than phosphoenol pyruvate. In PEP, the system is constrained to be in the less stable enol form due to the phosphate ester group. There is no such constraint in the pyruvate molecule. Therefore, PEP readily phosphorylates ADP in a reaction catalyzed by *pyruvate kinase*:

$$\underset{\underset{CH_2}{||}}{\overset{\overset{CO_2^-}{|}}{C}}-OPO_3^{2-} + ADP \quad \underset{}{\overset{Mg^{2+},\; K^+}{\rightleftharpoons}} \quad \underset{\underset{CH_3}{|}}{\overset{\overset{CO_2^-}{|}}{C}}=O + ATP \qquad \Delta G^{0\prime} = -7.5 \text{ kcal/mole}$$

PEP   Pyruvate

For each glucose molecule the system forms two molecules of PEP. Therefore, our net yield of ATP from anaerobic glycolysis is now two ATP molecules per molecule of glucose.

Pyruvate anion is a central molecule in metabolism, as you can see by reference to the chart at the back of the book. In addition to being a direct product of glycolysis, it can also be produced from certain amino acids and dicarboxylic acids. Pyruvate is the primary fuel for the processes of respiration in aerobic metabolism, as we will see in the next chapter. In the absence of oxygen, pyruvate substitutes as a terminal electron acceptor so that glycolysis proceeds on a continuous basis. The reoxidation of the NADH produced in the glyceraldehyde 3-phosphate dehydrogenase step (Step 5) is the basis for the last step of glycolysis.

Step 10   Formation of lactate   In this reaction, pyruvate is reduced to lactate by NADH in a reaction catalyzed by *lactate dehydrogenase* (LDH):

$$\underset{\text{Pyruvate}}{\underset{\overset{|}{CH_3}}{\overset{\overset{CO_2^-}{|}}{C=O}}} + NADH + H^+ \rightleftharpoons \underset{\text{L-Lactate}}{\underset{\overset{|}{CH_3}}{\overset{\overset{CO_2^-}{|}}{HO-C-H}}} + NAD^+ \qquad \Delta G^{0\prime} = -6.0 \text{ kcal/mole}$$

The lactate produced anaerobically diffuses out of the cell as a waste product. In microorganisms, such a waste product is excreted into the environment to serve as food for aerobic heterotrophs. (Think of this next time you eat sour cream or yogurt!) In higher animals, lactate is transported to the liver, where part of it is resynthesized into glucose and part is oxidized to generate energy. During heavy exertion, muscle tissue accumulates lactate when the energy demand exceeds the ability of the blood to supply oxygen to the tissues. This is termed a state of *oxygen debt*. Muscle fatigue and soreness is due in part to the acidity of this end product of anaerobic glycolysis. Since muscle tissue cannot resynthesize glucose from lactate, lactate produced in muscle tissue and in other tissues which function anaerobically must by transported by the blood to the liver, where **gluconeogenesis** (glucose biosynthesis) takes place. Figure 8.6 shows the response of blood lactate level to oxygen debt and subsequent recovery.

Figure 8.6

Graph of blood lactate concentration before, during and after a short period of intense exercise. During the period of exercise, the muscle tissues go into oxygen debt and metabolize glucose anaerobically, releasing lactate to the blood and increasing its concentration. During recovery, excess blood lactate is reconverted to pyruvate in the liver.

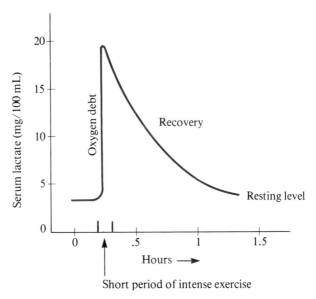

Alcoholic Fermentation

In organisms exhibiting alcoholic fermentation, the steps of glycolysis leading to pyruvate are identical to those just discussed. However, these organisms lack pyruvate dehydrogenase, containing instead the two enzymes *pyruvate decarboxylase* and *alcohol dehydrogenase*. It is certainly fortunate for our general level of sobriety that these two enzymes are not part of the metabolism of higher animals.

Pyruvate decarboxylase is a thiamine pyrophosphate-requiring enzyme that catalyzes the formation of acetaldehyde from pyruvate:

$$
\underset{\begin{array}{c}\\ \text{CH}_3\end{array}}{\overset{\begin{array}{c}\text{O}\diagup\text{O}^-\\ \text{C}\\ |\\ \text{C}=\text{O}\\ |\end{array}}{}} + \text{H}^+ \;\rightleftarrows\; \underset{\begin{array}{c}\\ \text{CH}_3\end{array}}{\overset{\begin{array}{c}\text{H}\diagup\text{O}\\ \text{C}\\ |\end{array}}{}} + \text{CO}_2\,(g)
$$

Acetaldehyde

This CO_2-producing enzyme is responsible for the head of foam on a glass of beer, a result of the CO_2 created by the catabolism of sugar during fermentation. The acetaldehyde formed serves the same function as pyruvate, namely that of a terminal electron acceptor. This reaction is analogous to that of lactate dehydrogenase and is catalyzed by alcohol dehydrogenase:

$$
\underset{\begin{array}{c}\\ \text{CH}_3\end{array}}{\overset{\begin{array}{c}\text{H}\diagup\text{O}\\ \text{C}\\ |\end{array}}{}} + \text{NADH} + \text{H}^+ \;\rightleftarrows\; \underset{\begin{array}{c}\\ \text{CH}_3\end{array}}{\overset{\begin{array}{c}\text{OH}\\ |\\ \text{CH}_2\\ |\end{array}}{}} + \text{NAD}^+
$$

Ethanol

Role of Glycolysis and Related Pathways

Glycolysis is a simple and direct means of conserving a modest amount of energy from the breakdown of glucose into two smaller molecules. The energetics of anaerobic glycolysis were summarized in Figure 8.2. From the overall free-energy change

glucose $\longrightarrow$ 2 lactate $\Delta G^{0\prime} = -47.0$ kcal/mole

only two ATP's are generated, and therefore only -14.6 kcal/mole of the total energy available are conserved. This is adequate for the simple needs of most microorganisms. In higher animals, anaerobic glycolysis provides a mechanism whereby certain tissues can operate for varying periods of time in the absence of oxygen. Glycolysis also serves to provide pyruvate as a fuel for respiration when oxygen is available. And finally, some of the glycolytic intermediates are important precursors for the biosynthesis of other biomolecules.

8.3

Alternate Fates of Glucose 6-Phosphate: Pentose Phosphate Pathway

Glucose 6-phosphate is another central biomolecule in carbohydrate metabolism. The importance of this substance is shown by Figure 8.7, which summarizes some of the transformations of glucose 6-phosphate we have already discussed and some which we will discuss presently. The liver, kidneys, and intestinal epithelium can synthesize and release free glucose. The formation of glucose from glucose 6-phosphate is accomplished by the enzyme *glucose 6-phosphatase*, found only in these three tissues in mammals. It may seem odd at first that an investment of ATP is negated by hydrolysis of the glucose 6-phosphate produced as a result of that investment. You must bear in mind, however, that the primary means the cell has for controlling the intracellular concentration of a metabolite such as glucose is **rate control**. This means balancing the rate at which the cell produces the metabolite with the rate at which the cell uses it up. When the concentration of glucose 6-phosphate is high, due, for example, to the accelerated breakdown of glycogen, the rate of hydrolysis of glucose 6-phosphate increases due to the increased concentration of the

Figure 8.7

Various fates of glucose 6-phosphate.

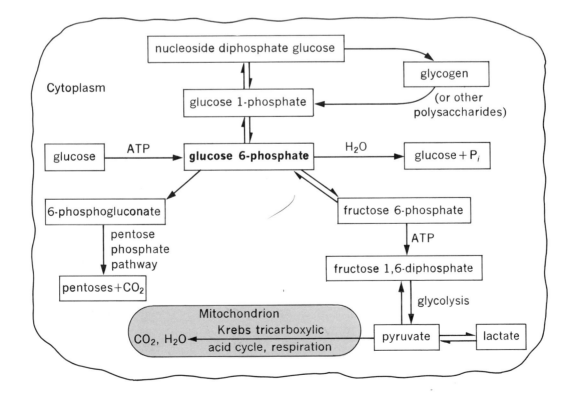

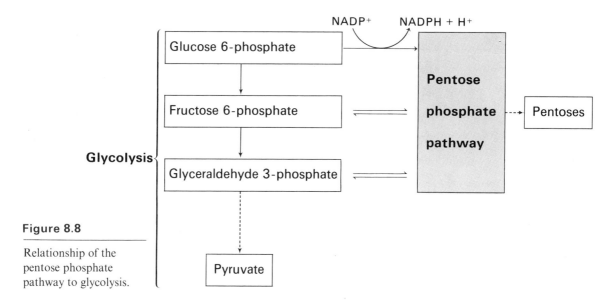

Figure 8.8

Relationship of the
pentose phosphate
pathway to glycolysis.

substrate for glucose 6-phosphatase. Ordinarily, the rate of glucose 6-phosphate hydrolysis in the liver and other tissues is low.

An important alternate fate of glucose 6-phosphate is the **pentose phosphate pathway**, also known as the phosphogluconate pathway and the pentose shunt, which takes place in the cytoplasm. The relationship between this pathway and "mainline" glycolysis can be appreciated by referring to Figures 8.4 and 8.8.

The pentose phosphate pathway accomplishes three crucial things in a living cell:

1. It provides the primary means for the generation of NADPH, a reducing agent required for the biosynthesis of fatty acids and steroids.
2. It provides a pool of ribose 5-phosphate that supplies the required five-carbon sugar component for DNA and RNA biosynthesis.
3. It provides for the non-oxidative interconversions of three-, four-, five-, six- and seven-carbon sugar phosphates.

The reactions comprising the pentose phosphate pathway are shown in Figure 8.9. It is apparent from this figure that the pentose phosphate pathway really doesn't "lead" anywhere. However, if ribose 5-phosphate is consumed in DNA or RNA synthesis, then the various equilibrium systems in Figure 8.9 shift accordingly to replenish ribose 5-phosphate; if NADPH is required to supply "reducing power" for the biosynthesis of highly reduced fatty acids and steroids, then this pathway provides a means of degrading glucose to CO_2 with the concurrent formation of NADPH. The pentose phosphate pathway can be subdivided into oxidative reactions and non-oxidative reactions as shown in Figure 8.9. Let us consider in detail some of the steps of this important pathway.

Figure 8.9

(*Opposite page*)
Reactions of the pentose
phosphate pathway.

Oxidative steps

Non-oxidative steps

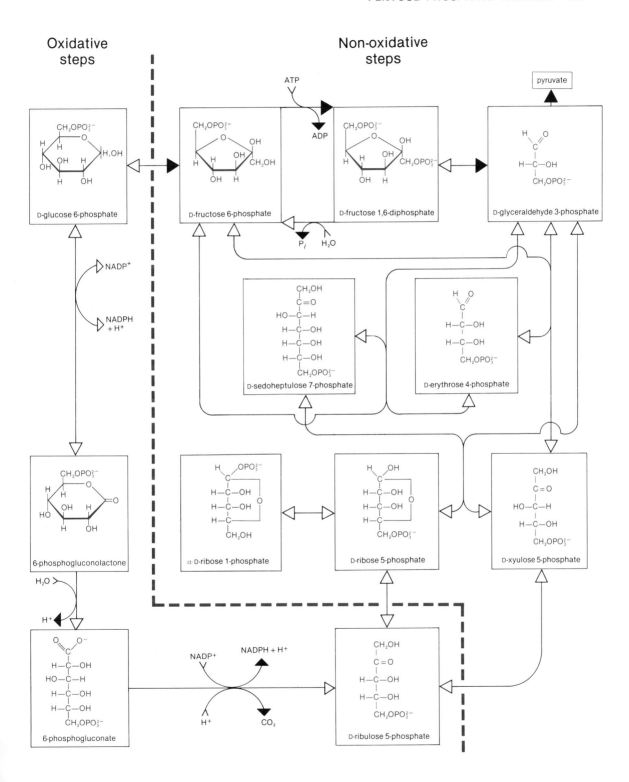

Oxidative Steps of the Pentose Phosphate Pathway

Oxidation of glucose 6-phosphate   Glucose 6-phosphate is "shunted" into this pathway by the action of glucose 6-phosphate dehydrogenase. This enzyme uses $NADP^+$ as a coenzyme and, as we saw in Chapter 3, $NADP^+/NADPH$ is a coenzyme redox system associated with biosynthesis, while the $NAD^+/NADH$ redox system is associated with catabolism.

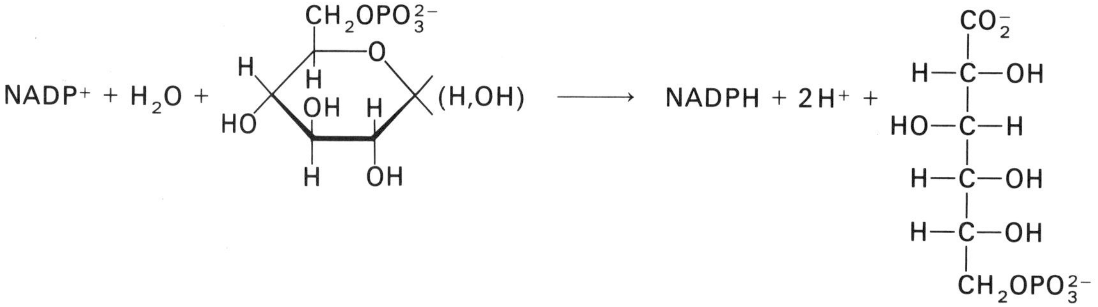

$$NADP^+ + H_2O + \text{(glucose 6-phosphate)} \longrightarrow NADPH + 2H^+ + \text{(6-Phosphogluconate)}$$

6-Phosphogluconate

Oxidation of 6-phosphogluconate   The oxidation of 6-phosphogluconate is catalyzed by 6-phosphogluconate dehydrogenase, yielding a five-carbon sugar phosphate and CO_2. This enzyme also requires $NADP^+$; therefore, this reaction provides a second step for the generation of NADPH for biosynthetic purposes.

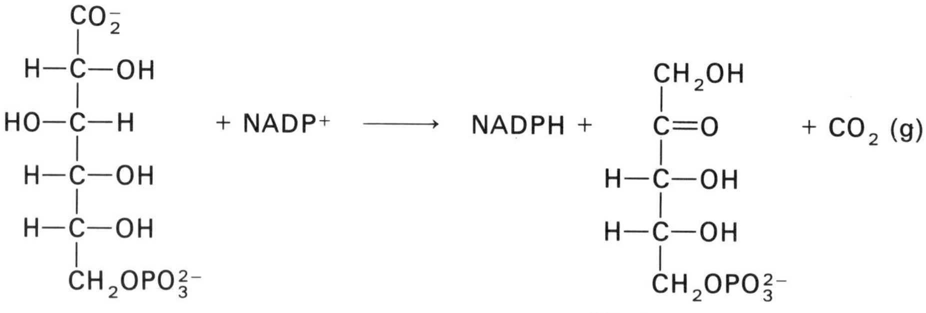

$$\text{(6-phosphogluconate)} + NADP^+ \longrightarrow NADPH + \text{(D-Ribulose 5-phosphate)} + CO_2\ (g)$$

D-Ribulose 5-phosphate

Non-oxidative Steps of the Pentose Phosphate Pathway

Formation of ribose 5-phosphate   The ribulose 5-phosphate produced in the step above can undergo a variety of interconversions, ultimately leading back to the glycolytic intermediates fructose 6-phosphate (and thus glucose 6-phosphate) and glyceraldehyde 3-phosphate. It can also produce

D-ribose 5-phosphate by an isomerization equilibrium catalyzed by phospho-pentose isomerase:

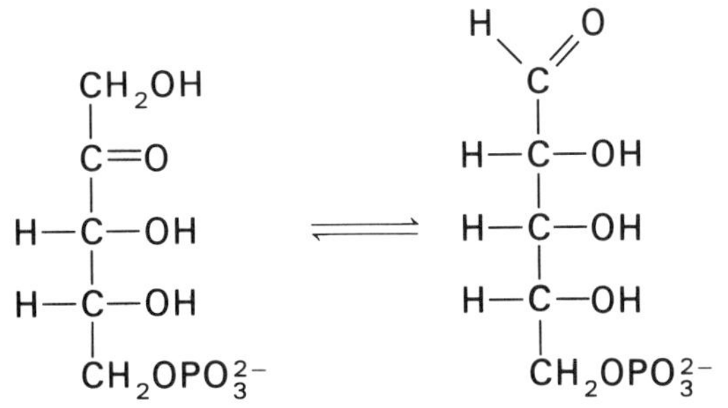

$$
\begin{array}{ccc}
\text{CH}_2\text{OH} & & \text{H}\diagdown\diagup\text{O} \\
| & & \text{C} \\
\text{C}=\text{O} & & | \\
| & & \text{H}-\text{C}-\text{OH} \\
\text{H}-\text{C}-\text{OH} & \rightleftharpoons & \text{H}-\text{C}-\text{OH} \\
| & & | \\
\text{H}-\text{C}-\text{OH} & & \text{H}-\text{C}-\text{OH} \\
| & & | \\
\text{CH}_2\text{OPO}_3^{2-} & & \text{CH}_2\text{OPO}_3^{2-} \\
\text{D-Ribulose 5-phosphate} & & \text{D-Ribose 5-phosphate}
\end{array}
$$

Other sugar interconversions   The remainder of the steps forming the pentose phosphate pathway are reactions involving equilibria between various three-, four-, five-, six-, and seven-carbon sugar phosphates. It is important to realize that the pentose phosphate pathway supplies the metabolites in these equilibria on demand. By applying our understanding of Le Châtelier's Principle (Chapter 7), we can see that if either ribose 5-phosphate or NADPH is consumed, the equilibria in the pentose phosphate pathway will produce more. Depending on the nature of the change in concentration, the equilibria will shift to maintain the proper equilibrium relationships between the metabolite concentrations. Therefore, the pentose phosphate pathway acts as a self-replenishing reservoir for pentoses and NADPH.

The proportion of incoming glucose diverted from glycolysis to the pentose phosphate pathway depends on the particular tissue. In liver tissue, an important site of biosynthesis, approximately 30% of the glucose metabolized goes through the pentose phosphate pathway. In adipose tissue, the mammary glands, testes, and adrenal cortex, where lipid and steroid biosynthesis is important, over 30% of the glucose used by the tissue is oxidized by this pathway. Muscle cells, however, utilize glucose almost exclusively by the glycolytic pathway.

The Pentose Phosphate Pathway and Red Blood Cells

The pentose phosphate pathway is essential for the normal functioning of red blood cells (erythrocytes). Mature erythrocytes have no mitochondria or nuclei, and cannot synthesize proteins. These specialized cells obtain energy by anaerobic glycolysis and rely on the pentose phosphate pathway as a source of NADPH for small molecule biosynthesis and reducing power. Erythrocytes use NADPH as a reducing agent in two major ways:

1.   To maintain the iron of hemoglobin in the Fe^{2+} state.
2.   To maintain glutathione (Chapter 3) in a reduced state.

Currently, glutathione is thought to act as a protective agent against the detrimental effects of H_2O_2, by means of the following glutathione peroxidase-catalyzed reaction:

$$2\,\text{Glutathione}-\text{SH} + H_2O_2 \longrightarrow 2\,H_2O + \begin{array}{c}\text{Glutathione}-\text{S} \\ | \\ \text{Glutathione}-\text{S}\end{array}$$

This is an important reaction because H_2O_2 can damage red blood cells by oxidizing hemoglobin (Fe^{2+}) to methemoglobin (Fe^{3+}) and by oxidizing unsaturated membrane lipids, thus weakening the membrane structure. Glutathione reductase catalyzes the reaction between oxidized glutathione and NADPH to regenerate glutathione.

The key enzyme in the pentose phosphate pathway is glucose 6-phosphate dehydrogenase. It is the primary regulatory factor in shunting material into the pathway. Synthesis of this enzyme is induced by increased amounts of carbohydrate in the diet. The enzyme is inhibited by one of its products, NADPH, and is activated by oxidized glutathione. The overall function of these regulatory factors is to maintain a constant ratio of $NADPH/NADP^+$ and to ensure increased activity when excessive concentrations of oxidized glutathione are present.

Glucose 6-phosphate dehydrogenase deficiency is a genetic defect of metabolism affecting over 100 million people, mainly in Mediterranean and tropical areas. The condition is common in these regions because it appears to make the individual a poor host for the malaria parasite, thus favoring the survival of persons with the defect. (In Chapter 2 we learned that sickle cell anemia has the same effect.) This deficiency causes a condition called *hemolytic anemia* wherein the erythrocyte membrane is fragile and ruptures easily. Thus, the average life span of the erythrocytes is shortened, resulting in anemia. Like the hemoglobin mutants discussed earlier (see problem 14, Chapter 2), a number of different glucose 6-phosphate dehydrogenase mutants have been identified.

8.4

Regulation of Glycolysis

Earlier we stated that the allosteric enzyme phosphofructokinase is important in controlling the rate of glycolysis in the forward direction. Now we will see how the rate at which glycolysis produces pyruvate is controlled with respect to other energy-conserving processes.

A few general observations relating to the control of glycolysis were known for a long time before the enzymatic basis of this control was understood.

These observations include:

1. **The Pasteur effect.** Louis Pasteur observed that yeast cells (facultative anaerobes) use much less glucose when grown in the presence of oxygen rather than under anaerobic conditions. The overall apparent effect is that glycolysis is inhibited under aerobic conditions.
2. **The Resynthesis of Glycogen.** Aerobic conditions accelerate the resynthesis of glycogen and other carbohydrates.
3. **The Accumulation of Lactate.** When oxygen is made available to the facultative anaerobe, the accumulation of lactate or other end products of anaerobic glycolysis ceases.

Clearly, any proposal concerning the control of glycolysis must be consistent with these important observations.

The control of glycolysis in higher animals is integrated with the overall control of the five processes shown in Figure 8.10. The five processes are:

I and II. Glycogenesis and Glycogenolysis The formation and breakdown of glycogen is discussed in detail in Chapter 12. These two processes are under both allosteric and hormonal control. Excess glucose and

Figure 8.10

The major processes of glucose metabolism which are subject to control. The five overall rate processes shown (I–V) are selectively inhibited or activated with the primary purpose of maintaining a constant ratio of ATP/ADP in the cell. Therefore, ATP, ADP, and AMP are among the primary modulating substances in the control of metabolism.

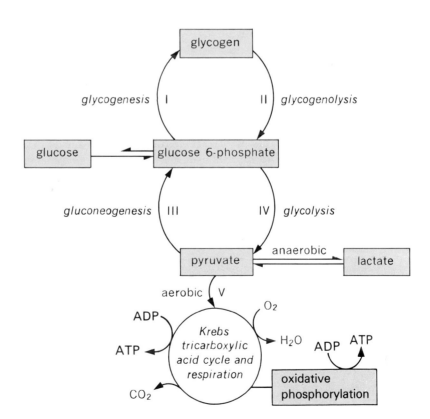

ATP are channeled, in part, into the synthesis of glycogen. Insufficient glucose and ATP stimulates the breakdown of glycogen to provide a ready source of fuel.

III. Gluconeogenesis This is the formation of glucose 6-phosphate from pyruvate (Chapter 12). This process is inhibited when glycolysis is active, and stimulated when glycolysis is sluggish.

IV. Glycolysis This is the transformation of glucose 6-phosphate to pyruvate and subsequent products, depending on conditions. In aerobic organisms, glycolysis and oxidative catabolism are linked in terms of both flow of material and control.

V. Tricarboxylic Acid Cycle and Respiration In the absence of oxygen these processes cannot occur. As Figure 8.10 shows, under aerobic conditions pyruvate enters this pathway and is oxidized to CO_2 and H_2O. The ATP produced by respiration is a major controlling factor in glycolysis.

Glycolysis is modulated by three regulatory enzymes: *hexokinase*, *phosphofructokinase*, and *pyruvate kinase*. Table 8.1 shows some of the modulators for these regulatory enzymes. In a respiring cell, the rate of glycolysis must be inhibited because most of the ATP generated comes from the aerobic catabolism of pyruvate. Since glycolysis generates pyruvate, it would be disastrous if pyruvate were to accumulate as rapidly under aerobic conditions as under anaerobic conditions because the subsequent oxidation of pyruvate yields so much energy. Therefore, the inhibition of glycolysis by increased concentrations of ATP, citrate, and fatty acids (Table 8.1), all end products of pyruvate catabolism, is a logical control mechanism. Under normal physiologi-

Table 8.1 Some Modulating Substances for Glycolysis.

Control enzyme	Activators	Inhibitors
Hexokinase	—	Glucose 6-phosphate (product inhibition)
Phosphofructokinase (PFK)	ADP, AMP	ATP, citrate
Pyruvate kinase	Fructose 1,6-diphosphate, glucose 6-phosphate, glyceraldehyde 3-phosphate	Alanine, ATP, fatty acids

cal conditions, PFK is the pacemaker for glycolysis. It is activated by increased levels of AMP and ADP and inhibited by increased levels of ATP and citrate. The net effect is a balance between two opposing forces, which tends to maintain a given ratio of ATP/ADP within narrow limits.

The case of pyruvate kinase control is interesting, since it is integrated with the control of both glycolysis and gluconeogenesis. The activators for this enzyme (Table 8.1) lead us to conclude that the system represents a mechanism to prevent the accumulation of glycolytic intermediates, a situation that might occur, for example, from the ingestion of a large amount of carbohydrate (e.g., candy). The reason for the inhibition of pyruvate kinase by alanine and fatty acids is not so obvious unless we recognize that the "reversal" of glycolysis (i.e., gluconeogenesis) proceeds under conditions wherein glycolysis is not active. If excessive amounts of fats or amino acids are present, the demand for pyruvate as an energy source diminishes. In this situation, gluconeogenesis becomes an important process, leading to a storable form of metabolic fuel or to the direct excretion of excess glucose. Thus, the inhibition of pyruvate kinase by alanine and by fatty acids prevents glycolysis and gluconeogenesis from competing.

The absence of glycolysis control is not normally observed as a hereditary disorder of metabolism (see Chapter 13), since a genetic abnormality in a mainline metabolic pathway such as glycolysis is ordinarily lethal. Figure 8.11 shows what happens to an organism when the regulatory properties of PFK are absent. In the mouse with abnormal PFK, glycolysis occurs at a maximal rate under all conditions. The excessive accumulation of pyruvate and subsequent products lead to the formation of fat. This is the only means the mouse has of continuously coping with excess pyruvate and with excessive amounts of the initial product of pyruvate catabolism, acetyl CoA (see Chapter 9).

Figure 8.11

The result of lack of glycolysis control due to abnormal phosphofructokinase. Both mice are the same age and have been reared in the same manner. The smaller mouse is normal, the larger mouse is genetically obese. Researchers have shown that PFK from the genetically obese mouse is catalytically normal but lacks all regulatory ability. This figure demonstrates the interrelationship of carbohydrate and fat metabolism and the importance of control. (Courtesy John L. Howland, Bowdoin College.)

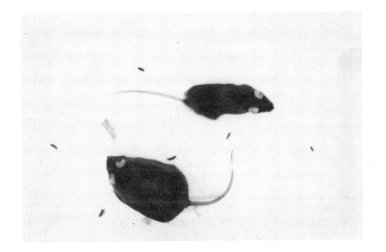

8.5

Transition from Anaerobic to Aerobic Metabolism

In this chapter we discussed anaerobic glycolysis and related processes. In the following chapter we will learn about the oxidative catabolism of glucose. The primary difference between aerobic and anaerobic glycolysis is that O_2 is the terminal electron acceptor in aerobic glycolysis, and therefore, under aerobic conditions, lactate does not form. The evolutionary advantage of respiration is tremendous, since under aerobic conditions a much smaller amount of fuel is needed to generate the required amount of ATP.

Suggestions for Further Reading

Howland, J. L. *Cell Physiology*. New York: Macmillan, Inc., 1973. A concise treatment of cell metabolism from the standpoint of integration and control.

Lehninger, A. L. *Biochemistry*. 2d ed. New York: Worth Publishers, 1975. This text should be the first place to look for the student wanting to learn more about glycolysis.

Stanier, R. Y.; Doudoroff, M. and Adelberg, E. A. *The Microbial World*. 3rd ed. Englewood Cliffs, N.J.: Prentice-Hall, 1970. A superb treatment of microbiology. Chapter 6 provides an excellent discussion of microbial fermentation.

Problems

1. For each of the terms below, provide a brief definition or explanation.

   Glycolysis   Pentose phosphate pathway
   Pasteur effect   Terminal electron acceptor

2. For each enzyme given below, write out the reaction catalyzed, giving the structures of reactants and products.

   Alcohol dehydrogenase   Lactate dehydrogenase
   Aldolase   Phosphofructokinase
   Enolase   Phosphoglucoisomerase
   Fructokinase   6-Phosphogluconate
   Glucose 6-phosphatase   dehydrogenase
   Glucose 6-phosphate   Phosphoglycerate kinase
     dehydrogenase   Phosphoglyceromutase
   Glutathione peroxidase   Pyruvate decarboxylase
   Glyceraldehyde 3-phosphate   Pyruvate kinase
     dehydrogenase   Triose phosphate isomerase
   Hexokinase

3. Using word equations, show how the following metabolites enter glycolysis:

   (a) maltose   (b) sucrose   (c) glycerol

4. Using equations, show how glucose might be converted to pyruvate without having fructose 1,6-diphosphate as an intermediate.

5. State the similarities and differences between anaerobic glycolysis and alcoholic fermentation.

6. Muscle aldolase cleaves not only fructose 1,6-diphosphate (its natural substrate) but also other ketose 1-phosphates. However, in the reverse direction, it is absolutely specific for dihydroxyacetone phosphate as the ketone reactant, but will accept a wide variety of aldehydes and aldehyde phosphates. Armed with this information, give the structures of the compounds present at equilibrium when pure muscle aldolase acts on a mixture of fructose 1,6-diphosphate, D-glyceraldehyde, and acetaldehyde.

7. The enzyme fructose 1-phosphate aldolase catalyzes the interconversion between fructose 1-phosphate and the trioses glyceraldehyde and dihydroxyacetone phosphate. Draw the structures of the reactants and products involved and show how this reaction is an aldol condensation.

8. Arsenate anion, $HAsO_4^{2-}$, is a competitive inhibitor for phosphate in many enzyme reactions involving phosphate. While the kinases of glycolysis are not "fooled" by arsenate, glyceraldehyde 3-phosphate dehydrogenase will utilize arsenate instead of phosphate. In the presence of arsenate, an acyl arsenate derivative forms instead of an acyl phosphate. This arsenate derivative hydrolyzes spontaneously with the liberation of heat.

   (a) Write out the steps for the conversion of fructose 1,6-diphosphate to pyruvate in the presence of arsenate (assume the presence of metabolites essential to glycolysis).
   (b) What is the net yield of ATP from glycolysis in the presence of arsenate?

9. Referring to this chapter and to the chart at the back of the book, write out the series of steps for the conversion of

   (a) D-fructose into ethanol and CO_2;
   (b) glycerol (from the digestion of dietary fat) into lactate;
   (c) glucose into α-D-ribose.

10. Does glucose 6-phosphate dehydrogenase deficiency interfere with the availability of pentoses for nucleic acid biosynthesis? Justify your answer.

CHAPTER 9

The Tricarboxylic Acid Cycle and Respiration

In Chapter 8 we studied glycolysis, a fundamental catabolic sequence capable of extracting a modest amount of the energy available in foodstuffs under either aerobic or anaerobic conditions. A cell relying only on glycolysis for energy production releases a waste product, such as ethanol or lactate, which is still a good fuel. For a primitive free-living procaryote, this inefficient but uncomplicated catabolic sequence is adequate for meeting the modest energy demands of such simple cells. But in more highly evolved and specialized eucaryotic cells of higher organisms, glycolysis, *by itself*, is not efficient enough to supply the energetic needs of such higher cells. Thus, nature has evolved pathways that utilize the waste products such as lactate which more primitive cells discard. Aerobes and facultative anaerobes have evolved the means of oxidizing pyruvate to CO_2 and H_2O in the presence of oxygen and, in the process, make complete utilization of the energy available in a foodstuff such as glucose. This efficiency of energy utilization means that a multicellular organism can increase in size and complexity without requiring a correspondingly large increase in food intake. In this chapter we will learn how aerobic cells oxidize pyruvate to CO_2 and H_2O and how this oxidation generates ATP.

9.1

Oxidation of Pyruvate: An Overview

Pyruvate is a central molecule in aerobic metabolism. In a respiring cell, the major fate of pyruvate is oxidation to *acetyl CoA* and CO_2. Pyruvate can be transformed into other metabolites as well; some of these processes are summarized in Figure 9.1.

The oxidation of pyruvate and subsequent steps require the action of three groups of enzymes:

1. The **pyruvate dehydrogenase** complex oxidizes and decarboxylates pyruvate to a form of acetate, namely, thioester acetyl CoA.
2. The Krebs tricarboxylic acid cycle (TCA cycle) oxidizes the two-carbon acetate unit to CO_2 and forms NADH and $FADH_2$.
3. The respiratory chain of electron-transferring enzymes reoxidizes the NADH and $FADH_2$ coenzymes produced in the dehydrogenation reactions of catabolism. In respiration, electrons and protons, originally

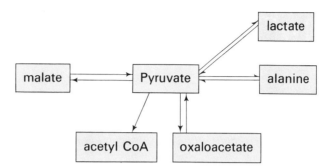

Figure 9.1

Various fates of pyruvate in cells of higher animals.

derived from food molecules, ultimately react with O_2 to produce H_2O. The respiratory chain of enzymes is located in the inner mitochondrial membrane and the terminal electron acceptor is oxygen. The redox energy gained by these electron-transfer (redox) reactions is partially conserved by *coupling* electron transfer to the phosphorylation of ADP. In addition to acting as a terminal electron acceptor for $FADH_2$ and NADH co-enzymes, produced in mitochondrial dehydrogenations, the respiratory chain can utilize certain pathways to act as a terminal electron acceptor for NADH produced in the cytoplasm during, for example, aerobic glycolysis.

These three groups of enzymes are the major subject of this chapter. The relationships among these groups are shown in Figure 9.2 (p. 272).

All the processes in Figure 9.2 occur in the **mitochondrion**, a subcellular organelle which has been called "the powerhouse of the eucaryotic cell" because it is where most of the ATP is generated during aerobic catabolism. Before we discuss the oxidation of pyruvate in more detail, it is appropriate to take a close look at the organelle in which this occurs.

9.2

Mitochondrial Structure and Function

In procaryotic cells, electron transport and related redox reactions and respiration occur in the cell membrane surrounding the cytoplasm (see Chapter 1). In eucaryotic cells, oxidative enzymes and respiratory systems are located in the mitochondrion, which is about the size of a typical bacterium (1 micron × 2 microns). Some biologists believe that the eucaryotic cell evolved from a symbiotic relationship between a primitive procaryotic anaerobe and a primitive procaryotic aerobe. The hypothesis that the mitochondrion developed from a primitive bacterium in such a relationship has some attractive features. For example, the mitochondrion has retained many of the functions of a separate

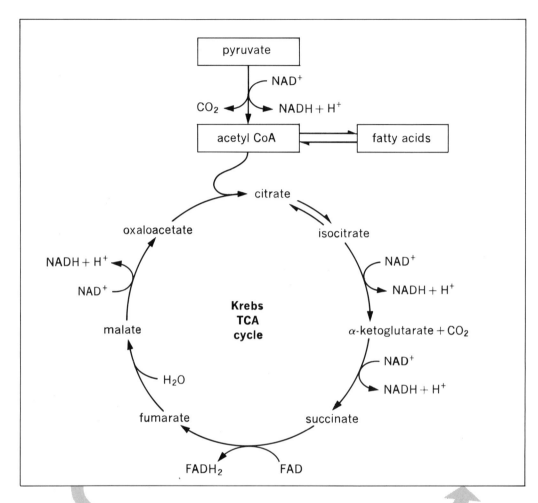

Figure 9.2

Relationships between
various steps in the
oxidation of pyruvate.

living cell, including self-replication and the synthesis of its own proteins. However, the mitochondrion has evolved into an organelle far removed from a free-living viable cell since it requires processes of the cytoplasm to provide necessary metabolites.

All aerobic and facultative anaerobic eucaryotes contain mitochondria. The number of mitochondria in a given cell can range from one in the cells of certain microorganisms, to a few dozen in yeast cells, to 1000 or more in liver cells. The egg cells of vertebrates can possess as many as 200,000 mitochondria. The number of mitochondria in a specific cell is related directly to the energy requirements of that cell.

We can best describe the structure of the mitochondrion by considering two deflated balloons, one large and one small. If you stuffed the large balloon into the small one, extensive folding of the large balloon would occur. If you then took a cross section, the result would be somewhat analogous to the diagrams of mitochondrial structure shown in Figures 9.3 and 9.4. The overall

Figure 9.3

(a) Structure of a mitochondrion, illustrating mitochondrion configuration as found in pancreatic cells. Note that the *cristae* greatly enlarge the surface of the inner membrane. The granules shown could contain any one of a number of materials. In some instances they appear to contain high-molecular-weight polymers of sugars; under other circumstances they may be formed from insoluble salts of calcium. The matrix contains many enzymes, as well as DNA and RNA.
(b) Structure of the inner mitochondrial membrane. The knob-like stalks, identified with ATP synthesis, have been observed under the electron microscope. (Both figures courtesy of J. L. Howland. From J. L. Howland, *Cell Physiology*, Macmillan, 1973.)

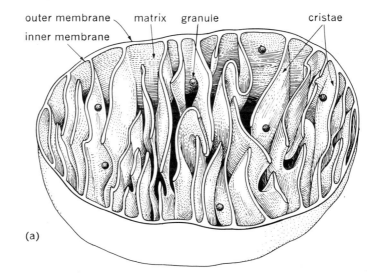

(a)

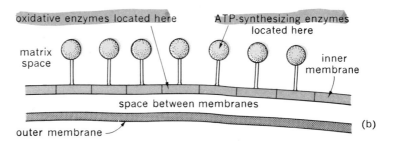

(b)

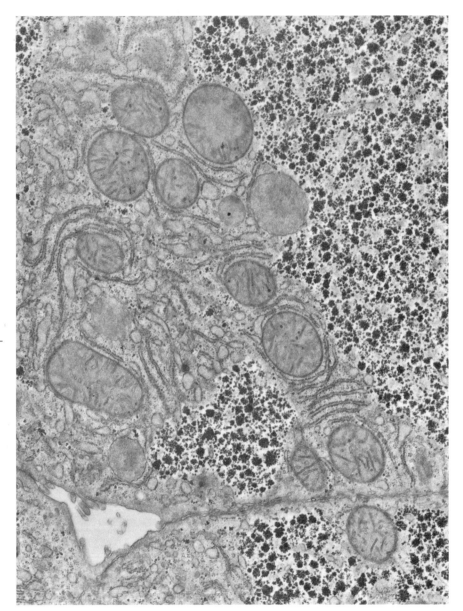

Figure 9.4

Mitochondria in human hepatocytes. The electron micrograph shows the detail of a human liver cell with numerous mitochondria, ×24,000. The cristae are evident as flattened vacuole-like structures in the mitochondrial envelope. (Figure courtesy of Solon Cole, M.D., Director of Electronmicroscopy, Department of Pathology, Hartford Hospital, Hartford, Conn.)

appearance of the mitochondrion is highly dependent on its source. While the basic structural features do not vary, the extent of cristæ formation and the shape (round vs oblong) can differ depending on the particular tissue or organism from which the mitochondrion is isolated.

The outer mitochondrial membrane consists of a typical trilaminar lipid membrane of the sort discussed in Chapter 6 and is permeable to most small molecules. It contains various membrane-bound enzymes, including those activated by fatty acids (Chapter 11). The enfoldings, or cristæ, of the inner membrane make the area of the inner mitochondrial membrane as much as 10^4 times the area of the outer membrane. The various functions of the mitochondrion are associated with the different parts of the structure shown in Figure 9.3. Table 9.1 summarizes the properties and functions of the structural features of the mitochondrion.

Table 9.1   Properties of the Major Structural Features of the Mitochondrion.

Part	Structural features	Physical features	Some enzymes localized in this part
Outer membrane	Trilaminar phospholipid membrane	Permeable to most molecules up to mol. wt. of 10^4	Fatty acyl thiokinases
Inner membrane	Phospholipid-protein membrane, much more complex and highly structured [see Figure 9.3(b)]	Permeable only to H_2O, short-chain fatty acids, and neutral molecules; impermeable to ions.	Repiratory chain; oxidative phosphorylation; α-keto acid dehydrogenases; succinate dehydrogenase; metabolite permease and transport systems
Matrix	Solution of proteins and metabolites	Contains granules of precipitated material such as calcium phosphate, ribosomes, and nucleic acids	Soluble enzymes of the Krebs TCA cycle; fatty-acid oxidation; urea biosynthesis

9.3

Conversion of Pyruvate to Acetyl CoA in the Mitochondrion

The predominant fate of pyruvate in most mammalian cells is the oxidative decarboxylation of the three-carbon α-keto acid to the elements of a two-carbon acetyl group. This is a highly exergonic reaction ($\Delta G^{0'} < -8$ kcal/mole). Part of the large free-energy change is conserved in the form of a reduced NADH coenzyme molecule, which serves as fuel for the respiratory chain upon re-oxidation. Another part of the free-energy change is conserved in a high-energy thioester link in the final product of this reaction, acetyl CoA. The net reaction and the structure of acetyl CoA are:

$$CH_3-\overset{\overset{O}{\|}}{C}-CO_2^- + NAD^+ + CoASH \longrightarrow CH_3-\overset{\overset{O}{\|}}{C}-SCoA + NADH + CO_2$$

Pyruvate $\qquad\qquad\qquad\qquad\qquad\qquad\qquad \Delta G^{\circ\prime} < -8$ kcal/mole

Acetyl $\qquad\qquad\qquad$ CoA

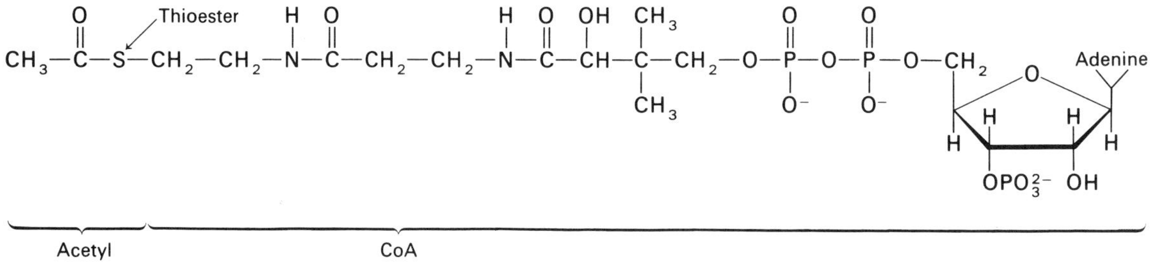

The enzyme system that carries out this reaction is the pyruvate dehydrogenase multienzyme complex, located on the inner surface of the inner mitochondrial membrane in eucaryotes and in the cell membrane in bacteria. The enzyme complex from *E. coli* has been studied extensively. Table 9.2 lists the major features of this enzyme.

Table 9.2 Pyruvate Dehydrogenase Complex of *E. coli.*

Part	Mol. wt. of smallest functional subunit	Number of subunits in complex	Coenzyme required
Pyruvate decarboxylase	90,000	24	Thiamine pyrophosphate (TPP)
Dihydrolipoyl transacetylase	70,000	24	Dihydrolipoic acid (DHL)
Dihydrolipoyl dehydrogenase	112,000	6	FAD, NAD$^+$

The basic enzymatic units of the pyruvate dehydrogenase complex are precisely arranged in a definite pattern, since the overall reaction involves several steps requiring closely coordinated action of the parts of the complex. The steps of the reaction are shown at the top of the next page.

Pyruvate decarboxylase (E_1—TPP):

$$E_1-TPP + CH_3-\overset{\overset{\textstyle O}{\|}}{C}-CO_2^- + H^+ \longrightarrow E_1-TPP-\overset{\overset{\textstyle OH}{|}}{CH}-CH_3 + CO_2$$

Dihydrolipoyl transacetylase $\left(E_2\text{—}\underset{S}{\overset{S}{\diagup}}\right)$:

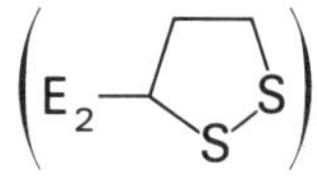

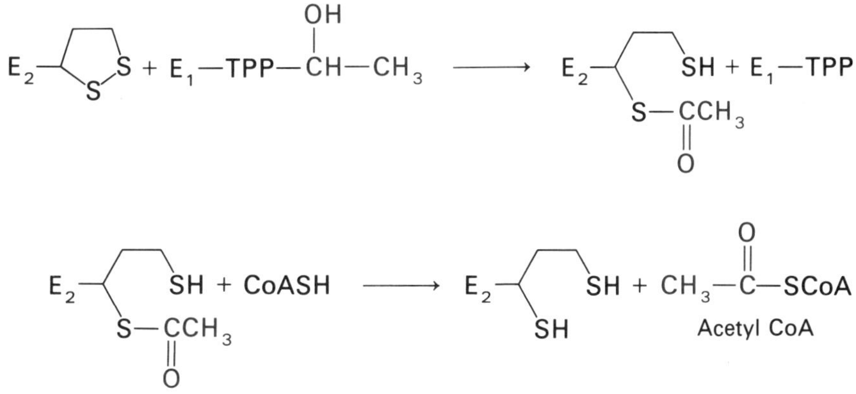

Dihydrolipoyl dehydrogenase (E_3—FAD):

$$E_3-FADH_2 + NAD^+ \longrightarrow E_3 FAD + NADH + H^+$$

The product of this *irreversible* oxidative decarboxylation reaction is acetyl CoA, which can also be produced by the degradation of fatty acids or by the catabolism of certain amino acids. Acetyl CoA can enter the Krebs TCA cycle and be degraded to CO_2, H_2O, and CoASH, or it can serve as the building block for the biosynthesis of fatty acids and other lipids, depending on metabolic needs.

9.4

The Krebs TCA Cycle

The Krebs TCA cycle and related processes constitute a central part of intermediary metabolism. It has been termed an **amphibolic** pathway because it not only provides energy by degrading the two-carbon acetyl unit of acetyl CoA, but it also makes the intermediate metabolites of the cycle available for bio-synthetic purposes. Since the Krebs TCA cycle is a closed loop, so to speak, there is *no net synthesis of material* in the cycle. Therefore, other pathways replenish the intermediates of the Krebs TCA cycle to make up for material drawn off for other uses. The extent to which this central cycle is integrated with other pathways can be appreciated by consulting the chart at the back of the book.

The overall stoichiometry of the Krebs TCA cycle is

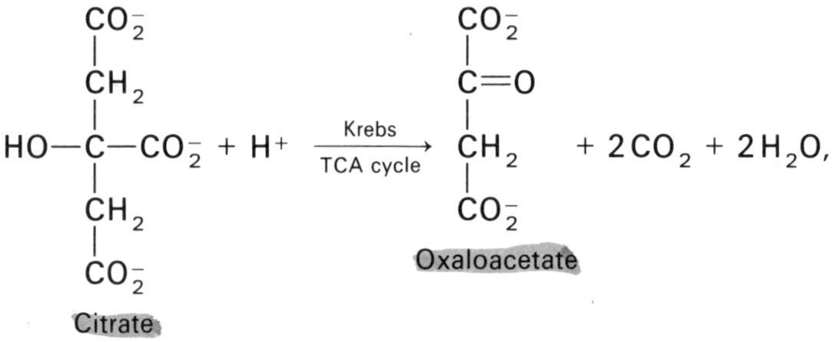

whereupon another mole of acetyl CoA reacts with oxaloacetate to produce citrate, permitting the continued "turning" of the cycle. Figure 9.5 shows the steps of the Krebs TCA cycle, which we will now discuss in more detail.

Steps of the Krebs TCA Cycle

Formation of citrate The formation of citrate is carried out by *citrate synthetase*, also known as citrate-condensing enzyme or citrate oxalo-acetate lyase. While the actual mechanism of this enzyme is not completely understood, the organic chemistry involved is a condensation reaction between a carbonyl group and a nucleophile:

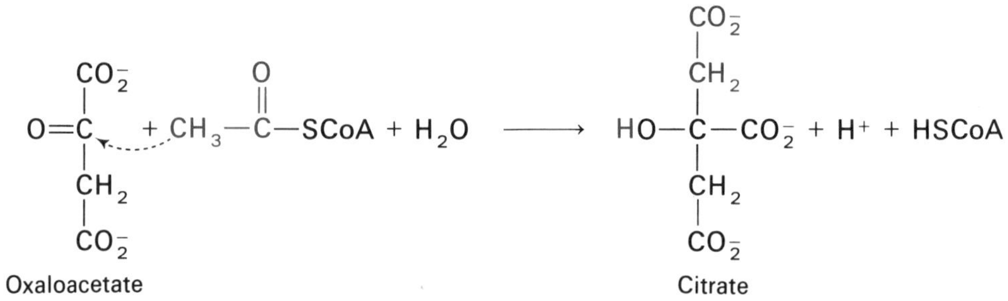

We do know that citrate synthetase is a regulatory enzyme that controls the rate of the Krebs TCA cycle. For example, ATP is a negative modulator of citrate synthetase. Inhibition by ATP decreases the affinity of the enzyme for

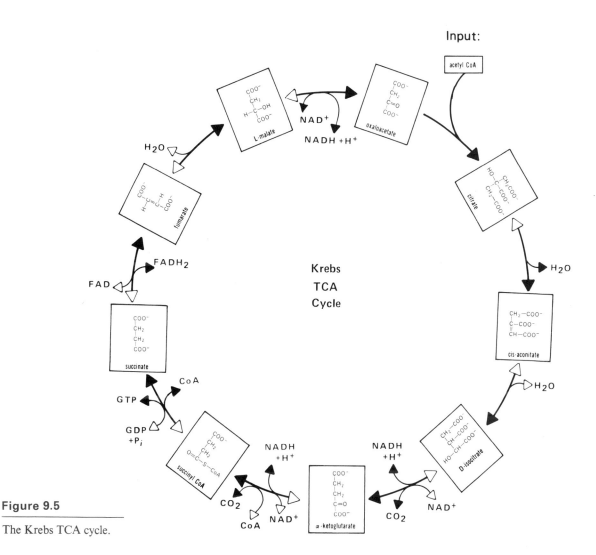

Figure 9.5

The Krebs TCA cycle.

acetyl CoA when the need for acetyl CoA oxidation by the Krebs TCA cycle is low. Acetyl CoA is then shunted into other channels, such as lipid biosynthesis. Even though the product of this reaction is a symmetrical molecule, the enzyme reaction occurs as if it were an asymmetric substrate (see Problem **9**, Chapter 3).

Formation of isocitrate The enzyme *aconitase* catalyzes the reversible hydration of the unsaturated tricarboxylic acid anion *cis*-aconitate, yielding either citrate or isocitrate depending on the stereochemistry of the hydration of *cis*-aconitate. In the normal operation of the cycle, aconitase catalyzes the overall conversion of citrate to isocitrate.

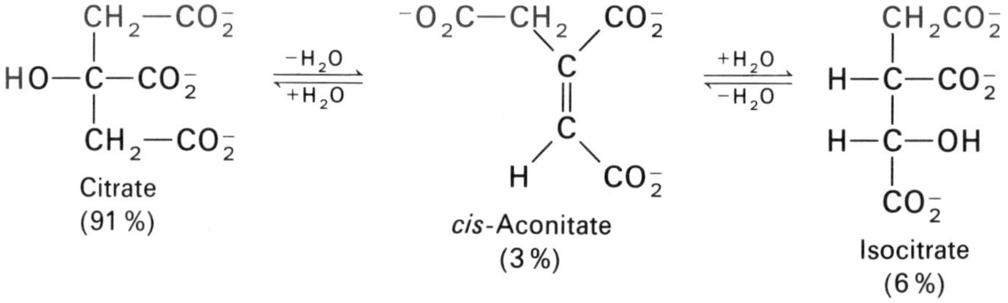

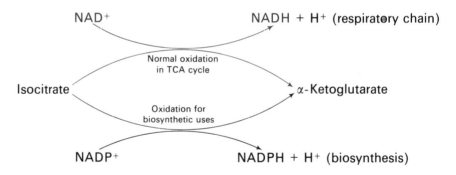

Citrate (91%) — cis-Aconitate (3%) — Isocitrate (6%)

Dehydrogenation of isocitrate There are two forms of *isocitrate dehydrogenase*, one form requiring NAD^+ and the other requiring $NADP^+$. This reaction is the rate-limiting step of the Krebs TCA cycle and therefore controls the overall rate of respiration. The NAD^+ isocitrate dehydrogenase is a regulatory enzyme activated by ADP and NAD^+ and inhibited by ATP and NADH. The catabolic NAD^+ enzyme "shuts off" when the level of ATP (and thus acetyl CoA) is high. Isocitrate oxidized by the $NADP^+$ isocitrate dehydrogenase yields NADPH, which is required for the conversion of acetyl CoA into fatty acids and other biosynthetic uses.

The product of the reaction with either enzyme is α-ketoglutarate, which can be further broken down in the Krebs TCA cycle or can enter amino acid metabolism by conversion to glutamate:

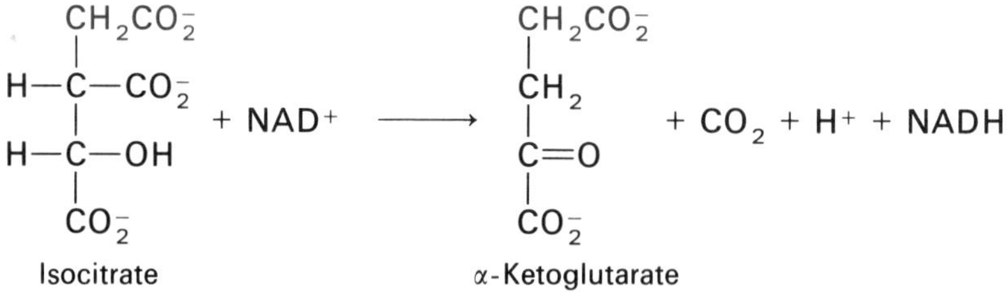

Formation of succinate The oxidative decarboxylation of α-ketoglutarate is formally identical to the oxidation of pyruvate by pyruvate dehydrogenase. It should not be surprising, then, that the enzyme α-keto-glutarate dehydrogenase is similar in structure and properties to the pyruvate dehydrogenase multienzyme complex we discussed earlier. Just as the pyruvate complex forms acetyl CoA, α-ketoglutarate dehydrogenase leads to the formation of *succinyl CoA*:

$$
\begin{array}{c}
CH_2CO_2^- \\
| \\
CH_2 \\
| \\
C{=}O \\
| \\
CO_2^-
\end{array}
\quad + NAD^+ + CoASH \longrightarrow \quad
\begin{array}{c}
CH_2CO_2^- \\
| \\
CH_2 \\
| \\
C{=}O \\
| \\
SCoA
\end{array}
\quad + NADH + H^+ + CO_2
$$

α-Ketoglutarate   Succinyl CoA

Succinyl CoA can transform into succinate in the continuation of the Krebs TCA cycle, or it can serve as a biosynthetic percursor of fatty acid and heme (porphyrin) biosynthesis.

The thioester bond in succinyl CoA has a higher free energy of hydrolysis than ATP:

$$
\text{Succinyl-SCoA} + H_2O \rightleftharpoons \text{Succinate} + \text{CoASH} \qquad \Delta G^{0'} = -9.5 \text{ kcal/mole}
$$

The cell conserves this energy change by coupling the formation of succinate with a phosphorylation reaction catalyzed by the enzyme succinyl thiokinase. The generation of one ATP molecule per succinate formed is the only place in the Krebs TCA cycle where ATP is generated directly. The process is called *substrate-level phosphorylation*. The phosphorylated intermediate produced is **guanosine triphosphate (GTP)**, which is similar to ATP except that the purine base guanine is present in the structure of GTP instead of adenine (see Chapter 13).

$$
\text{Succinyl-SCoA} + \text{GDP} + P_i \xrightleftharpoons{Mg^{2+}} \text{Succinate} + \text{GTP} + \text{CoASH}
$$

Guanosine triphosphate readily phosphorylates ADP in the presence of the enzyme nucleoside diphosphate kinase:

$$
\text{GTP} + \text{ADP} \xrightleftharpoons{Mg^{2+}} \text{GDP} + \text{ATP}
$$

Dehydrogenation of succinate   Succinate dehydrogenase is the only FAD-requiring dehydrogenase in the Krebs TCA cycle, and is tightly associated with the elements of the respiratory chain of enzymes in the inner mitochondrial membrane.

$$
\begin{array}{c}
\text{CO}_2^- \\
| \\
\text{CH}_2 \\
| \\
\text{CH}_2 \\
| \\
\text{CO}_2^-
\end{array}
\quad + \text{ FAD} \quad \longrightarrow \quad
\begin{array}{c}
\text{H}\diagdown\quad\diagup\text{CO}_2^- \\
\text{C} \\
\| \\
\text{C} \\
{}^-\text{O}_2\text{C}\diagup\quad\diagdown\text{H}
\end{array}
\quad + \text{ FADH}_2
$$

Succinate   Fumarate

$$
\begin{array}{c}
\text{CO}_2^- \\
| \\
\text{CH}_2 \\
| \\
\text{CO}_2^-
\end{array}
$$

Malonate

The enzyme is competitively inhibited by such substances as malonate and oxaloacetate, which are similar in structure to the substrate succinate.

$$
\begin{array}{c}
\text{CO}_2^- \\
| \\
\text{C}{=}\text{O} \\
| \\
\text{CH}_2 \\
| \\
\text{CO}_2^-
\end{array}
$$

Oxaloacetate

Formation of L-malate The stereospecific hydration of the carbon-carbon double bond of fumarate yields the α-hydroxy dicarboxylic acid anion L-malate. The enzyme catalyzing this reaction is fumarase, which is stereospecific for the formation of L-malate:

$$
\begin{array}{c}
\text{H}\diagdown\quad\diagup\text{CO}_2^- \\
\text{C} \\
\| \\
\text{C} \\
{}^-\text{O}_2\text{C}\diagup\quad\diagdown\text{H}
\end{array}
\quad + \text{ HOH} \quad \longrightarrow \quad
\begin{array}{c}
\text{CO}_2^- \\
| \\
\text{HO}{-}\text{C}{-}\text{H} \\
| \\
\text{CH}_2\text{CO}_2^-
\end{array}
$$

Fumarate   L-Malate

Dehydrogenation of L-malate This reaction is the oxidation of a secondary alcohol functional group to produce a ketone function (see Appendix B), and is therefore similar to the reaction catalyzed by lactate dehydrogenase (LDH).

$$
\begin{array}{c}
\text{CO}_2^- \\
| \\
\text{HO}{-}\text{C}{-}\text{H} \\
| \\
\text{CH}_2\text{CO}_2^-
\end{array}
\quad + \text{ NAD}^+ \quad \longrightarrow \quad
\begin{array}{c}
\text{CO}_2^- \\
| \\
\text{C}{=}\text{O} \\
| \\
\text{CH}_2 \\
| \\
\text{CO}_2^-
\end{array}
\quad + \text{ NADH} + \text{H}^+
$$

L-Malate   Oxaloacetate

One of the exciting results of recent X-ray crystallographic studies on the three-dimensional structures of LDH and malate dehydrogenase (MDH) is the discovery that the tertiary and quaternary structures of these enzymes are almost identical. It is probable that both LDH and MDH evolved from a common ancestral protein. In addition to mitochondrial MDH, there is a cytoplasmic form of MDH important in aerobic metabolism in the cytoplasm, as we will see shortly.

With the formation of oxaloacetate, the Krebs TCA cycle has come full circle. Continued operation of the cycle requires only a supply of acetyl CoA; a sufficient steady-state level of each of the intermediates of the cycle; activation of the regulatory enzymes of the cycle; and, of course, oxygen to act as a terminal electron acceptor for the reduced coenzymes produced in the cycle. The net result of one turn of the Krebs cycle is the generation of

1 ATP from substrate level phosphorylation;
3 NADH from NAD-linked dehydrogenations;
1 $FADH_2$ from FAD-linked dehydrogenation.

Control of the Krebs TCA Cycle

Control of the Krebs TCA cycle affects not only the steady-state concentrations of the intermediates in the cycle, but also the rates at which related processes such as glycolysis, lipogenesis, gluconeogenesis, and glycogenolysis take place. In an aerobic cell, the Krebs TCA cycle leads to the formation of most of the ATP generated in the cell. From what we have seen in our study of glycolysis, it should be apparent that ATP is the major allosteric modifier for the regulation of many related metabolic pathways. Therefore, the rate of the Krebs TCA cycle indirectly controls the rate of many other processes.

The major controlling enzymes in the self-regulation of the Krebs TCA cycle are

1. **Citrate synthetase.** This regulatory enzyme is strongly inhibited by ATP and NADH, which are both end products of the cycle.

2. **Isocitrate dehydrogenase.** This regulatory enzyme is the primary controlling factor in the Krebs TCA cycle. The NAD^+-requiring form is activated by ADP and inhibited by NADH and ATP.

3. **Succinate dehydrogenase.** This enzyme is strongly inhibited by low concentrations of oxaloacetate, and therefore seems to be "self-pacing," since the ultimate product of its action is oxaloacetate.

9.5

Replenishment of Krebs TCA Cycle Intermediates

The substrates of the Krebs TCA cycle have high turnover rates. That is, material is constantly withdrawn from the Krebs TCA cycle for biosynthetic purposes and then replenished with newly produced intermediates. This replenishment is essential because the Krebs TCA cycle does not, in itself, produce any new material. Its action only provides a continuous means for the degradation of successive molecules of acetate in the form of acetyl CoA. The term **anaplerosis** refers to those processes serving to replenish the Krebs TCA cycle. It is clear that the addition of any *one* of these intermediates suffices to replenish the entire cycle. There are two major anaplerotic reactions:

Pyruvate carboxylase Pyruvate and oxaloacetate can be interconverted by the action of the biotin-requiring enzyme *pyruvate carboxylase*:

$$CO_2 + CH_3-\overset{\overset{\displaystyle O}{\|}}{C}-CO_2^- \rightleftharpoons \ ^-O_2C-\overset{\overset{\displaystyle O}{\|}}{C}-CH_2-CO_2^- + H^+$$

Pyruvate   Oxaloacetate

The reaction in the forward direction is stimulated by acetyl CoA, an allosteric activator for pyruvate carboxylase. Thus, high levels of acetyl CoA, stimulating this enzyme, enhance the means for utilizing acetyl CoA.

Malic enzyme While pyruvate carboxylase is the primary anaplerotic enzyme in liver and kidney tissue, malic enzyme is important in heart and skeletal muscle. Here, pyruvate is carboxylated and *reduced* to form L-malate:

$$CH_3-\overset{\overset{\displaystyle O}{\|}}{C}-CO_2^- + CO_2 + NADPH + H^+ \rightleftharpoons \begin{array}{c} CO_2^- \\ | \\ HO-C-H \\ | \\ CH_2CO_2^- \end{array} + NADP^+$$

Pyruvate   L-Malate

In addition to the two reactions above, phosphoenol pyruvate can be reversibly dephosphorylated and carboxylated by the action of phosphoenol pyruvate carboxykinase, forming oxaloacetate. These reactions are not only important for anaplerosis, but, in the reverse direction, they are important in gluconeogenesis.

9.6

The Glyoxylate Cycle

Many organisms, such as germinating seeds and microorganisms growing in simple media, must satisfy all biosynthetic needs with very simple precursors. In such cases, carbohydrates and lipids must serve as the major raw materials

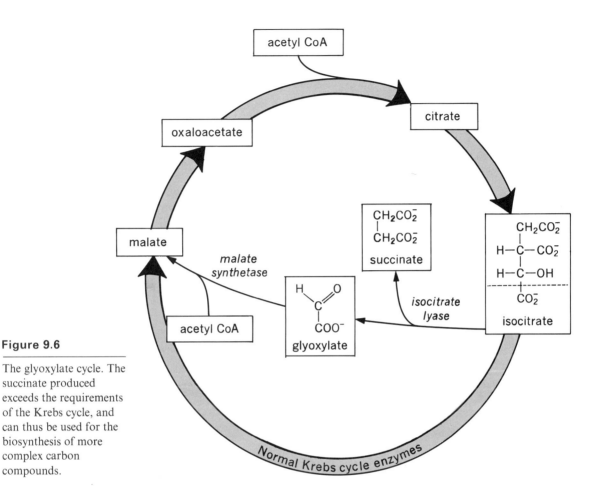

Figure 9.6

The glyoxylate cycle. The succinate produced exceeds the requirements of the Krebs cycle, and can thus be used for the biosynthesis of more complex carbon compounds.

for biosynthesis as well as for energy. Since both of these raw materials lead to the formation of acetyl CoA, a pathway must exist to utilize acetyl CoA as a biosynthetic precursor for substances other than lipids.

In such organisms we find a subsidiary to the Krebs TCA cycle known as the **glyoxylate cycle**, which permits the formation of one mole of succinate from two moles of acetyl CoA. The glyoxylate cycle is thus a form of internal anaplerosis, since the formation of "new" succinate means that the concentration of any of the Krebs TCA cycle intermediates can be increased. The overall sequence of the glyoxylate cycle is shown in Figure 9.6. The key enzymes are isocitrate lyase and malate synthetase, which are not present in the cells of most higher organisms. These two enzymes are *inducible* (see Chapter 3) in germinating seeds and are associated with subcellular particles in eucaryotes called *glyoxysomes*. Even when the glyoxylate cycle is working, the Krebs TCA cycle continues to function to supply energy.

9.7

Electron Transport and Oxidative Phosphorylation

Oxidation-reduction is the fundamental basis for all energy production in biological systems. For example, even though the net oxidation state of the system in anaerobic glycolysis remains the same, an oxidation reaction is necessary to produce the high-energy phosphate compound 1,3-diphospho-glycerate (see p. 253). Subsequent ATP generation in anaerobic glycolysis merely capitalizes on the overall energetics of this oxidation step. In the case of aerobic catabolism, the role of redox reactions is more obvious.

The energetics of a redox reaction are based on the tendencies of the oxidant to accept electrons and of the reductant to donate electrons. We can quantitatively express this tendency to donate or accept electrons by an overall free-energy change for the redox reaction, measured in kilocalories or, by means of a simple conversion factor, in volts.

The electron-accepting ability of an oxidant can be expressed as its **standard reduction potential** $E^{0'}$, measured in volts. The conventions with respect to the standard free-energy change for a reaction, discussed in Chapter 6, also apply to reduction potentials. An important redox couple for our purposes is the H_2O/O_2 couple:

$$\underset{\text{Oxidant}}{\tfrac{1}{2}O_2 + 2H^+ + 2e^-} \rightleftharpoons \underset{\text{Reductant}}{H_2O} \qquad E^{0'} = +0.82 \text{ volts} \quad (1)$$

We can relate electrical potential energy (volts) to chemical potential energy (ΔG) by the relation:

$$\Delta G = -nF \quad \text{(electrical potential in volts)}$$

where n is the number of electrons transferred ($n = 2$ in Eq. 1) and F is a quantity called Faraday's constant and equals 23,062 calories/volt–equivalent. The free-energy change in Eq. 1 can easily be calculated as:

$$\Delta G^{0'} = -nFE^{0'} = -2(23,062)(+0.82) = -37.8 \text{ kcal/mole}$$

The important thing to note here is that a seemingly small voltage translates into what is, in fact, a large difference in chemical energy. Equation (1) is called a *half-cell reaction*. We can combine this with another half-cell reaction, such as the $NADH/NAD^+$ redox couple shown in Eq. (2), obtaining an overall redox reaction given in Eq. (3).

$$\text{NAD}^+ + 2\text{H}^+ + 2\text{e}^- \rightleftharpoons \text{NADH} + \text{H}^+ \qquad E^{0\prime} = -0.32 \text{ V} \qquad (2)$$

The $E^{0\prime}$ value of the $\text{H}_2\text{O}/\text{O}_2$ couple is more positive (and thus more exergonic) than the NADH/NAD^+ couple. This means that thermodynamically speaking, O_2 will oxidize NADH:

$$
\begin{array}{lll}
\tfrac{1}{2}\text{O}_2 + 2\text{e}^- + 2\text{H}^+ \rightleftharpoons \text{H}_2\text{O} & & E^{0\prime} = +0.82 \text{ V} \\
\text{NADH} + \text{H}^+ \rightleftharpoons \text{NAD}^+ + 2\text{H}^+ + 2\text{e}^- & & E^{0\prime} = +0.32 \text{ V} \\
\hline
\text{Net:} \quad \tfrac{1}{2}\text{O}_2 + \text{NADH} + \text{H}^+ \rightleftharpoons \text{H}_2\text{O} + \text{NAD}^+ & & \Delta E^{0\prime} = 1.14 \text{ V}
\end{array}
$$

$$G^{0\prime} = -2(23{,}062)(1.14) = -52.3 \text{ kcal/mole}$$

These manipulations should be familiar to you from general chemistry. The important thing here is that redox reactions can involve free-energy changes that are massive in a biological context. The reaction in Eq. (3) describes what occurs overall when oxygen serves as a terminal electron acceptor for NADH. The overall energy change is more than enough to drive the phosphorylation of seven ADP molecules per NADH oxidized. However, cells cannot make use of such large amounts of energy at one time, and so they contain a mechanism that breaks up the large decrease in chemical (redox) potential energy into more manageable steps. This is done by the respiratory chain of redox enzymes, which accepts electrons from good reducing agents (NADH and FADH_2) and lowers the energy of these electrons stepwise by means of coupled redox reactions, conserving the optimal amount of energy as ATP. This concept is illustrated in Figure 9.7.

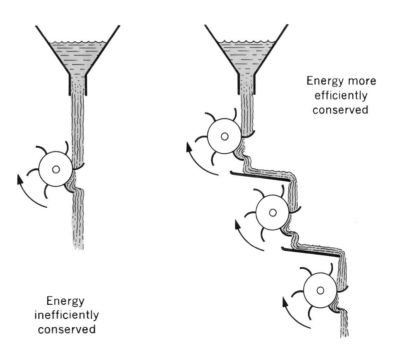

Figure 9.7

Stepwise conservation of large overall free-energy changes compared with a single conservation step. Note that though the water runs faster in the left-hand figure, the water wheel can turn at only a finite maximum rate.

Energy more efficiently conserved

Energy inefficiently conserved

Respiratory Chain

The respiratory chain is a common pathway taken by all electrons removed from substrate molecules during dehydrogenation steps in aerobic metabolism. The proteins of the respiratory chain are embedded in the mitochondrial inner membrane as discrete respiratory assemblies. In bacteria, we find respiratory assemblies in the cell membrane, where they play a role similar to that of the mitochondrial respiratory chain.

The coenzyme groups associated with the respiratory chain include those redox coenzymes discussed in Chapter 3. Table 9.3 summarizes the carriers and their coenzymes. The coenzymes shown in the table have all been discussed before except for the nonheme iron (NHI) prosthetic group of the NADH dehydrogenase complex. NHI proteins are a broad class of redox proteins that do not contain the Fe^{2+}/Fe^{3+} cofactor associated with a heme group. Instead, in many NHI proteins the metal ion is associated with sulfur in an iron-sulfur redox prosthetic group. The trend of $E^{0'}$ values in Table 9.3 is significant because it illustrates how the respiratory chain works. Electrons flow spontaneously from a high potential (negative voltage) to a low potential. In the course of this electron flow, work can be performed. The work obtained from electron transport in the respiratory chain is the biological analog of the work obtained from electrons flowing through the field and armature coils of an electric motor, causing the armature to turn and do mechanical work.

Table 9.3 Electron Carriers of the Respiratory Chain.

Protein	Prosthetic group	Number of electrons transferred	$E^{0'}$ (V)
NADH dehydrogenase	FMN; NHI	2	-0.11
Coenzyme Q (not a protein)	Quinone	2	$+0.10$
Cytochrome b	Heme	1	$+0.06$
Cytochrome c_1	Heme	1	$+0.22$
Cytochrome c	Heme	1	$+0.25$
Cytochrome a, a_3	Heme	1	$+0.28$
(Oxygen)	—	4	$+0.82$

Steps of the Respiratory Chain

The overall scheme of the respiratory chain is shown in Figure 9.8. The order of electron carriers in the respiratory chain and the relationships of the various electron donors to the chain are fairly well understood. The actual mechanism of the electron transfers, particularly the means by which the energy changes in electron transport drive the phosphorylation of ADP, is the subject of intense current research.

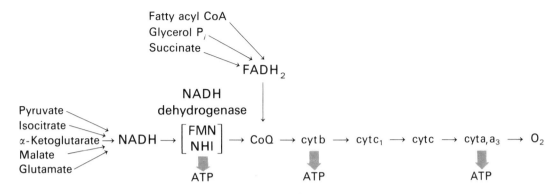

Figure 9.8

Path of electron transfer
in the respiratory chain.
The large arrows
represent probable
points at which energy is
conserved by coupling
electron transfer to the
phosphorylation of
ADP.

We can consider the respiratory chain in two parts: hydrogen-transfer
steps, where pairs of protons and electrons are transferred concurrently, and
electron-transfer steps, where only electrons are transferred.

NADH dehydrogenase   The substrate for this membrane-bound
enzyme complex is NADH, which is produced by NAD-linked dehydrogenations. The action of this complex is not well understood, but we do know that
both FMN (Chapter 3) and a nonheme iron [NHI] prosthetic group are
involved. The reaction catalyzed by NADH dehydrogenase is

$$\text{NADH} + \text{H}^+ + \text{E}\!-\!\begin{bmatrix}\text{FMN}\\\text{NHI}\end{bmatrix} \;\rightleftharpoons\; \text{NAD}^+ + \text{E}\!-\!\begin{bmatrix}\text{FMNH}_2\\\text{NHI}\end{bmatrix} \quad \Delta G^{0\prime} = -9.7 \text{ kcal/mole}$$

A large free-energy change occurs at this step, and the cell conserves part of
this energy change by coupling this exergonic reaction with the phosphorylation
of ADP.

Coenzyme Q   Coenzyme Q, or ubiquinone, is currently thought to
act as an acceptor for electrons from reduced flavin coenzymes. Reduced
NADH dehydrogenase ($FMNH_2$), reduced succinate dehydrogenase ($FADH_2$),
and other reduced flavoproteins donate electrons to coenzyme Q. This quinone
coenzyme is a derivative of benzoquinone, which is highly soluble in the lipid
environment of the mitochondrial inner membrane. The reaction between
CoQ (ox) and a reduced flavin coenzyme, FlH_2 (red), can be given as

Both of these redox steps transfer protons and electrons together. We can also see that NADH is a stronger reducing agent than $FADH_2$ or $FMNH_2$, since the reduced flavin coenzyme of NADH dehydrogenase is formed in an exergonic reaction coupled to ATP generation. Therefore, the reoxidation of NADH by the respiratory chain will produce *one more ATP* than the reoxidation of reduced flavin coenzymes.

The Cytochrome System

Electrons from reduced CoQ enter the chain of **cytochromes**, which are electron-transferring heme proteins with different redox properties. From Table 9.3 you can see that a net change in redox potential of 0.22 V occurs on going from cytochrome b to cytochrome a, a_3—a change of 10.1 kcal/mole! Yet the prosthetic group for the cytochromes is basically the same Fe^{2+}/Fe^{3+} heme system discussed in Chapter 3. Therefore, the protein part of each cytochrome must provide a sufficiently different environment for the heme group to produce the markedly different $E^{0'}$ values observed.

The sequences of reactions in the cytochrome system is shown diagrammatically in Figure 9.9. The net result of the sequential redox steps is to break up a large decrease in free energy into two relatively manageable free-energy decreases, enabling the generation of two moles of ATP per two moles of electrons transferred to oxygen to form H_2O.

Respiratory Inhibitors

The respiratory chain electron carriers can be inhibited by specific compounds. These respiratory inhibitors have been particularly helpful in deducing the overall process of electron transport by allowing investigators to "turn off" the chain at specific points. Table 9.4 summarizes the structures and

Figure 9.9

The cytochrome system.

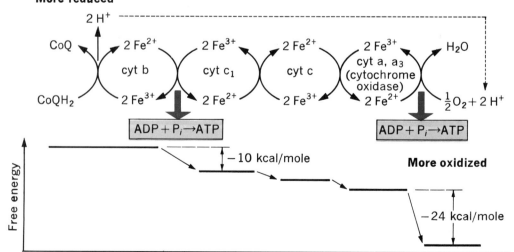

actions of some important inhibitors. Some of the compounds in this table are familiar poisons. For example, cyanide and carbon monoxide both cause death by complexing with the heme iron of cytochrome (a + a$_3$), thus directly preventing the reaction with O_2. In this case all of the respiratory chain carriers preceding cytochrome (a + a$_3$) would be in reduced forms. Some inhibitors such as *rotenone* do not totally shut down the respiratory chain. In the case of rotenone, electrons can still enter the chain at the CoQ level.

Table 9.4   Some Important Respiratory Inhibitors.

Inhibitor	Site of action
Cyanide ion: CN$^-$ Azide ion: N$_3^-$ Carbon monoxide: CO	Acts on heme iron of cytochrome (a + a$_3$)
Antimycin A	A fungal antibiotic that blocks the transfer of electrons from cytochrome b to cytochrome c$_1$
Thenoyl trifluoracetone	Prevents the reduction of cytochrome b by succinate
Piericidin A	Prevents the reduction of cytochromes by NADH by acting as an inactive structural analog of CoQ
Amytal	A barbiturate that prevents the reduction of CoQ
Rotenone	A fish poison isolated from the root of a South American plant that inhibits NADH dehydrogenase. Rotenone is a widely used insecticide.

Oxidative Phosphorylation

The process of oxidative phosphorylation is best described by Figure 9.10. Although the mechanism of oxidative phosphorylation is not well understood, it has been known experimentally for some time that electron transfer causes ATP to be generated. As scientists began to understand the stoichiometry of this respiration-driven phosphorylation, they introduced the "P:O ratio":

$$P:O \text{ ratio} = \frac{\text{moles } P_i \text{ used up}}{\text{moles "O" used up}} \equiv \frac{\text{number of ATP generated}}{\frac{1}{2}O_2 \text{ consumed}}$$

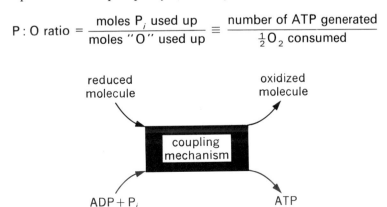

Figure 9.10

The black box of oxidative phosphorylation.

Table 9.5 P:O Ratios of Some Respiratory-Chain Substrates.

Substrate	P:O ratio
NADH + H$^+$	3
FADH$_2$	2
Citrate $\longrightarrow$ oxaloacetate	12
α-Ketoglutarate $\longrightarrow$ oxaloacetate	9
Succinate $\longrightarrow$ oxaloacetate	5
Malate $\longrightarrow$ oxaloacetate	3

To determine this quantity, we can experimentally measure the consumption of O_2 and the intake of P_i by respiring mitochondria *in vitro*. By adding different substrates to the medium we can determine the P:O ratios, as shown in Table 9.5. In view of the P:O ratio value for NADH, we can see that the energetic efficiency of the electron-transport chain is quite high:

$$\text{NADH} + \text{H}^+ + \tfrac{1}{2}\text{O}_2 \xrightarrow[\text{chain}]{\text{Respiratory}} \text{NAD}^+ + \text{H}_2\text{O} \qquad \Delta G^{0\prime} = -52.7 \text{ kcal/mole}$$

3 moles of ATP are generated in this process,

$$\text{so the cell recovers } 3 \times 7.3 \text{ kcal} = +21.9 \text{ kcal/mole}$$
$$\text{Net loss as heat:} \quad \overline{-30.8 \text{ kcal/mole}}$$

The net efficiency is at least $(21.9/52.7) \times 100 = 42\%$.

The mechanism by which oxidative phosphorylation is coupled to respiration is unclear. Scientists have proposed several hypotheses to explain this coupling, including:

1. **Conformational coupling.** Conceptually this is similar to the reverse of muscle contraction. We know that mitochondria swell and contract, depending on the degree of respiratory activity. We do not know how such gross alterations in mitochondrial structure affect processes at the molecular level.

2. **High-energy intermediate coupling.** Many workers feel that electron transfer leads to the generation of a high-energy intermediate, which in turn drives the formation of ATP. Such an intermediate has not yet been observed, but the hypothesis can be illustrated schematically by Figure 9.11.

3. **Chemiosmotic coupling.** In 1963, Peter Mitchell proposed that pH gradients set up as a result of electron transfer actually drive ATP formation by a reverse active-transport type of process (see Chapter 7). For this proposal he was awarded a Nobel Prize in 1978.

There is much experimental evidence to support any one of these coupling hypotheses. For example, a great deal has been learned from studies of *un-*

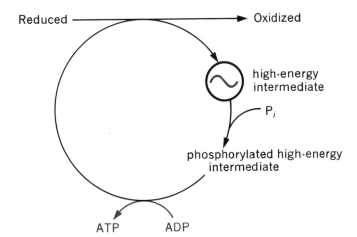

Figure 9.11

A proposed coupling
mechanism involving a
high-energy intermediate.

coupled respiration, where researchers use chemical treatment to sever the thermodynamic ties between respiration and oxidative phosphorylation. In the presence of an uncoupling agent, mitochondria respire but do not generate ATP. Many uncoupling agents act either as lipid-soluble proton carriers, such as 2,4-dinitrophenol, or as ionophores, such as gramicidin (see Chapter 2), which makes membranes permeable to ions. These experimental observations might suggest that the chemiosmotic hypothesis is valid, but the uncouplers might just as well be altering the mechanical properties of the inner membrane or inhibiting the formation of a high-energy intermediate.

9.8

Entry of Metabolites into the Mitochondrion

The mitochondrion is like a little cell within a bigger cell, and like the plasma membrane of the cell, the mitochondrial inner membrane is impermeable to most solutes:

permeable: H_2O, small fatty acids, glycerol, urea
impermeable: ions, sugars, amino acids, NAD^+, FAD, acetyl CoA, etc.

Transport systems exist for many metabolites, as shown in Table 9.6 (p. 294).
During aerobic catabolism, NADH is generated in the cytoplasm. The cell must have a way of permitting the indirect reoxidation of this extra-mitochondrial NADH by the respiratory chain, because we know from the Pasteur effect (Chapter 8) that lactate accumulation stops when oxygen becomes available to a facultative anaerobe. The reoxidation of cytoplasmic NADH is

Table 9.6 Some Solutes for which Transport Systems Exist in the Mitochondrial Inner Membrane.

ADP, ATP, P*	Glutamate
Succinate, malate, pyruvate	Na^+, K^+, Mg^{2+}, Ca^{2+}
Isocitrate, citrate, *cis*-aconitate	

* In the case of ADP/ATP, the transport system is called *exchange diffusion* because one ATP is carried out for every ADP carried in.

Figure 9.12

The glycerol-phosphate shuttle and the malate shuttle. Because mitochondrial glycerol phosphate dehydrogenase is an FAD-enzyme, the glycerol shuttle yields one less ATP per extramitochondrial NADH than the malate shuttle.

performed by the shuttle systems. There are two primary shuttle systems operating: the *glycerol-phosphate shuttle* and the *malate shuttle*. Both serve to regulate the amount of NADH in the cytoplasm, as shown in Figure 9.12. In addition to the shuttle systems and respiration, there is evidence that NADH can be reoxidized to NAD^+ by the *transhydrogenase reaction*:

$$NADP^+ + NADH \rightleftharpoons NADPH + NADP^+$$

This reaction is thought to form additional NADPH when biosynthesis needs are high.

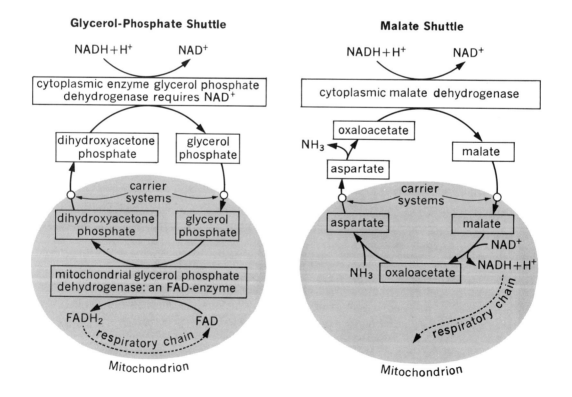

Glycerol-Phosphate Shuttle

Malate Shuttle

9.9

Control of Respiration

Figure 9.13

Respiratory control: the stimulation of mitochondrial respiration by ADP. This is a trace produced by a polarographic electrode sensitive to oxygen. A downward deflection of the line indicates an increase in the rate at which oxygen is used up. The reaction mixture at the beginning of the experiment (left side of the chart) contains buffer, magnesium, phosphate, succinate and water and is saturated with oxygen. When a suspension of mitochondria is introduced, oxygen uptake occurs as succinate is oxidized to fumarate. The addition of ADP stimulates oxygen consumption, which then returns to the original rate when the ADP has all been phosphorylated to form ATP. The amount of oxygen that is used at the higher rate may serve as the basis for estimating the efficiency of ATP synthesis—i.e., the ATP/oxygen ratio. (Text and figure from J. L. Howland, *Introduction to Cell Physiology*, Macmillan, 1968.)

As we have seen in our discussion of catabolism, the controlling elements of the rates of glycolysis and the Krebs TCA cycle center around the ATP/ADP ratio. This is also the case with the control of respiration rate. The respiratory chain is responsive to ATP demand expressed as a higher or lower ATP/ADP concentration ratio. The rate of respiration is sensitive to the concentration of ADP, as shown in Figure 9.13. As this figure shows, the respiration rate (oxygen consumption rate) of resting mitochondria is constant. When the concentration of ADP increases, either by cellular ATP demand *in vivo* or by direct addition of ADP *in vitro*, the rate of respiration increases sharply until the ATP/ADP ratio returns to the previous (resting) value, at which point the respiration rate falls back to the resting value.

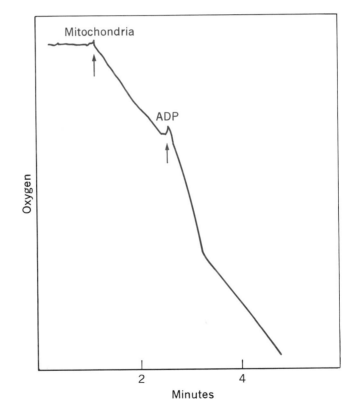

9.10

Overall Energetics of Glycolysis and Respiration

We can inventory the moles of ATP produced in aerobic carbohydrate catabolism by adding up the reduced coenzymes formed and taking into account the number of substrate-level phosphorylations, thus arriving at a total figure for ATPs generated per mole of glucose catabolized. This inventory is shown in Table 9.7. It is obvious that respiration makes efficient and complete use of glucose as a fuel molecule.

Table 9.7 Energy Inventory for Glycolysis and Respiration of One Mole of Glucose

Reaction	Reduced coenzyme	ATP formed by oxidative phosphorylation	ATP formed at substrate level
Glycolysis (aerobic)	2 NADH	4*	2
2 pyruvate $\longrightarrow$ 2 acetyl CoA	2 NADH	6	
2 isocitrate $\longrightarrow$ 2 α-ketoglutarate	2 NADH	6	
2 α-ketoglutarate $\longrightarrow$ 2 succinate	2 NADH	6	2
2 succinate $\longrightarrow$ 2 fumarate	2 FADH$_2$	4	
2 malate $\longrightarrow$ 2 oxaloacetate	2 NADH	6	
Total ATP generated:		36 ATP/glucose	

* This is 4 and not 6 because we are assuming that the glycerol-phosphate shuttle is bringing electrons from extramitochondrial NADH into the mitochondrion.

Respiration is a process common to a vast number of heterotrophic cells. Glucose is gained directly or indirectly from autotrophic cells, which do not use high-energy electrons derived from substrate molecules to drive electron transport and oxidative phosphorylation, but instead use *light* energy by a process known as photosynthetic electron transport. We will discuss photosynthesis in the next chapter and learn that there are many similarities between mitochondrial electron transport and photosynthetic electron transport.

Suggestions for Further Reading

Lehninger, A. L. *Biochemistry.* 2d ed. New York: Worth Publishers, 1975. This comprehensive text treats respiration and the Krebs TCA cycle in Chapters 17–19.

Metzler, D. E. *Biochemistry—The Chemical Reactions of Living Cells.* New York: Academic Press, 1977. This is a detailed and interesting advanced text. Chapter 10 gives a thorough discussion of respiration.

Scientific American

Goodenough, U. W. and Levine, R. P. "The Genetic Activity of Mitochondria and Chloroplasts." November 1970.

Hinkle. P. C. and McCarty, R. E. "How Cells Make ATP." March 1978.

Lehninger, A. L. "Energy Transformation in the Cell." May 1960.

———. "How Cells Transform Energy." September 1961.

Racker, E. "The Membrane of the Mitochondrion." February 1968.

Siekevitz, P. "Powerhouse of the Cell." July 1957.

Problems

1. For each of the terms below, provide a brief definition or explanation.

Amphibolic	Krebs TCA cycle
Anaplerosis	Oxidative phosphorylation
Cytochromes	Respiratory chain
Glyoxylate cycle	Substrate level phosphorylation

2. For each enzyme given below, write out the reaction catalyzed, showing the structures of reactants and products.

Aconitase	Malate dehydrogenase
Citrate synthetase	Malate synthetase
Fumarase	Malic enzyme
Isocitrate dehydrogenase	Pyruvate dehydrogenase
Isocitrate lyase	Succinate dehydrogenase
α-Ketoglutarate dehydrogenase	

3. Show the steps involved in each of the overall conversions given below. Draw the structures of all reactants and products. (Note that you may be able to write more than one pathway to accomplish the conversion, and you may have to consult Chapter 8.)

    (a) Pyruvate → malate
    (b) Succinate → pyruvate
    (c) Glucose 6-phosphate → ribose 5-phosphate
    (d) Acetyl CoA → net increase in concentration of succinate

4. How many ATPs are generated during the complete oxidative degradation to CO_2 and H_2O of each of the following compounds?

    (a) Fructose 6-phosphate
    (b) Acetyl CoA
    (c) Sucrose
    (d) Glycerol

5. How many ATPs are generated when glucose oxidizes to CO_2 and H_2O in a eucaryotic cell treated with the uncoupling agent 2,4-dinitrophenol?

6. Using equations, show the fate of the two electrons ($2e^- + 2H^+$) of the reduced NAD-prosthetic group of glyceraldehyde 3-phosphate dehydrogenase produced during glycolysis in:

   (a) Human connective-tissue cells functioning anaerobically.
   (b) Human hepatocytes functioning aerobically.
   (c) Brewer's yeast cells metabolizing glucose anaerobically.

7. Citric acid is a symmetrical molecule. However, when it is produced by the reaction of the Krebs TCA cycle catalyzed by citrate synthetase, the two-carbon fragment from acetyl CoA is added in a nonsymmetrical manner. Referring to the discussion of stereospecificity of enzyme action in Chapter 3, explain how citrate synthetase can condense acetyl CoA with oxaloacetate in a stereospecific manner.

8. Draw a diagram for respiration similar to Figure 9.8. On this diagram note the processes affected by each of the respiratory inhibitors given in Table 9.4.

9. The plant poison *atractyloside* specifically inhibits the mitochondrial ATP/ADP transport system. Describe the effects of this substance on the Krebs cycle, oxidative phosphorylation and respiration.

10. Write out the overall sequence of steps for the oxidation of extramitochondrial NADH via the respiratory chain, assuming the glycerol phosphate shuttle is operating.

11. When acetate must serve as the only carbon source for an organism, obviously the biosynthetically crucial substances of the Krebs TCA cycle cannot be synthesized from glucose. Here acetate is first converted to acetyl CoA. How would a bacterium living in dilute acetic acid convert acetate to succinate?

CHAPTER 10

Photosynthesis

The basic form of heterotrophic catabolism is the degradation of a reduced molecule such as a sugar or a fat, producing energy and substances that are ultimately more oxidized than the original foodstuffs. The situation is quite similar to our use of coal and petroleum products as fuel. In each case, we use energy which came originally from sunlight. In biological terms, the energy content of sunlight is enormous. In this chapter we will learn how this sunlight energy is conserved by photosynthetic cells.

10.1

Light Energy and Living Systems

We can visualize the basic unit of light radiation, the *photon* or *light quantum*, as a wave packet of electromagnetic radiation. A photon is to a beam of light as a molecule is to a sample of a chemical substance. Einstein developed a simple relationship between the energy of light and its wavelength:

$$E \equiv kcal/mole \equiv Nh\nu = \frac{Nhc}{\lambda} = \frac{286{,}000}{\lambda(\text{Å})} \; kcal/mole$$

where N = Avogadro's number = 6.02×10^{23} particles/mole, h = Planck's constant = 1.6×10^{-34} calories-sec, c = velocity of light = 3×10^{10} cm/sec, λ = wavelength of light in Angstroms ($1 \text{ Å} = 10^{-8}$ cm), and ν = frequency of light in Hertz ($1 \text{ Hz} = sec^{-1}$). Figure 10.1 shows the energy content per Einstein of light (6.02×10^{23} photons) as a function of wavelength.

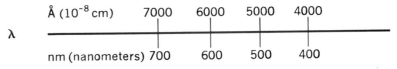

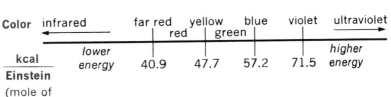

Figure 10.1

Energy content of different wavelengths of visible light.

299

Figure 10.2

Electronic excitation of a pigment molecule, showing how the electronically excited molecule can reduce an electron-acceptor molecule and oxidize an electron-donor molecule. The levels in the figure represent electronic energy levels in the molecules. We will use the concept shown in this figure often in our discussion of photosynthetic electron transport in Section 10.3.

The absorption of light by a chemical substance involves a process called *electronic excitation*. The wavelengths of light absorbed by a substance depend on the electronic structure of the molecule. A substance which appears white absorbs no light in the visible region of the spectrum; it reflects all colors, which is why it looks white. A yellow substance absorbs purple light and reflects the remaining wavelength components of visible light, thus appearing yellow. The process of electronic excitation by light absorption is a fundamental part of photosynthesis. Figure 10.2 illustrates how a substance can absorb light and enter an energy-rich electronically excited state.

The energy values of visible light shown in Figure 10.1 are of the same magnitude as the overall energy changes in respiration. It should thus be possible to design an apparatus that converts light energy into chemical energy and drives the overall reactions of the respiratory chain in reverse, producing O_2 from H_2O, NADPH from $NADP^+$, and energy to synthesize sugars from CO_2 and H_2O. Nature, of course, did this long ago—the apparatus is called a chloroplast and the process is photosynthesis.

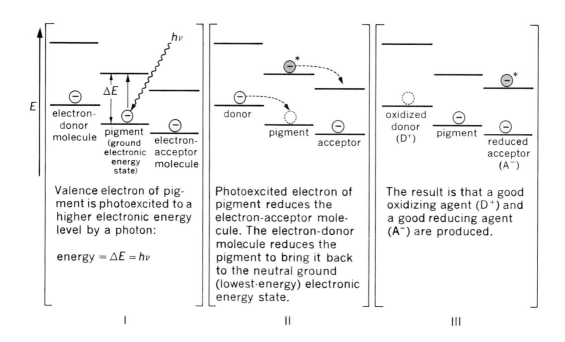

Valence electron of pigment is photoexcited to a higher electronic energy level by a photon:

energy $= \Delta E = h\nu$

Photoexcited electron of pigment reduces the electron-acceptor molecule. The electron-donor molecule reduces the pigment to bring it back to the neutral ground (lowest-energy) electronic energy state.

The result is that a good oxidizing agent (D^+) and a good reducing agent (A^-) are produced.

I   II   III

10.2

The Chloroplast and Photosynthetic Organisms

Photosynthesis is conducted by both procaryotes and eucaryotes. The distribution of photosynthetic organisms includes:

    algae, plants, and diatoms;
    protozoa;
    photosynthetic bacteria, including purple sulfur bacteria, green
        sulfur bacteria, and purple nonsulfur bacteria.

In procaryotes, the photosynthetic pigments are located in the cell membrane. In eucaryotes, the photosynthetic apparatus is located in the organelle known as the **chloroplast**.

The number of chloroplasts per cell varies from one to many, depending on the organism. The size of the chloroplast is of the same order of magnitude or larger than the mitochondrion. In fact, the lamellar (layered) lipid membrane structure of the chloroplast closely resembles the enfolded membrane structure of the mitochondrion. Figure 10.3 shows the major features of chloroplast structure. The **thylakoid disks** shown in this figure are the basic photosynthetic units of the chloroplast and contain the pigment systems responsible for absorbing and transforming light energy into chemical energy. The presence of the large starch granule in the chloroplast electron micrograph implies that carbohydrate formation using CO_2 and NADPH takes place in the stroma, or soluble matrix, of the chloroplast. The presence of mitochondria adjacent to the chloroplast shows that in the absence of sunlight, plant cells function as heterotrophs (see also Figure 1.5, page 10).

The photosynthetic pigments include the green chlorophylls, the yellow–orange carotenoids, and the blue–red phycobilins. The structures of these organic molecules are shown in Figure 10.4. The chlorophyll a and b molecules in the figure look similar to the heme macrocyclic system, but the ring is actually quite different from the porphyrin ring (see Problem **2** at the end of this chapter). Moreover, the metal ion in chlorophyll is Mg^{2+}, rather than iron as in heme systems. In addition to pigments, the photosynthetic unit contains the components of two electron-transport chains. These chains include cytochromes and nonheme iron proteins.

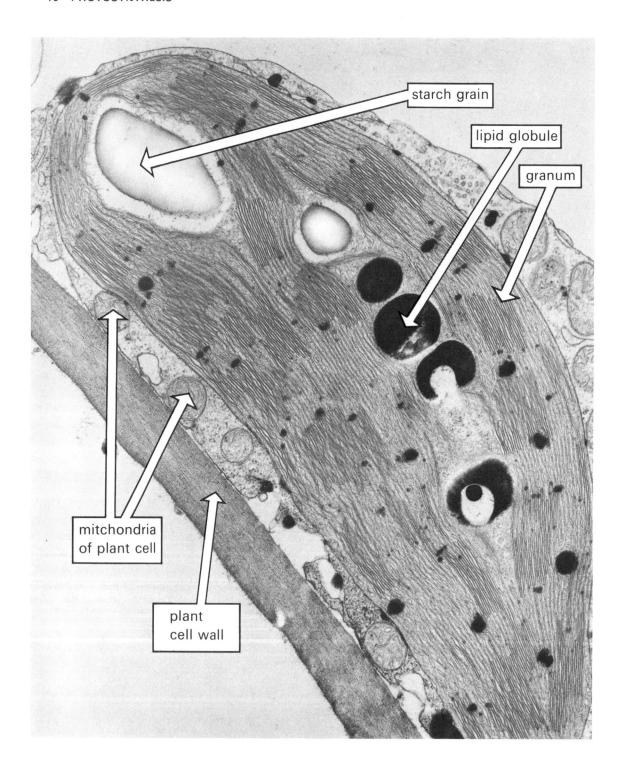

starch grain

lipid globule

granum

mitchondria
of plant cell

plant
cell wall

Figure 10.3

Chloroplast structure. The electron micrograph is similar to the chloroplasts of Figure 1.5. Note the starch grains, which represent conserved sunlight energy. In the dark, the photosynthetic cell shown here becomes a heterotroph, as indicated by the presence of mitochondria. (Courtesy of Dr. Myron Ledbetter, Biology Dept., Brookhaven National Laboratory. Prior publication in M. C. Ledbetter and K. Porter, *Introduction to the Fine Structure of Plant Cells*, Springer-Verlag, 1970.) The drawing shows the structural features of a chloroplast of higher plants.

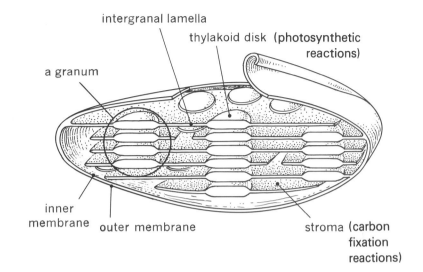

intergranal lamella

thylakoid disk (photosynthetic reactions)

a granum

inner membrane   outer membrane

stroma (carbon fixation reactions)

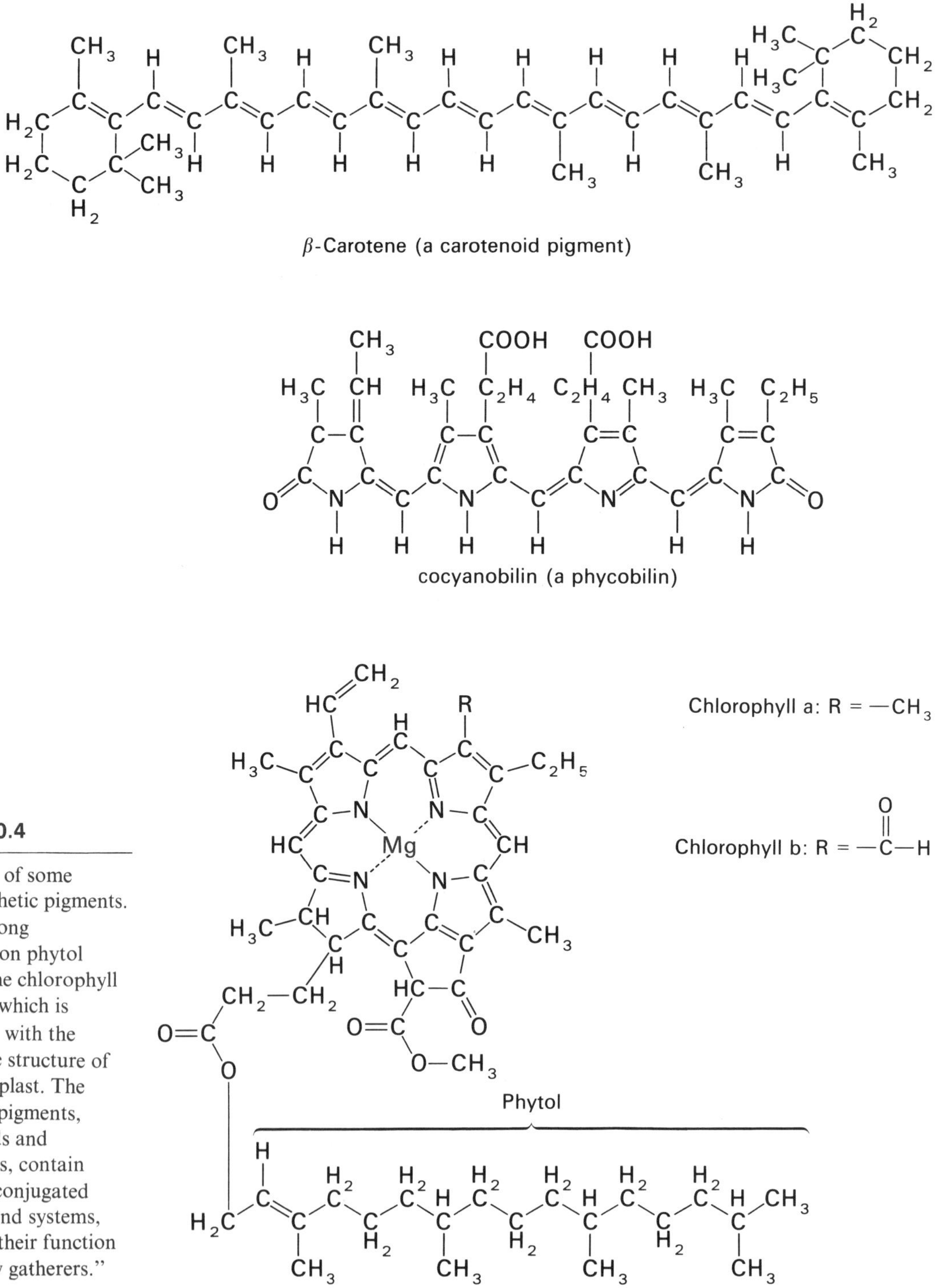

β-Carotene (a carotenoid pigment)

cocyanobilin (a phycobilin)

Figure 10.4

Structures of some photosynthetic pigments. Note the long hydrocarbon phytol chain of the chlorophyll molecule, which is associated with the membrane structure of the chloroplast. The accessory pigments, carotenoids and phycobilins, contain extensive conjugated double–bond systems, related to their function as "energy gatherers."

Chlorophyll a: R = —CH₃

Chlorophyll b: R = —C(=O)—H

Phytol

Light absorption by the pigments directly parallels photosynthetic activity. If we measure the light absorption of the pigments as a function of the wavelength of the light (the absorption spectrum) and compare this with the wavelength dependence of photosynthetic activity (action spectrum), we find that the two curves coincide as shown in Figure 10.5. The light-gathering apparatus of the chloroplast is closely integrated with the apparatus that converts the light energy into chemical energy, and with good reason. The average lifetime of the electronically excited state of a molecule is only about 10^{-9} seconds. This means that the chloroplast must transfer energy from the photoexcited pigment (Figure 10.2) to the rest of the photosynthetic unit in less than 10^{-9} seconds; otherwise the excited pigment molecule will emit a photon and go back into the ground electronic energy state.

Figure 10.5

The wavelength dependence of photosynthetic activity and the absorption spectrum of the photosynthetic pigments.

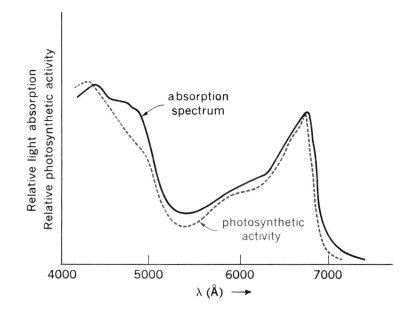

10.3

Reactions of Photosynthesis

The overall photosynthetic reaction for aerobic photosynthetic autotrophs is:

$$n\,H_2O \;+\; n\,CO_2 \quad \xrightarrow{\;h\nu\;} \quad (CH_2O)_n \;+\; n\,O_2 \qquad (1)$$

Electron donor   Electron acceptor   Carbohydrate

The O_2 produced in photosynthesis comes from H_2O. Here H_2O is acting as an electron donor.

For those photosynthetic anaerobes and bacteria which do not use H_2O as an electron donor, Eq. (2) gives the general photosynthetic reaction:

$$n\,H_2X + n\,CO_2 \xrightarrow{\ h\nu\ } (CH_2O)_n + n\,X \qquad (2)$$

where X = S, organic compounds, etc. This general representation is an over-simplification because there are actually two sets of reactions taking place in the photosynthetic unit during photosynthesis: *light reactions* and *dark reactions*.

The light reactions conserve sunlight energy in the form of ATP and NADPH, providing the reducing power needed for carbohydrate biosynthesis:

$$NADP^+ + n h\nu + H_2O + ADP + P_i \longrightarrow NADPH + H^+ + \tfrac{1}{2}O_2 + ATP \qquad (3)$$

The dark reactions are concerned with *carbon fixation*, converting CO_2 to glucose. The overall stoichiometry of this process is shown in Eq. (4) and is discussed in detail later in this chapter.

6 ribulose 1,5-diphosphate $+\ 6\,CO_2 + 18\,ATP + 12\,NADPH + 12\,H^+$

$\longrightarrow$ 6 ribulose 1,5-diphosphate $+$ hexose phosphate $+ 18\,ADP + 17\,P_i + 12\,NADP^+$ $\qquad (4)$

We can perform the light and dark reactions separately in the laboratory. Hill studied the light reactions using isolated chloroplasts and various synthetic dyes in place of $NADP^+$ to act as electron acceptors. The *Hill reaction* and the normal light reaction are:

Normal light reaction:

$$2\,H_2O + 2\,NADP^+ \xrightarrow[\text{Chloroplasts}]{h\nu} 2\,NADPH + 2\,H^+ + O_2 \qquad (5)$$

Hill reaction:

$$2\,H_2O + \underset{\text{(Redox dye)}}{2\,A} \xrightarrow[\text{Chloroplasts}]{h\nu} \underset{\text{(Changes color)}}{2\,AH_2} + O_2 \qquad (6)$$

In both cases, light energy causes electron flow from H_2O, either to the $NADP^+$ or to the dye.

Studies on systems like the Hill reaction have provided much information about the mechanism of photosynthesis. One important finding is that there are *two* sets of light-driven reactions: Photosystem I and Photosystem II. Photosystem I acts with red light (700 nm) and can either reduce $NADP^+$ or phosphorylate ADP to yield ATP. Photosystem II acts with light of wavelengths less than 680 nm and is coupled to Photosystem I as the terminal electron acceptor and to H_2O as an electron donor to yield oxygen. The relationship of

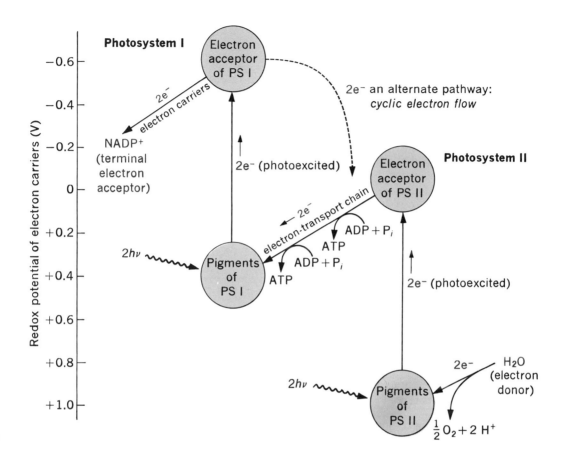

Redox potential of electron carriers (V)

Photosystem I

Electron acceptor of PS I

−0.6

−0.4

2e⁻ electron carriers

−0.2

NADP⁺ (terminal electron acceptor)

0

2e⁻ (photoexcited)

2e⁻ an alternate pathway: *cyclic electron flow*

Electron acceptor of PS II

Photosystem II

+0.2

2hν

2e⁻ electron-transport chain

ADP + P$_i$

ATP

Pigments of PS I

ADP + P$_i$

ATP

+0.4

2e⁻ (photoexcited)

+0.6

+0.8

2hν

Pigments of PS II

2e⁻

H$_2$O (electron donor)

+1.0

$\frac{1}{2}$O$_2$ + 2 H⁺

Figure 10.6

A general scheme for electron flow and phosphorylation during photosynthesis in chloroplasts. The arrows depict the direction of electron flow.

PS I to PS II and the path of electron flow from H_2O to $NADP^+$ is shown in Figure 10.6.

You can see from this figure that there are two modes of photosynthetic electron flow: cyclic electron flow and noncyclic electron flow. In the case of cyclic electron flow, or **cyclic photophosphorylation**, only PS I is operating. Electrons become photoexcited and, in falling back to the ground state, pass through part of the electron transport chain in a manner analogous to mitochondrial electron transport. The net reaction is

$$2\,h\nu + P_i + ADP \longrightarrow ATP + H_2O$$

In cyclic photophosphorylation, oxygen is not evolved since only PS I is acting. Cyclic photophosphorylation is an important way of generating ATP *only* when the demand for NADPH is less than the demand for ATP.

Noncyclic electron flow requires the action of both PS I and PS II. The overall reaction is the one we normally associate with the overall light reactions of photosynthesis:

$$4\,hv + 2\,P_i + 2\,ADP + H_2O + NADP^+ \longrightarrow NADPH + \tfrac{1}{2}O_2 + 2\,ATP + H^+ + 2\,H_2O$$

Referring to Figure 10.6, we can see that a photoexcited electron pair in PS II passes through an electron-transport chain and acts as an electron donor for PS I. Photoexcitation of PS I causes either the reduction of $NADP^+$ or the entry of the electrons into an electron-transport chain back to the pigments of PS I, along with the generation of ATP. This final path of electrons, either to $NADP^+$ or back through the electron-transport chain, depends on the relative cellular demand for ATP vs NADPH.

In Chapters 8, 9 and 10 we have learned about three means by which ADP can be phosphorylated to ATP: *substrate level phosphorylation*; *oxidative phosphorylation*; and *photophosphorylation*.

10.4

Carbon Fixation: The Dark Reactions

Carbon dioxide is transformed into carbohydrate, or fixed, in leaves by a series of reactions discovered by Calvin and Bassham in the late 1940s. These researchers incubated algae in the presence of light and radioactive $^{14}CO_2$. After a short time they disrupted the cells and analyzed the extract for radioactive metabolites. They found that one of the first radioactive compounds formed in the cells was *3-phosphoglycerate*, an intermediate of glycolysis. This three-carbon or "C_3" compound is produced in most plants, including fruits, vegetables, grains and legumes. Such plants are thus termed "C_3 plants." Further experiments by Calvin, Bassham and other workers led to the discovery of a cyclic process for CO_2 fixation now known as the *Calvin cycle*. The major details of this process are shown in Figure 10.7.

The initial product of the reaction of ribulose 1,5-diphosphate and CO_2 in the Calvin cycle is two molecules of 3-phosphoglycerate. The reaction is catalyzed by a key enzyme in carbon fixation, *diphosphoribulose carboxylase*. The stoichiometry becomes rather complicated because part of the 3-phosphoglycerate is used for the biosynthesis of sugars and part is used for the resynthesis of ribulose 1,5-diphosphate, so that more CO_2 can be fixed [see Eq. (4)]. The formation of sugars from 3-phosphoglycerate takes place essentially by a reversal of glycolysis. Since ribulose 5-phosphate is an intermediate of the pentose phosphate pathway, the resynthesis of ribulose 1,5-diphosphate requires only one additional enzyme, *phosphoribulokinase*, which phosphorylates ribulose 5-phosphate to regenerate the original reactant in the Calvin cycle. Here we have another close metabolic interrelationship illustrating the general rule that cellular metabolism has been evolving toward an overall scheme which requires the minimum number of different enzymes. In the Calvin cycle, the addition of two key enzymes to a pathway we have already studied (pentose phosphate pathway, Chapter 8) accounts for the process of CO_2 fixation in plants.

Figure 10.7

The essential features of the Calvin cycle for photosynthetic carbon fixation. These reactions take place in the stroma or soluble part of the chloroplast (see Figure 9.3). The monosaccharide phosphates are shown in their open-chain forms for clarity. Most of the compounds in this figure are already familiar from our study of glycolysis and related processes in Chapter 8. Note that the formation of new carbohydrates from CO_2 can proceed continuously in the presence of NADPH and ATP without using up the key starting compound D-ribulose 1,5-diphosphate.

In the middle 1960s a group of Australian workers showed that the first product of carbon fixation in such plants as corn, sugar cane, millet, crabgrass and certain other plants of tropical origin was not a three-carbon metabolite but rather a four-carbon compound. These organisms are termed "C_4 plants" and have fundamental differences from C_3 plants not only in their carbon-fixation biochemistry but also in the structure of their leaves (see Suggestions for Further Reading). The method of CO_2 fixation used by C_4 plants is called the Hatch–Slack Pathway. The initial reaction is:

$$CO_2 + \underset{\text{Phosphoenol pyruvate}}{H_2C=\overset{\overset{\displaystyle OPO_3^{2-}}{|}}{C}-CO_2^-} \xrightarrow{\text{PEP carboxylase}} \underset{\text{Oxaloacetate}}{^-O_2C-\overset{\overset{\displaystyle O}{\|}}{C}-CH_2-CO_2^-} + P_i$$

The oxaloacetate is formed in the cells near the leaf surface, transformed into malate, and then transported to a group of specialized inner cells of the leaf

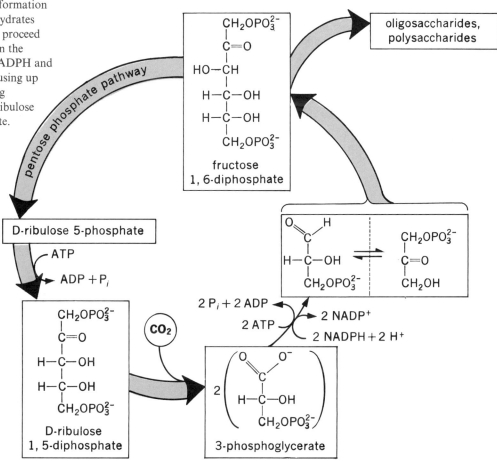

Figure 10.8

The Hatch–Slack pathway for CO_2 fixation in C_4 plants. These reactions take place in two types of leaf cells. This division of labor actually increases the rate at which CO_2 can be fixed, even though more energy is needed to do so. Because most C_4 plants are tropical in origin, the energy requirement is not a problem due to increased rates of photosynthesis.

called the *bundle sheath cells*. Here, malate is converted to pyruvate and CO_2. The CO_2 is converted to hexose by the Calvin cycle and pyruvate is converted back into phosphoenol pyruvate by phosphorylation with ATP, thus completing the cycle. The scheme of carbon fixation in C_4 plants is shown diagrammatically in Figure 10.8. Because AMP is produced in the step of the C_4 scheme that leads to phosphoenol pyruvate, the overall pathway from CO_2 to hexose requires 2 moles more of ATP per mole of CO_2 than does the Calvin cycle in C_3 plants. While more energy is needed to fix carbon by the C_4 pathway, the division of labor between cells in the leaf enables C_4 plants to assimilate

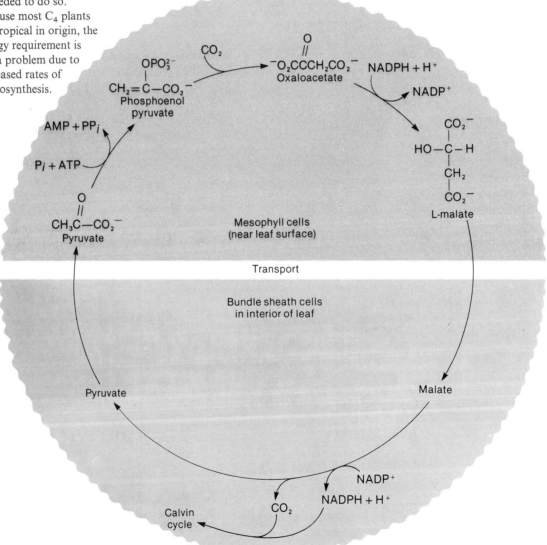

CO_2 approximately twice as rapidly as C_3 plants. Because of this, C_4 plants produce two to three times as much vegetable matter per acre as C_3 crops. Breeding food crops that have an increased rate of CO_2 assimilation is an important goal of plant scientists.

Recently, researchers have shown that animal cells can fix CO_2 by reactions similar to those in plants. CO_2 fixation is not a major process in most cases in animals (heterotrophs) because they do not have the photosynthetic apparatus to produce ATP and NADP independently from the catabolism of foodstuffs. One example of an important carbon fixation reaction in animals is the group of anaplerotic reactions associated with the Krebs TCA cycle (see Section 9.5).

Photorespiration

If you look back at the micrograph in Figure 10.3, you can observe mitochondria adjacent to the chloroplast of the green plant cell. This means that in addition to being autotrophs, green plant cells can function heterotrophically. Obviously, plants must function as heterotrophs in the dark. However, workers have noticed respiratory activity in green plant cells in the light as well. This process of **photorespiration** consumes a portion of the reducing power produced by photosynthetic electron transport and therefore acts in competition with it. Since the oxidation process is not coupled with ATP generation, photorespiration represents a net "waste" of reducing power.

The details of the origin of photorespiration have only recently become understood, although considerable research is still being conducted on this process by plant biochemists. It was observed by the German biochemist Warburg as early as 1920 that oxygen inhibits CO_2 assimilation by plants at low CO_2 concentrations. This is called the *Warburg effect* and occurs in algae and in C_3 plants. It does not occur in C_4 plants. Recently it was found that the key enzyme of the Calvin cycle, diphosphoribulose carboxylase, can also act as an *oxygenase*. The carboxylase activity of this enzyme is inhibited by oxygen and the oxygenase activity is inhibited by CO_2. The products of the oxygenase reaction are *phosphoglycolate acid* and *3-phosphoglycerate*, an intermediate of glycolysis. The phosphoglycolate formed is hydrolyzed and oxidized by the FAD enzyme *glycolate oxidase* to produce glyoxylate (see Figure 9.6). The key steps of photorespiration are shown in Figure 10.9.

While the role of photorespiration is not clearly understood, it is generally considered detrimental to crop yields in C_3 food plants. In C_3 plants that are staple food crops, such as wheat, soybeans and potatoes, photorespiration can consume over 50% of the reduced molecules produced by photosynthesis. Plant scientists are currently trying to cross C_3 and C_4 plants to improve yields by genetically limiting photorespiration. In another approach, plant biochemists are experimenting with specific inhibitors for various steps of the photorespiration process, in order to find an agricultural chemical that can be used to improve crop yields through increased efficiency in CO_2 fixation.

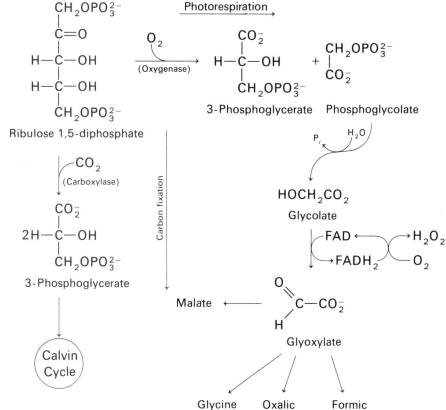

Figure 10.9

The major steps of the photorespiration process. In the conversion of one mole of ribulose 1,5-diphosphate to one mole of glyoxylate, two moles of O_2 are consumed. This represents a considerable loss of reducing power and substrates that could have been used to biosynthesize new plant tissue.

10.5

Plant Growth Regulators

Plants, like animals, have a complex physiology in which hormones, or regulator substances, play a crucial role in the development of the individual. In Chapter 6 we saw how steroid hormones control such cycles as ovulation, pregnancy and sexual development in humans and other mammals. Within the annual cycle, plants must also do certain things at certain times. Basic functions such as

sprouting	leaf fall
fruit development	fruit fall
flowering	growth

are all under the control of chemical substances called *plant growth regulators.* These remarkable substances fall into four main classes, which are summarized in Table 10.1. While the mechanism by which these plant growth regulators act

Table 10.1 The Types and Properties of Plant Growth Regulators.

Class of compound	Function
Auxins	Cause enlargement of plant cells
Gibberellins	Stimulate cell division or cell enlargement, or both
Cytokinins	Stimulate cell division
Inhibitors	A diverse group of substances that inhibit some biochemical processes in plants.

is not yet known, one current theory proposes that members of all four classes of regulators are present in normal plant tissue and that changes in their concentration ratios elicit given physiological responses. Such changes could be caused by variations in ambient temperature or daily light duration.

Today, millions of pounds of synthetic plant growth regulators are used each year in modern agriculture. For example, the auxin *indole butyric acid* is used to promote the rooting of cuttings. *Maleic hydrazide* is a synthetic inhibitor used to prevent the sprouting of onions and potatoes in storage. The inhibitor *2-chloroethyl trimethylammonium chloride* reduces the length of the plant stalk, especially in cereal grains. The use of this plant growth regulator greatly reduces the tendency of wheat and other cereal grains to "lodge" or become flattened by heavy winds and rain. This substance has saved millions of bushels of grain all over the world.

3-Indole butyric acid   Maleic hydrazide   2-Chloroethyl trimethyl ammonium chloride

One of the great challenges of plant biochemistry is to gain a molecular understanding of how plant physiology is regulated. The results of such an understanding will lead to a tremendous increase in world crop production (see Suggestions for Further Reading).

Suggestions for Further Reading

Clayton, R. K. *Light and Living Matter.* 2 vols. New York: McGraw-Hill, 1971. Volume 2 of this paperback set discusses photobiology in a clear and concise manner.

Metzler, D. E. *Biochemistry—The Chemical Reactions of Living Cells.* New York: Academic Press, 1977. This text discusses the role of light in biology in detail in Chapter 13.

Nickell, L. G. "Plant Growth Regulators." *Chemical and Engineering News*, October 9, 1978. A review of the structure and applications of plant growth regulators.

Zelitch, I. "Photosynthesis and Plant Productivity." *Chemical and Engineering News,* February 5, 1979. This is a comprehensive and readable review of the current state of applied research in photosynthesis and photorespiration.

Scientific American
Bassham, J. A. "The Path of Carbon in Photosynthesis." June 1962.
Bjorkman, O. and Berry, J. "High Efficiency Photosynthesis." October 1973.
Govindjee and Govindjee, R. "The Primary Events of Photosynthesis." December 1974.
Levine, R. P. "The Mechanism of Photosynthesis." December 1969.
Rabinowitch, E. I. "Photosynthesis." August 1948.
Rabinowitch, E. I. and Govindjee. "The Role of Chlorophyll in Photosynthesis." July 1965.
van Overbeek, J. "The Control of Plant Growth." July 1968.
Wald, George. "Light and Life." October 1959.

Problems

1. For each of the terms below, provide a brief definition or explanation.

    Cyclic photophosphorylation   Light reactions
    C_3 plant   Noncyclic electron flow
    C_4 plant   Photorespiration
    Dark reactions   Thylakoid disk

2. Draw the structures of chlorophyll a and of ferroprotoporphyrin IX (heme coenzymes—see Chapter 3). Compare the two structures and list at least three differences between the two molecules.

3. Refer to Figure 10.6 for this problem.

    (a) What is oxidized by photosystem II?
    (b) What is the terminal electron acceptor of a photoexcited electron pair in photosystem II?
    (c) What is the terminal electron acceptor in photosystem I during cyclic photophosphorylation?
    (d) What is the terminal electron acceptor in photosystem I during noncyclic electron flow?
    (e) If you were going to select a redox dye to use in place of $NADP^+$ for the Hill reaction, what general criterion would the redox potential of the dye have to satisfy? (For a discussion of oxidation-reduction, see Chapter 9.)

4. Using structures, show how fructose 1,6-diphosphate can be converted to ribulose 1,5-diphosphate.

5. Glyoxylate, a product of photorespiration, can be converted to malate, glycine, oxalate and formic acid. Show how each of these conversions can occur. Give the type of enzyme and the probable prosthetic group involved in each case. You may wish to refer to Appendix B and Chapter 3.

6. Compare the structure of Vitamin A (Section 3.4) with the structure of β-carotene (Figure 10.4). How are these two structures related?

CHAPTER 11

Catabolism of Fatty Acids and Nitrogen Compounds

In Chapter 7 we outlined the overall relationship among the catabolisms of amino acids, lipids and carbohydrates. In this chapter we will study the details of the catabolism of fatty acids, amino acids and other nitrogen-containing metabolites. We will see how these processes relate to the general system of energy-generating processes in the cell.

11.1

Fatty Acid Catabolism

Approximately 10% of the body weight in mammals is in the form of triacylglycerols (Chapter 6), or fat. These highly reduced substances represent an important form of stored energy and are subject to a continual turnover in the organism. Depot fat, or stored lipid, is quite mobile. For example, the half-life of depot fat in rats is about 10 days, which means that the formation and breakdown of depot fat occur concurrently at a fairly rapid rate.

The triacylglycerol molecule, containing many —CH_2— groups in the fatty acid chains, represents an ideal storage form for surplus metabolic energy. It is highly reduced and thus yields over *twice* the energy per gram of carbohydrate when broken down to CO_2 and H_2O:

	$\Delta G^{0\prime}$ kcal/mole	$\Delta G^{0\prime}$ kcal/g
$C_6H_{12}O_6 + 6O_2 \longrightarrow 6CO_2 + 6H_2O$ (Glucose)	−686	−3.8
$CH_3(CH_2)_{14}CO_2H + 23O_2 \longrightarrow 16CO_2 + 16H_2O$	−2338	−9.3

Since lipid is not hydrated, as is carbohydrate, it is much more compact and thus occupies less volume than carbohydrate of equal weight *in vivo*. In vertebrate metabolism, nearly half the total metabolic energy comes from fat breakdown in most tissues. The major exception is brain tissue, which derives energy solely from glucose. As an energy storage form, lipid is important even in microorganisms in which we can observe fat globules in the cytoplasm.

The catabolism of fats begins with free fatty acids. These are produced either by the *intra*cellular hydrolysis of depot triglycerides or by the hydrolysis of dietary fat in the small intestine by lipase enzymes, followed by absorption of the free fatty acids through the intestinal epithelium. Fatty acids are primarily transported by the blood in a form bound to the protein plasma albumin, which constitutes a large fraction of the protein in blood plasma. Irrespective of the source, fatty acids are oxidized in the mitochondrion. Therefore, it is useful to begin our discussion of fatty acid oxidation by learning how fatty acids in the cytoplasm enter the mitochondrion. Figure 11.1 presents an overview of this process as it relates to pyruvate catabolism in the mitochondrion.

Figure 11.1 not only illustrates the close relationship between carbohydrate and fatty acid metabolism, but also shows the relationship between fatty acid catabolism, which is a mitochondrial process, and fatty acid biosynthesis, which occurs in the cytoplasm. Examination of the figure also reveals that the substance **carnitine** plays a central role in this overall process as the agent which "permits the entry" of the fatty acyl group into the mito-

Figure 11.1

An overview of the relationship between the free fatty acid pool in the cytoplasm and mitochondrial fatty acid oxidation. (←O→ denotes a specific transport system.)

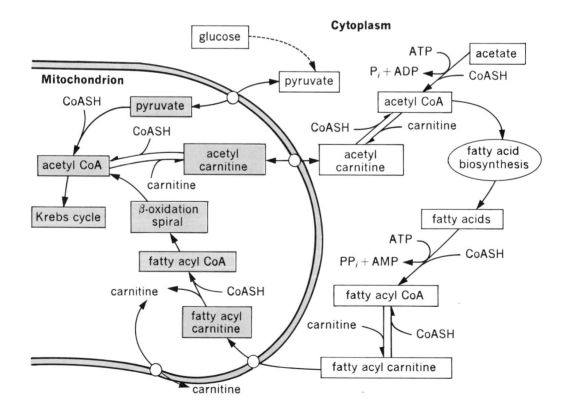

chondrion. This means that the mitochondrial inner membrane must contain a transport system (see Chapter 7) for fatty acyl carnitines.

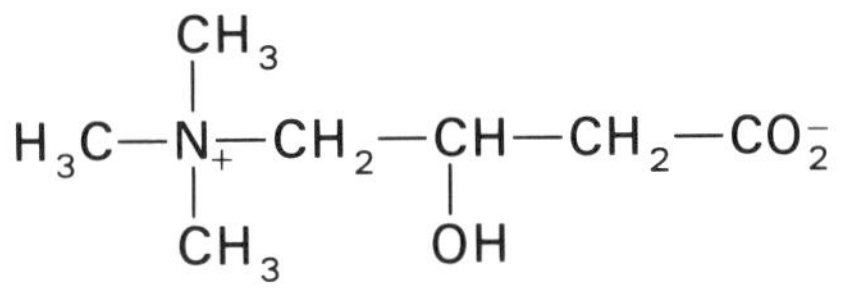

$$H_3C-\overset{\overset{\textstyle CH_3}{|}}{\underset{\underset{\textstyle CH_3}{|}}{N^+}}-CH_2-\overset{\overset{}{\underset{\underset{\textstyle OH}{|}}{CH}}}{}-CH_2-CO_2^-$$

Carnitine

We can trace the sequence of steps from the degradation of a fatty acid molecule in the cytoplasm to the formation of acetyl CoA in the mitochondrion by beginning with the activation of free fatty acids by reaction with acetyl CoA.

Activation of Free Fatty Acids

As Figure 11.1 shows, free fatty acids in the cytoplasm must react with coenzyme A and ATP to produce an activated fatty acyl CoA species:

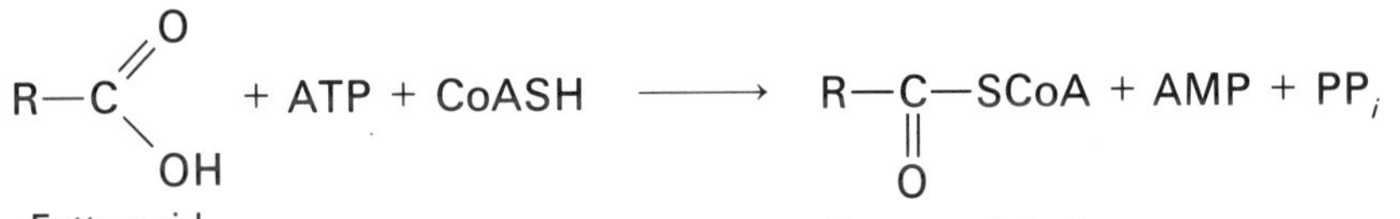

$$R-\overset{\overset{\textstyle O}{\|}}{C}\diagdown_{OH} + ATP + CoASH \longrightarrow R-\overset{\overset{}{\underset{\underset{\textstyle O}{\|}}{C}}}{}-SCoA + AMP + PP_i$$

Fatty acid   Fatty acyl CoA

This reaction is catalyzed in the cytoplasm by a group of enzymes known collectively as *thiokinases*. The specificity of the enzymes is determined by the chain length (long, medium, short) of the fatty acid substrates. This reaction is an example of **pyrophosphoryl cleavage** of ATP (see Chapter 7). Much of the driving force for this reaction is due to the fact that pyrophosphate ion, PP_i, which is one of the products, is rapidly broken down by a pyrophosphatase enzyme:

$$PP_i + H_2O \xrightarrow{\text{Pyrophosphatase}} 2 P_i$$

Entry of Fatty Acyl Groups into the Mitochondrion

The mitochondrial inner membrane contains transport systems specific for fatty acyl derivatives of the amino hydroxy acid carnitine. Fatty acyl CoA molecules react with free carnitine, yielding fatty acyl carnitine derivatives which pass through the mitochondrial inner membrane:

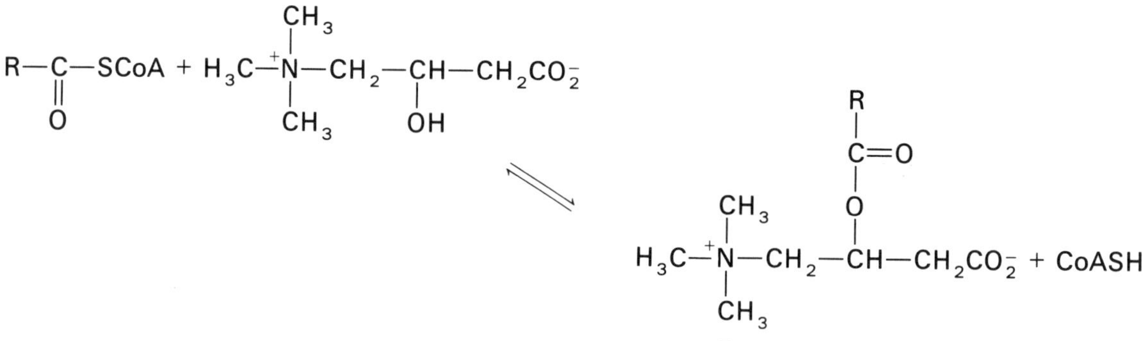

$$R-\underset{\underset{O}{\|}}{C}-SCoA + H_3C-\overset{+}{\underset{\underset{CH_3}{|}}{N}}-CH_2-\underset{\underset{OH}{|}}{CH}-CH_2CO_2^-$$

$$\rightleftharpoons$$

$$H_3C-\overset{+}{\underset{\underset{CH_3}{|}}{N}}-CH_2-\underset{\underset{\underset{\underset{R}{|}}{C=O}}{|}}{CH}-CH_2CO_2^- + CoASH$$

Fatty acyl carnitine

Inside the mitochondrion, the process is reversed: coenzyme A reacts with fatty acyl carnitine to produce intramitochondrial fatty acyl CoA and free carnitine, which passes back out to the cytoplasm and permits the passage of additional fatty acyl groups into the mitochondrion. The net effect is the entry of a fatty acyl CoA into the mitochondrion.

β-Oxidation Spiral

The next set of reactions degrades the fatty acyl CoA molecules into two-carbon units (acetyl CoA), proceeding two carbons at a time. We term this process a spiral instead of a cycle because each "turn" shortens the fatty acyl chain by two carbons. Figure 11.2 shows the overall scheme of this process in detail. The organic chemistry of this pathway is quite straightforward.

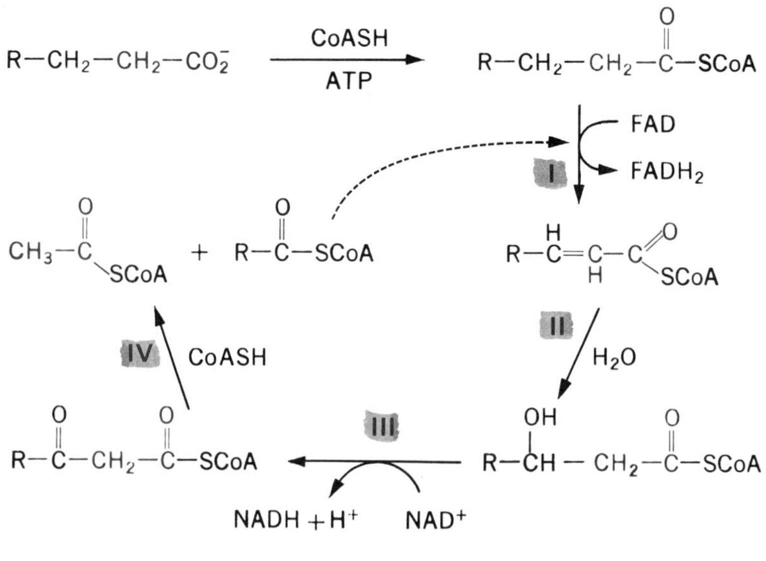

Figure 11.2

The β-oxidation spiral for fatty acid catabolism. The Roman numerals in the figure refer to the numbered steps in the text.

I. **Oxidation of** $-CH_2-CH_2-$ **to** $-CH=CH-$ The initial step of the β-oxidation spiral oxidizes fatty acyl CoA using a group of enzymes called *fatty acyl CoA dehydrogenases*. Like the thiokinases, different members of this group of similar enzymes are specific for various ranges of chain length (long, medium, short) in the fatty acyl CoA substrates. These are all FAD-requiring enzymes, and the $FADH_2$ produced is reoxidized by the respiratory chain at the CoQ level:

$$R-CH_2-\overset{\beta}{C}H_2-\overset{\alpha}{C}H_2-\overset{\overset{\displaystyle O}{\|}}{C}-SCoA + FAD \longrightarrow R-CH_2-\overset{\overset{\displaystyle H}{|}}{C}=\overset{\overset{\displaystyle }{}}{\underset{\underset{\displaystyle H}{|}}{C}}-\overset{\overset{\displaystyle }{}}{\underset{\underset{\displaystyle O}{\|}}{C}}-SCoA + FADH_2$$

$$\Delta^{2,3}\text{-}trans\text{-enoyl CoA}$$

II. **Addition of HOH to** $-CH=CH-$ The double bond of enoyl CoA is hydrated to produce a secondary alcohol. This reaction is catalyzed by *enoyl hydratase*:

$$R-CH_2-\overset{\overset{\displaystyle H}{|}}{C}=\underset{\underset{\displaystyle H}{|}}{C}-\underset{\underset{\displaystyle O}{\|}}{C}-SCoA + HOH \rightleftharpoons R-CH_2-\underset{\underset{\displaystyle H}{|}}{\overset{\overset{\displaystyle OH}{|}}{C}}-CH_2-\underset{\underset{\displaystyle O}{\|}}{C}-SCoA$$

$$\text{L-}\beta\text{-hydroxyacyl CoA}$$

III. **Oxidation of** $-\overset{\overset{\displaystyle OH}{|}}{C}H-$ **to** $-\overset{\overset{\displaystyle O}{\|}}{C}-$ The secondary alcohol functional group of the L-β-hydroxyacyl CoA is dehydrogenated to a ketone by the NAD^+-enzyme hydroxyacyl CoA dehydrogenase. For each mole of substrate oxidized, one mole of NADH is produced, which is reoxidized by the respiratory chain.

$$R-CH_2-\underset{\underset{\displaystyle O}{\|}}{\overset{\overset{\displaystyle OH}{|}}{C}H}-CH_2-\underset{\underset{\displaystyle O}{\|}}{C}-SCoA + NAD^+ \rightleftharpoons R-CH_2-\underset{\underset{\displaystyle O}{\|}}{C}-CH_2-\underset{\underset{\displaystyle O}{\|}}{C}-SCoA + NADH$$
$$+ H^+$$

$$\beta\text{-ketoacyl CoA}$$

IV. **Removal of acetyl CoA** If we removed the elements of acetyl CoA from the product of the preceding step we would have a fatty acyl group with two fewer carbons than we had in the first place. This reaction is catalyzed by the enzyme *thiolase*, which reacts CoASH with the β-ketoacyl CoA molecule to produce acetyl CoA and a shortened fatty acyl CoA molecule:

$$R-CH_2-\underset{\underset{O}{\|}}{C}-CH_2-\underset{\underset{O}{\|}}{C}-SCoA + CoASH \;\rightleftharpoons\; R-CH_2-\overset{\overset{O}{\|}}{C}-SCoA + CH_3-\overset{\overset{O}{\|}}{C}-SCoA$$

Shortened fatty acyl CoA

This shortened fatty acyl chain can be further degraded by the β-oxidation spiral, repeating the steps just discussed.

The overall β-oxidation spiral is summarized by Eq. (1):

$$CH_3-(CH_2)_n-CH_2-\overset{\overset{O}{\|}}{C}-SCoA + CoASH + FAD + NAD^+ + H_2O \tag{1}$$

$$\longrightarrow CH_3-(CH_2)_{n-2}-CH_2-\overset{\overset{O}{\|}}{C}-SCoA + CH_3-\overset{\overset{O}{\|}}{C}-SCoA + FADH_2 + NADH + H^+$$

The potential for generation of ATP by this process is great. If the acetyl CoA produced is oxidized via the Krebs TCA cycle, the net yield of ATP per turn of the β-oxidation spiral is:

2 ATP from oxidation of 1 FADH$_2$ (step I above; reoxidation by respiratory chain)
3 ATP from oxidation of 1 NADH (step III above; reoxidation by respiratory chain)
12 ATP from oxidation of acetyl CoA via TCA cycle

Net: 17 ATP generated per "turn" of the β-oxidation spiral

We can construct an energy inventory for the oxidation of a C_6-fatty acid to compare this process with the oxidation of glucose ($C_6H_{12}O_6$). Breaking down a six-carbon fatty acid requires two turns of the spiral and generates 3 acetyl CoA, 2 FADH$_2$, and 2 NADH. However, we require *one* ATP to activate the fatty acid. Thus, the balance for a C_6-fatty acid is

6 ATP from 2 NADH
4 ATP from 2 FADH
36 ATP from 3 acetyl CoA
minus   1 ATP for activation of fatty acid

Net: 45 ATP

Compare the 45 ATP generated from the oxidation of a C_6-fatty acid with the 36 ATP generated from the oxidation of glucose (see Section 9.10).

Other Oxidation Schemes

While β-oxidation is the predominant pathway for fatty acid oxidation in living organisms, there are others which are important in a localized context. For example, in brain tissue, *α-oxidation* is an important pathway for fatty acid oxidation. This is particularly true for fatty acids with substituents in positions that block β-oxidation (see Figure 11.3). In this case, fatty acids are oxidized to α-hydroxyacids, which are then oxidized further to give an α-keto acid. The scheme is completed when the α-keto acid is oxidatively decarboxylated to a fatty acid having one fewer carbon than the original fatty acid. *Refsum's disease* is an inherited defect in α-oxidation characterized by severe neurological disorder. This disease causes large quantities of *phytanic acid* to accumulate in blood plasma and body tissues, particularly the liver. This accumulation is due to the body's inability to catabolize phytanic acid by α-oxidation. Phytanic acid is a common branched-chain fatty acid derived from the phytol side chain of chlorophyll (Figure 10.4). Its primary dietary sources are cow's milk and animal fat. Figure 11.3 shows the steps of α-oxidation using phytanic acid as a substrate. Note that the structure of phytanic acid prevents β-oxidation from occurring until the branch point moves from the β to the α position as a result of α-oxidation (see Problem **8** at the end of this chapter).

Figure 11.3

The α-oxidation scheme using phytanic acid as a substrate. In the case of Refsum's disease, a defect in the α-oxidation system causes phytanic acid to accumulate in the blood plasma and tissues. In the example below, the product can be converted to the acyl CoA derivative and oxidized by the β-oxidation scheme.

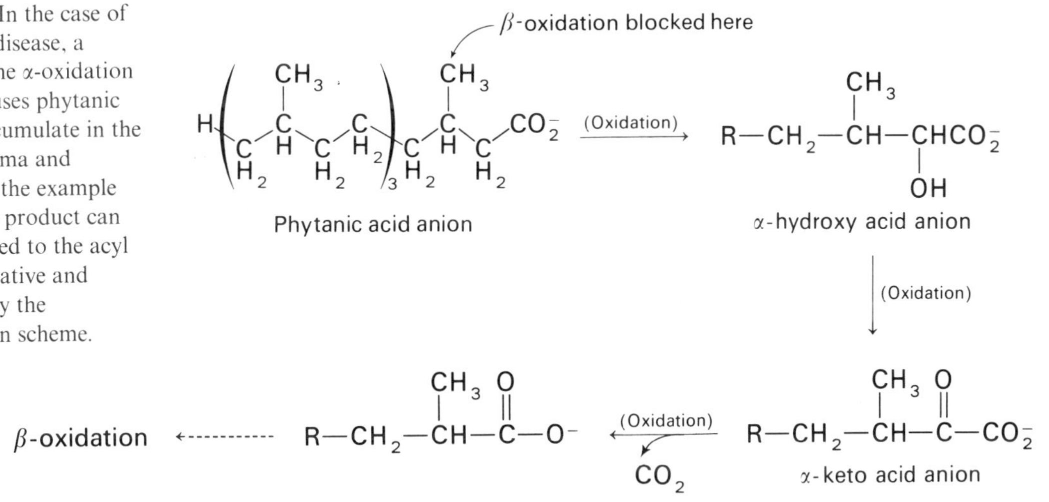

Propionate Metabolism

The substance propionyl CoA is produced from several processes, including the oxidation of fatty acids with odd numbers of carbons and the catabolism of certain amino acids. This molecule is converted to succinyl CoA by a series of reactions, one of which involves a coenzyme B_{12}-requiring enzyme. The sequence is shown in Figure 11.4. Note that the general type of

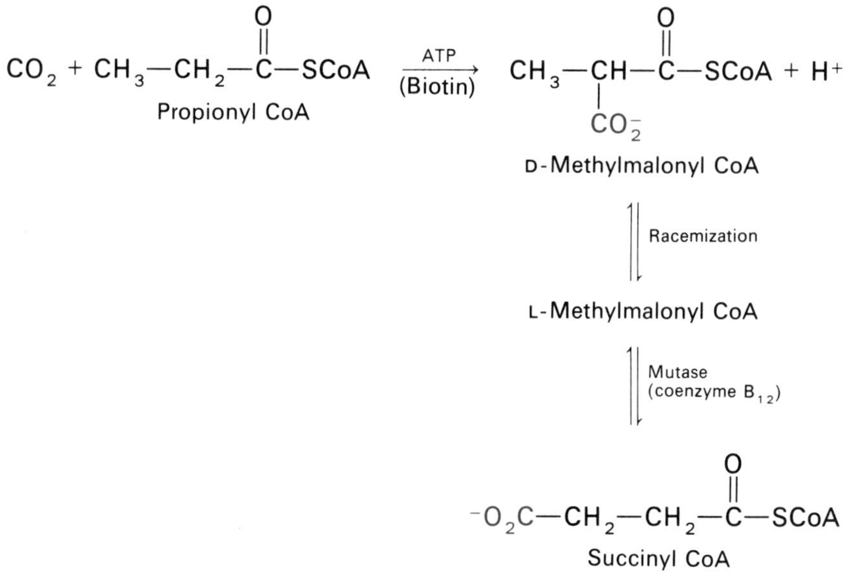

$$CO_2 + CH_3-CH_2-\overset{\displaystyle O}{\overset{\|}{C}}-SCoA \xrightarrow[\text{(Biotin)}]{\text{ATP}} CH_3-\underset{\underset{\displaystyle CO_2^-}{|}}{CH}-\overset{\displaystyle O}{\overset{\|}{C}}-SCoA + H^+$$

Propionyl CoA   D-Methylmalonyl CoA

$\updownarrow$ Racemization

L-Methylmalonyl CoA

$\updownarrow$ Mutase (coenzyme B_{12})

$$^-O_2C-CH_2-CH_2-\overset{\displaystyle O}{\overset{\|}{C}}-SCoA$$

Succinyl CoA

Figure 11.4

Scheme of propionate metabolism. In cases of severe vitamin B_{12} deficiency, both propionate and methylmalonate are excreted in the urine.

isomerization catalyzed by coenzyme B_{12}-enzymes is well illustrated here by the reaction catalyzed by methyl malonyl CoA mutase (see Chapter 3).

Two conditions can interfere with this step of propionate metabolism and cause methyl malonate to accumulate and be excreted in the urine. In the case of vitamin B_{12} deficiency (see Section 3.4), both methyl malonate and propionate appear in the urine. In the case of the hereditary disease *methyl malonic aciduria*, which causes mental retardation and stunted growth in children, there may be a defect in either the mutase protein or the vitamin B_{12} metabolism. In either case, methyl malonate accumulates and is excreted when the amount exceeds the renal threshold.

11.2

Ketone Bodies

Acetoacetyl CoA is a central molecule in lipid metabolism. You can recognize it as the product of the next-to-last step in the oxidation of a four-carbon fatty

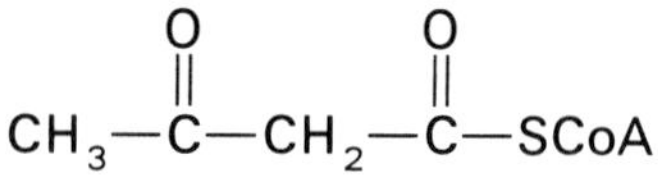

$$CH_3-\overset{\displaystyle O}{\overset{\|}{C}}-CH_2-\overset{\displaystyle O}{\overset{\|}{C}}-SCoA$$

Acetoacetyl CoA

acyl CoA. It is therefore an intermediate produced during the breakdown of any even-chain fatty acid. Acetoacetyl CoA is primarily formed directly from acetyl CoA by the *ketothiolase* reaction:

$$2\,CH_3-\overset{\displaystyle O}{\overset{\displaystyle \|}{C}}-SCoA \;\rightleftharpoons\; CH_3-\overset{\displaystyle O}{\overset{\displaystyle \|}{C}}-CH_2-\overset{\displaystyle O}{\overset{\displaystyle \|}{C}}-SCoA + CoASH$$

If the steady-state level of acetyl CoA is high, formation of acetoacetyl CoA is favored. Figure 11.5 summarizes some of the major fates of acetoacetyl CoA.

Two products of acetoacetate metabolism are the **ketone bodies, acetone** and β-hydroxybutyric acid. Ordinarily, the level of these two substances in blood plasma is very low. Higher than normal concentrations of them result in the excretion of ketone bodies in the urine. The presence of acetone or β-hydroxybutyrate in the urine is consistent with several possible metabolic abnormalities outlined on the next page.

Figure 11.5

Various fates of acetoacetyl CoA

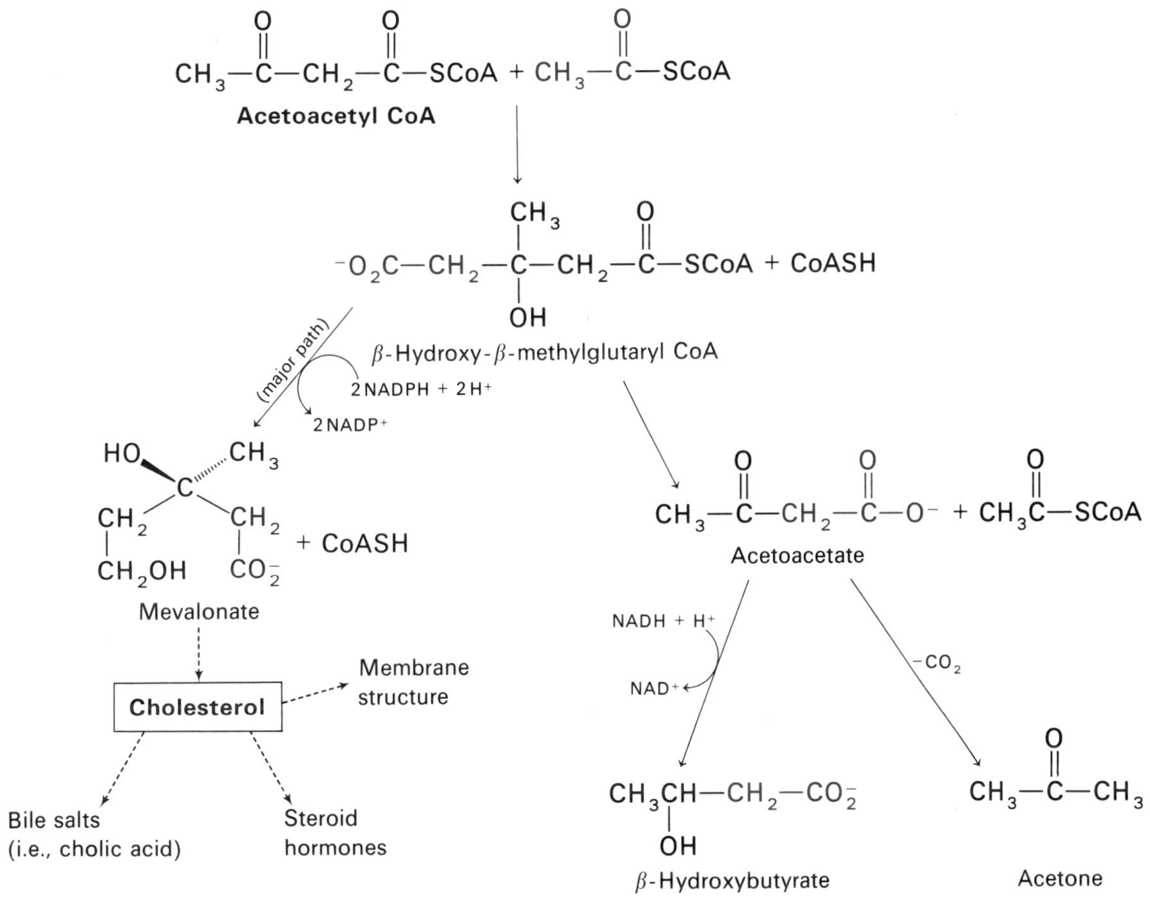

1. **Starvation.** Depletion of carbohydrate reserves and overcatabolism of fat occurs during starvation. The lack of carbohydrate stimulates fatty acid degradation in the liver. This in itself causes increased levels of acetoacetyl CoA, acetyl CoA, and their products. Because there is no carbohydrate entering glycolysis, the levels of pyruvate and other glycolytic intermediates fall off. As we learned in our discussion of anaplerosis and the Krebs TCA cycle in Chapter 9, pyruvate is in a steady state with oxaloacetate and malate (see p. 284). Therefore, in starvation, or in the case of a deliberate restriction of carbohydrate in the diet, the levels of the Krebs TCA cycle intermediates are also depressed. This unfortunately diminishes the ability of the starving cell to catabolize acetyl CoA, which accentuates the accumulation of acetyl CoA and leads to increased formation of ketone bodies. This metabolic abnormality illustrates well the close interrelationships among the metabolic processes we have discussed.

   In cases of acute starvation, once fat reserves are depleted, protein must be catabolized to provide energy. Under these conditions, as much as 6% of the body protein may be lost per day.

2. **Abnormal diet.** The total exclusion of carbohydrate from the diet results in the overcatabolism of fat and a resulting increase in ketone body formation, as in starvation (see discussion above). This is a typical example of the close integration of fat and carbohydrate metabolism. Some popular "fad" diets for losing weight take advantage of this principle for fat loss by eliminating carbohydrate from the diet.

3. **Diabetes.** Insulin is apparently required by cells for the uptake of blood glucose. In one form of diabetes, insulin deficiency prevents glucose from being catabolized at a normal rate. Then, since blood glucose cannot serve as a fuel, the cell mobilizes depot fat to an increased extent. High levels of fat catabolism again result in high levels of acetoacetyl CoA. Both acetone and β-hydroxybutyrate are produced in excess. In severe cases, high concentrations of β-hydroxybutyrate (a carboxylic acid species) can exceed the buffer capacity of the blood plasma, and the resultant lowering of the blood pH (acidosis) can lead to diabetic coma and death.

11.3

Amino Acid Catabolism

In the first half of the 1900s, many experiments were performed on the nutritional functions of the various amino acids in the diets of animals. In the course of studies by W. S. Rose and others, two general facts were recognized:

1. Some amino acids are not required (**nonessential amino acids**), since the animal could biosynthesize them from carbohydrate or other carbon sources and a nitrogen source. Other amino acids are indispensable: the animal could not biosynthesize these **essential amino acids** from other precursors, and therefore they had to be supplied in the diet. For example, either whole wheat or soybeans offer a complete source of essential amino acids, but corn, an important staple in many societies, is ordinarily lacking in lysine. Plant scientists are actively trying to breed new feed grain varieties that offer a complete source of protein, and recently agronomists at Purdue University were able to breed a "high-lysine" corn which has already had an important dietary impact in many areas of the world.

2. Some dietary amino acids are **ketogenic**: excess amounts of these in the diet give rise to increased levels of acetyl CoA and thus ketone bodies. Other amino acids are **glycogenic**: excess glycogenic amino acids cause increased glucose or glycogen biosynthesis.

Table 11.1 summarizes the details of these observations. It is readily apparent that *amino acid metabolism is closely integrated with carbohydrate and fat metabolism.*

Table 11.1 Amino Acids and Their Relationship to Diet and Metabolism.

Essential amino acids*	Nonessential amino acids	Glycogenic	Ketogenic	Both
Arginine	Alanine	Alanine	Leucine	Isoleucine
Histidine	Aspartic acid	Arginine		Lysine
Isoleucine	Cysteine	Aspartic acid		Phenylalanine
Leucine	Glutamic acid	Cysteine		Tryptophan
Lysine	Glycine	Glutamic acid		Tyrosine
Methionine	Proline	Glycine		
Phenylalanine	Serine	Histidine		
Threonine	Tyrosine†	Methionine		
Tryptophan		Proline		
Valine		Serine		
		Threonine		
		Valine		

* These were determined for white rats and correspond to the case for many mammals.

† Tyrosine is synthesized from phenylalanine, and might therefore be considered an essential amino acid as well.

General Scheme of Amino Acid Catabolism

Amino acid metabolism in higher animals takes place mainly in the liver. Much is known concerning the intermediary metabolism of amino acids. In this text we can only sketch the general outlines of the pathways involved; you may obtain further information by consulting one of the suggested readings at the end of the chapter.

Removal of the $-NH_2$ group is often the first step in the catabolism of an amino acid. This occurs primarily by either of two processes: **transamination** or **oxidative deamination**.

Transamination is catalyzed by enzymes which require pyridoxal phosphate as a coenzyme. The action of pyridoxal phosphate in transamination reactions was discussed in Section 3.5. Transamination is involved in the catabolism of the following amino acids: alanine, arginine, asparagine, aspartic acid, cysteine, isoleucine, lysine, phenylalanine, tryptophan, tyrosine, and valine. There are two broad types of transaminases, alanine transaminase and the glutamate transaminases:

Alanine transaminase:

$$\underset{\text{Pyruvate}}{CH_3-\overset{\overset{\displaystyle O}{\|}}{C}-CO_2^-} + \alpha\text{-amino acid} \;\rightleftharpoons\; \underset{\text{L-Alanine}}{CH_3-\overset{\overset{\displaystyle NH_3^+}{|}}{CH}-CO_2^-} + \alpha\text{-keto acid}$$

Glutamate transaminases:

$$\underset{\alpha\text{-Ketoglutarate}}{^-O_2C-CH_2-CH_2-\overset{\overset{\displaystyle O}{\|}}{C}-CO_2^-} + \alpha\text{-amino acid} \;\rightleftharpoons\; \underset{\text{L-Glutamate}}{^-O_2C-CH_2-CH_2-\overset{\overset{\displaystyle NH_3^+}{|}}{CH}-CO_2^-} + \alpha\text{-keto acid}$$

There are several glutamate transaminases that are specific for various amino acids, including one which is specific for L-alanine. The net effect of the transaminases is to "collect" the nitrogens of different amino acids onto *glutamate*, which is a central source of nitrogen for both biosynthesis and excretion. Note also that both pyruvate and α-ketoglutarate are key metabolites in the action of transaminases, and thus serve as important links between amino acid and carbohydrate metabolism.

In oxidative deamination the amino group of an amino acid is oxidized to an imine function which then hydrolyzes to a ketone functional group:

$$-CH_2-\overset{\overset{\displaystyle \overset{+}{N}H_3}{|}}{CH}- \;\xrightarrow{-(2e^-\,+\,2H^+)}\; \underset{\text{Imine}}{-CH_2-\overset{\overset{\displaystyle \overset{+}{N}H_2}{\|}}{C}-} \;\xrightarrow{H_2O}\; -CH_2-\overset{\overset{\displaystyle O}{\|}}{C}- + NH_4^+$$

The major oxidative path is the reaction catalyzed by L-glutamate dehydrogenase, although other amino acid oxidases and dehydrogenases are present in metabolism.

The reaction catalyzed by glutamate dehydrogenase is another vital connection between amino acid and carbohydrate metabolism and again illustrates the central role of glutamate in nitrogen metabolism. The enzyme is allosteric and is subject to inhibition by ATP, GTP, and NADH and activation by ADP, certain amino acids, and certain hormones. It can use either NAD^+ or $NADP^+$ as a coenzyme:

$$\begin{Bmatrix} NAD^+ \\ or \\ NADP^+ \end{Bmatrix} + \text{L-glutamate} + H_2O \rightleftharpoons \alpha\text{-ketoglutarate} + \begin{Bmatrix} NADH \\ or \\ NADPH \end{Bmatrix} + NH_4^+$$

The NH_4^+ produced is an extremely toxic substance in a cell, particularly in cells of higher animals. Normally, blood NH_4^+ concentration is about $3 \times 10^{-5}M$. Adding excess NH_4^+ drives the glutamate dehydrogenase redox equilibrium to the left, thereby depleting the crucial Krebs cycle intermediate α-ketoglutarate.

Elimination of NH_4^+: Urea Biosynthesis

There are three major routes in the liver for removing NH_4^+ or NH_3 from the cell:

1. **Reversal of glutamate dehydrogenase**

$$NADPH + \alpha\text{-ketoglutarate} + NH_4^+ \rightleftharpoons \text{glutamate} + H_2O + NADP^+$$

2. **Synthesis of glutamine**

$$\underset{\text{Glutamate}}{{}^-O_2C-CH_2CH_2\overset{\overset{\displaystyle NH_3^+}{|}}{C}H-CO_2^-} + ATP + NH_3 \rightleftharpoons \underset{\text{Glutamine}}{H_2N-\overset{\overset{\displaystyle O}{||}}{C}-CH_2CH_2\overset{\overset{\displaystyle NH_3^+}{|}}{C}H-CO_2^-} + ADP + P_i$$

Glutamine synthesis is a nontoxic way to transport NH_3 from the liver to other parts of an organism. The blood plasma concentration of glutamine is about twice that of any other amino acid. The amide group readily hydrolyzes to yield glutamate and ammonia. The enzyme which catalyzes this reversible reaction is *glutamine synthetase*, a complex regulatory enzyme of the cytoplasm.

3. **Synthesis of carbamoyl phosphate.** Mammals, which excrete excess nitrogen primarily as urea, form carbamoyl phosphate as part of the biosynthetic pathway from NH_3 to urea:

$$NH_4^+ + HCO_3^- + 2ATP \longrightarrow H_2N-\overset{\overset{\displaystyle O}{||}}{C}-OPO_3^{2-} + 2ADP + P_i$$

This reaction is catalyzed by *carbamoyl phosphate synthetase*, an important regulator of urea biosynthesis. In one form of *congenital hyperammonemia*, this enzyme is defective, thus causing increased blood levels of NH_4^+ (see Problem **10** at the end of this chapter).

Even though we should properly consider the biosynthesis of urea in Part III of this text, we will discuss it here in the context of nitrogen excretion and amino acid catabolism.

The bulk of excess nitrogen is excreted into the environment through three general modes. In aquatic organisms, ammonia is eliminated directly. Such organisms are called **ammonotelic**. **Uricotelic** animals excrete excess nitrogen primarily as *uric acid*, a much less toxic substance than NH_4^+. Uric acid is excreted as a semiliquid paste in a minimum of water. These uric acid-excreting organisms include birds and terrestrial reptiles, which must use a minimum amount of H_2O due to either weight (birds) or availability (snakes, etc.). Mammals excrete excess nitrogen primarily in the form of urea and are therefore called **ureotelic** organisms. Humans ordinarily excrete 5–10 grams of urea a day in the urine.

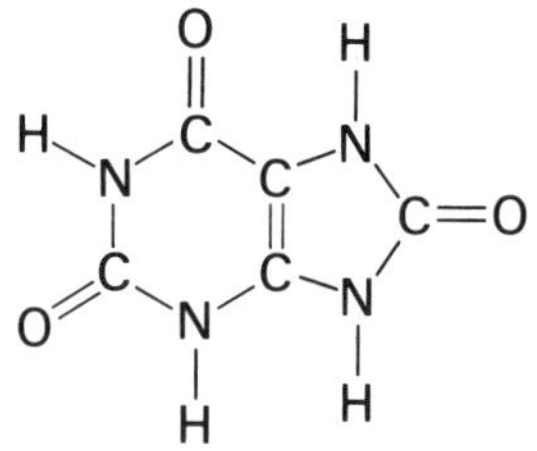

Uric acid

Urea synthesis occurs in the liver and the urea thus formed is carried to the kidneys for excretion. The details of the metabolic pathway for the formation of urea were worked out by Krebs and Henseleit in the 1930s. They observed that the rate of urea formation caused by "feeding" different amino acids to liver tissue slices was greatest when arginine was used. Like so many other biochemical processes which must carry out a continuous conversion of one metabolite into another, the steps of urea synthesis are cyclical in nature. The crucial step of the **Krebs Urea Cycle** is the reaction catalyzed by the enzyme *arginase*, which is present only in ureotelic organisms:

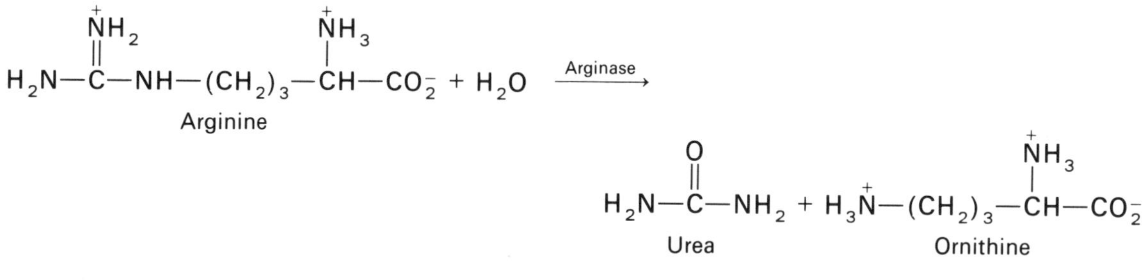

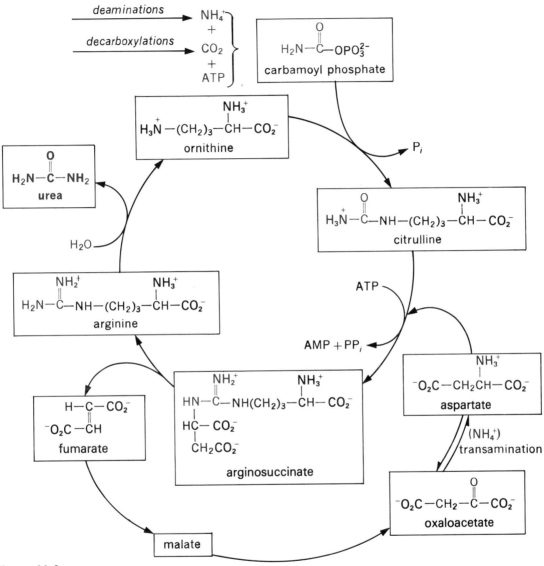

Figure 11.6

The Krebs urea cycle. The net effect is to remove two moles of NH_4^+ per turn. Note the metabolic interrelationship between this cycle and the Krebs TCA cycle (Chapter 9).

Arginase readily hydrolyzes arginine to ornithine and urea. Therefore, the job of the urea cycle is to convert ornithine back into arginine on a continuous basis. The conversion, of course, involves incorporating the elements of urea back into the ornithine molecule. The steps of this pathway are shown in Figure 11.6.

While urea is the major form in which nitrogen is excreted in humans, other nitrogen compounds are also normally present. These are summarized in Table 11.2.

Table 11.2 Nitrogen-containing compounds found in normal urine. These are forms in which nitrogen is excreted.

Compound	Structure	% of excreted N
Urea	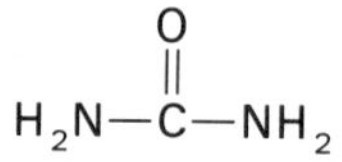 $H_2N-\overset{\overset{O}{\|\|}}{C}-NH_2$	86.0
Creatinine	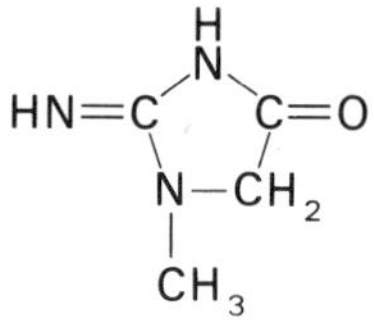 $HN=C\underset{N-CH_2}{\overset{H\atop N}{\diagup}}C=O$, $N-CH_3$	4.5
Ammonium ion	NH_4^+	2.8
Uric acid	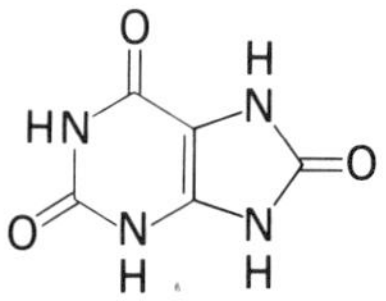 (structure)	1.7
Others		5.0

Overview of Steps of Amino Acid Catabolism

It is beyond the scope of this text to detail all the steps of amino acid metabolism. We have seen that some amino acids are ketogenic and some are glycogenic. In addition, we have noted that the initial event in the breakdown of many amino acids involves deamination, and that the $-NH_2$ groups thus removed can become involved in nitrogen-compound biosynthesis or can be excreted in mammals as urea. Figure 11.7 summarizes the overall catabolism of amino acids as related to the catabolism of pyruvate and acetyl CoA.

Glycogenic amino acids produce either pyruvate or a Krebs TCA cycle intermediate. The increase in the level of any Krebs cycle intermediate causes a net increase in the rate of gluconeogenesis (hence, the origin of the term "glycogenic"). The ketogenic amino acids fall into two classes (Table 11.1). Leucine is degraded directly to acetyl CoA and acetoacetyl CoA only. Those amino acids which are both ketogenic and glycogenic produce both acetyl CoA and either pyruvate or a Krebs TCA cycle intermediate.

Hereditary Defects in Amino Acid Catabolism

The details of amino acid metabolism have been studied extensively because many hereditary errors of amino acid metabolism have been observed in humans (see Chapter 13). In fact, the study of individuals with defects in

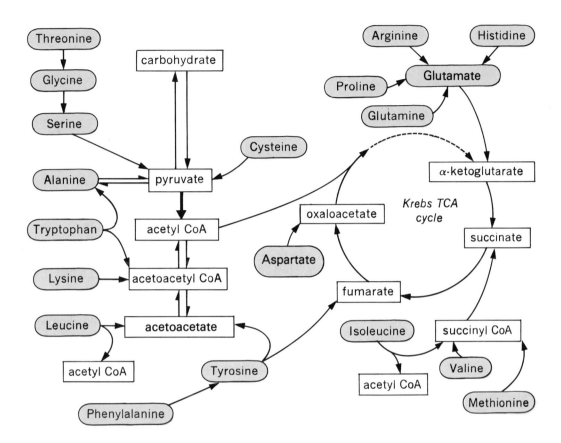

Figure 11.7

General relationships of amino acid catabolism. In many cases in this figure, the arrows represent a sequence of steps.

amino acid metabolism has provided researchers with a powerful means of determining the sequence of reactions in normal pathways. We have already seen numerous examples of inborn errors in carbohydrate and lipid metabolism. Now let's consider some of the many known hereditary defects in amino acid metabolism.

Phenylketonuria (PKU) is an inborn error of metabolism causing severe mental retardation if not detected and controlled in infancy. This is an example where an alternate pathway, usually of minor importance, leads to an abnormally high concentration of a minor metabolite. In PKU, the enzyme *phenylalanine hydroxylase* is inactive. Ordinarily, an important fate of phenylalanine is oxidation to tyrosine:

$$\text{Phenylalanine} \xrightarrow{[O]} \text{Tyrosine}$$

Phenylalanine   Tyrosine

In about 0.01% of the population, a genetic defect results in the absence or low activity of this enzyme. Phenylalanine accumulates in these individuals and is converted by an alternate pathway to phenylpyruvate:

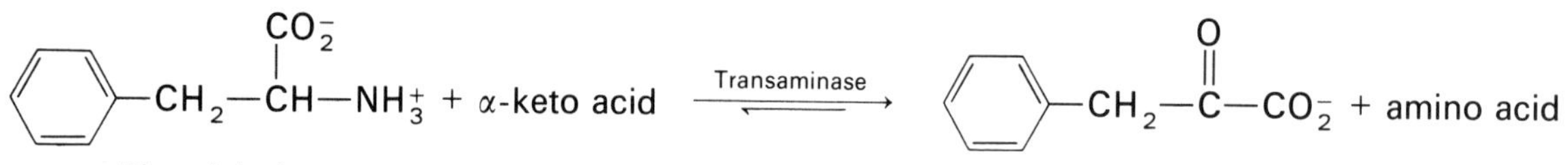

In individuals with PKU, as much as 2 grams of phenylpyruvate can accumulate daily. This excessive amount of phenylpyruvate is excreted in the urine. The undesirable effects of PKU can be controlled by early recognition of the condition followed by restriction of phenylalanine in the diet. In many states, newborn infants are normally screened for PKU shortly after birth.

Alkaptonuria is one of the four inborn errors of metabolism described by Garrod in 1908. In the stepwise degradation of tyrosine (and phenylalanine) to acetoacetyl CoA and fumarate, *homogentisate* is formed as an intermediate metabolite. In the case of alkaptonuria, the enzyme *homogentisate oxidase* is defective. This causes its substrate, homogentisate, to accumulate and be excreted in the urine. On exposure to air under alkaline conditions, homogentisate is oxidized to form a dark pigment which discolors the urine. Aside from abnormal pigmentation of the urine and connective tissue, alkaptonuria does not appear to cause the severe consequences that many other inborn errors of metabolism do.

Maple syrup urine disease is a rare hereditary defect of amino acid metabolism in which the urine smells like maple syrup due to the presence of large amounts of branched-chain α-keto acids. This condition is generally fatal and affects about 1/200,000 individuals. It is caused by defective enzymes involved in the catabolism of the three-branched-chain aliphatic amino acids, valine, isoleucine and leucine. Figure 11.8 shows the overall scheme of the steps where the disorder occurs (see also Problem **11** at the end of this chapter). As the figure shows, the dehydrogenases that oxidize and decarboxylate the branched-chain α-keto acids are defective.

Figure 11.8

The overall steps of valine, isoleucine and leucine catabolism, showing the process blocked in maple syrup urine disease.

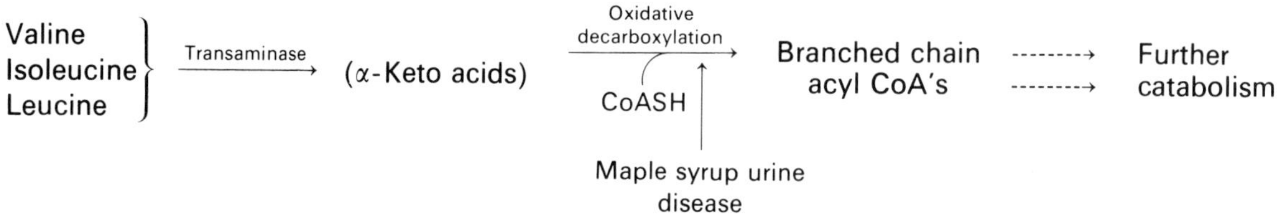

11.4

Catabolism of Nucleotides

Nucleotides obtained from either the diet or from the breakdown of cells in the body are hydrolyzed to yield the corresponding purine or pyrimidine base. Some of these bases can be "salvaged" by a family of enzymes called *phosphoribosyl transferases* which convert the bases into nucleotide monophosphates. However, most of the purines and pyrimidines derived from the diet are catabolized. The pyrimidines (thymine, cytosine and uracil) are oxidized to soluble non-cyclic products. The catabolism of purines (adenine and guanine) is more complicated; Figure 11.9 shows an overview of this process. One

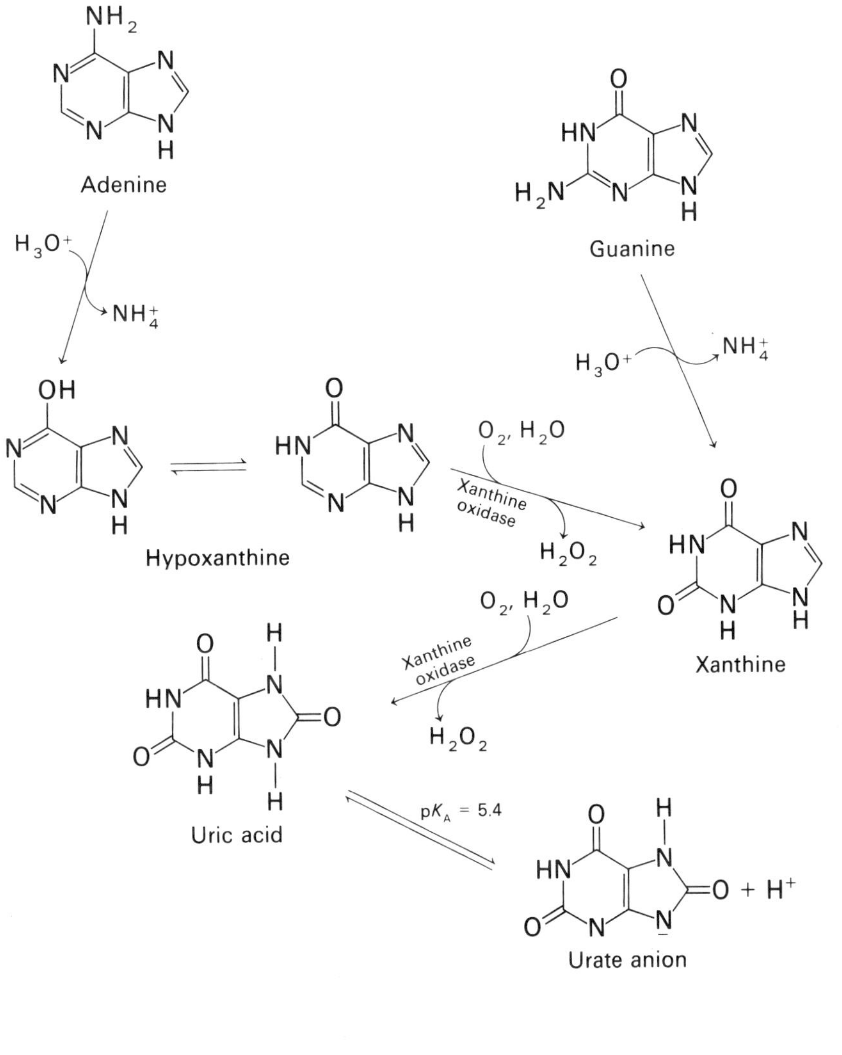

Figure 11.9

Overall scheme for the catabolism of purine bases (adenine and guanine). The common hereditary disease gout results from the accumulation of uric acid. Substances that specifically inhibit xanthine oxidase alleviate this condition. The equilibrium between tautomeric forms that exists for purine and pyrimidine bases is illustrated here for hypoxanthine.

interesting facet of purine catabolism is its connection with the common disease *gout*. This is a hereditary defect of metabolism that affects about three people per thousand. In humans and other primates, uric acid is excreted in the urine (see Table 11.2). Gout can arise from two sources: an increase in the rate of uric acid formation or a defect in the process which causes its excretion. Either way, the end is the same; increased amounts of uric acid in the blood cause insoluble urate salts to be deposited in the joints. This results in the familiar and painful gouty arthritis. Gout can be controlled by specifically inhibiting the formation of uric acid, using the drug *allopurinol*, a competitive inhibitor of xanthine oxidase. Note the similarity between the structure of allopurinol shown in the margin and the structure of the natural substrate for xanthine oxidase, hypoxanthine (see Figure 11.9).

Allopurinol

11.5

Concluding Remarks on Catabolism

We have now completed our discussion of the general ways in which cells extract energy from food molecules. Since the cell is a steady-state chemical system, **processes of catabolism and anabolism occur concurrently**. Our separation of these two broad types of pathways is merely for the sake of convenient discussion. Within the cell we find a steady-state concentration, or *pool*, of each metabolite. Whether a given metabolite is broken down to generate energy or is used in biosynthesis is governed ultimately by the ratio of ATP/ADP in the cell.

In this chapter we have seen how amino acids can be broken down into the same central intermediates of lipid and carbohydrate catabolism. The alternative, of course, is that the cell can use an amino acid as a protein building block. In Part III we will discuss some of the ways in which the cell produces biomolecules and biopolymers.

Suggestions for Further Reading

Lehninger, A. L. *Biochemistry*. 2d ed. New York: Worth Publishers, Inc. 1975. Chapters 20 and 21 give a good, detailed discussion of fatty acid and amino acid catabolism.

White, A.; Handler, P.; Smith, E.; Hill, R. and Lehman, I. *Principles of Biochemistry*. 6th ed. New York: McGraw-Hill, 1978. This comprehensive text is strongly oriented toward human physiology. Chapters 17–23 provide a good discussion of lipid and amino acid metabolism; Chapter 27 contains an interesting analysis of hereditary disorders of metabolism.

Scientific American

Dawkins, M. J. R. and Hull, D. "The Production of Heat by Fat." August 1965.

Green, D. E. "The Metabolism of Fats." January 1954.

Young, V. R. and Scrimshaw, N. S. "The Physiology of Starvation." October 1971.

Problems

1. For each of the terms below, provide a brief definition or explanation.

Fatty acyl carnitine	α-oxidation
Glycogenic	β-oxidation
Ketogenic	Pyrophosphoryl cleavage
Ketone bodies	Transamination
Krebs urea cycle	Ureotelic

2. For each enzyme given below, write out the reaction catalyzed, giving the structures of reactants and products.

Arginase	Hydroxyacyl CoA dehydrogenase
Enoyl hydratase	Ketothiolase
Fatty acyl CoA dehydrogenase	Pyrophosphatase
L-Glutamate dehydrogenase	Thiokinase
Glutamate-oxaloacetate transaminase	

3. Why are only two "turns" of the β-oxidation spiral required to completely degrade a C_6-fatty acid?

4. Write out all of the steps for the complete oxidation of the C_{18}-fatty acid stearic acid by the β-oxidation spiral, starting with stearic acid in the cytoplasm. How many ATP's are generated per stearic acid molecule as a result of complete oxidation, assuming maximum energy conservation?

5. Refer to Figure 11.3, page 321. Give the structures and relative molar amounts of the acyl CoA's produced by the β-oxidation of pristanic acid, which is the result of the α-oxidation of phytanic acid.

6. Refer to Figure 11.4. In methylmalonic aciduria, methylmalonyl CoA accumulates. It is known that this substance is an inhibitor of pyruvate carboxylase (Chapter 9). Explain why ketosis develops in cases of methylmalonic aciduria.

7. Using structures, show the fate of the following metabolites when they are catabolized.

   (a) aspartate   (b) glutamine   (c) alanine

   (d) acetoacetate   (e) acetate

8. Suppose an experiment is done on the effect of adding exogenous ("from the outside") substrates to a suspension of well-aerated pieces of pigeon flight muscle. Which would lead to the generation of more ATP: addition of 1 mole of acetate ion or 1 mole of citrate ion? Justify your answer.

9. Give at least four ways in which pyruvate can form from precursors other than phosphoenol pyruvate. You will have to refer to Chapter 9 as well as Chapter 11.

10. What metabolite, besides NH_4^+, would most likely be present in elevated amounts in the blood of a person with congenital hyperammonemia? Where would this metabolite come from, and what mainline process of metabolism would be directly affected?

11. Refer to Figure 11.8. Write out the individual steps shown, giving the structures of reactants and products.

12. The Dutch biochemist Knoop was one of the first to study fatty acid catabolism (1905). He fed dogs synthetic fatty acids containing a phenyl group at the end with the general formula:

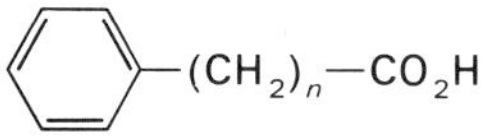

n = odd or even

The phenyl ring was not oxidized in the animals but excreted in the urine as an identifiable product. Depending on whether a dog was fed fatty acids with an even or odd number of $—CH_2—$ groups, Knoop isolated either a benzoic acid derivative or a phenylacetic acid derivative from the dog's urine. Which of the two final

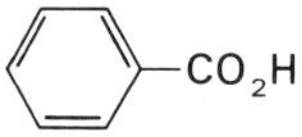

Benzoic acid

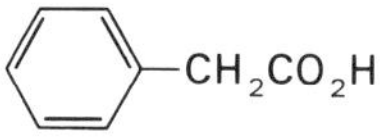

Phenylacetic acid

oxidation products isolated by Knoop resulted from the oxidation of the fatty acids having an even number of $—CH_2—$ groups?

PART III

Biosynthesis: Energy-Requiring Reactions

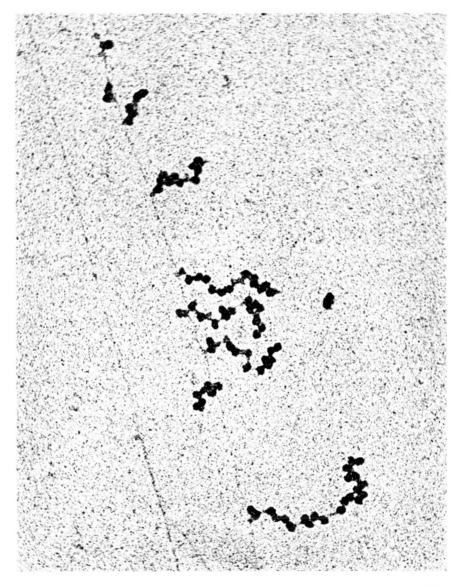

CHAPTER 12

Biosynthesis of Small Molecules

In Part I we learned about families of molecules that we might consider as "building blocks" for biosynthetic purposes. In Part II we learned how the cell conserves energy and how it can degrade biomolecules into a relatively small number of central substances such as pyruvate and acetyl CoA. In this chapter we will study some of the ways in which the cell synthesizes a number of important small molecules, including purines, pyrimidines, fatty acids and vital substances derived from amino acids. We will see why several coenzymes such as biotin and tetrahydrofolate are essential in biosynthetic pathways.

In addition to small-molecule biosynthesis, we will discuss the biosynthesis of glucose and the biosynthesis and breakdown of glycogen. Finally, we will review some of the methods the cell uses to keep the processes of catabolism and anabolism under control. Before going into these topics in detail, let's first review some general biosynthetic principles.

12.1

General Principles of Biosynthesis

There are three underlying requirements of biosynthesis:

1. **The Energy Problem.** Creating order by producing larger molecules from smaller molecules requires energy. In addition to providing the driving force needed to synthesize a product which is thermodynamically less stable than its precursors, energy is also expended in supplying a "thermodynamic push" sufficient to ensure that a process goes to completion with a high yield. As we saw in Chapter 7, this is the same as saying that ΔG is large and negative, or $K_{eq} \gg 1$. We know, of course, that the primary energy source for biosynthesis is based on ATP and other high-energy phosphate compounds.

2. **The Parts Problem.** Cells do not have free $-CH_3$ groups or free $-\overset{\displaystyle O}{\overset{\displaystyle \|}{C}}-CH_3$ groups floating around in the cytoplasm, even though many methylation

and acetylation reactions occur in the cell. In the laboratory, we might use acetic acid as a source of acetyl groups. But free acetic acid is toxic to the cell. Therefore, means must exist in the cell to provide such a group by the appropriate coenzyme. We know, for example, that in the case of citrate synthetase (Chapter 9), coenzyme A acts as a carrier for an activated $-\overset{\overset{\displaystyle O}{\|}}{C}-CH_3$ group in the citrate-forming reaction.

3. **The Pattern Problem.** Any given molecule has a unique structure. Therefore, the cell must contain a mechanism for ensuring that a particular metabolic pathway produces the proper product. There are two levels at which formation of the correct pattern or structure is verified. On the macromolecular level, the cell produces the correct sequence of amino acids in a protein molecule by using a nucleic acid as a template for its synthesis (Chapter 13). On the molecular level, in the synthesis of non-informational macromolecules and small biomolecules, the correct structure is assured by the specificities of the enzymes involved (see Section 3.7).

With these three underlying requirements of biosynthesis in mind, let's consider the energy problem and the parts problem in more detail.

Providing a Driving Force for Biosynthesis by Group Activation

In Chapter 7 we saw how an endergonic process could be coupled to the hydrolysis of ATP. In biosynthetic reactions involving $R-OH$, $R-NH_2$, and related types of compounds, activation generally involves either *phosphorylation* or the formation of *nucleoside phosphate derivatives* to produce a more reactive substrate.

Activation by phosphorylation   follows the general scheme

$$R-OH + ATP \;\rightleftharpoons\; R-OPO_3^{2-} + ADP$$

The activation group RO— can react with some acceptor molecule A to form a new bond. The action of the phosphate group makes this reaction more exergonic:

$$R-OH + A \;\rightleftharpoons\; RO-A + H^+ \qquad K_{eq} < 1$$
$$R-O-PO_3^{2-} + A \;\rightleftharpoons\; RO-A + P_i \qquad K_{eq} > 1$$

A few examples of this type of process were given in Chapter 7 when we discussed the coupling of an endergonic reaction to the hydrolysis of ATP. We can illustrate this process further by discussing another example.

Inosine (Section 12.8) is biosynthesized from ribose 1-phosphate and hypoxanthine:

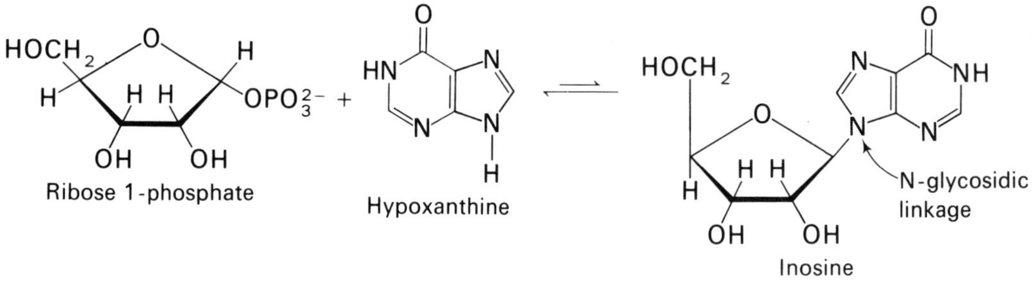

Ribose 1-phosphate   Hypoxanthine   Inosine

Here the number 1 —OH group of ribose is activated by prior formation of a phosphate ester.

Activation by formation of a nucleoside phosphate derivative is particularly important in the biosynthesis of proteins and carbohydrates. An important example of how this works is the formation of uridine diphosphate glucose (UDP glucose), which is the building block for the formation of glycogen:

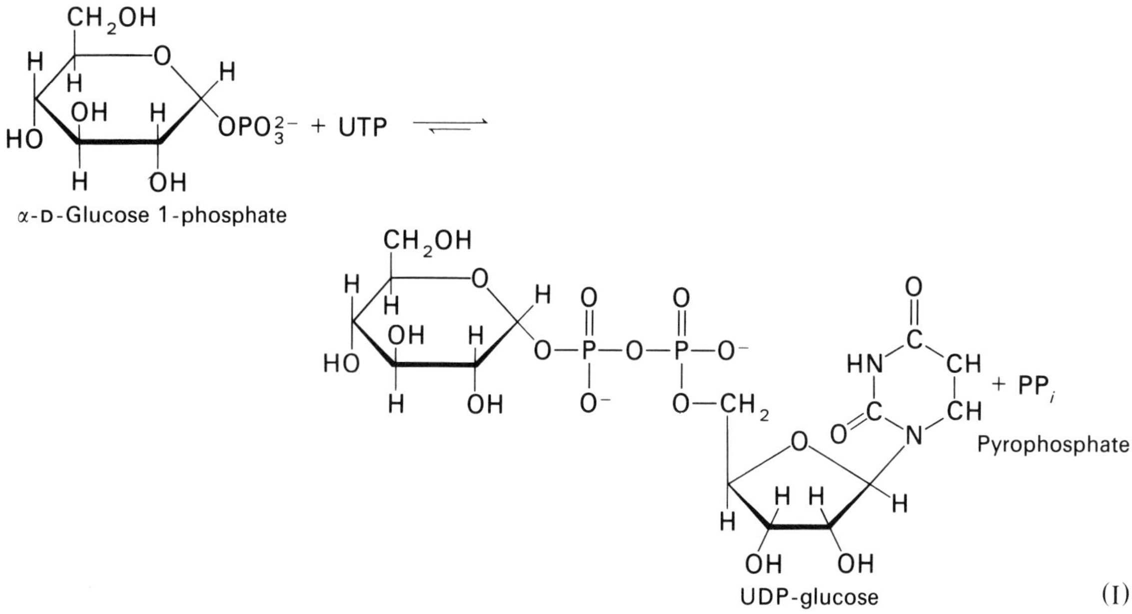

α-D-Glucose 1-phosphate

UDP-glucose   (I)

$$PP_i + H_2O \xrightarrow{\text{Pyrophosphatase}} 2P_i \qquad (II)$$

The sum of reactions (I) and (II) is strongly exergonic because the hydrolysis of pyrophosphate [reaction (II)] helps pull reaction (I) along by Le Châtelier's Principle (Section 7.5). This scheme of pyrophosphate cleavage and subsequent hydrolysis is quite general, and provides the driving force for many key activation reactions.

Nucleoside diphosphate sugars are important in the formation of glycoside linkages (Section 4.2). In Section 12.5 we will see some specific examples of UDP-glucose participating in the creation of glycoside links.

Group-Carrying Coenzymes

Many reactions in biosynthesis consist of a small activated group or "part" combining with an already existing molecule to produce a more complex product. This group addition involves certain coenzymes (Section 3.5) serving as specific carriers for certain groups. For example, we just saw how UDP acts as a specific carrier for a *glycosyl* group and activates that glucose-derived group with respect to addition to some acceptor molecule, as in the formation of a glycoside link. Table 12.1 lists a number of important group-carrying coenzymes, together with some of the biosynthetic processes in which they are involved.

Not shown in Table 12.1 is the large class of nucleoside phosphate derivatives of various functional groups, which are extremely important in biosynthesis. In this class, the "energy problem" and the "parts problem" are satisfied simultaneously because the nucleoside phosphate carries the group in an activated form. In Chapter 13 we will learn about an important example of this when we discuss the transfer-RNA-amino acid derivative as the true building block for protein biosynthesis.

Also not shown in Table 12.1 is the important methyl group carrier *S-adenosyl methionine (SAM)*. This substance can be considered as a "CH_3^+" donor in its reactions (see Figure 12.5).

$$NH_3^+$$
$$\mid$$
$$CH_2CH_2CHCO_2^-$$
$$/$$
$$H_3C-\overset{+}{S}-CH_2 \quad O \quad \text{Adenine}$$

S-adenosyl methionine

With these general biosynthetic principles in mind, let's see how they are applied by discussing some specific examples of small-molecule biosynthesis.

Table 12.1 Some Group-Carrying Coenzymes Involved in Biosynthesis. [An asterisk (*) marks the active part of the coenzyme molecule.]

Coenzyme	Group carried	Some biosynthetic uses
Coenzyme A; CoASH	Acyl group: $-\overset{\displaystyle O}{\underset{\displaystyle \parallel}{C}}-R$	Steroids; Fatty acids (Section 12.2); Triacylglycerols (Figure 12.2); Citrate (Chapter 9)
Tetrahydrofolate: 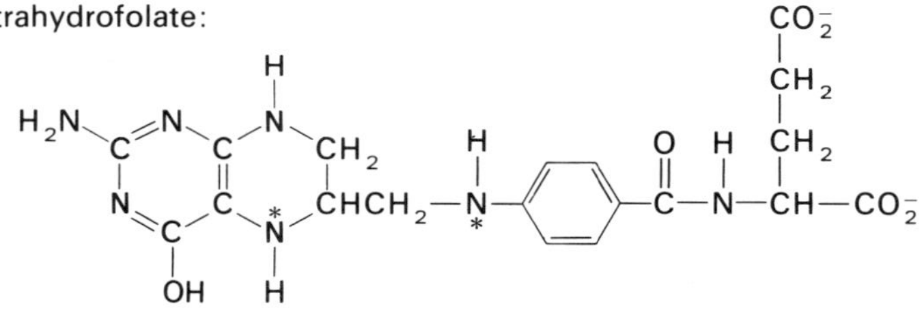 [structure]	One-carbon groups: $-CH_2OH$; $-CH_3$; $-\overset{\displaystyle O}{\underset{\displaystyle \parallel}{C}}-H$	Purines; Glycine (Figure 12.4); Methionine (Section 12.3)
Biotin: 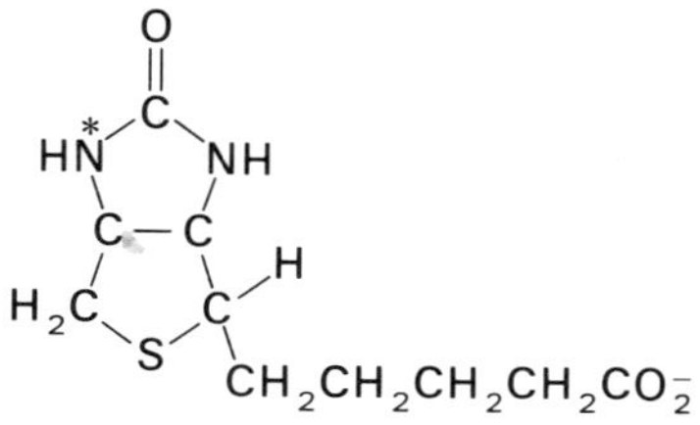 [structure]	Carboxylate group: $-\overset{\displaystyle O}{\underset{\displaystyle \parallel}{C}}-O^-$	Carboxylic acids (Section 9.5)
Thiamine pyrophosphate (TPP): 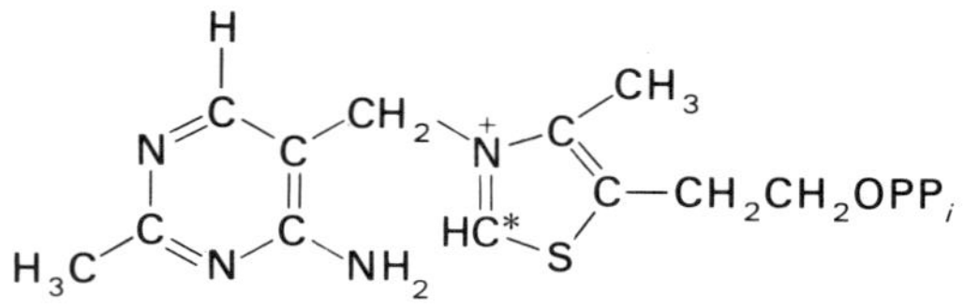 [structure]	"Activated aldehyde": $\left[R-\overset{\displaystyle O^-}{\underset{\displaystyle \parallel}{C}}- \right]$ (see Figure 3.6)	$\overset{\displaystyle O}{\underset{\displaystyle \parallel}{R-C}}-\overset{\displaystyle OH}{\underset{\displaystyle \mid}{C}}-R'$ (α-ketol derivatives) (Figures 8.9 and 11.4)
Nicotinamide adenine dinucleotide phosphate (NADP)	Hydrogen carrier in reduced form: $(H^+ + e^-)$	Serves as a reducing agent in biosynthetic reduction reactions

12.2

Fatty Acid Biosynthesis

In Chapter 11 we learned how fatty acids are degraded to acetyl CoA by the β-oxidation spiral. Fatty acid biosynthesis, or **lipogenesis**, is the opposite of this process. Before we discuss the details of lipogenesis, there are some overall features worth summarizing.

1. **Location.** Lipogenesis occurs in the cytoplasm through the action of a multienzyme complex called the *fatty acid synthetase* complex. As the name implies, this enzyme complex forms new fatty acid molecules, using acetyl CoA as starting material. While the synthesis of new fatty acid molecules does not occur in the mitochondrion, where fatty acid oxidation takes place, there are mitochondrial enzymes that catalyze the lengthening of preexisting medium- and long-chain fatty acids.

    Lipogenesis is an example of how the cell keeps degradative and biosynthetic processes separated. It is a general rule that the cell does not biosynthesize and degrade a given metabolite using exactly the same pathway. Were this the case, a situation of metabolic stasis, or "futile cycle," would result. By keeping biosynthetic and catabolic pathways separate, the cell can exert control over both processes in a coordinated manner. In the case of lipogenesis vs β-oxidation the cell achieves this separation by conducting each process in a different compartment (see Section 12.6). In the case of glycolysis vs gluconeogenesis (Section 12.4) the catabolic and anabolic pathways share many steps in common, except for key irreversible steps catalyzed by enzymes unique to each pathway.

2. **CO_2 requirements.** The cytoplasmic fatty acid synthetase system requires CO_2. This is worth noting because CO_2 is not a reactant or product in the β-oxidation spiral of fatty acid degradation.

3. **Raw materials.** The fatty acid synthetase system is not a simple reversal of the β-oxidation spiral. While the end product of the β-oxidation spiral is acetyl CoA, the key building block of fatty acid synthesis is *malonyl CoA*, a derivative of CO_2 and acetyl CoA:

$$^-O_2C-CH_2-\overset{\overset{\textstyle O}{\textstyle \|}}{C}-S-CoA$$

Malonyl CoA

The reason for this makes sense in terms of the driving force for lipogenesis. When a malonyl CoA unit is added to a growing fatty acid chain, CO_2 is released. This makes the net addition of a two-carbon unit to the

Figure 12.1

Diagram of the functional units of the yeast fatty acid synthetase complex. The numerals in this figure refer to the steps in lipogenesis discussed in the text. The arrangement of the complex enables the fatty acid chain elongation reactions to occur sequentially as the reactant-ACP thioester derivative "swings" from enzyme to enzyme. When the fatty acyl chain is sufficiently long (12–16 carbons) it is cleaved from the long side chain of ACP (Step VII).

fatty acid chain essentially irreversible and effectively solves the "energy problem" in lipogenesis.

The Enzymes of Lipogenesis

Workers have isolated and studied the fatty acid synthetase system from such sources as *E. coli*, yeast, and pigeon liver. The form of this enzyme complex varies from species to species, but the basic mode of action appears common to all. Fatty acid synthetase from *E. coli* has seven subunits and a total molecular weight of 2.3×10^6. Each subunit possesses an individual catalytic function, as illustrated in Figure 12.1. Together with the separate enzyme *acetyl CoA carboxylase*, which forms malonyl CoA, these seven subunits constitute the enzymes required for fatty acid biosynthesis.

Formation of Malonyl CoA

Acetyl CoA carboxylase, a biotin-requiring enzyme that catalyzes the addition of a carboxyl group to acetyl CoA, is a regulatory enzyme controlling

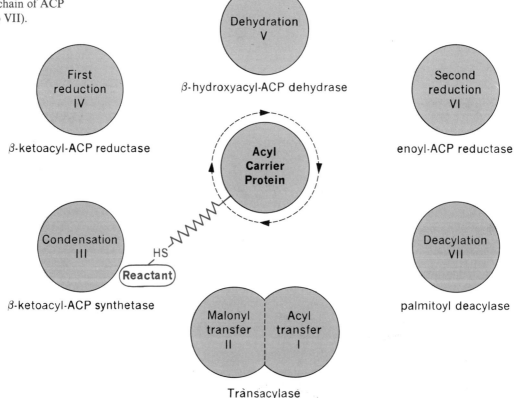

the overall rate of lipogenesis. Both citrate and isocitrate activate this enzyme. Increased concentrations of the Krebs cycle intermediates occur when excess fuel such as glucose, pyruvate, or acetyl CoA is present. Lipogenesis provides a means for storing this excess fuel. In the absence of citrate and isocitrate, acetyl CoA carboxylase remains inactive.

$$CH_3-\overset{\overset{\displaystyle O}{\|}}{C}-SCoA + CO_2 + ATP \xrightarrow[\text{(Biotin)}]{Mg^{2+}} HO_2C-CH_2-\overset{\overset{\displaystyle O}{\|}}{C}-SCoA + ADP + P_i$$

Malonyl CoA

Note that biotin is required as a coenzyme since this is a carboxylation reaction (see Table 12.1).

Formation of Fatty Acid Chain

The basic reaction forming carbon–carbon bonds in lipogenesis is

$$E-S-\overset{\overset{\displaystyle O}{\|}}{C}-R + {}^-O_2C-CH_2-\overset{\overset{\displaystyle O}{\|}}{C}-S-E' \longrightarrow E'-S-\overset{\overset{\displaystyle O}{\|}}{C}-CH_2-\overset{\overset{\displaystyle O}{\|}}{C}-R + HS-E + CO_2$$

Enzyme-bound fatty acyl group   Enzyme-bound malonyl group   Enzyme-bound longer chain

This reaction is much like the reverse of the last step of the β-oxidation spiral, the ketothiolase reaction (see Section 11.1), except that here the release of CO_2 provides the driving force needed to make the reaction proceed in the direction indicated.

Figure 12.1 shows how the functional units of the fatty acid synthetase complex in yeast are arranged. Note the "swinging arm" in the figure, called **acyl carrier protein** (ACP). As the figure shows, ACP acts as a conveyor, bringing the fatty acyl chain bound to its terminal —SH group from one stage of fatty acid chain elongation to the next in the fatty acid synthetase complex. The action of ACP eliminates the need for a random diffusion of reactant from one unit in the complex to the next.

Acyl carrier protein is a small (mol. wt. 10,000) protein containing a pantothenic acid group covalently linked to a serine residue of the protein through a phosphate bridge:

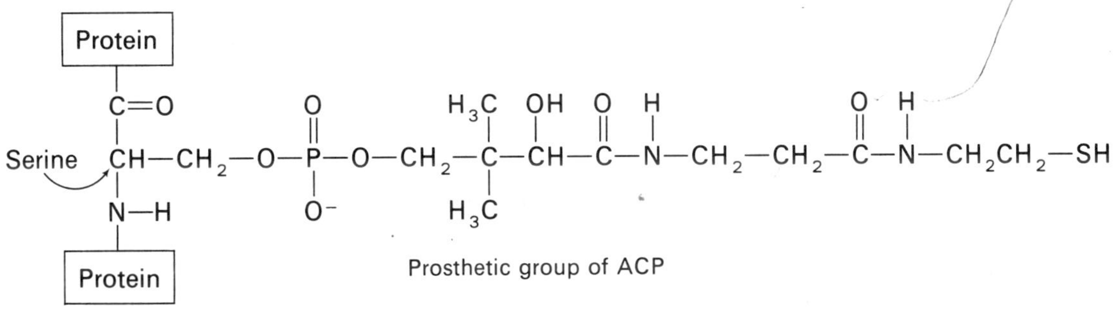

Prosthetic group of ACP

The terminal —SH group of the pantothenic acid part of ACP forms a thioester link with an acetyl group, a malonyl group, or a fatty acyl group; remember that the —SH group in coenzyme A has a similar function (see Section 3.5).

$$R-\overset{\overset{\displaystyle O}{\|}}{C}-S\diagdown\diagup\diagdown\diagup\diagdown\boxed{\text{ACP}}$$

We will follow the steps of lipogenesis, shown in Figure 12.1, as we did in Chapter 11 for the β-oxidation spiral.

I.   Initiation of lipogenesis   Lipogenesis is initiated by the reaction of acetyl CoA with the —SH group of ACP:

$$\underset{\text{Acetyl CoA}}{CoAS-\overset{\overset{\displaystyle O}{\|}}{C}-CH_3} + HS-ACP \rightleftharpoons CoASH + \underset{\text{Acetyl-S-ACP}}{CH_3-\overset{\overset{\displaystyle O}{\|}}{C}-S-ACP}$$

The "swinging arm" of ACP is now charged with an acetyl group which can accept additional two-carbon units.

II.   Initiation of chain elongation   The acetyl group of acetyl-S-ACP transfers temporarily to a cysteine residue —SH group, clearing ACP for the addition of a malonyl group. In subsequent turns of the lipogenesis cycle, a fatty acyl group transfers to the cysteine —SH. In either case, malonyl CoA forms a thioester link with the resulting free —SH group of ACP.

$$CH_3\overset{\overset{\displaystyle O}{\|}}{C}-S-ACP + HS-E_{\text{II}} \rightleftharpoons CH_3\overset{\overset{\displaystyle O}{\|}}{C}-S-E_{\text{II}} + HS-ACP$$

$$\underset{\text{Malonyl CoA}}{^-O_2C-CH_2\overset{\overset{\displaystyle O}{\|}}{C}-SCoA} + HS-ACP \rightleftharpoons \underset{\text{Malonyl-S-ACP}}{^-O_2C-CH_2\overset{\overset{\displaystyle O}{\|}}{C}-S-ACP} + CoASH$$

III.   Condensation   In this step the fatty acyl chain lengthens by two carbon atoms through the reaction of malonyl-S-ACP with the enzyme-bound acetyl (or fatty acyl) group of the previous step:

$$O_2C-CH_2\overset{\overset{\displaystyle O}{\|}}{C}-S-ACP + CH_3\overset{\overset{\displaystyle O}{\|}}{C}-S-E_{\text{II}} \longrightarrow \underset{\substack{\text{Acetoacetyl-S-ACP} \\ (\beta\text{-ketoacyl ACP})}}{CH_3\overset{\overset{\displaystyle O}{\|}}{C}-CH_2\overset{\overset{\displaystyle O}{\|}}{C}-S-ACP} + CO_2(g) + HSE_{\text{II}}$$

The driving force for this process is provided by the elimination of CO_2 from the malonyl group during the reaction.

IV. First reduction The β-keto acyl group undergoes reduction by NADPH, yielding a β-hydroxyacyl group:

$$CH_3\overset{\overset{\displaystyle O}{\|}}{C}-CH_2\overset{\overset{\displaystyle O}{\|}}{C}-S-ACP + NADPH + H^+ \;\rightleftharpoons\; CH_3\overset{\overset{\displaystyle OH}{|}}{CH}-CH_2\overset{\overset{\displaystyle O}{\|}}{C}-S-ACP + NADP^+$$

β-hydroxybutyryl-S-ACP

This is just the reverse of the corresponding step in the β-oxidation spiral, except that NADPH is the reductant in lipogenesis while NAD^+ is the oxidant in the step of β-oxidation. This is consistent with our understanding of NADPH as a reducing agent in biosynthesis.

V. Dehydration In this step an alcohol is dehydrated to an alkene:

$$CH_3\overset{\overset{\displaystyle OH}{|}}{CH}-CH_2\overset{\overset{\displaystyle O}{\|}}{C}-S-ACP \;\rightleftharpoons\; CH_3-CH{=}CH-\overset{\overset{\displaystyle O}{\|}}{C}-S-ACP + H_2O$$

Crotonyl-S-ACP

Finally reduction of the product of this step yields a fatty acyl group containing two more carbons than the one that started the sequence of steps.

VI. Second reduction Here an alkene undergoes reduction by NADPH to yield an alkane. In this case, the product is butyryl-S-ACP, which undergoes further chain elongation beginning with step II.

$$CH_3-CH{=}CH-\overset{\overset{\displaystyle O}{\|}}{C}-S-ACP + NADPH + H^+ \;\rightleftharpoons\; CH_3CH_2CH_2\overset{\overset{\displaystyle O}{\|}}{C}-S-ACP + NADP^+$$

Butyryl-S-ACP

VII. Release of completed fatty acid After seven "turns" of Steps I–VI, palmityl-S-ACP (C_{16}) is formed. At this point, further elongation does not occur. Instead, the thioester link is hydrolyzed, releasing palmitic acid:

$$CH_3-(CH_2)_{14}-\overset{\overset{\displaystyle O}{\|}}{C}-S-ACP + H_2O \;\rightleftharpoons\; CH_3-(CH_2)_{14}-\overset{\overset{\displaystyle O}{\|}}{C}-OH + HS-ACP$$

Palmityl-S-ACP   Palmitic acid

Formation of Higher Fatty Acids

Fatty acids with more carbons than palmitic acid are formed by elongation reactions that can occur in two principal ways. In both cases an acetate unit is added to the carboxyl end of the existing fatty acid chain, thus extending it by two carbon atom units at a time. In the *mitochondrial elongation system*, acetyl CoA is the carbon unit donor and NADPH is the reductant. In the *cytoplasmic elongation system*, enzymes associated with the endoplasmic reticulum (see Figure 1.4) use malonyl CoA as a two-carbon source and NADPH as a reductant.

Unsaturation is introduced into the fatty acid chain by a *fatty acid desaturase* enzyme system that requires O_2 and NADH. In mammals, this desaturase system specifically inserts a *cis* – C=C at a point no more than nine carbon atoms from the —COOH group in fatty acids having sixteen or more carbon atoms. Therefore, the *cis*-9,10 double bond in oleic acid can readily be formed from the saturated precursor, stearic acid. Because the mammalian fatty acid desaturase system cannot introduce double bonds in a carbon chain beyond the 9,10 position, the polyunsaturated linoleic and lino-

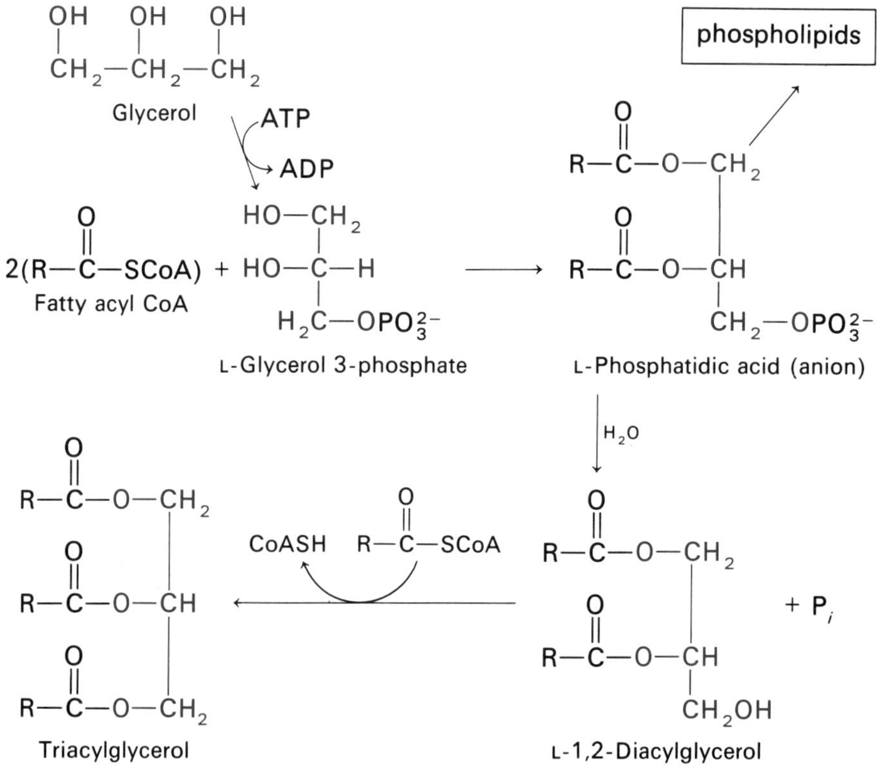

Figure 12.2

Formation of triacylglycerols and related substances. L-Glycerol 3-phosphate is derived primarily from glycolysis. The glycolytic intermediate dihydroxyacetone phosphate is reduced to L-glycerol 3-phosphate.

lenic acids cannot be produced in the body. These two fatty acids are thus an essential part of the diet, due to the specificity of the fatty acid desaturase system (see Problem **6** at the end of this chapter).

An important source for the NADPH required for fatty acid biosynthesis is the set of dehydrogenation reactions of the pentose phosphate pathway (Section 8.3). Both the biosynthesis and catabolism of fatty acids normally occur concurrently but separately in different locations in the cell. Therefore, we can say that fatty acids are in a constant state of turnover in the cell.

Free fatty acids are normally not found in animals in any significant amount. They are used to produce triacylglycerols and related lipids. The major features of triacylglycerol synthesis are shown in Figure 12.2. Adipose tissue and the liver are the major sites of triacylglycerol synthesis in animals. Note that phosphatidic acid is a central substance in lipid biosynthesis (see Chapter 6).

12.3

Biosynthesis of Nitrogen Compounds

In Section 11.3 we learned how urea is biosynthesized via the Krebs urea cycle. In addition to urea, which is synthesized to eliminate NH_4^+, there are several other nitrogen-containing compounds of functional importance whose biosynthetic pathways have been investigated. These include amino acids and their derivatives, oligopeptides, porphyrin, purines, and pyrimidines. It is beyond the scope of this book to discuss the biosynthesis of all these compounds in detail. However, we will consider some examples that illustrate important points relevant to the biosynthesis of nitrogen compounds.

Amino Acids

The biosynthetic pathways of amino acids have been studied extensively, particularly in bacteria. These simple organisms are used because it is much easier to isolate and investigate a given process in bacterial metabolism than to study the corresponding process in the vastly more complex cells of higher organisms.

In general, many processes of amino acid catabolism are common to amino acid biosynthesis. The net effect is to maintain a constant "pool" of each amino acid in the cell. Furthermore, the biosynthesis of most amino acids is under feedback control. In Section 3.9 we used the example of isoleucine biosynthesis in discussing feedback control. This concept is illustrated again in Figure 12.3 for the case of valine biosynthesis. The figure shows that formation of α-acetolactate from the starting material pyruvate requires thiamine pyrophosphate (see Table 12.1).

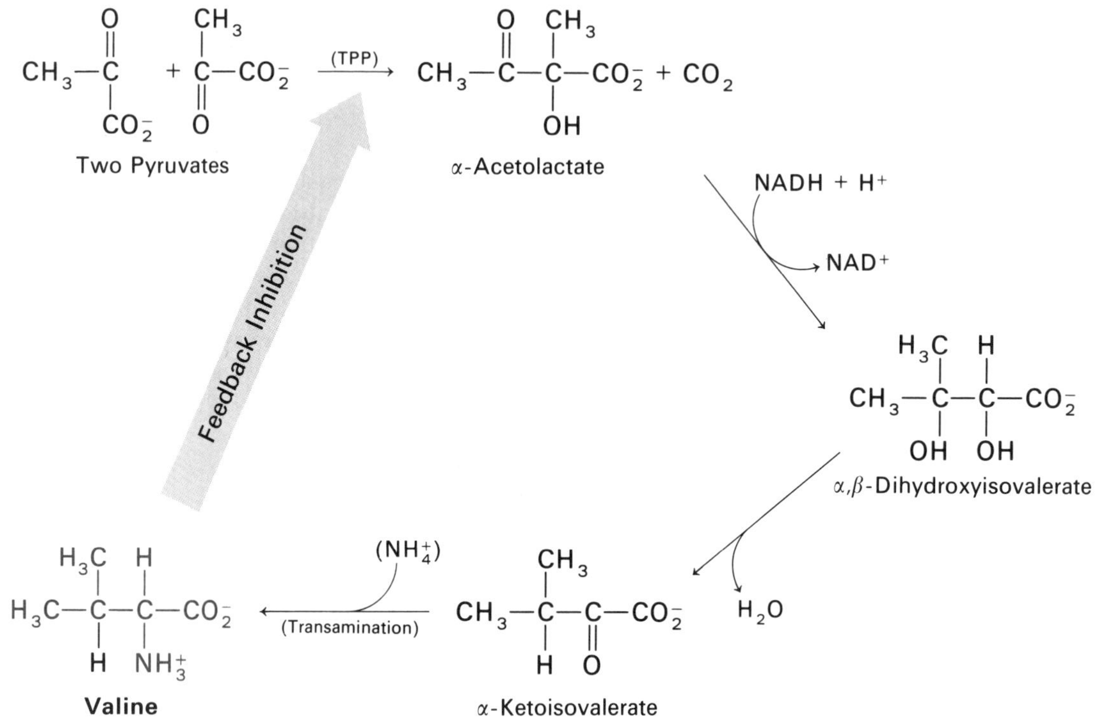

Figure 12.3

Valine biosynthesis, showing the feedback control common to many amino acid biosynthetic pathways.

In addition to being produced from precursors which are intermediates in carbohydrate metabolism, some amino acids are biosynthesized from other amino acids. An example of this is given in Figure 12.4, which shows the biosynthesis of serine from 3-phosphoglyceric acid, an intermediate in glycolysis, followed by the formation of glycine from serine. The reversible formation of glycine from serine also illustrates the action of tetrahydrofolate as a coenzyme involved in one-carbon-group transfers. The 5,10-methylene tetrahydrofolate produced in the biosynthesis of glycine can be oxidized or reduced, providing a source for a variety of one-carbon units.

Many amino acids are precursors for a wide variety of nitrogen-containing biomolecules. For example, tyrosine is a precursor for the skin pigment *melanin*. In specialized skin cells called *melanocytes*, tyrosine is oxidized by the action of the enzyme *tyrosinase* to give dihydroxyphenylalanine, an intermediate which is then further oxidized to give the polymeric pigment melanin. The inborn error of metabolism labeled *albinism*, expressed as a lack of skin pigmentation, has been traced to a genetic defect in tyrosinase. An interesting sidelight is the fact that tyrosinase is not only involved in skin pigment formation in animals, but also causes the darkening of such fruits as apples and peaches when they are exposed to air.

Two important classes of nitrogen-containing substances are the *amines* and the purine and pyrimidine bases of DNA and RNA. We will discuss the important features of each of these in turn.

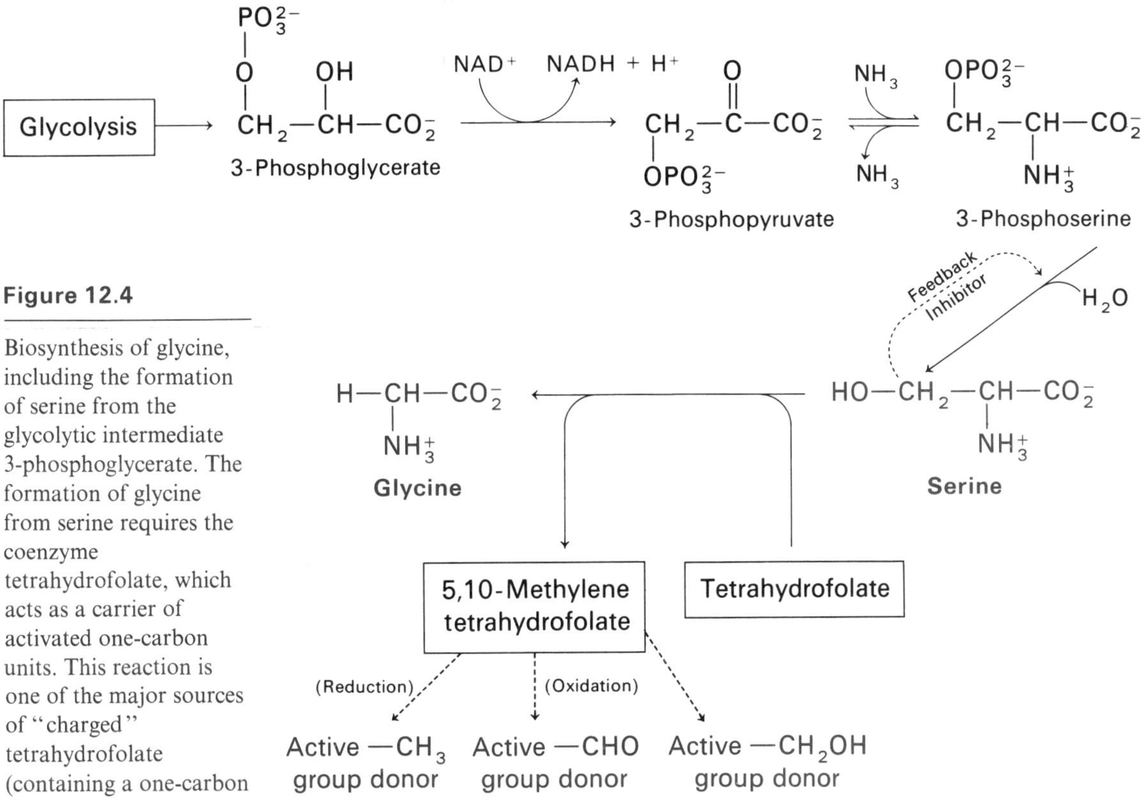

Figure 12.4

Biosynthesis of glycine, including the formation of serine from the glycolytic intermediate 3-phosphoglycerate. The formation of glycine from serine requires the coenzyme tetrahydrofolate, which acts as a carrier of activated one-carbon units. This reaction is one of the major sources of "charged" tetrahydrofolate (containing a one-carbon fragment). Tetrahydrofolate derivatives can donate an activated one-carbon unit in any of three oxidation states: $-CH_3$; $-CH_2OH$; $-CHO$.

Amino Acids as Precursors of Amines

There are a number of amines with strong physiological activity that are produced from the simple decarboxylation of amino acids. Some amines are important in the functioning of the nervous system, and others produce strong responses, such as swelling of tissues in the case of histamine.

Histamine   is produced by the decarboxylation of histidine as part of the response of the immune system to antigens. Histamine causes the capillaries to expand, producing edema (swelling) and lowering of blood pressure. This can create a great deal of discomfort for people with allergies, since their systems are constantly producing histamines. The discomfort can be eased by the consumption of *antihistamines*. Antihistamines are drugs that block the action of histamine and therefore alleviate its effects.

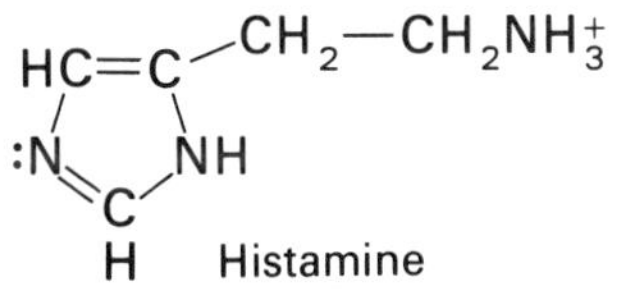

Amines are important substances in the biochemistry of the nervous system. While it is beyond the range of this text to treat these substances and their action in detail, a few examples will provide a brief glimpse of this exciting new branch of biochemistry.

Nerve action is basically a chemical process, wherein certain amines act as neurochemical transmitters in the transmission of nerve impulses from the sending end of one nerve cell (*neuron*) to the receiving end of another, as illustrated below:

Neuron
transmits
nerve
signal
⎬ ⟶ Neurochemical transmitter released ⟶ { Transmitter migrates to receptor site of another neuron to trigger continuation of the signal

One type of junction or *synapse* between nerve cells is called a *cholinergic synapse* because it functions by using *acetyl choline*. Figure 12.5 shows how acetyl choline is produced from serine and how it acts as a neurochemical transmitter.

The *catecholamines* are another important family of neurochemical transmitters which act in a different type of nerve synapse. All the compounds in this group are derived from tyrosine. Figure 12.6 shows the relationship between the catecholamines and their precursor, tyrosine.

One of the members of this family, dopamine, is produced in brain tissue from L-DOPA, a molecule that can cross the blood-brain barrier. L-DOPA has been used with great success in the treatment of the debilitating *Parkinson's*

Figure 12.5

The formation of choline from serine and its action in the neurochemical cycle. Choline is acetylated at the transmitter end of one neuron to produce acetyl choline. This migrates across the gap between nerve cells called the synapse, is bound by the cholinergic receptor of another nerve cell, and then hydrolyzed back to choline. In the process, a nerve impulse is transmitted. Acetyl cholinesterase is a serine hydrolase which is irreversibly inhibited by nerve gases (see Section 3.7).

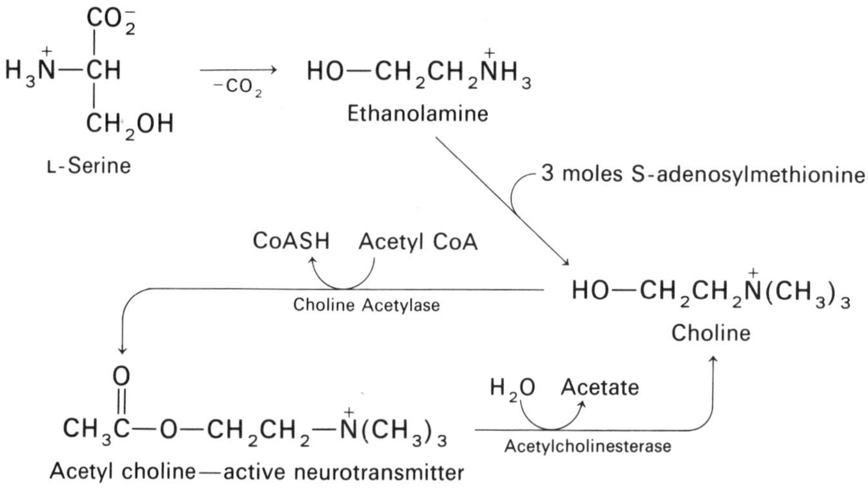

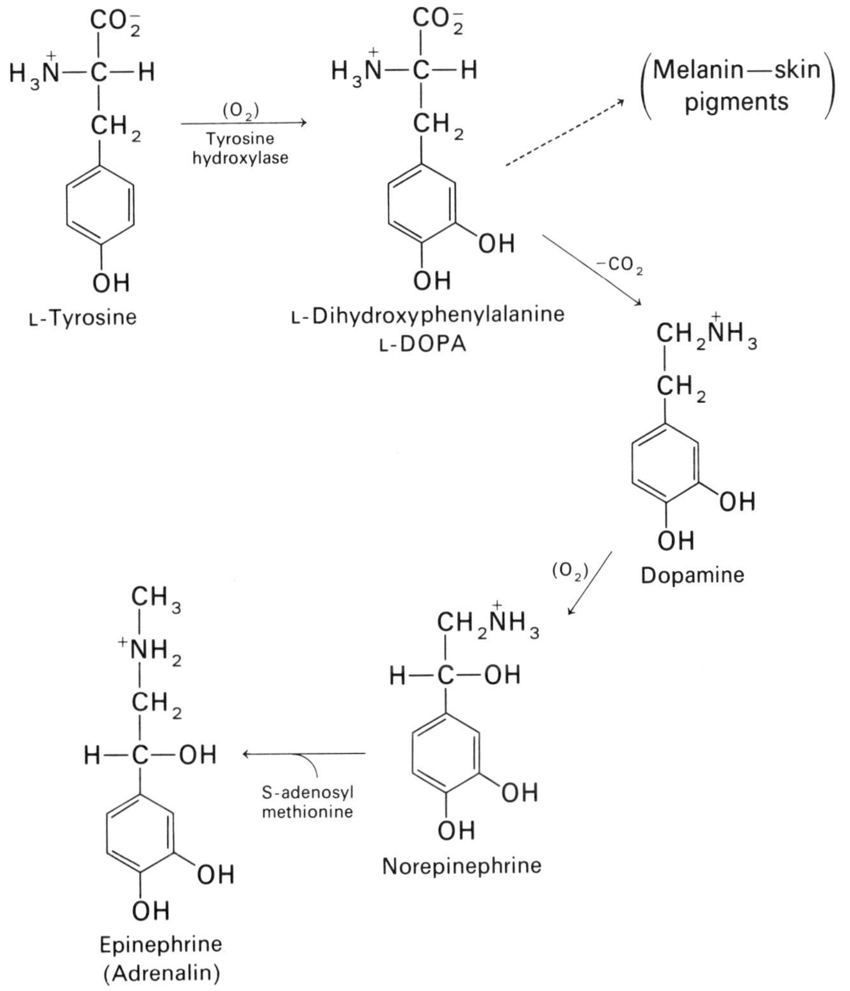

Figure 12.6

The formation of the catecholamines from tyrosine. In humans and other mammals, dopamine and norepinephrine are important neurotransmitters and epinephrine is an important hormone.

disease. Currently, Parkinson's disease is thought to arise from a lack of dopamine in certain parts of the brain.

The catecholamines are broken down by a family of enzymes called *monoamine oxidases.* It is interesting to note that specific inhibitors for monoamine oxidase have proven to be powerful antidepressant drugs.

Tryptophan is the precursor for many neuroactive substances found in plants and animals. One important tryptophan product is *serotonin*, a key neurochemical substance in brain function. There is evidence that serotonin is required for normal sleep and appetite response. Many potent psychoactive substances are structurally related to serotonin. Figure 12.7 shows the relationship between tryptophan, serotonin and some well-known psychoactive compounds. The examples given are just a few of the many known substances that are related to tryptophan.

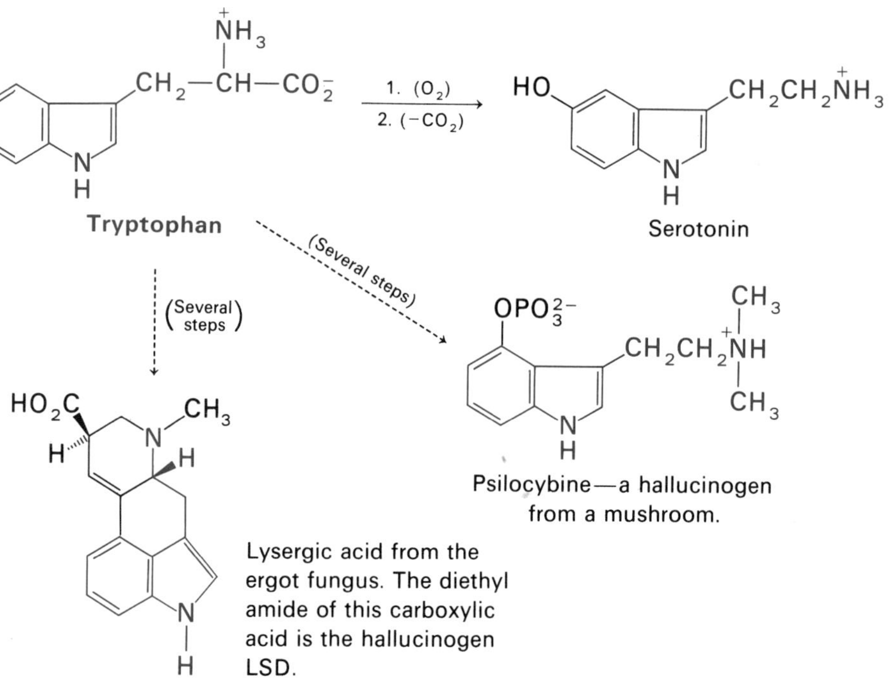

Figure 12.7

Neuroactive substances derived from tryptophan. Serotonin is an important neurochemical in brain function, and psilocybine and LSD are both hallucinogens. All three have structures which are closely related to the structure of their common precursor.

Purine and Pyrimidine Biosynthesis

In Chapter 13 we will see how the purine and pyrimidine bases of DNA and RNA play pivotal roles in the storage and transfer of genetic information and in the expression of this information by protein biosynthesis. It is only appropriate, therefore, that we briefly outline the biosynthetic origins of these remarkable molecules.

The general features of purine biosynthesis are illustrated in Figure 12.8. The initial purine formed is inosinic acid, a precursor for both adenylic acid and guanylic acid. The formation of inosinic acid from ribose 5-phosphate is regulated by the ultimate products of the pathway: GMP, GDP, GTP, ATP, ADP, and AMP. The relative amounts of adenylic acid (AMP) and guanylic acid (GMP) produced are controlled by two mechanisms. The first is feedback inhibition by AMP on the conversion of inosinic acid to AMP plus feedback inhibition by GMP on the conversion of inosinic acid to GMP. The second mechanism is the activation of one process by the product of the other. This means that an accumulation of GTP stimulates the formation of ATP, and vice versa. The net result is a balance in the amount of adenine and guanine formed from inosinic acid.

The biosynthetic origins of the atoms in the purine nucleus were established by the use of isotopically labeled precursors, as shown in Figure 12.9. Among the isotopes useful for tracer studies are ^{3}H, ^{14}C, ^{32}P, ^{35}S, ^{55}Fe, and ^{131}I, which are radioactive, and ^{2}H (deuterium) and ^{15}N, which are not.

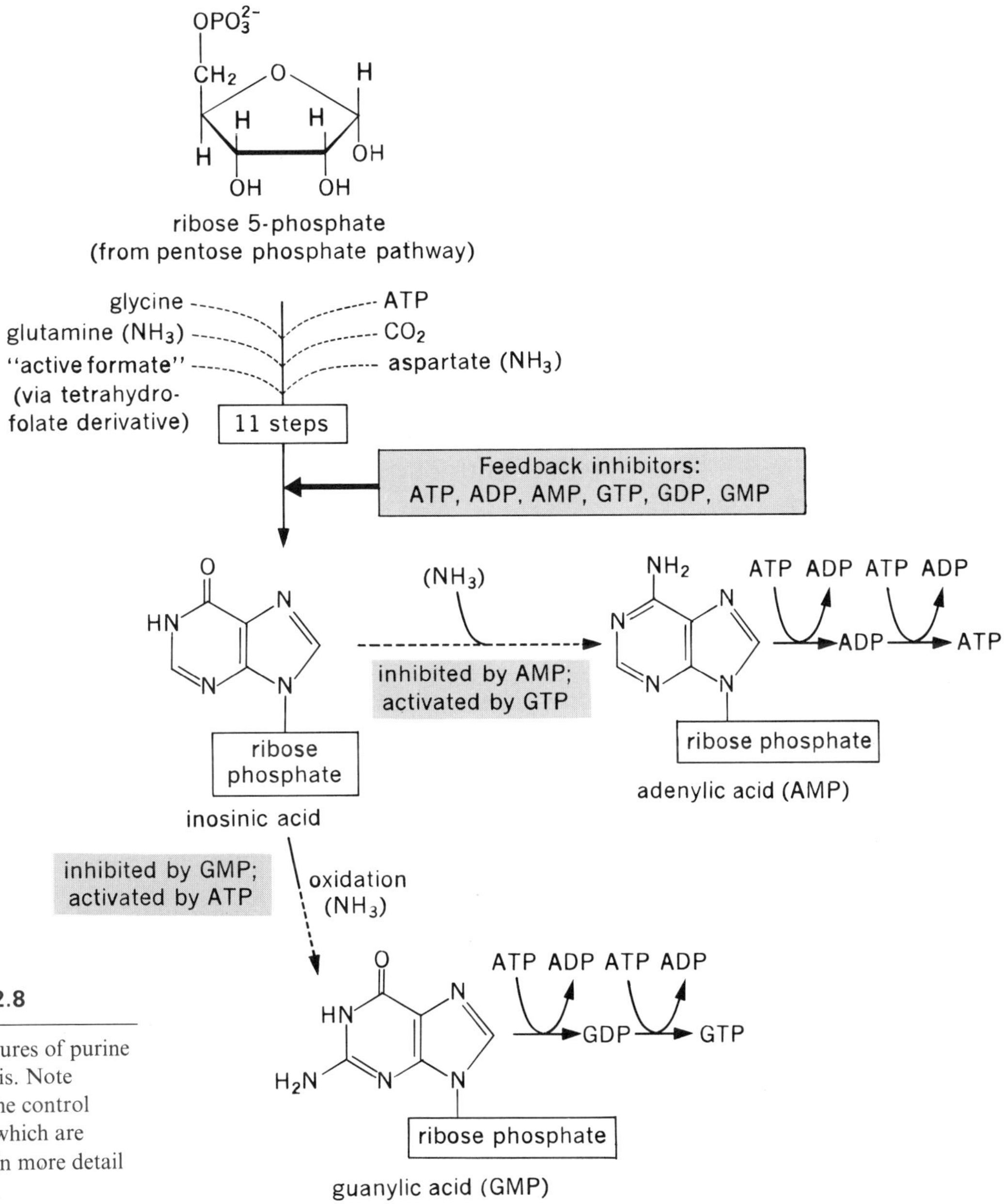

Figure 12.8

Major features of purine biosynthesis. Note carefully the control elements, which are discussed in more detail in the text.

The pyrimidine nucleus is much simpler than the purine ring and its biosynthesis is also less complex. The pathway for the biosynthesis of pyrimidines is summarized in Figure 12.10. Here carbamoyl phosphate, which plays a central role in urea biosynthesis (Section 11.3), also serves as a primary precursor for pyrimidines. The enzyme carbamoyl phosphate synthetase, which

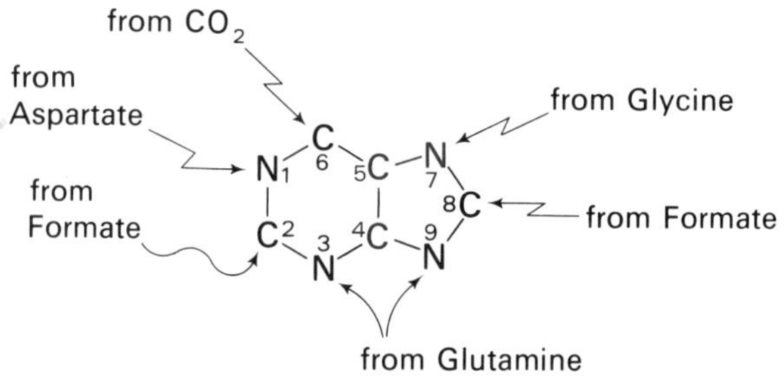

	Compound fed	Purine ring position where label is found
	$H_3^+N{-}CH_2{-}^{14}CO_2^-$	4
	$H_3^+N{-}^{14}CH_2{-}CO_2^-$	5
	$H_3{}^{15}N^+{-}CH_2{-}CO_2^-$	7*
	$^{14}CO_2$	6
	$^-O_2C{-}CH{-}CH_2{-}CH_2{-}\overset{\displaystyle O}{\overset{\|}{C}}{-}^{15}NH_2$ $\quad\quad\;\; \underset{NH_3^+}{\|}$	3 and 9*
	$H{-}^{14}CO_2^-$	2 and 8
	$\underset{\displaystyle ^-O_2C{-}CH{-}CH_2{-}CO_2^-}{\overset{\displaystyle ^{15}NH_3^+}{\|}}$	1*

Figure 12.9

Origins of the carbon and nitrogen atoms in the purine nucleus, as determined by isotopic tracer experiments. The isotopically labeled precursor is fed to an organism; then, after a suitable period, the purine-containing compounds are isolated and analyzed.

* ^{15}N is not radioactive and is analyzed by a technique known as mass spectrometry.

produces this metabolite in the cytoplasm, is different from the mitochondrial enzyme associated with urea biosynthesis. The reaction between carbamoyl phosphate and aspartate is catalyzed by aspartate transcarbamoylase, a regulatory enzyme that controls the rate of pyrimidine biosynthesis. The products shown in Figure 12.10, UMP and CTP, serve as precursors for the biosynthesis of other pyrimidine derivatives.

12.4

Biosynthesis of Carbohydrates: Gluconeogenesis

In Chapters 8 and 9 we briefly discussed gluconeogenesis, the formation of glucose from pyruvate. In Chapter 10 we described the Calvin cycle, by which carbohydrate is biosynthesized through the dark reactions of photosynthesis.

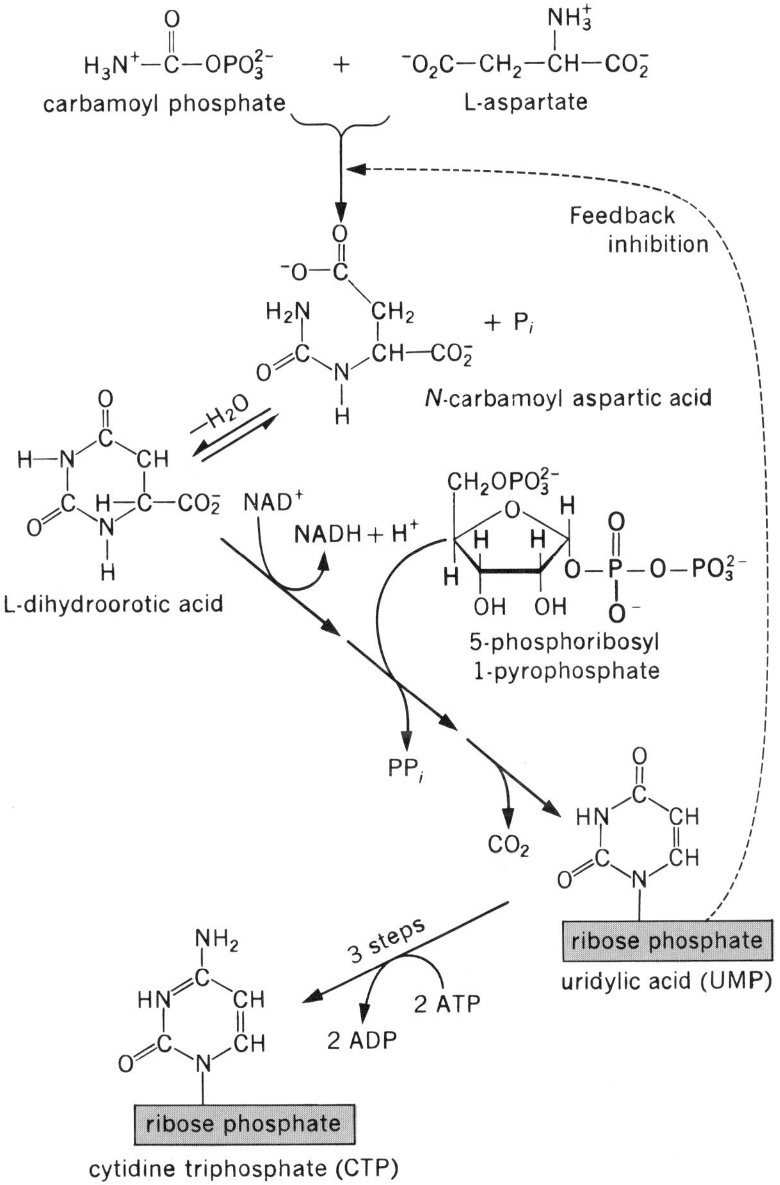

Figure 12.10

A summary of
pyrimidine biosynthesis.

In this section we will examine gluconeogenesis in more detail. In addition, we will see how glycogen is formed and broken down and how higher organisms exert hormonal control over these processes.

Physiological Role of Glycogen

Animals, plants, and many microorganisms produce storage forms of carbohydrates which act as a reserve of glucose or other monosaccharides, available

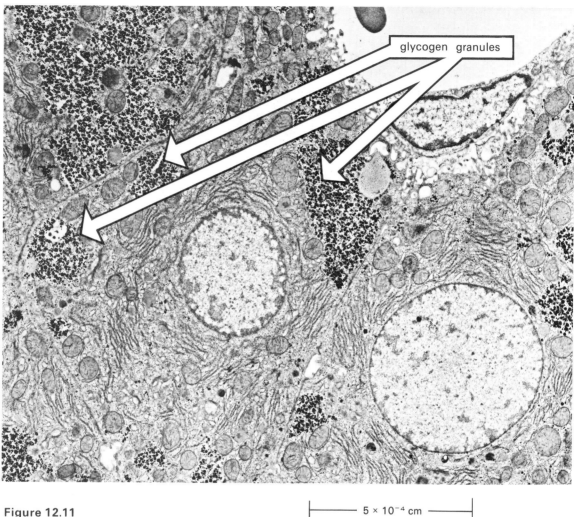

glycogen granules

Figure 12.11

⊢——— 5 × 10⁻⁴ cm ———⊣

An electron micrograph of a section through human hepatocytes. The dark granules clustered throughout the cells are stored glycogen. (Electron micrograph courtesy of Solon Cole, M.D., Director of Electron Microscopy, Department of Pathology, Hartford Hospital.)

to the organism on demand. This is particularly important for higher animals which do not feed continuously but rather consume food at various intervals. Since nervous tissue must have a continuous supply of glucose in order to function, the animal must have a way of maintaining a constant level of blood glucose between meals. This supply of glucose comes from gluconeogenesis in liver and kidney tissue, and, most importantly, from the breakdown of stored glycogen in the liver and elsewhere in the body. Figure 12.11 shows granules of stored glycogen in a human liver cell.

In mammals, the amount of glycogen in the liver ranges from about 2% to 8% of the wet weight of the liver, depending on metabolic circumstances. The turnover of glycogen in liver tissue is continuous: conditions promoting the formation of glycogen inhibit its breakdown, and vice versa. The control of these opposing processes is a major factor in maintaining the concentration of blood glucose within a normal range (65–90 mg per 100 mL blood plasma for humans).

Although glycogen is present in most tissues, its metabolism has been studied most extensively in liver and in skeletal muscle of mammals and birds. The major difference between glycogen metabolism in liver and muscle tissue is that only the liver can release free glucose into the blood, as shown in Figure 12.12. The enzymatic basis for this situation is that, unlike liver tissue, muscle tissue lacks glucose 6-phosphatase. This enzyme hydrolyzes glucose 6-phosphate (see Section 8.3), which can be produced either from glycogen breakdown or by gluconeogenesis. Therefore, muscle glycogen acts as an energy reserve for periods of strenuous muscular activity. During such periods, the blood may not be able to supply oxygen to the muscle rapidly enough, so that the muscle glycogen undergoes anaerobic glycolysis, producing lactate. In such a situation, the lactate enters the bloodstream and is carried to the liver, where it is metabolized aerobically to produce either glucose, which is released to the blood, or glycogen.

The formation of glucose from lactate in liver tissue is an important process in the overall regulation of glucose metabolism in higher animals.

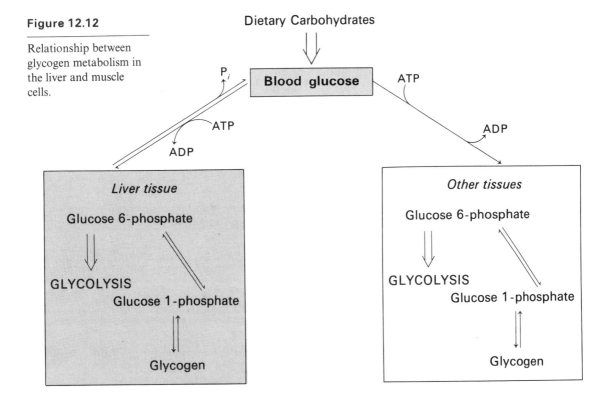

Figure 12.12

Relationship between glycogen metabolism in the liver and muscle cells.

Before we learn how glycogen is produced from glucose, we will discuss the biosynthesis of glucose in the liver. As we will see, glucose biosynthesis utilizes many familiar steps of glycolysis (Chapter 8) and some of the reactions related to the Krebs TCA cycle (Chapter 9).

Steps of Gluconeogenesis

In addition to the availability of exogenous (outside the cell) glucose, certain cells can produce glucose endogenously (inside the cell) through **gluconeogenesis** ("formation of new sugar"). Figure 12.13 shows the general pathway for certain metabolites leading back to glucose 6-phosphate. As you can see, this overall scheme permits the cell to synthesize glycogen or other polysaccharides from glycogenic amino acids, lactate, and Krebs TCA cycle intermediates.

As Figure 12.13 shows, many steps in gluconeogenesis are simply reversals of glycolysis reactions. However, two steps in glycolysis cannot be reversed: these are the pyruvate kinase reaction and the phosphofructokinase reaction. They provide another illustration of how the cell keeps biosynthetic and degradative pathways separate. Pyruvate kinase catalyzes the formation of pyruvate from phosphoenol pyruvate (PEP). The $\Delta G^{0\prime}$ for this reaction is large and negative, which means that the reverse reaction is highly unfavorable. Phosphofructokinase catalyzes the formation of fructose 1,6-diphosphate from fructose 6-phosphate. It is a regulatory enzyme with a very low affinity for fructose 1,6-diphosphate, so the reverse reaction is extremely slow. Therefore, the pathway from pyruvate (or lactate) back to glucose must circumvent these two hurdles.

An important function of gluconeogenesis in the liver is the conversion of exogenous and endogenous lactate, produced during anaerobic catabolism,

Figure 12.13

An overview of gluconeogenesis.

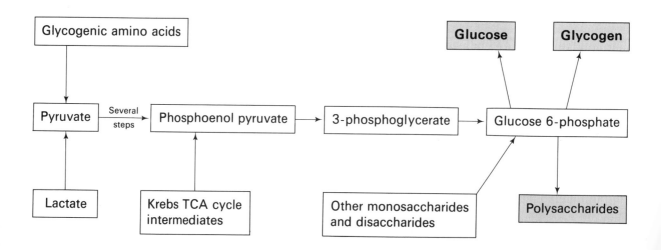

back to glucose 6-phosphate. Exogenous lactate is carried to the liver in the blood. The lactate dehydrogenase reaction

$$\text{Lactate} + \text{NAD}^+ \rightleftharpoons \text{Pyruvate} + \text{NADH} + \text{H}^+$$

is readily reversible, so there is no difficulty in the interconversion of lactate and pyruvate.

The conversion of pyruvate to PEP proceeds in four steps:

I.   **Formation of oxaloacetate from pyruvate in the mitochondrion.**

$$CO_2 + \text{Pyruvate} + \text{ATP} \rightleftharpoons \text{Oxaloacetate} + \text{ADP} + P_i$$

This reaction is one of the anaplerotic reactions discussed in Chapter 9. It is catalyzed by pyruvate carboxylase, a biotin-requiring enzyme. Pyruvate carboxylase is a regulatory enzyme which requires acetyl CoA for activity. Clearly, the presence of excess acetyl CoA corresponds to a metabolic situation favorable for gluconeogenesis.

II. **Formation of malate in the mitochondrion.**

$$\text{NADH} + \text{H}^+ + \text{Oxaloacetate} \rightleftharpoons \text{NAD}^+ + \text{malate}$$

This reaction is the reverse of the malate dehydrogenase reaction of the Krebs TCA cycle. Excess NADH and oxaloacetate, by Le Châtelier's Principle, shifts the malate/oxaloacetate equilibrium in favor of malate. The malate formed is transported out of the mitochondrion into the cytoplasm.

III. **Formation of oxaloacetate in the cytoplasm.**

$$\text{Malate} + \text{NAD}^+ \rightleftharpoons \text{Oxloacetate} + \text{NADH} + \text{H}^+$$

This step, catalyzed by cytoplasmic malate dehydrogenase, was discussed in Chapter 9 in connection with shuttle systems.

IV. **Formation of phosphoenol pyruvate in the cytoplasm.**

$$\text{Oxaloacetate} + \text{GTP} \longrightarrow \text{Phosphoenol pyruvate} + CO_2 + \text{GDP}$$

The phosphorylation and decarboxylation of oxaloacetate is catalyzed by phosphoenol pyruvate carboxykinase and completes the transformation of pyruvate to phosphoenol pyruvate.

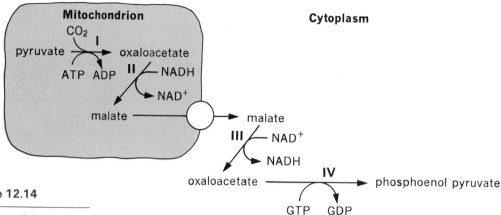

Figure 12.14

Summary of the conversion of pyruvate in the mitochondrion to phosphoenol pyruvate in the cytoplasm. This involves the reverse of the malate shuttle (Section 9.8). The roman numerals refer to the numbered steps in the text.

In sum, the overall formation of phosphoenol pyruvate consumes two high-energy phosphate bonds per phosphoenol pyruvate molecule produced. These steps are summarized in Figure 12.14.

At this point you might wonder why PEP doesn't just react to form pyruvate by the pyruvate kinase reaction of glycolysis. This futile cycle of material is prevented by the regulatory properties of pyruvate kinase. This enzyme is turned off by elevated levels of ATP (see Section 8.8), alanine, or fatty acids, while the same conditions activate gluconeogenesis. Therefore, PEP formed from pyruvate and from Krebs TCA cycle intermediates transforms to fructose 1,6-diphosphate by a reversal of the steps of glycolysis. In doing so, the next hurdle in gluconeogenesis is reached.

Formation of Fructose 6-Phosphate from Fructose 1,6-Diphosphate

In our discussion of glycolysis in Chapter 8, we saw that phosphofructokinase is a regulatory enzyme which acts irreversibly and is turned off by elevated levels of ATP (see Figure 8.5). The hydrolysis of fructose 1,6-diphosphate (the reverse of the phosphofructokinase reaction) is catalyzed by another regulatory enzyme, *diphosphofructose phosphatase*. It is strongly inhibited by AMP and ADP and activated by ATP. Thus, it is active under conditions favoring gluconeogenesis:

$$\text{Fructose 1,6-diphosphate} + H_2O \longrightarrow \text{Fructose 6-phosphate} + P_i$$

Fructose 6-phosphate readily converts to glucose 6-phosphate by the reversal of the phosphoglucoisomerase reaction of glycolysis. There are two possible fates for glucose 6-phosphate formed in this manner:

I. Glucose 6-phosphate $\rightleftharpoons$ Glucose 1-phosphate $\rightleftharpoons$ UDP-glucose

II. Glucose 6-phosphate + H_2O $\rightleftharpoons$ Glucose + P_i

In the first of these two reactions, UDP-glucose, the building block for glycogen formation, is synthesized from glucose 6-phosphate. In the second reaction, the enzyme glucose 6-phosphatase carries out the hydrolysis of glucose 6-phosphate, forming free glucose which is rapidly released by the liver to the blood.

It is apparent that gluconeogenesis is an important factor in an animal's ability to maintain a constant range of blood glucose concentration. In cases of starvation, this is the route by which glycogenic amino acids are converted to glucose. Of even greater importance in this regard is the biosynthesis and breakdown of stored glycogen.

12.5

Glycogen Metabolism

Raw Material for Glycogen Synthesis

The building block for glycogen synthesis (glycogenesis) is Uridine DiPhosphate-Glucose (Section 12.1). This is one of a number of nucleoside diphosphate-sugar molecules involved in many hexose interconversions and in the biosynthesis of oligo- and polysaccharides. The formation of UDP-glucose is catalyzed by *uridine diphosphoglucose pyrophosphorylase*, and proceeds as follows:

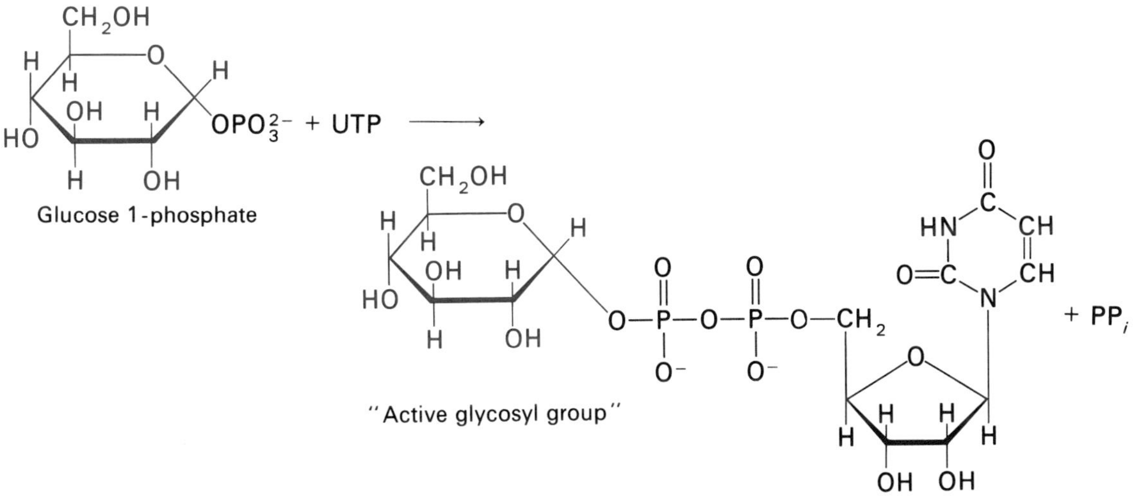

Glucose 1-phosphate

"Active glycosyl group"

UDP-glucose acts as a high-energy glycosyl-group donor. The way this works in the formation of sucrose is

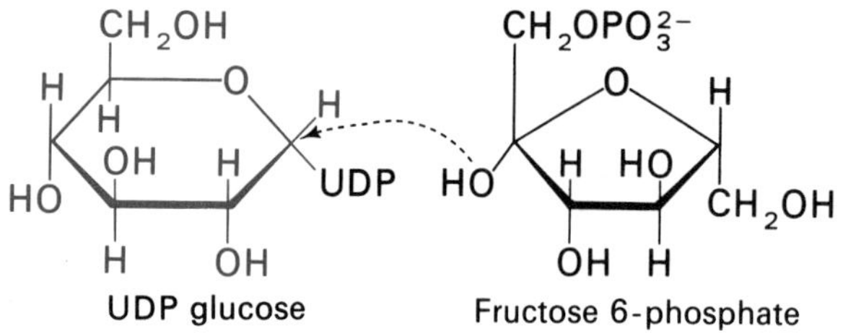

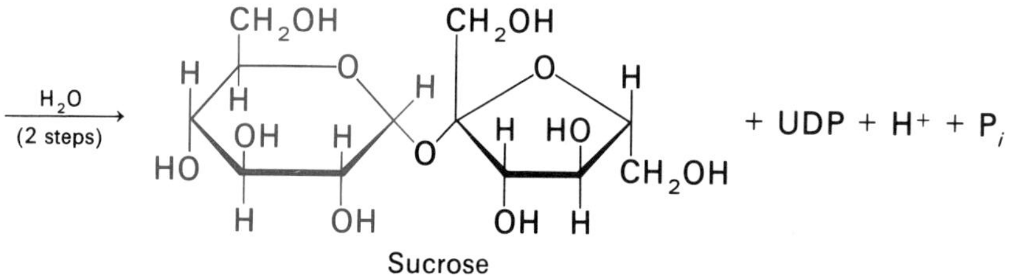

UDP glucose   Fructose 6-phosphate

Sucrose

This example illustrates a general way of forming glycoside links in biosynthesis.

Glycogen Formation and Breakdown

Glycogenesis requires the action of two enzymes, *glycogen synthetase* and *glycogen branching enzyme*. Glycogen synthetase, in its activated form, carries out the reaction

$$n\,\text{UDP-glucose} + \text{short primer strand (glucose)}_x \longrightarrow (\text{glucose})_{n+x} + n\,\text{UDP}$$

By itself *in vitro*, glycogen synthetase acts upon UDP-glucose to produce only an α-1,4-linked homopolysaccharide identical with amylose (Chapter 5). The branching of the polysaccharide chain is brought about by the action of glycogen branching enzyme, which works in a concerted manner with glycogen synthetase. The way a branch point is introduced in a growing glycogen molecule is illustrated schematically in Figure 12.15.

Glycogenolysis, the breakdown of glycogen, involves three enzymes acting to break the molecule primarily into glucose 1-phosphate units. This **phosphorolysis** actually conserves the free energy of the glycoside links in glycogen by transferring the elements of phosphoric acid to the glycoside link being broken rather than to the elements of HOH. This phosphorolysis reaction is shown in Eq. (1):

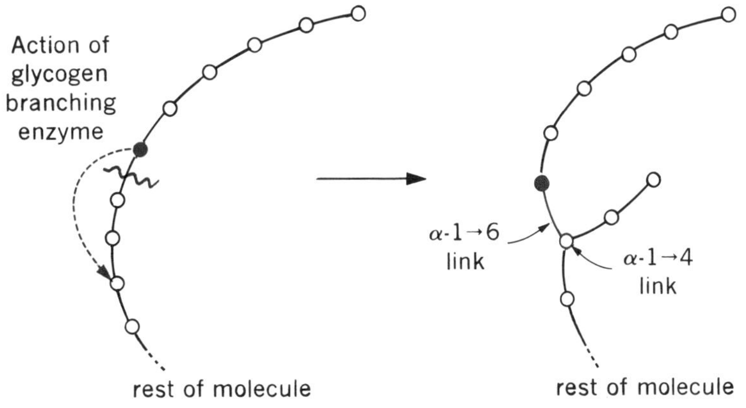
Action of
glycogen
branching
enzyme
rest of molecule
α-1→6
link
α-1→4
link
rest of molecule

Figure 12.15

A schematic diagram of the action of glycogen branching enzyme. The ends of the chain are free to continue growth by the action of glycogen synthetase and additional UDP-glucose. The figure abbreviates the polysaccharide chain as (—O—O—O—).

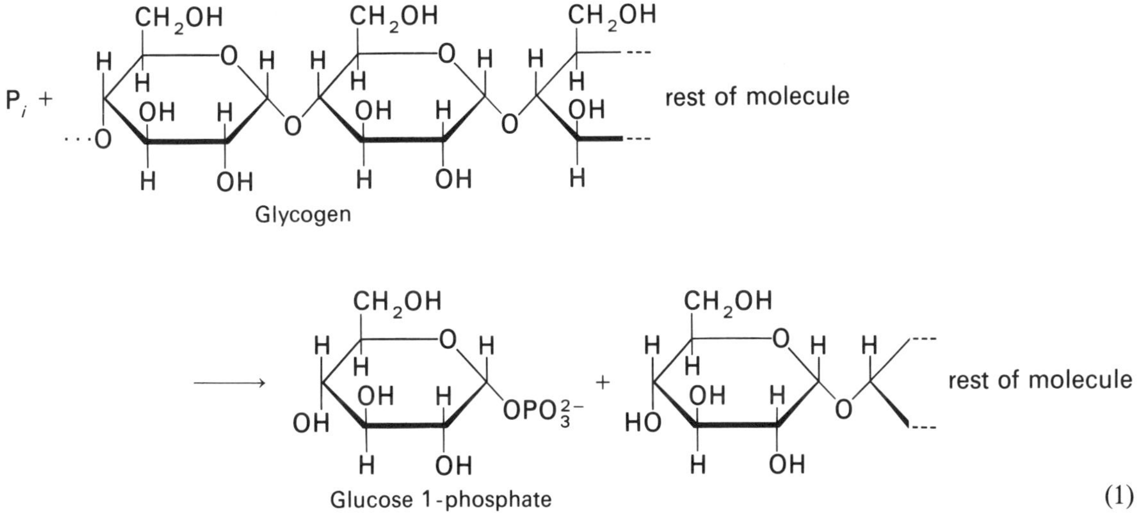
P_i +
Glycogen
rest of molecule
Glucose 1-phosphate
rest of molecule

(1)

Equation (1) is catalyzed by *glycogen phosphorylase*, the key enzyme in the breakdown of glycogen.

Both glycogen synthetase and glycogen phosphorylase are regulatory enzymes which regulate their respective processes. Under metabolic conditions activating glycogen phosphorylase, glycogen synthetase becomes inhibited, and vice versa. This prevents glycogenesis and glycogenolysis from acting in opposition to each other in a futile cycle.

Control of Glycogen Metabolism

There are two levels of control for glycogenesis and glycogenolysis: control at the cellular level and control by the action of certain hormones. Before

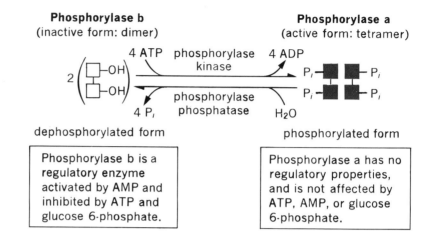

Figure 12.16

Active and inactive forms of glycogen phosphorylase. The activation/inactivation processes require two regulatory enzymes, as shown.

discussing the control of these two processes, we must first learn about the regulatory properties of the two key enzymes, glycogen synthetase and glycogen phosphorylase.

Glycogen phosphorylase initiates glycogenolysis and exists in both an active and a nearly inactive form, as shown in Figure 12.16. In the absence of external control, the breakdown of glycogen is inhibited by an excess of its raw materials. Under these conditions, the formation of glycogen to store these excess carbohydrates becomes the preferred process. Formation of the active tetrameric phosphorylase *a* form involves the phosphorylation of one unique serine —OH group in each subunit of the inactive dimeric phosphorylase *b* form, as shown in Figure 12.16. This phosphorylation and dephosphorylation in controlling the activity of glycogen phosphorylase is, itself, under close regulation, as we will soon see.

Glycogen synthetase initiates glycogenesis and, like its opposite counterpart, glycogen phosphorylase, exists in an active and an inactive form, as shown in Figure 12.17. The phosphorylated form of this enzyme, synthetase D, remains inactive until an excess of raw material, glucose 6-phosphate, is present. Then glycogen synthetase converts to the active synthetase I, its dephosphorylated form. This is the converse of what happens in the case of glycogen phosphorylase.

The self-regulation of glycogen synthesis and breakdown is evident from a consideration of the regulatory properties of the inactive forms of the two enzymes. Those allosteric modifiers that favor glycogenesis inhibit glycogenolysis, and vice versa. The phosphorylation of synthetase I and of phosphorylase *b* activates glycogenolysis. This occurs as a result of a metabolic demand for additional carbohydrate, as in contracting muscle. The dephosphorylation of synthetase D and of phosphorylase *a* activates glycogenesis. This occurs as a

Figure 12.17

Active and inactive
forms of glycogen
synthetase. Synthetase
phosphatase
and synthetase kinase
are both regulatory
enzymes.

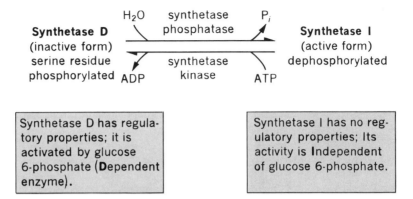

Synthetase D has regulatory properties; it is activated by glucose 6-phosphate (**Dependent** enzyme).

Synthetase I has no regulatory properties; Its activity is Independent of glucose 6-phosphate.

result of low metabolic demand for carbohydrate, as in the case of resting muscle. Both of these processes are under hormonal control, a mechanism presenting a marvelous means of using a purely biochemical system to amplify a small stimulus by orders of magnitude.

Hormones are a diverse class of substances secreted by endocrine glands that affect the rate of cellular reactions. The term "hormone" comes from the Greek "to excite," because at one time hormone action was thought to involve only a stimulation of metabolic processes. We now know that hormones can either decrease or increase the rate of physiological processes. Hormones generally produce a large physiological response when present in minute concentrations. All hormones have relatively short physiological half-lives; they are produced and broken down rapidly. Furthermore, hormone action is extremely specific. For example, the endocrine hormones *glucagon* and *epinephrine* (adrenalin) both stimulate glycogenolysis in liver tissue, but of the two hormones, only epinephrine stimulates glycogenolysis in muscle tissue.

Glucagon (a peptide hormone)
(N-terminal)   His-Ser-Glu-Gly-Thr-Phe-Thr-Ser-Asp-Tyr-Ser-Ser-Lys-Tyr-Leu-
            Asp-Ser-Arg-Arg-Ala-Gln-Asp-Phe-Val-Gln-Trp-Leu-Met-Asn-Thr (C-terminal)

Epinephrine

This suggests that the cells of the various tissues "recognize" specific hormone molecules.

Since epinephrine, glucagon and other hormones act as physiological "messengers" outside the cells of the "target tissue" to which their physiological activity is directed, there must be a second messenger that transmits this activity within the target cell. Research by Sutherland and others has shown that cyclic nucleotides, particularly **cyclic AMP**, act as messengers inside the cell.

Cyclic AMP is formed by the action of the adenyl cyclase system, located in the plasma (cell) membrane of the cells of many animal tissues. The formation and structure of cyclic AMP (c-AMP) are

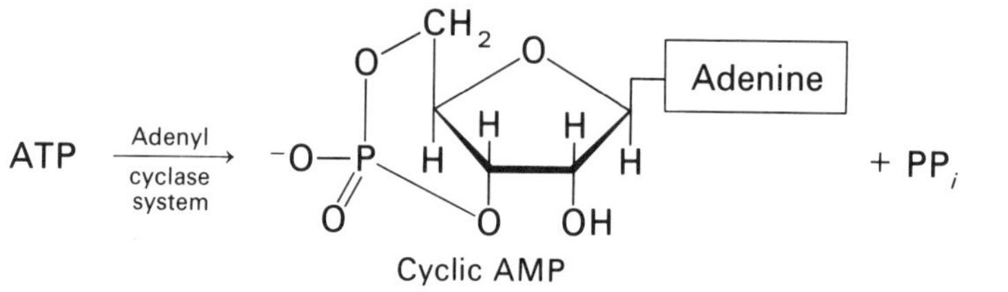

Cyclic AMP

$$\text{ATP} \xrightarrow[\text{cyclase system}]{\text{Adenyl}} \text{Cyclic AMP} + \text{PP}_i \qquad (2)$$

The action of c-AMP as a second messenger in hormone action is illustrated in Figure 12.18. As shown in the figure, the *hormone receptor site* on the surface of the cell has a high affinity and specificity for a given hormone molecule. A cell may have many receptors for different hormones. The high affinity enables the organism to govern processes within the cells of a given tissue with minute quantities of hormones. The high specificity permits the action of the hormone to be translated inside the cells of various target tissues by essentially the same mechanism: namely, the formation of the second messenger c-AMP. We will now see how this general scheme works for the hormonal control of glycogenolysis and glycogenesis in liver tissue.

The binding of either glucagon or epinephrine by their specific hormone receptor sites activates the adenyl cyclase system, which in turn catalyzes the synthesis of c-AMP in the cell. Cyclic AMP promotes the formation of synthetase D (inactive form) by activating *synthetase kinase*. This also promotes the formation of phosphorylase *a* (active form) by causing the activation of phosphorylase kinase. The net result is the stimulation of glycogenolysis and the inhibition of glycogenesis.

The details of these activation processes are shown in Figure 12.19. You should study this figure carefully. A cascade effect is apparent that brings about a considerable amplification in the effect of binding, say, a single hormone molecule. Binding one hormone molecule leads to the formation of many c-AMP molecules. Cyclic AMP, in turn, activates many phosphorylase kinase kinase molecules, activating many more phosphorylase kinase molecules, and so on. The situation is the biochemical analog to a chain letter.

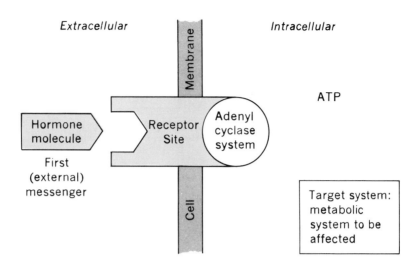

Target system not activated

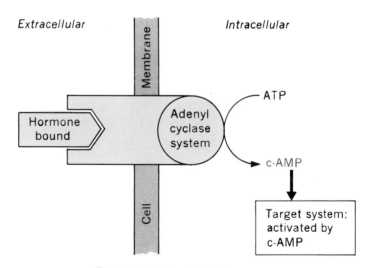

Target system activated

Figure 12.18

A general scheme for hormonal control of a given process inside a cell—the second-messenger concept. The binding of one hormone molecule to a specific receptor site activates the adenyl cyclase system and causes the formation of many c-AMP molecules, thereby effecting a considerable amplification of the hormone action. Cyclic AMP activates the target system, thus causing the physiological response characteristic of the hormone in the target tissue. The lifetime of c-AMP in the cell is very short due to the presence of phosphate-ester hydrolyzing enzymes. In the absence of hormone, the amount of c-AMP in the cell is negligible.

We said earlier that glucagon does not promote glycogenolysis in muscle tissue. Now we know the explanation: muscle cells lack adenyl cyclase receptor sites for glucagon, and therefore c-AMP cannot be produced in muscle in response to this hormone.

Glycogenolysis (turned ON by hormonal control)

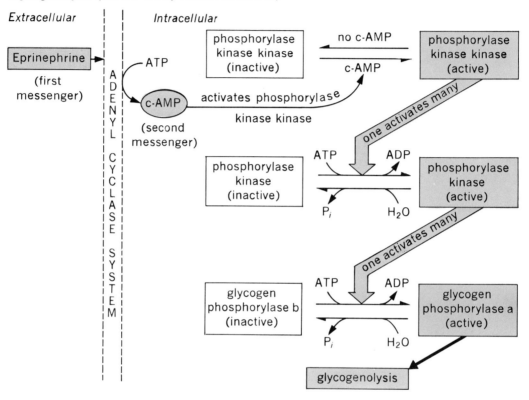

Glycogenesis (turned OFF by hormonal control)

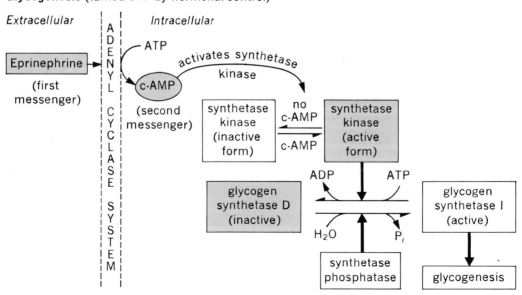

Inherited Defects in Glycogen Metabolism

There are nine known hereditary defects in glycogen metabolism. All of these defects severely limit an individual's chances for survival. Table 12.2 lists these glycogen storage diseases along with the defective enzymes involved. For all cases except glycogen synthetase deficiency (Type IX), glycogen breakdown is affected, either directly or indirectly (see Problem **12** at the end of this chapter). For example, in *McArdle's disease*, muscle glycogen phosphorylase is affected and glycogen accumulates in the muscle tissue until the action of the muscle is impaired. In the case of *Von Gierke's disease*, the ultimate product of glycogenolysis, glucose 6-phosphate, cannot be hydrolyzed and released to the blood. Because of the regulatory properties of the enzymes involved in glycogenolysis and glycogenesis, Von Gierke's disease effectively ensures that glycogen is formed but not broken down.

Table 12.2   Inherited Defects in Glycogen Metabolism. In All Cases Listed Here, Except for Type IX, Glycogen Accumulates in the Affected Tissue.

Type	Name	Principal tissues affected	Defective enzyme
I	Von Gierke's disease	liver, kidney	glucose 6-phosphatase
II	Pompe's disease	generalized	an enzyme in the glycogen de-branching system in glycogenolysis
III	Cori's disease	liver, muscle	same as type II
IV	Anderson's disease	liver	glycogen branching enzyme
V	McArdle's disease	muscle	glycogen phosphorylase
VI	Her's disease	liver	glycogen phosphorylase
VII	———	muscle	phosphofructokinase
VIII	———	liver	phosphorylase kinase
IX	———	liver	glycogen synthetase

Figure 12.19

(*Opposite page*)
The action of cyclic AMP in the hormonal control of glycogen metabolism, using epinephrine as an example. This general scheme describes the action of many different hormones on different tissues in evoking various physiological responses. Note the amplification effect here: one molecule of epinephrine can cause the formation of many c-AMP molecules, turning off glycogenesis and ultimately bringing about the phosphorolysis of many glycogen molecules.

In all of these glycogen storage diseases, the affected individuals are *hypoglycemic*. That is, blood glucose level is depressed in the absence of dietary sugars because an essential part of the blood glucose regulatory mechanism is defective. In normal individuals, glycogen is a constantly replenished reservoir for the maintenance of blood glucose levels.

This completes our discussion of the metabolism of small molecules. Throughout this discussion, we have made particular note of control mechanisms at the enzyme level. In addition, we have seen how external control can be effected through the participation of an internal second messenger. Before we go on to see how proteins are biosynthesized, we will review some of the primary ways by which a cell keeps its various metabolic processes organized.

12.6

Review of Metabolic Regulation

Since metabolic control is such a vital feature of the biochemistry of all intact cells, let's summarize the various types of control mechanisms we have seen up to now.

Metabolic Regulation at the Genetic Level

In Section 3.9 we saw how protein synthesis is regulated, enabling the concentrations of certain inducible enzymes to be controlled. Since the reactions of metabolism are catalyzed by enzymes, regulation of a given enzyme concentration is an effective means of metabolic control. Only certain key enzymes are inducible or repressible, and can thus serve to turn on or turn off entire metabolic pathways. Induction and repression have been demonstrated most clearly in bacterial metabolism.

Metabolic Control at the Enzyme Level

There are several ways for the cell to regulate the activity of an enzyme.

1. **Allosteric control.** Throughout this book we have seen examples of *pacemaker* enzymes, which control the rate of metabolic sequences and thus the steady-state concentrations of the metabolites involved. These allosteric or regulatory enzymes have properties permitting their activity to be regulated by the binding of certain metabolites. It is apparent that in catabolism the ratio of ATP/ADP is a key regulating factor for many regulatory enzymes. In most cases, allosteric regulation is a *negative feedback* type of control. An increase in the ultimate product decreases its rate of formation, and vice versa. This feature of allosteric control was particularly apparent in our discussion of amino acid biosynthesis earlier in this chapter.

2. **Control by enzyme modification.** In some ways this process is similar to allosteric control, because the reversible binding of a negative or positive modulator does modify the structure of a regulatory enzyme. However,

we can make the distinction clear by using the activation of glycogen phosphorylase as an example. Here, a single amino acid residue (serine) is modified to form a phosphate ester, changing the enzyme from inactive to active form. This is a distinct chemical modification, as opposed to a non-covalent modification induced by the binding of a biomolecule.

3.   **Control by enzyme specificity.** Substrate specificity in enzymes is the basis for the cell's ability to carry out many chemical reactions in the same solution simultaneously. The binding of a substrate by an enzyme molecule is the key event in the biochemistry of cellular metabolism.

Metabolic Control by Compartmentalization

We have seen that various metabolic processes occur in separate locations within the cell. For example, mitochondrial processes do not interfere with similar processes in the cytoplasm because of the permeability barrier of the mitochondrial membrane. This kind of spatial control of metabolism is important in preventing competing processes occurring in the same cell from interfering directly with each other. For example, liver cells secrete free glucose into the bloodstream, but can also metabolize glucose for energy. These two processes can take place simultaneously in the same cell without competing directly and producing a stalemate situation. The enzyme that hydrolyzes glucose 6-phosphate, glucose 6-phosphatase, is associated with the endoplasmic reticulum in liver cells. Any free glucose formed by the action of this enzyme is thus rapidly transported outside the cell. Hexokinase, which forms glucose 6-phosphate from glucose, is a cytoplasmic enzyme and therefore does not engage in a futile "tug of war" with glucose 6-phosphatase.

Metabolic Control by Hormone Action

In general, metabolic processes appear to be affected by hormone action at two levels. At the first level, the binding of a hormone molecule to a receptor site on the surface of a target tissue cell triggers the synthesis of cyclic AMP or some other cyclic nucleotide, which then functions as a "second messenger" inside the cell. The cyclic nucleotide thus produced binds to a specific regulatory enzyme or protein in the metabolic pathway affected. We saw how this worked in the case of the hormones glucagon and epinephrine acting on glycogen metabolism.

At the second level, the hormone is taken up by the cell and controls the actual biosynthesis of protein molecules. For example, the steroid hormones can regulate metabolic processes by regulating the biosynthesis of specific proteins.

Throughout the text we have discussed a number of hormones and their effects on various metabolic processes. Now that we have concluded our discussion of intermediary metabolism, it would be worthwhile reviewing the

Table 12.3 A Summary of Hormone Substances Discussed in the Text, with Sites of Action and Section References.

Hormone	Major site of action	Section
Peptide hormones:		
Oxytocin	uterus, mammary gland	2.11
Vasopressin	kidney, arteries	2.11
Insulin	generalized	2.13, 7.8, 11.2
Glucagon	liver, fat tissue	12.5
Lipid hormones:		
Steroids	various tissues	6.7
Prostaglandins	smooth muscle, other tissues	6.9
Amino acid derivatives:		
Epinephrine	generalized	12.3, 12.5
Norepinephrine	generalized	12.3
Serotonin	central nervous system	12.3

earlier discussions of these various hormones and relating their action to the overall scheme of catabolism and anabolism. To help you do that, Table 12.3 lists these hormones along with their principal sites of action in humans and the section in the text where each can be found.

The systems of metabolic control do not act separately, but rather in a concerted manner, to keep all processes of metabolism functioning normally. The way in which these control mechanisms interact and interrelate is only beginning to be understood and in the years to come will be a productive and exciting area of biochemical research.

In this chapter we made some general remarks about biosynthetic reactions and illustrated these points with specific examples. In addition, we reviewed some of the important aspects of metabolic control. The suggested reading at the end of this chapter should be a first step if you wish to learn more about biosynthesis and control. There remains one last subject concerning biosynthesis which is crucial to our study of biochemistry—this is the biosynthesis of proteins, the subject of Chapter 13.

Suggestions for Further Reading

Howland, John L. *Cell Physiology*. New York: Macmillan, 1973. This very readable book emphasizes the regulation of cellular processes at the molecular and cellular levels. It is a good starting point for learning more about control mechanisms.

Lehninger, A. L. *Biochemistry*. 2d ed. New York: Worth Publishers, 1975. This text treats the subject material of this chapter in greater detail. The chapters on amino acid, lipid, and carbohydrate biosynthesis are excellent.

Scientific American

Axelrod, J. "Neurotransmitters." June 1974.

Fernstrom, J. D. and Wurthman, R. J. "Nutrition and the Brain." February 1974.

Green, D. E. "The Synthesis of Fat." February 1960.

McEwen, B. S. "Interactions Between Hormones and Nerve Tissues." July 1976.

Nathanson, J. A. and Greengard, P. "Second Messengers in the Brain." August 1977.

Neutra, M. and Leblond, C. P. "The Golgi Apparatus." February 1969.

O'Malley, B. W. and Schrader, W. T. "The Receptors of Steroid Hormones." February 1976.

Pastan, I. "Cyclic AMP." August 1972.

Problems

1. For each of the terms below, provide a brief definition or explanation.

Catecholamine	Glycogenolysis
Compartmentalization	Hormone
Feedback control	Lipogenesis
Gluconeogenesis	Phosphorolysis
Glycogenesis	Second-messenger concept

2. Draw the structures of the metabolites given below.

Biotin	Serotonin
Cyclic AMP	Tetrahydrofolate
Epinephrine	UDP-glucose
Malonyl CoA	

3. For each of the enzymes given below, give the reaction catalyzed, showing the structures of reactants and products.

Acetyl CoA carboxylase	Glycogen phosphorylase
Adenyl cyclase system	Glycogen synthetase
Diphosphofructose phosphatase	Pyrophosphatase
Glucose 6-phosphatase	Pyruvate carboxylase

4. Using the formation of UDP-glucose as an example, show how pyrophosphate cleavage can help provide a driving force for the formation of a nucleoside-sugar derivative.

5. Using chemical equations, show the sequence of steps for the conversion of acetyl CoA to butanoyl CoA by the fatty acid synthetase complex (see Figure 12.1). In each step give the structures of the organic groups undergoing chemical change.

6. Fatty acid desaturases in mammals have specificity requirements such that certain polyunsaturated fatty acids cannot be formed in the body, but must instead be part of the diet. Refer to the discussion on fatty acid biosynthesis in Section 12.2 and to the structures shown in Table 6.2, page 186, when answering parts (a) and (b).

    (a) Give the structure of the product of fatty acid desaturase acting on palmitic acid and stearic acid, where the carbon-carbon double bond is introduced as far as possible from the carboxyl group.

(b) Explain why linoleic acid cannot be biosynthesized from palmitic acid in humans.

7. Write balanced equations for the steps in the synthesis of tripalmitin (Chapter 6) from glycerol and palmitic acid. (You might wish to refer to Chapter 11.)

8. A hereditary defect in diphosphofructose phosphatase (Section 12.4) has been described in which the activity of this enzyme is absent.

   (a) What is the effect of this disease on the control of blood glucose and blood lactate levels?
   (b) Would you expect to find excess fat in the liver of an individual with this disease? Justify your answer.

9. In our discussion of gluconeogenesis we traced the pathway from lactate to glucose 6-phosphate, using word-reactions. Give the complete series of steps for this transformation, showing the structures of all reactants and products.

10. Why can't acetyl CoA serve as a substrate for gluconeogenesis in animals? What must a cell contain to permit this conversion to occur? (Refer to Chapter 9.)

11. Describe the states of glycogen synthetase and glycogen phosphorylase under the following circumstances:

    (a) Resting muscle: [ATP]/[ADP] high; [Glucose 6-phosphate] high.
    (b) Contracting muscle: [ATP]/[ADP] low; [glucose 6-phosphate] low.
    (c) Glucagon is bound to liver cell glucagon receptor site.

12. Why can't glucagon stimulate glycogenolysis in muscle tissue?

13. Refer to Table 12.2 and the accompanying text on inherited diseases of glycogen metabolism when answering the following questions:

    (a) In Her's disease (type VI), what is the effect on the body's control of blood glucose level?
    (b) Under what conditions could glycogenolysis proceed to a limited extent in the case of glycogen storage disease type VIII?
    (c) In glycogen storage disease type VII, the enzyme phosphofructokinase is defective. Why does this enhance the rate of glycogen formation?

13

Protein Biosynthesis

One of the unifying features of biochemistry is the fact that all chemical processes in a cell are catalyzed and controlled by proteins. Up until now, we have taken for granted that the cell can biosynthesize the proteins it needs. In Chapter 3 we learned that the amino acid sequence of a protein is the all-important factor in determining its identity and activity. In this chapter we will see how the cell translates genetic information into the final primary structure of a protein molecule.

13.1

The General Scheme of Information Transfer in a Cell

We now know that the carrier of genetic information in a cell is **deoxyribonucleic acid (DNA)**, which contains, in coded form, all the information needed to direct the synthesis of all proteins and nucleic acids in a cell. An interesting account of how DNA came to be identified as the carrier of genetic information is given by G. Stent in "Prematurity and Uniqueness in Scientific Discovery" (see Suggestions for Further Reading at the end of this chapter). DNA was first isolated from cell nuclei in 1869, but its function was not understood until 1944, when Avery and his coworkers published the results of their experiments. The organism they studied was the *pneumococcus* bacterium responsible for pneumonia in humans. The infectious form of this organism has a slimy mucopolysaccharide outer coat called a *capsule* that is required for its disease-producing capacity. These infectious types form smooth colonies and are thus called *S-forms*. Non-infectious mutants of pneumococcus lack the capsule coat and form rough colonies when grown on a nutrient plate. They are therefore termed *R-forms*. Avery and his colleagues found that when they made a pure preparation of the DNA from the infectious S-form and allowed the DNA to be taken up by a culture of R-pneumococci, the non-infectious forms were *transformed* into infectious ones. That is, the introduction of the S-form DNA made R-pneumococci take on the hereditary characteristics of the S-pneumococci. This discovery firmly established DNA as the carrier of genetic information, even though Avery's findings were the subject of hot debate for several years.

Further evidence of the role of DNA as the carrier of genetic information came from the work of Chargaff and others, who made a careful study of the base composition of DNA from various species. They discovered that the molar amounts of the four major bases found in DNA—adenine, guanine, cytosine and thymine—have the following relationship:

$$[adenine] = [thymine] \quad and \quad [guanine] = [cytosine]$$

They also found that the base composition of DNA was the same for all cells in a given species, and varied from one species to another. The base composition of DNA was discussed in detail in Section 5.2.

Since cells multiply by a process of division, DNA must reproduce in exactly the same form in each cell from generation to generation of cells. In addition, the normal functioning of an individual cell requires that the genetic information carried by DNA be used to direct the biosynthesis of enzyme proteins. These two points define the role of genetic material in the cell and give rise to the *central dogma* of molecular genetics. This proposal outlines the roles of DNA and **ribonucleic acid (RNA)** in the transmission of genetic information from a stored form to the final primary structure of a protein molecule, as shown in Figure 13.1.

Three major processes are apparent from Figure 13.1: *Replication, Transcription* and *Translation*.

Replication involves the linear assembly of DNA monomer units, forming an exact replicate, or copy, of the sequence of the original DNA structure. This process permits the formation of two daughter DNA molecules during cell division, each an exact copy of the parent DNA.

Figure 13.1

The central dogma of molecular genetics. The arrows indicate the direction of flow of genetic information. The dotted lines show special cases that deviate from this scheme.

Transcription involves the linear assembly of RNA monomer units, or ribonucleotides, using a small specific portion (gene) of the DNA strand as a model. RNA molecules not only provide a working template for protein biosynthesis, but also serve as specific carriers for amino acids, as well as furnishing the site where protein synthesis occurs.

Translation involves the linear assembly of amino acid monomers, using one specific type of RNA as a template and another specific type of RNA as an amino acid carrier and adapter. This corresponds to the actual process of protein synthesis.

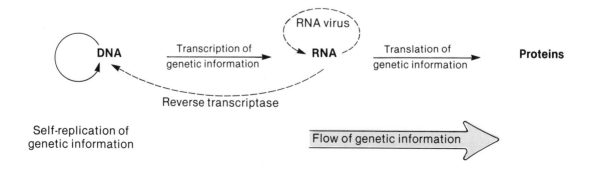

The scheme shown in Figure 13.1 also includes a minor process, discovered by recent research: under certain conditions, RNA can act as a template for DNA biosynthesis. This process is termed *reverse transcription* and will be discussed later in more detail.

At this point it is clear that nucleic acids play a central role in protein biosynthesis. Before continuing on in this chapter, you should review the chemistry of nucleotides and nucleic acids discussed in Chapter 5. This is important because we will be applying this material to our study of the function of nucleic acids in protein biosynthesis.

13.2

Base Pairing and the Principle of Complementarity

In Chapter 5 we discussed the structures of purine and pyrimidine bases. These bases are central to the function of nucleic acids because in the covalent structure of a nucleic acid, the bases extend from the pentose-phosphodiester backbone much like the R groups of a protein. Figure 13.2 is a review of the structures of the purine and pyrimidine bases and their relationships to the covalent structure of a nucleic acid.

These nitrogen bases form the chemical basis for the action of nucleic acids. The —NH and —C=O groups of these molecules are capable of hydrogen bond formation (Chapter 1). In Chapter 5 we learned that the geometry of the purine and pyrimidine bases is such that only certain pairs of bases will form strong hydrogen bonds with each other (see Figure 5.5, p. 165). This specificity of base pairing forms the chemical basis for the structure and action of DNA and RNA. Table 13.1 summarizes the major features of base-pairing specificity in DNA and RNA.

Table 13.1 Purine and Pyrimidine Bases of DNA and RNA and their Base-Pairing Specificity.

Name	Symbol	Found in	Pairs with
Purines			
Adenine	A	DNA, RNA	Thymine (DNA) Uracil (RNA)
Guanine	G	DNA, RNA	Cytosine
Pyrimidines			
Thymine	T	DNA (certain types of RNA)	Adenine
Uracil	U	RNA	Adenine
Cytosine	C	DNA, RNA	Guanine

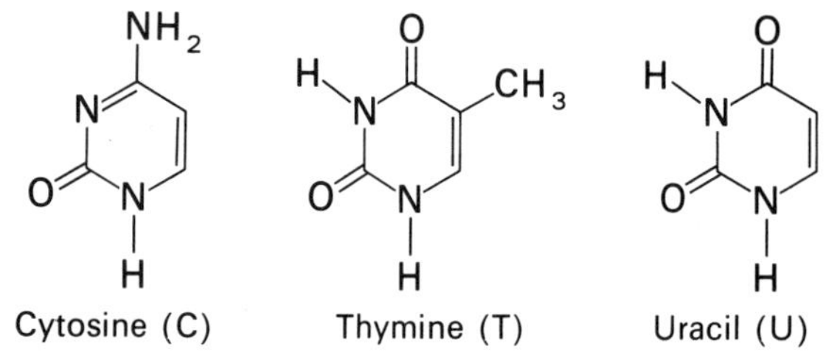

Cytosine (C)   Thymine (T)   Uracil (U)   Adenine (A)   Guanine (G)

The pyrimidine bases cytosine, thymine, and uracil. 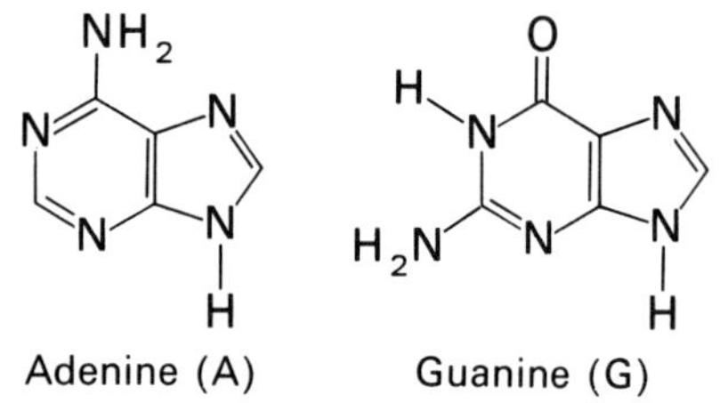   The purine bases adenine and guanine.

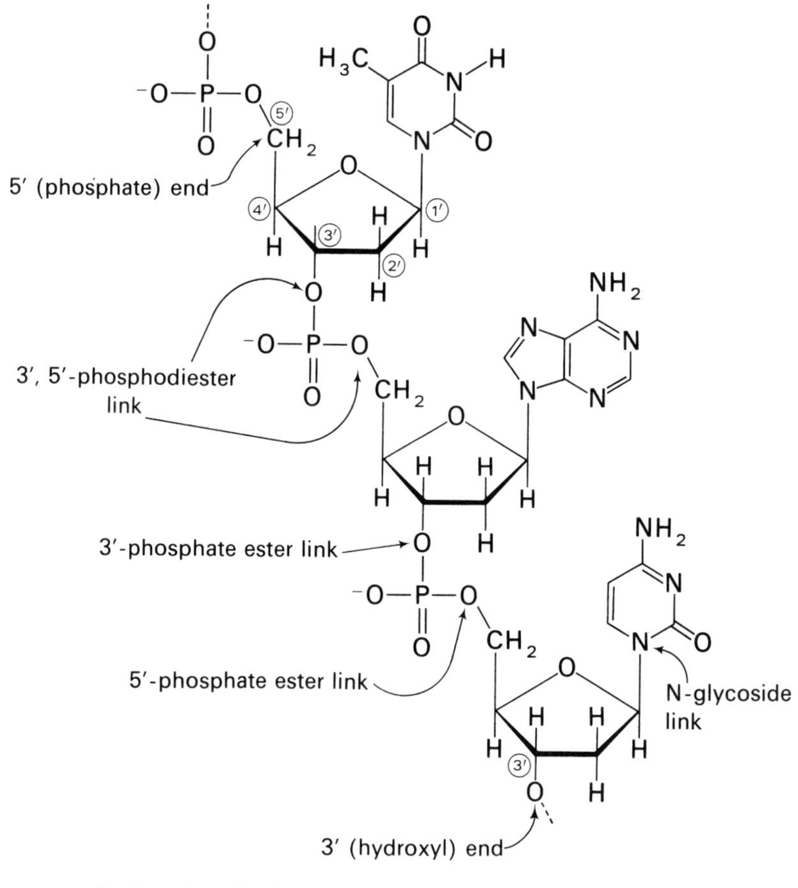

Figure 13.2

The major purine and pyrimidine bases and an illustration of the general covalent structure of a nucleic acid. The trinucleotide fragment demonstrates the important concept of polarity in which any linear nucleic acid chain has a 5′ (phosphate) end and a 3′ (hydroxyl) end.

A trinucleotide fragment of a deoxyribonucleic acid.

Before we discuss the properties of DNA and RNA in more detail, we should learn more about how the bases of a nucleic acid participate in its function. Consider a *tri*-ribonucleotide containing A, G and U, as shown in Figure 13.3. The trinucleotide which would bind to it most strongly, and

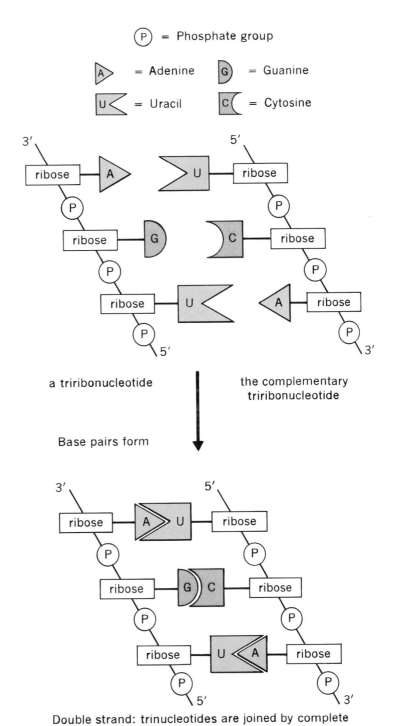

Figure 13.3

The principle of complementarity. For convenience, the trinucleotide fragment is represented schematically. Note that the two strands are anti-parallel; they have opposite polarity.

Double strand: trinucleotides are joined by complete base pairing of complementary strands.

therefore with the highest probability, would be the one in which each of the three bases forms a base pair with the corresponding base in the original trinucleotide, as shown in the figure. The two chains (or separate segments of the same chain) completely base-paired in this way are said to have **complementary nucleotide (or base) sequences**. The *principle of complementarity* thus provides the chemical basis for the action of nucleic acids and is founded on the specificity of base pairing between purine and pyrimidine bases. Now let's apply this principle to the structure of DNA.

13.3

DNA

Every living cell contains one or more genetic units called **chromosomes**, each containing a coiled DNA double strand. In procaryotic cells, the DNA is associated with small, positively charged polyamines. Eucaryotic chromosomes are more complex and contain RNA and proteins along with DNA. Table 13.2 gives the approximate DNA content for several types of organisms. It is important to remember that within a multicellular organism each somatic or diploid cell (any cell other than an egg or sperm cell) has an identical chromosome content. That is, the DNA molecules are the same in all cells of the organism. It is beyond the scope of this text to discuss the topic of cellular differentiation. However, it is clear that since the DNA molecules are the same in each cell of a higher organism, the cells of different tissues must translate different parts of this genetic information. Table 13.2 also shows that, in general, as the complexity of an organism increases, so does the amount of genetic material. The more extensive metabolism of higher organisms requires a greater number of enzymes, and therefore more genetic information in the form of DNA.

Rather than have a longer and longer double DNA strand, cells of more specialized and complex organisms divide their DNA into more than one chromosome. Table 13.3 lists the number of chromosomes found in cells of some higher organisms. Note the progression in the number of chromosomes on going from simpler to more complex organisms. Not counted as chromo-

Table 13.2 Approximate Cellular DNA Content for Several Types of Organisms (Diploid Cells).

Organism	Grams DNA per cell	Number of base pairs
Mammals	6×10^{-12}	5500×10^6
Birds	2×10^{-12}	2000×10^6
Fungi	ca. 2×10^{-14}	20×10^6
Bacteria	ca. 2×10^{-15}	2×10^6

Table 13.3   Number of Chromosomes in Cells of Some Organisms. (Each chromosome contains a single DNA molecule.)

Organism	Number of chromosomes
Procaryotes	
Bacteria	1
Eucaryotes	
Drosophila (fruit fly)	8
Corn	20
Frog	26
Cat	38
Rabbit	44
Human	46
Chicken	78

somes in Table 13.3 are the short, circular double-stranded DNA molecules called plasmids, which are found in most bacteria. Plasmid size can range from 7,000 to about 150,000 base pairs and a bacterial cell may have as many as 20 plasmids. While the exact function of these circular DNA molecules remains unclear, important use has been made of *E. coli* plasmids in experiments on gene manipulation, as we will see later in this chapter. Also not included in Table 13.3 are the small DNA molecules found in all mitochondria and chloroplasts.

In the bacterial cell, DNA is found associated with certain polyamines such as spermine and spermidine:

$$H_3\overset{+}{N}-CH_2CH_2CH_2-\overset{H}{\underset{\underset{N}{+}}{N}}-CH_2CH_2CH_2CH_2-\overset{H}{\underset{\underset{H}{+}}{N}}-CH_2CH_2CH_2-\overset{+}{N}H_3$$

Spermine

$$H_3\overset{+}{N}-(CH_2)_3-\overset{H}{\underset{\underset{H}{+}}{N}}-(CH_2)_4-\overset{+}{N}H_3$$

Spermidine

These positively charged molecules bind to the negatively charged phosphate groups in the DNA molecule, thus rendering it more flexible and amenable to folding within the confines of the cell.

Spermine and spermidine do not serve the same purpose in the eucaryotic chromosome. Instead, certain proteins called **histones** and **nonhistones** are found associated with DNA. The function of these proteins is more complex than that of the polyamines in bacteria. Recent studies have shown that the histone molecules play an important role in the folding of DNA in the eucaryotic chromosome. The histones and nonhistones are also a major factor in the regulation of gene transcription. The way these proteins function in the genetic activity of eucaryotic chromosomes is currently a very active area of research.

DNA Structure

In Chapter 5 we discussed some of the features of the Watson–Crick model for the secondary structure of DNA. First proposed in 1953, this model has withstood the test of further experimentation. Let's review some of the features of the Watson–Crick model, along with some important facts about DNA structure.

1. The overall structure of DNA consists of two polynucleotide chains coiled about a common axis, producing a double helix. This is shown schematically in Figure 13.4.

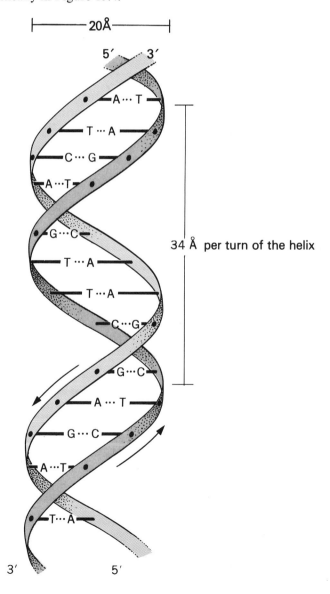

Figure 13.4

The overall configuration of the DNA double helix. Note that the two strands are complementary and antiparallel.

2. The two chains are joined by base pairing and by hydrophobic interactions involving the purine and pyrimidine rings of the bases. The base sequence of one strand is complementary to the base sequence of the other.

3. The purine and pyrimidine bases are located on the inside of the helix and the pentose-phosphodiester backbone forms the outer surface of the helix. The planar purine and pyrimidine rings are perpendicular to the long axis of the helix, ensuring more effective base pairing.

4. The individual DNA strands in the double helix are antiparallel; that is, one strand runs from the 3′ to the 5′ end, while the other strand runs from the 5′ to the 3′ end. This is equivalent to saying that the strands have opposite polarity.

5. The two strands of the double helix can only be separated by unwinding the coils—they cannot simply be pulled apart. The uncoiling of DNA occurs during replication and partially during transcription. It can also be brought about by heating DNA in solution, causing the orderly helical structure to denature. In denatured DNA, the strands are largely separated.

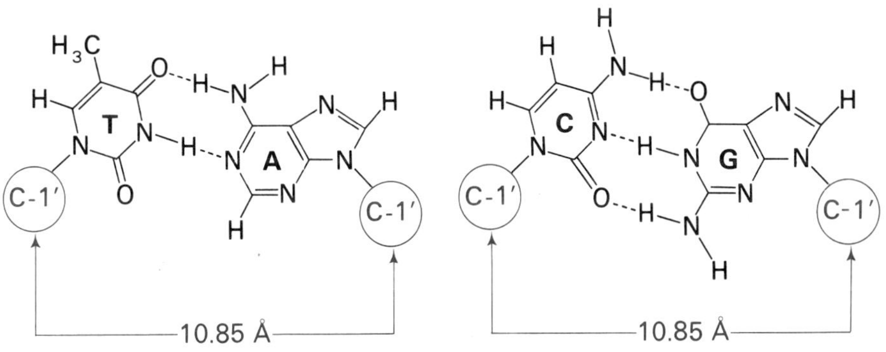

Figure 13.5

Models for the A–T and G–C base pairs in DNA, showing the distance between the 1′ carbons of the deoxyribose groups. In the double helix structure, the 1′ carbons of the deoxyribose units attached to any hydrogen-bonded pair of bases are always 10.85 Å apart. A purine-pyrimidine base pair just fills the 10.85 Å gap. A purine–purine base pair is too large for this space, irrespective of the geometric considerations of the bases themselves; pyrimidine–pyrimidine base pairs are too small to fill the space. Thus, the overall dimensions of the double helix help to ensure the specificity of base pairing.

6. The total length of DNA in a cell can range from about 1.2 mm in *E. coli* to a total of about 2 meters in mammals.

7. The single chromosome of double-stranded DNA in bacteria has been shown to be circular (see Figure 13.7). Higher organisms contain more than one chromosome and each DNA molecule is correspondingly longer. It is probable that the chromosomes of higher organisms are also circular.

Figure 13.5 shows how the interior dimensions of the double helix help to ensure that only the "right" bases form base pairs. It is remarkable that the overall distance across an A–T base pair in the double helix is the same as the distance across a G–C pair.

Now that we have established the major features of DNA structure, let's discuss how DNA replicates during cell division so that each daughter cell includes an exact copy of the parent DNA.

13.4

DNA Replication

The process of DNA replication has been studied most extensively in bacterial systems. Implicit in this research is the idea that the results obtained from such simple systems are also generally valid for cells of higher organisms, which are more difficult to study experimentally.

In 1958 Meselson and Stahl showed that the replication of DNA is **semiconservative** in character, meaning that each daughter DNA double helix contains one strand from the parent DNA molecule and one newly synthesized strand. The general scheme of semiconservative replication is shown in Figure 13.6.

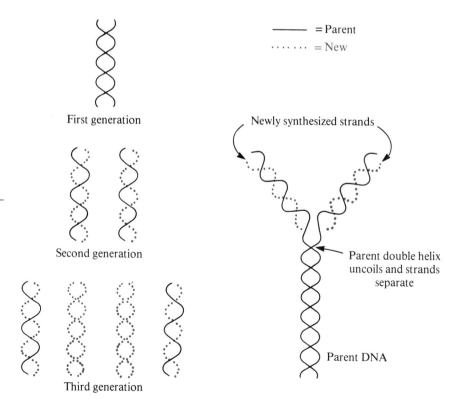

Figure 13.6

Overall scheme of DNA replication, showing its semiconservative character; every daughter DNA double helix consists of one newly synthesized strand and one strand from the previous generation. Each parent strand acts as a template for the formation of a new DNA double helix.

We stated earlier that the bacterial chromosome is circular. This is readily apparent from the remarkable photograph shown in Figure 13.7(a), which clearly shows the circular chromosome of *E. coli* in the process of

Figure 13.7(a)

Replication of circular DNA. Autoradiograph of a DNA molecule isolated from *E. coli*. The cells were grown in a medium containing radioactive thymidine. The isolated DNA was then spread on a photographic plate and the radioactive thymine nucleotide units caused the plate to be exposed along the outline of the DNA molecule. The molecule of DNA in this figure was clearly undergoing replication just prior to its isolation. (Courtesy John Cairns, Imperial Cancer Research Fund, Mill Hill Laboratories, London, England. Prior publication: *Scientific American*, Jan. 1966.)

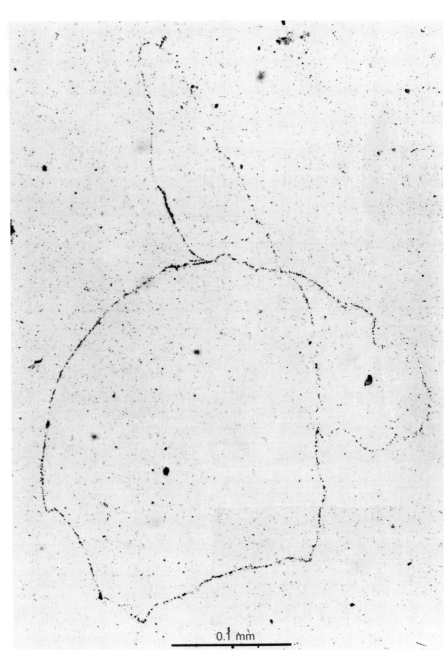

0.1 mm

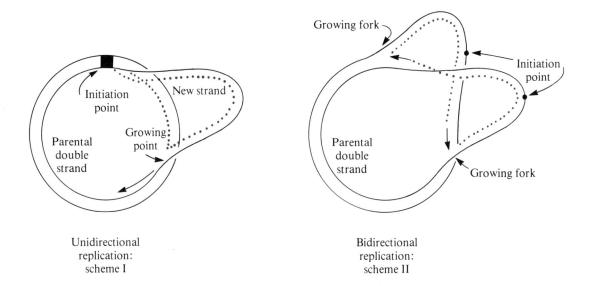

Unidirectional
replication:
scheme I

Bidirectional
replication:
scheme II

Figure 13.7(b)

A schematic
representation of the
replication of circular
DNA. Originally, it was
thought that replication
took place in only one
direction around the
parent double helix, as
shown in Scheme I. We
now know that the
replication of DNA is
bidirectional, as shown in
Scheme II. In Scheme II
the two growing forks
move in opposite
directions and ultimately
meet as replication is
completed. There is still
much debate on how the
parent DNA strand is
able to unwind without
causing the remaining
part of the double helix
to twist up during the
replication process.

replicating. A schematic interpretation of this process is given in Figure 13.7(b) and (c). Replication commences at a specific *initiation point* on the chromosome. The double helix unwinds and daughter strands form as DNA synthesis travels around the chromosome. Ultimately, two identical circular DNA molecules are formed and the two new chromosomes separate.

Enzymes of DNA Replication

In 1958, A. Kornberg and his colleagues began studying the enzymes responsible for carrying out the DNA replication process in *E. coli*. They were soon able to isolate an enzyme from *E. coli* that catalyzes the formation of phosphodiester links, using preformed single-strand DNA as a template and dNTP's (see Figure 5.6, p. 165) as substrates. This enzyme is now called *DNA polymerase I* or *Pol I*. Initially it was thought that *Pol I* was solely responsible for DNA replication. Now there is strong evidence that this is not the case. In 1969, DeLucia and Cairns discovered an *E. coli* mutant having less than 1% of the normal *DNA polymerase I* activity. This mutant strain of *E. coli* was observed to grow and multiply at the same rate as the normal cells. The major difference was that the mutant cells were highly sensitive to mutagenic factors such as ultraviolet light, and they lacked the capacity to repair breaks in the DNA strand. This suggests that *Pol I* is important in DNA repair. Two new DNA polymerases were isolated from these mutants and were later found in normal *E. coli* as well. These enzymes are designated *DNA polymerase II (Pol II)* and *DNA polymerase III (Pol III)*.

Figure 13.7(c)

DNA replication in eucaryotic cells. Electron micrographs have shown that eucaryotic DNA is replicated bidirectionally from many initiation points, thus ensuring the rapid replication of the large eucaryotic chromosome. This process is shown schematically in the figure.

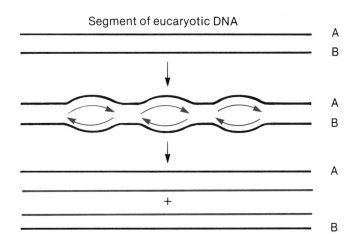

Segment of eucaryotic DNA

All DNA polymerases catalyze the formation of a polynucleotide from a single DNA template strand and a short complementary DNA or RNA primer strand. This process is illustrated schematically in Figure 13.8.

The chemistry of the DNA polymerase-catalyzed formation of a phosphodiester link is quite straightforward, as shown in Figure 13.9. Note that **the growth of the DNA chain is in the 5′ → 3′ direction**, and also that

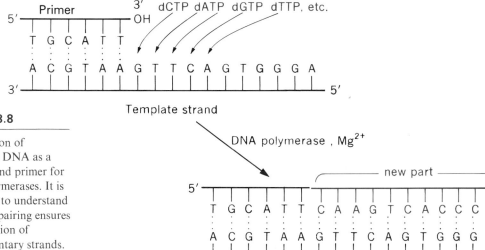

Figure 13.8

The function of preformed DNA as a template and primer for DNA polymerases. It is important to understand how base-pairing ensures the formation of complementary strands. Note that the growth of the DNA strand is in the 5′ → 3′ direction.

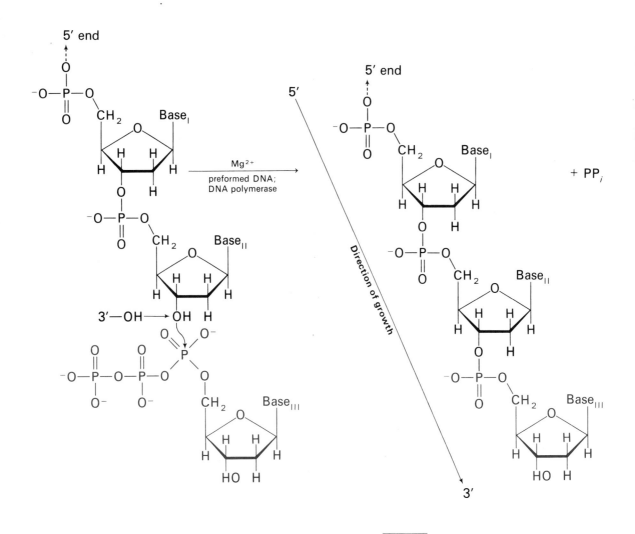

Overall Reaction:

Figure 13.9

Formation of a
phosphodiester link in
the 5′ → 3′ direction and
the overall reaction
catalyzed by the DNA
polymerase.

$$
\left.
\begin{array}{l}
n_1 \text{ dATP} \\
n_2 \text{ dCTP} \\
n_3 \text{ dGTP} \\
n_4 \text{ dTTP}
\end{array}
\right\}
\xrightarrow[\text{Mg}^{2+};\ \text{DNA polymerase}]{\text{DNA template–primer}}
\boxed{
\begin{array}{l}
\text{DNA} \\
\text{dAMP}_{n_1} \\
\text{dCMP}_{n_2} \\
\text{dGMP}_{n_3} \\
\text{dTMP}_{n_4}
\end{array}
}
+ (n_1 + n_2 + n_3 + n_4)\text{PP}_i
$$

pyrophosphate cleavage occurs. The subsequent hydrolysis of the pyrophosphate by pyrophosphatase enzymes provides a strong driving force for the
reaction *in vivo.*

Figure 13.10

The "proofreading" or editing of a newly formed phosphodiester link in DNA in which an incorrect base has been introduced. Unpaired (incorrect) nucleotides are hydrolyzed as shown by the $3' \rightarrow 5'$ exonuclease activity of the DNA polymerases. The high level of fidelity in DNA replication is therefore thought to be due in large part to the self-correcting ability of the DNA polymerases and the inability of the polymerase to extend a new strand that is improperly base-paired.

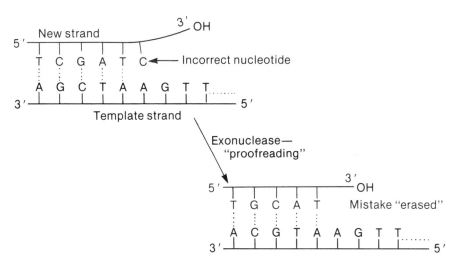

All three DNA polymerases are large proteins having multiple catalytic activities. In addition to catalyzing the formation of new DNA, all three exhibit $3' \rightarrow 5'$ *exonuclease activity*. There is evidence that this $3' \rightarrow 5'$ exonuclease activity fulfills a *proofreading* or editing function in which an incorrectly base-paired nucleotide can be removed prior to the addition of the next nucleotide. The editing function of the DNA polymerases is illustrated in Figure 13.10. Pol I also exhibits $5' \rightarrow 3'$ exonuclease activity, unlike Pol II and Pol III. This is probably involved with a repair function of Pol I in which short segments of a damaged or incorrectly base-paired DNA strand can be excised and replaced.

The relative activity levels of the three DNA polymerases differ greatly. Pol III is about 15 times more active on a molar basis than Pol I, and about 300 times more active than Pol II. Because of the difference in activity levels, it is thought that Pol I and Pol III produce most of the newly formed DNA strands. The actual role of Pol II is not yet understood. In addition to the DNA polymerases, other proteins are also required for DNA replication. They include *unwinding proteins* that promote the separation of the double helix to provide two single-strand segments as templates for DNA replication. Another prerequisite is the key enzyme *DNA ligase*.

DNA ligase catalyzes the formation of a $3',5'$-phosphodiester link between the ends of two DNA strands or between the $3'$ end and the $5'$ end of a single DNA strand. This enzyme serves to repair breaks or "nicks" in an existing DNA molecule and to join the ends of a linear DNA double helix, producing a circular molecule. The action of DNA ligase is illustrated in Figure 13.11. We will see shortly how this enzyme is believed to participate in DNA replication.

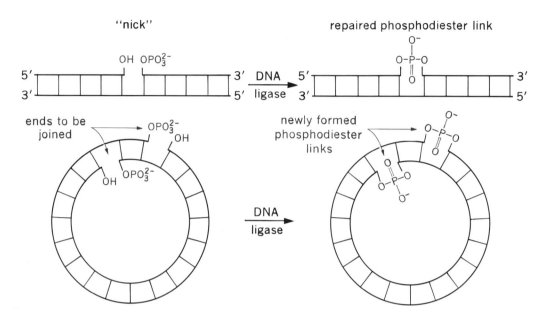

Figure 13.11

The action of DNA ligase. The examples show some phosphodiester links explicitly and the rest in schematic form.

A Model for DNA Replication

Currently, DNA replication in *E. coli* is thought to proceed by five major steps. While there are differences from one species to another, replication in eucaryotic cells probably follows a similar procedure. These steps are consistent with the following known features of DNA replication:

1. DNA replication is semiconservative.
2. DNA replication is bidirectional.
3. DNA replication starts at specific points on the chromosome.
4. DNA replication is initiated by a short RNA primer segment produced on the template DNA strand.
5. DNA is biosynthesized in the $5' \rightarrow 3'$ direction on a template strand in short segments that are later joined together to produce the completed daughter strand.

Let's consider the steps of DNA replication in more detail:

Step I. Initiation of Replication   Replication starts at a specific initiation point on the *E. coli* chromosome. By a process not yet completely understood, *DNA-directed RNA polymerase* binds at this point, thus beginning DNA replication. The uncoiling of the helix through the action of unwinding proteins also commences at the initiation point. In eucaryotic cells, replication of the larger DNA is started simultaneously at a number of points on the chromosome.

Step II.   Formation of RNA Primer Strands   In *E. coli*, a short RNA strand of about 100 nucleotides serves as a primer for the initiation of DNA biosynthesis. In eucaryotic cells, this strand is only about 10 nucleotides long. The RNA primer strand is biosynthesized in the $5' \rightarrow 3'$ direction by DNA-directed RNA polymerase. This enzyme has the capacity to bind to the initiation point on the DNA strand and, after DNA strand separation, to produce a complementary RNA primer strand. This occurs at the initiation site and at other points on each DNA strand, leaving long gaps between the short primer strands that are filled in by DNA in the next step.

Step III. Formation of DNA Segments The free $3'$-OH end of the RNA primer strand serves as the point from which the active form of *Pol III* catalyzes the formation of DNA in the $5' \rightarrow 3'$ direction. Newly formed DNA is complementary to the template strand. The DNA segments produced in this way in *E. coli* are from 1000–2000 nucleotide units long. In eucaryotic cells, these segments are approximately 100–150 nucleotide units long.

Step IV. Removal of RNA Primers Once the gaps between the RNA primer strands are filled in with DNA, the RNA strands are hydrolytically cleaved away, leaving segments of DNA on a template strand separated by small gaps. These portions of DNA have been isolated and are called *Okazaki fragments*. Their existence is evidence that the replication of DNA is discontinuous in nature.

Step V. Completion of the DNA Strand The gaps between the Okazaki fragments are filled by the action of Pol I functioning in the $5' \rightarrow 3'$ direction. The result is a series of adjacent segments of DNA on the template strand. The final joining of the $5'$ end of one segment and the $3'$ end of the neighboring one is catalyzed by DNA ligase.

Because each strand of the parent DNA undergoes this process, replication proceeds in both directions around the chromosome simultaneously. The proposed steps of replication just discussed are shown in Figure 13.12 (p. 394).

Each eucaryotic chromosome is much larger than the *E. coli* chromosome. There is evidence that replication in eucaryotes is initiated at many points at the same time to ensure that the process is completed in a reasonably short time. For example, in an egg cell of the fruit fly, *Drosophila*, replication is completed in about 3 minutes. This means that initiation must take place at thousands of points on each chromosome simultaneously.

RNA-Directed DNA Synthesis

Originally it was thought that only DNA could serve as a template for the formation of new DNA. This is now known not to be universally true. For example, a number of viruses having only RNA as genetic material can transform normal cells in animals into cancer cells. One such virus is the *Rous sarcoma virus*, a widely studied RNA virus that causes cancer in chickens. These tumor viruses have their RNA transcribed into DNA in the host cell, and the DNA then becomes a permanent part of the host cell's genetic material. These viruses are able to affect the host cell DNA because they contain

I. Initiation of replication:

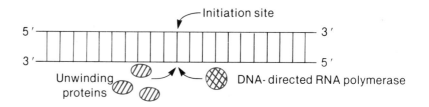

II. Forming RNA primers

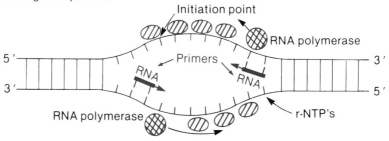

III. Formation of DNA segments

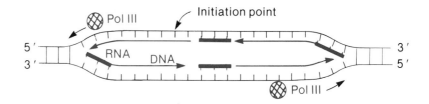

IV. Removal of RNA primers

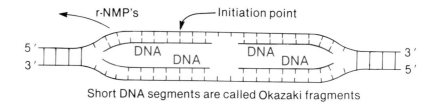

Short DNA segments are called Okazaki fragments

V. Completion of DNA strand:

Figure 13.12

A model for DNA
replication in *E. coli*.
The steps shown are
consistent with what is
known about this
process.

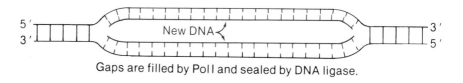

Gaps are filled by Pol I and sealed by DNA ligase.

RNA-directed DNA polymerases. This family of enzymes is also called *reverse transcriptases* because the flow of genetic information is from RNA to DNA—the reverse of the original "central dogma" of molecular biology. Reverse transcriptases were originally thought to be unique to RNA viruses, but they were subsequently found in normal cells as well. The function of these enzymes in normal cells is not clearly understood.

Gene Manipulation: Recombinant DNA

We have discussed cases where cells were transformed by the incorporation of additional genetic material into existing DNA. Recently it has become possible to transplant a segment of DNA from one organism into the chromosome of another unrelated organism. These segments, each of which directs the formation of a specific RNA molecule, are called **genes**. Currently, much work is being done on the manipulation of readily accessible bacterial DNA such as *E. coli* plasmids. It is now possible to engineer bacteria that produce some human proteins, including insulin. Since human insulin would be much more effective for treating diabetes than the bovine insulin currently used, genetic engineering holds great promise for the production of insulin and other proteins for the treatment of disease. On the other hand, irresponsible workers could insert genes for antibiotic resistance into strains of pathogenic (disease-producing) bacteria, thus creating a potentially deadly organism. Much of the experimentation on recombinant DNA has used strains of *E. coli*, the most commonly studied bacteria. There is considerable concern that an altered strain of *E. coli* could have widespread and possibly grave health consequences if it were released into the environment. Because of these anxieties, the National Institutes of Health (NIH) and other research institutions have enacted rules governing the types of DNA recombination experiments that can be done and the precautions that must be taken in each case.

The major tool for genetic engineering is a family of enzymes called *restriction endonucleases*, which cleave DNA molecules only at certain nucleotide sequences. This characteristic makes possible the production of a mixture of specific DNA fragments from a donor organism. (The action of these enzymes is analogous to the function of the various proteases that are so important in protein sequencing; see Table 2.10, p. 50). The restriction endonucleases have another significant property: they produce "sticky" ends on the DNA fragments by cleaving each strand of the DNA duplex at a slightly different point. Because of the specificity of restriction endonucleases, the overlapping strands of the donor DNA fragments are complementary to those of the host DNA. Under the proper conditions, the two can couple and be sealed at the "nicks" in the new DNA duplex by DNA ligase. Once the modified DNA is inserted into a suitable host cell (ordinarily *E. coli*), it can be replicated and perpetuated by normal cellular processes. The overall scheme of recombinant DNA technique is illustrated in Figure 13.13 for the manipulation of *E. coli* plasmids. For a more detailed discussion of the scientific and ethical features of gene manipulation, consult the Suggestions for Further Reading at the end of this chapter.

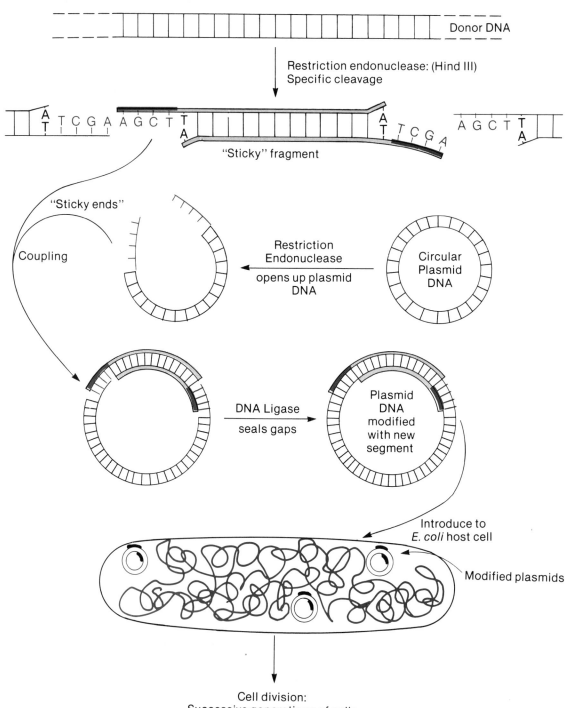

Donor DNA

Restriction endonuclease: (Hind III)
Specific cleavage

"Sticky" fragment

"Sticky ends"

Coupling

Restriction
Endonuclease
opens up plasmid
DNA

Circular
Plasmid
DNA

DNA Ligase
seals gaps

Plasmid
DNA
modified
with new
segment

Introduce to
E. coli host cell

Modified plasmids

Cell division:
Successive generations of cells
with modified DNA

Base Triplets and the Genetic Code

We have now seen how the genetic information carried in the base sequence of DNA is replicated from generation to generation. The code relationship between the DNA base sequence and the amino acid sequence in a protein is summarized in Figure 13.14. Although we will discuss the **genetic**

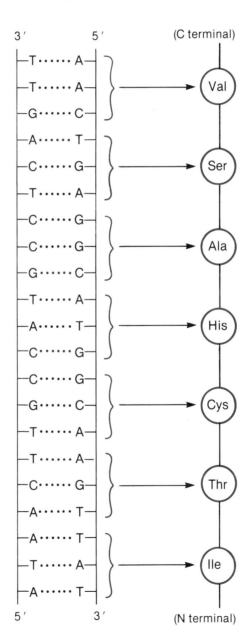

Figure 13.14

The genetic code is a
triplet code. Shown here
in schematic form is the
relationship between the
sequence of base triplets
in DNA and the amino
acid sequence of a
corresponding
oligopeptide.

code in detail later in this chapter (Section 13.9), it is important at this point to recognize that the genetic code is a *three-letter* (*triplet*) *code* involving any combination of the four purine and pyrimidine bases. This arrangement yields $4^3 = 64$ possible combinations of bases, or code words. Clearly, this is more than adequate to provide genetic code words for the 20 amino acids. As we will see, many amino acids have more than one possible base triplet code word, or **codon**.

13.5

Transcription of Genetic Information: RNA Biosynthesis

In Figure 13.1 we saw how the flow of genetic information in a cell runs from DNA to RNA. The *transcription* of a specific unit of genetic information in DNA results in the formation of a single-stranded RNA molecule with a base sequence complementary to that portion of the DNA strand transcribed. We can think of a DNA strand as being divided into many short interconnected segments. Each segment, or **gene**, consists of a sequence of bases which codes for a unique RNA molecule. The RNA molecule corresponding to a given gene may be any of the three types of RNA:

1.   *Messenger RNA* (*m-RNA*), which acts as a working template for the synthesis of a protein chain.

2.   *Ribosomal RNA* (*r-RNA*), which acts as the nucleic acid component in the structure of ribosomes, where protein synthesis is carried out.

3.   *Transfer RNA* (*t-RNA*), which acts as a specific amino acid carrier for the formation of a polypeptide chain.

The vast majority of genes in a chromosome code for m-RNA molecules, thus providing the primary direction for protein synthesis. Genetic maps showing the locations of many genes corresponding to particular proteins have been deduced for several bacteria, including *E. coli* (see the text by J. M. and E. M. Barry cited at the end of this chapter).

The process of synthesizing an RNA molecule by the transcription of the corresponding DNA template can be divided into several steps.

Step 1.   The enzyme RNA polymerase attaches to a specific sequence of bases, or start signal, at the beginning of the gene undergoing transcription. These initiation sites are pyrimidine-rich base sequences having about 10 nucleotides. The attachment of RNA polymerase to the initiation site causes the uncoiling of a short segment of the DNA double helix. For any given gene, only *one strand* of the double helix serves as a template for transcription.

The RNA polymerase of *E. coli* produces all three types of cellular RNA. In mammalian cells, there is evidence that there are several different RNA polymerases. The *E. coli* RNA polymerase has a molecular weight of approximately 5×10^5 and consists of five subunits. Four of these, collectively labeled $\alpha_2\beta\beta'$, are responsible for binding to the DNA template and the catalytic action of the enzyme. The fifth subunit, σ, recognizes the start signals or *promoter regions* on the DNA template and initiates transcription.

Step 2. The substrates for the RNA polymerase reaction, ATP, GTP, UTP, and CTP, base-pair to their complementary bases on *one* of the uncoiled sections of the DNA. The specificity of base-pairing permits the DNA to act as a template for the addition of ibonucleoside triphosphates in the proper order to the growing RNA strand.

RNA polymerase catalyzes the formation of a phosphodiester link between a ribonucleoside triphosphate and the 3'-OH end of the growing RNA strand. Cleavage and subsequent hydrolysis of pyrophosphate help provide a driving force for this reaction. The action of RNA polymerase is analogous to that of DNA polymerase I. RNA strand growth, as in the case of DNA, is in the $5' \to 3'$ direction.

Step 3. As the RNA polymerase moves down the DNA strand, the resulting hybrid RNA/DNA duplex uncoils and the DNA template strand reforms the more stable DNA/DNA double helix with the complementary strand of the chromosome. At the termination of the gene, a specific base sequence, or *stop signal*, causes transcription to cease and the RNA polymerase detaches from the DNA molecule. In some cases, there is evidence that a specific protein, the ρ *factor*, may be involved in the termination process.

Step 4. After an RNA molecule is synthesized, it may be chemically modified. For example, it is known that the 18 S and 28 S r-RNA of mammalian ribosomes are the products of the methylation and cleavage of a single 45 S precursor. This is reminiscent of the formation of zymogens or inactive precursors of certain enzyme proteins (Chapter 3). There is evidence that t-RNA molecules are produced by the selective cleavage of larger RNA molecules. Furthermore, the minor bases, particularly common in t-RNA, are probably the result of post-transcriptional chemical modification of the t-RNA precursor.

The transcription process is shown schematically in Figure 13.15. Study this figure carefully, referring to the steps outlined above.

Figure 13.15

RNA biosynthesis by transcription of DNA by RNA polymerase. There are several key points to note:

1. Only one strand of the DNA duplex serves as a template.

2. The building blocks for RNA synthesis are ribonucleoside triphosphates.

3. Base-pairing of the NTP's with the complementary bases in the DNA strand ensures accuracy of transcription.

4. The chemistry of transcription involves the reaction of the 3'—OH end of the growing RNA molecule with a ribonucleoside triphosphate. Pyrophosphate is cleaved and hydrolyzed to provide additional driving force for the formation of the phosphodiester link.

(a) Overview of Transcription

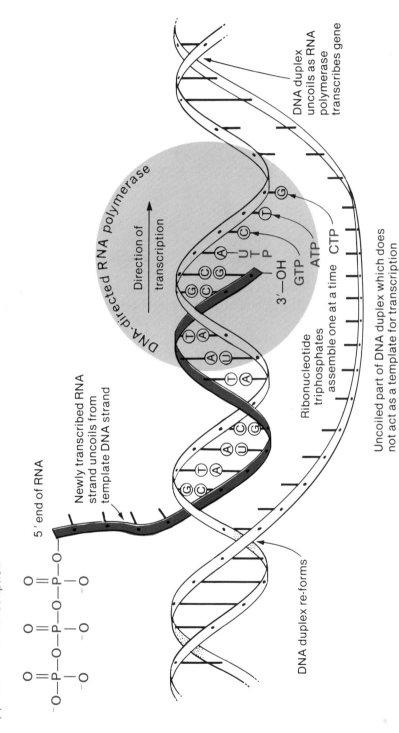

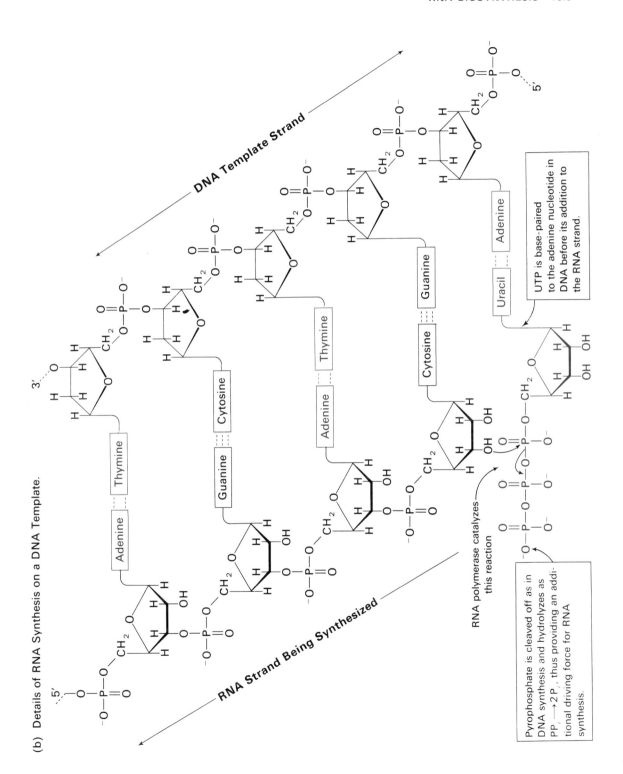

(b) Details of RNA Synthesis on a DNA Template.

DNA Template Strand

RNA Strand Being Synthesized

UTP is base-paired to the adenine nucleotide in DNA before its addition to the RNA strand.

RNA polymerase catalyzes this reaction

Pyrophosphate is cleaved off as in DNA synthesis and hydrolyzes as PP$_i$ → 2 P$_i$, thus providing an additional driving force for RNA synthesis.

Thymine — Adenine
Cytosine — Guanine
Adenine — Thymine
Guanine — Cytosine
Uracil — Adenine

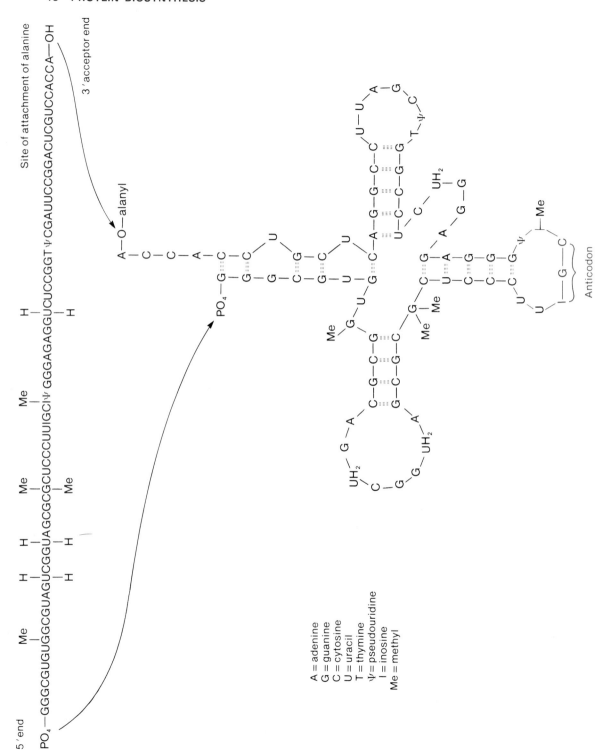

13.6

Properties and Functions of RNA

Transfer RNA is the smallest form of RNA. Because of its size, you may find it referred to in older literature as (*s*) *RNA* due to the fact that it remained in the supernatant solution when the other (heavier) forms of RNA were precipitated by ultracentrifugation. Each of the 20 amino acids has at least one unique t-RNA molecule that serves to transport it to the site of protein synthesis and ensure its proper placement in the amino acid sequence of the protein being synthesized. Figure 13.16 gives a schematic representation of a t-RNA molecule. The *anticodon* loop shown in this figure contains a base triplet (anticodon) that is complementary to one of the codons for alanine. The anticodon plays a key role in protein synthesis.

The "cloverleaf" model of t-RNA structure shown in Figure 13.16 is only a two-dimensional approximation of the actual shape of such a molecule. Using X-ray diffraction analysis, the three-dimensional structures of a number of t-RNAs have been determined. Figure 13.17 gives a perspective representation of the three-dimensional structure of yeast phenylalanine t-RNA.

Note from Figure 13.16 that one anticodon for alanine is the triplet $_3$CGI$_{5'}$, which is complementary to the codon $_{5'}$GCC$_{3'}$. (**I** stands for inosine, discussed below.) Researchers possess a great deal of evidence that the third position in the anticodon (the base at the 5′ end of the anticodon) has much more freedom of motion than the first two bases. This observation is called the

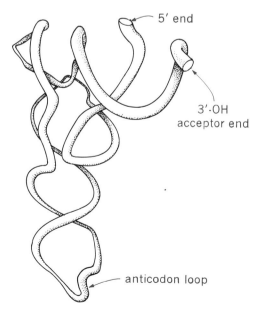

5′ end

3′-OH
acceptor end

anticodon loop

"**wobble concept**" and explains why the same anticodon triplet in a given t-RNA molecule can base-pair with several different codon triplets. As we will see when we discuss the genetic code in detail, a given amino acid can have more than one codon triplet. In the case of the aminoacyl t-RNA molecule shown in Figure 13.16, the codons GCU, GCC, GCA, and GCG all code for alanine. Of these four codons, the anticodon shown in Figure 13.16 will base-pair with GCU, GCC, and GCA. Note that the first two bases in these three codons are the same (GC), and only the third position varies. The interactions between inosine, the "wobble base" in the alanine anticodon, and each of the three bases it can pair with are shown in Figure 13.18. It is important to note that this multiple base-pairing potential in the third anticodon position means that the same aminoacyl t-RNA molecule can base-pair with up to three codons, all specifying the same amino acid.

Figure 13.18

Base pairs formed by the "wobble" base inosine. Although it is not apparent from this figure, inosine must be allowed considerable freedom of motion in the t-RNA molecule to permit this base-pairing to occur.

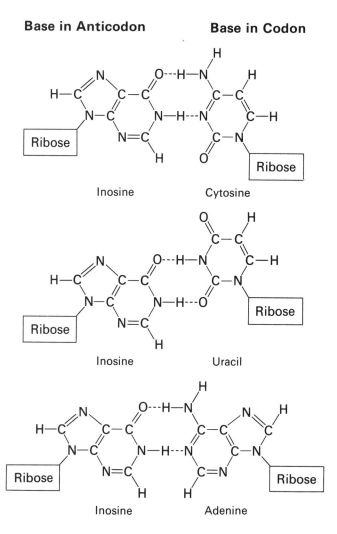

The efficiency of binding between, for example, CGI and each of its three corresponding codons is not the same for each. This variation in binding ability may serve as a basis for controlling the rate of protein synthesis. It is possible, therefore, that the rate at which a given amino acid adds onto a growing protein chain is determined by which of the possible codons is used.

Formation of Aminoacyl t-RNA

The formation of aminoacyl t-RNA involves the creation of an ester link between the 3′—OH end of the t-RNA and the α—COOH group of the amino acid. This reaction is catalyzed in two steps by a group of enzymes called *aminoacyl t-RNA synthetases*. For each amino acid there corresponds at least one unique aminoacyl t-RNA synthetase, specifically linking the amino acid to the appropriate t-RNA molecule(s).

An aminoacyl t-RNA synthetase must carry out two processes. First it must specifically bind and activate its particular amino acid. Secondly, it must specifically bind the correct t-RNA for that amino acid and catalyze the formation of aminoacyl t-RNA.

Amino acids are activated by the formation of *aminoacyl adenylates*. This is an activation process similar to the formation of nucleoside diphosphate sugars in the synthesis of polysaccharides (see Section 12.1).

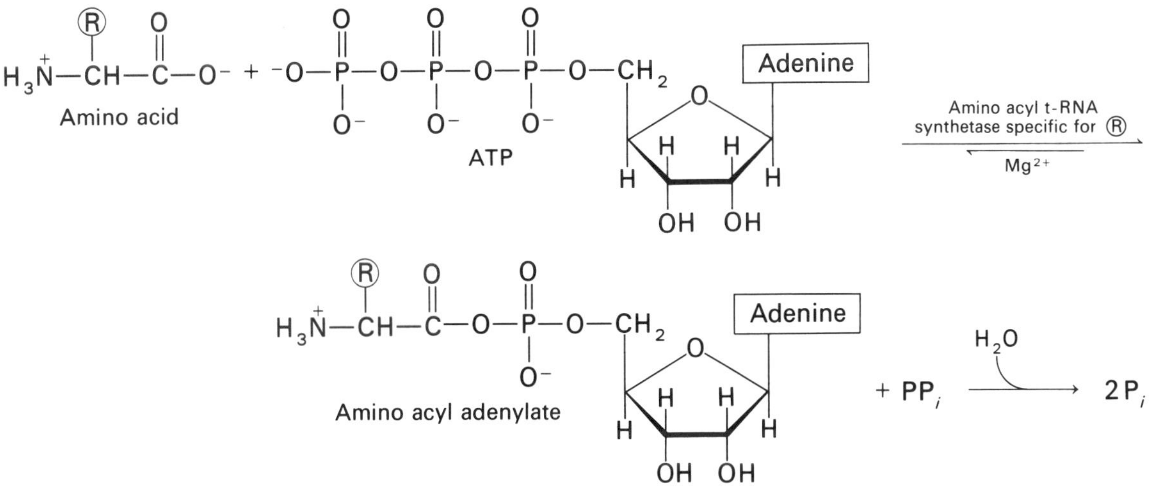

In this activation reaction we can see two ways in which the cell overcomes the "energy problem" in biosynthesis. The cleavage of pyrophosphate and its subsequent hydrolysis provide a strong driving force for the formation of the aminoacyl adenylate. The aminoacyl adenylate derivative is also a much more reactive aminoacyl-group donor than the original amino acid, by virtue of the mixed carboxylic-phosphoric acid anhydride linkage in the molecule (see Appendix B).

The aminoacyl adenylate remains bound to the aminoacyl t-RNA synthetase molecule and reacts with the appropriate t-RNA molecule, also bound to the same enzyme. The result is a completed aminoacyl t-RNA molecule, which then detaches from the enzyme:

$$\boxed{\begin{array}{c} \text{t-RNA specific} \\ \text{for } \circledR \end{array}} - C - C - \overset{3'}{A} - OH + H_3\overset{+}{N} - \underset{}{\overset{\overset{\circledR}{|}}{C}H} - \overset{\overset{O}{\parallel}}{C} - AMP$$

Aminoacyl adenylate

$$\xrightarrow[\quad]{Mg^{2+}} \boxed{\begin{array}{c} \text{t-RNA specific} \\ \text{for } \circledR \end{array}} - C - C - A - O - \overset{\overset{O}{\parallel}}{C} - \underset{}{\overset{\overset{\circledR}{|}}{C}H} - \overset{+}{N}H_3 + AMP$$

Aminoacyl t-RNA

The aminoacyl t-RNA synthetases have a very high degree of specificity toward their particular amino acid and t-RNA substrate molecules. It has been estimated that the error rate in substrate binding is less than 1 in 10,000.

The aminoacyl group in aminoacyl t-RNA is in an activated form. The reaction of this aminoacyl group with an amino group, producing a peptide link, is an exergonic (spontaneous) reaction.

We have just described how amino acid units are carried to the site of protein synthesis. Let's now discuss the working template for protein synthesis, messenger RNA.

Messenger RNA

The size of the m-RNA molecule depends on the number of amino acid residues in the protein for which it serves as a template. Clearly, synthesis of a protein containing 500 amino acid residues must be directed by an m-RNA molecule having at least 1500 (3 × 500) bases.

In our discussion of t-RNA structure we saw that a unique anticodon in the t-RNA molecule corresponds to the particular amino acid carried. In the base sequence of the working template for protein synthesis, m-RNA, we find base triplets, or codons, complementary to the t-RNA anticodon. The location of each codon in the m-RNA strand corresponds to the location of the corresponding amino acid in the protein primary structure for which the m-RNA serves as a template.

The turnover of m-RNA in bacteria is very rapid, exhibiting an average lifetime of about two minutes. While the transcription of a given gene yields only one m-RNA molecule at a time, this one m-RNA molecule can direct the biosynthesis of many protein molecules simultaneously, as we shall soon see.

The relationship of aminoacyl t-RNA to m-RNA is shown schematically in Figure 13.19. This figure summarizes what we have said about the role of codon–anticodon interactions in the proper placement of amino acids in the protein chain. Now that we have discussed the "parts" and the template for protein synthesis, all that remains to discuss is the site of protein synthesis, the ribosome.

Figure 13.19

Aminoacyl t-RNA molecules and their relationship to m-RNA. The high specificity of codon-anticodon base-pairing ensures the correct location of the aminoacyl t-RNA molecule along the m-RNA template.

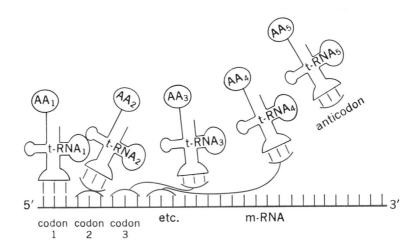

Figure 13.20

Structure of the *E. coli* ribosome, showing the dissociation of each subunit into its r-RNA and protein components. Note that an intact 70 S ribosome is formed when the subunits bind with m-RNA. (Note also that the sedimentation coefficient depends on both molecular weight and the shape of a molecule. Therefore, the S-values of the ribosomal subunits do not add up to give the S-value of the intact ribosome.)

Ribosomal RNA and Ribosomes

Protein synthesis occurs on the surface of an RNA-protein complex known as the **ribosome**. The overall function of the ribosome is to ensure proper orientation between the m-RNA template and the aminoacyl t-RNA molecules being bound to the template. Therefore, the ribosome must specifically bind m-RNA, incoming aminoacyl t-RNA's, and part of the growing protein chain, all in the correct stereochemical orientation. Furthermore, the ribosome contains certain enzymes called *translocases* that cause the ribosome to move along the m-RNA strand as the synthesis of the protein proceeds.

In a procaryote such as *E. coli*, about 15,000 ribosomes are distributed throughout the cytoplasm. In Chapter 5 we saw that the intact *E. coli* ribosome has a particle weight of 3×10^6 Daltons and is termed a 70 S ribosome due to its sedimentation properties in the ultracentrifuge (see Table 5.3, p. 177). The approximate overall shape of the 70 S ribosome is shown in Figure 13.20.

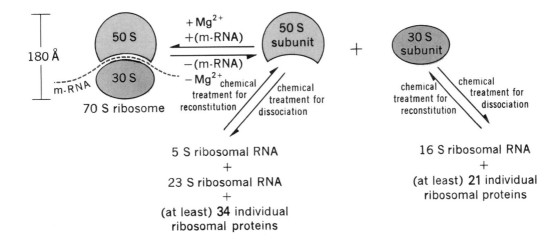

In the absence of m-RNA and at low Mg^{2+} concentrations, the 70 S ribosome dissociates into two subunits: a 50 S subunit (particle weight about 2×10^6 Daltons) and a 30 S subunit (particle weight about 1×10^6 Daltons). The ribosomal RNA and the protein components of each subunit can be dissociated and isolated by appropriate chemical treatment and fractionation. The result of separating out the constituents of each of the *E. coli* ribosomal subunits is shown in Figure 13.20. It is remarkable that under the right conditions it is possible to spontaneously reconstitute active 30 S and 50 S ribosomal subunits. Thus it is clear that the complex and highly specific ordering of proteins and nucleic acids in the ribosome results from the spontaneous self-assembly of the components.

In eucaryotic cells, protein synthesis occurs not only in the cytoplasm, but also to a limited extent in mitochondria and chloroplasts. The ribosomes of chloroplasts and mitochondria are similar to the 70 S ribosomes of procaryotes, while the ribosomes in the cytoplasm of eucaryotic cells are larger and more complex. Like the 70 S ribosome of procaryotes, the 80 S ribosome of eucaryotes dissociates into a large (60 S) subunit and a small (40 S) subunit. The 60 S subunit contains three RNA molecules: 5 S, 7 S and 23 S. The 40 S subunit has a single 18 S RNA molecule. In addition, there are ribosomal proteins in the nucleoprotein complex structure of the eucaryotic ribosome.

In a eucaryotic cell, such as the hepatocyte shown in Figure 1.4, p. 8, ribosomes are ordinarily found in association with the *endoplasmic reticulum*, a structure made up of many channels extending in all directions throughout the cytoplasm. Figure 13.21 shows an electron micrograph of a human hepatocyte in which the endoplasmic reticulum and its associated ribosomes are clearly visible.

Figure 13.21

The endoplasmic reticulum of a human hepatocyte, $\times 24,000$. Note the ribosomes, which appear as granules lined up along the passageways of the endoplasmic reticulum. (Electron micrograph courtesy of Solon Cole, M.D., Director of Electronmicroscopy, Department of Pathology, Hartford Hospital, Hartford, Conn.)

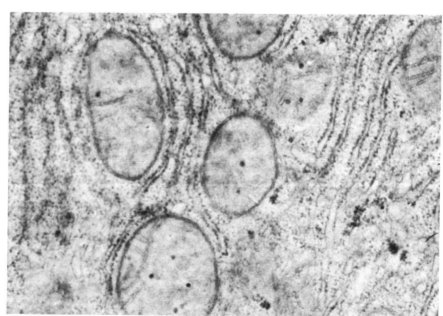

Irrespective of the type of cell, the action of the ribosome in protein synthesis is universal. The two ribosomal subunits form a complete ribosome when bound to m-RNA. The ribosome-m-RNA complex represents an active protein synthesizing unit. The relationship among the ribosome, m-RNA and t-RNA is shown in Figure 13.22. Note that normally more than one ribosome binds to an m-RNA strand. This permits the simultaneous formation of several proteins from the same template in an "assembly line" manner. This multi-ribosome/m-RNA complex is called a **polysome**, and has been observed directly with the electron microscope, as in the remarkable photograph of Figure 13.23. You should study this figure carefully since it summarizes our discussion of transcription and translation of genetic information. Note that the formation

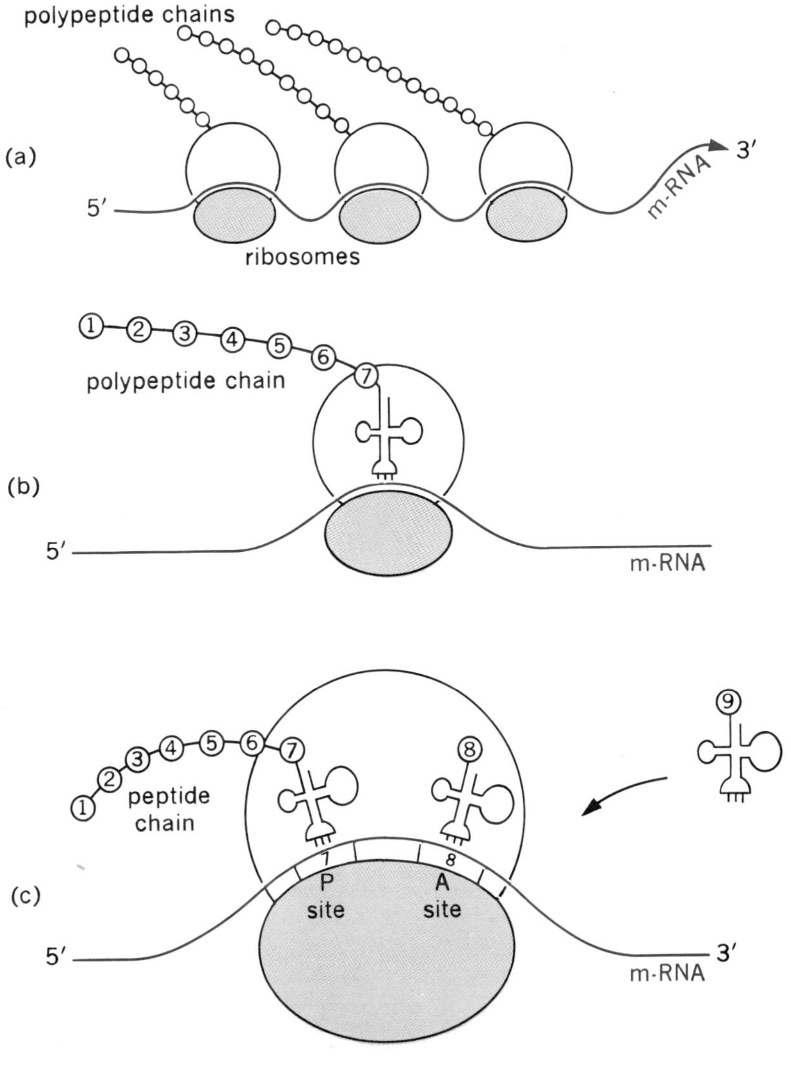

Figure 13.22

Relationship among ribosomes, m-RNA, and t-RNA. (a) A polysome, consisting of three ribosomes "scanning" the m-RNA strand, with polypeptide chains in various stages of completion. (b) Amino acid 7 has been carried to the ribosome by its t-RNA molecule and positioned on the m-RNA strand by codon/anticodon base-pairing. It is shown already attached to the growing polypeptide chain. (c) A closer look at the ribosome showing the A (*aminoacyl*) *site*, which binds the incoming aminoacyl t-RNA molecules, and the P (*peptidyl*) *site*, which binds the peptidyl t-RNA.

of m-RNA by transcription from DNA and the translation of the base sequence of m-RNA into protein structure by ribosomal protein synthesis are closely coordinated with respect to time and place, at least in *E. coli*. As the 5′ end of the newly formed m-RNA peels off from the DNA template (see Figure 13.15), the ribosomal subunits bind and protein synthesis commences even as the m-RNA template is still being made!

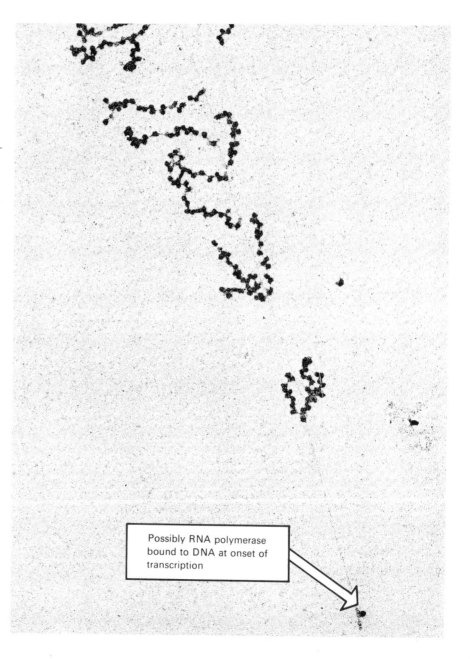

Possibly RNA polymerase bound to DNA at onset of transcription

Figure 13.23

(a and b) Genetically active and inactive portions of *E. coli* chromosomes. In both electron micrographs, the key factors in transcription and translation are apparent. The DNA duplex is the continuous thin line, and the polysomes of varying length and numbers of ribosomes are also clearly visible. Note that the transcription and translation of genetic information is closely synchronized. These remarkable electron micrographs summarize in real terms the information in Figures 13.20 and 13.22. The degree of magnification can be estimated from the fact that the diameter of each ribosome is approximately 200 Å. [Electron micrographs courtesy O. L. Miller, Jr.; prior publication in *Science*, **169**, 392 (1970).]

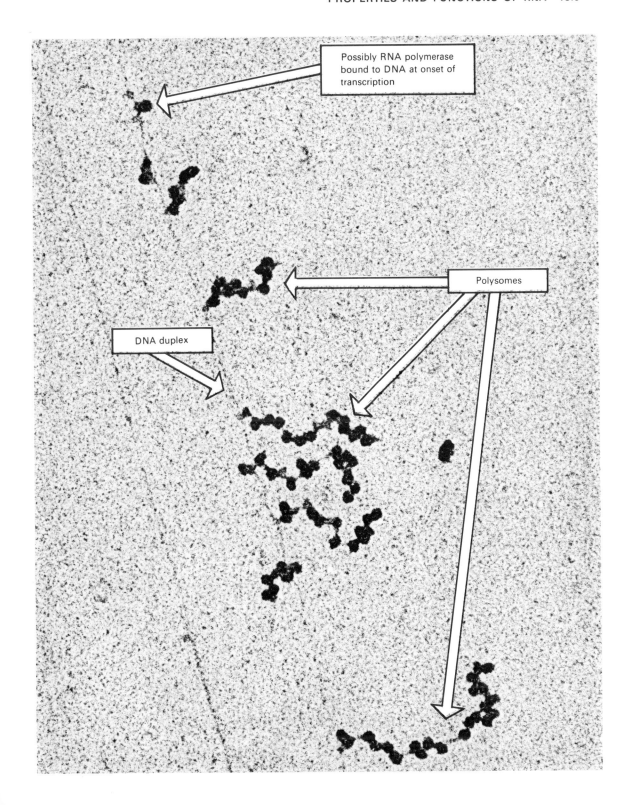

Possibly RNA polymerase bound to DNA at onset of transcription

Polysomes

DNA duplex

We have now described all the units required for protein biosynthesis: the building blocks, the template, and the location. Now we can take a closer look at the process itself.

13.7

Translation of Genetic Information: Protein Biosynthesis

The process of protein biosynthesis can be divided into three main steps, summarized in Table 13.4. Looking at this table we can see that the cell requires several types of substances for protein synthesis. The building blocks are the aminoacyl t-RNA molecules. The template and site of synthesis is the m-RNA/ribosome complex. A number of proteins ("factors") whose properties are not yet fully understood are also essential. Finally, phosphate-bond energy is required in the form of ATP and GTP. Let's examine the steps of protein biosynthesis in more detail.

Table 13.4 Major Steps in Protein Synthesis in *E. coli.*

Step	Substances required
Initiation of polypeptide chain	Initiating aminoacyl t-RNA; m-RNA; 30 S and 50 S ribosomal subunits; GTP; Mg^{2+}; initiation factors (proteins): IF_1, IF_2 and IF_3.
Elongation of polypeptide chain	Aminoacyl t-RNA's; GTP; elongation factors (proteins): EF-Tu, EF-Ts and EF-G.
Termination of polypeptide chain	Termination codon in m-RNA strand; release factors (proteins): RF 1 and RF 2.

Initiation of Protein Synthesis

In procaryotes and possibly in eucaryotes, all polypeptide formation is initiated by N-formylmethionyl t-RNA (fMet-$tRNA_f$):

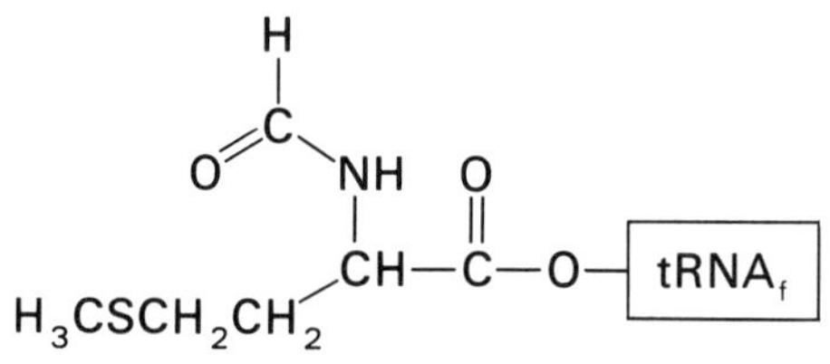

N-formylmethionyl t-RNA

The structure of the N-formylmethionyl group of fMet-$tRNA_f$ is interesting because the N-formyl portion is similar to a peptide. Therefore, it may be that the intact peptide link of fMet-tRNA$_f$ is needed for protein chain initiation. Once the protein is fully formed, the N-formyl group may be cleaved or several terminal residues including the methionine residue may be removed. Some proteins may retain the formyl group at their N-terminus; many bacterial proteins do have methionine as the N-terminal amino acid.

There is a unique t-RNA molecule associated with N-formyl methionine. It differs from the t-RNA that is the carrier for methionine residues in the internal positions in the polypeptide chain. When methionine is attached to t-RNA$_f$, it can be formylated by the action of a specific transformylase enzyme, yielding fMet-tRNA$_f$. When methionine is attached to the regular t-RNA$_m$, it cannot be formylated by the transformylase and therefore cannot initiate protein synthesis.

The process of polypeptide chain formation commences with the assembly of an intact ribosome/m-RNA complex, including N-formylmethionyl tRNA$_f$ bound to the initiating codon of m-RNA at the aminoacyl (A) site within the 70 S ribosome (see Figure 13.22).

The steps of initiation are shown schematically in Figure 13.24. It is important to note that the polypeptide chain grows from the *N-terminal end*. That is, the C-terminal amino acid residue is added last. Also the "reading" of the m-RNA strand runs from the 5′ to the 3′ end. This is vital because it makes it possible to begin the translation process of m-RNA before its transcription is complete (see Figure 13.23).

The Elongation Cycle

The elongation cycle in *E. coli* consists of three basic processes:

1. **Aminoacyl t-RNA binds at the aminoacyl (A) site of the 70 S ribosome,** using specific codon/anticodon base-pairing. The "lining up" of the next amino acid residue to be added to the polypeptide chain requires GTP and a protein, *EF-Tu*. The elongation factor EF-Tu is present as a mixed dimer, *EF-Tu/EF-Ts*, unless activated by reaction with GTP. The part of the polypeptide chain already formed is attached to the peptidyl (P)

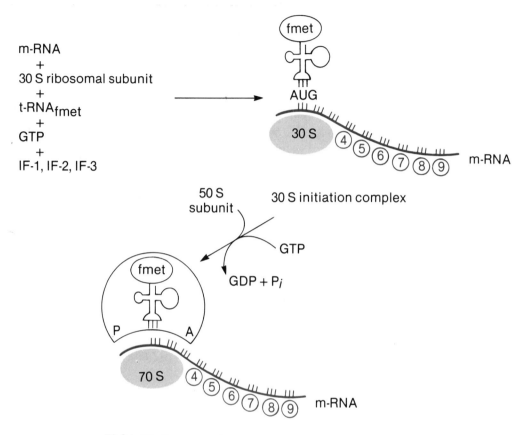

m-RNA
+
30 S ribosomal subunit
+
t-RNA$_{fmet}$
+
GTP
+
IF-1, IF-2, IF-3

30 S initiation complex

50 S subunit

GTP

GDP + P$_i$

70 S initiation complex

Figure 13.24

Initiation of protein
synthesis by the
formation of an
initiation complex,
consisting of a functional
70 S ribosome, m-RNA,
and amino acyl
t-RNA$_{fmet}$. Translation
of m-RNA begins at the
5′ end. It has been
shown that t-RNA$_{fmet}$
and m-RNA will bind to
the isolated 30 S subunit
in the presence of the
initiation factors and
GTP.

site through the still covalently linked t-RNA molecule of the preceding
amino acid residue (see Figure 13.25).

2.   **The peptide link is formed.** The free amino group of the aminoacyl t-RNA
bound at the A site reacts with the activated ester linkage of the peptidyl
t-RNA attached to the P site, as shown in Figure 13.25. This reaction
requires an enzyme protein called *peptidyl transferase* which appears
to be one of the 34 proteins of the 50 S ribosomal subunit.

3.   **The ribosome translocates,** or moves along the m-RNA strand a distance
corresponding to one codon. Since this is an actual movement, it is not
surprising that this step requires energy, in the form of GTP. A specific
protein called EF-G is also essential for translocation. This step of the
elongation cycle is included in Figure 13.25.

The entire elongation cycle is analogous to a hypothetical automobile
assembly line, where the cars in various stages of assembly remain stationary
and the entire factory moves over them during the assembly process. The cycle

Figure 13.25

Steps of the elongation cycle. (a) The binding of aminoacyl t-RNA at the A site.
(b) The peptidyl transferase reaction; this reaction consists mainly of the interchange between an ester (peptidyl t-RNA) and an amide (new peptide link).
(c) The translocation step, in which the newly elongated peptidyl t-RNA/m-RNA is moved from the A site to the P site of the 70 S ribosome.

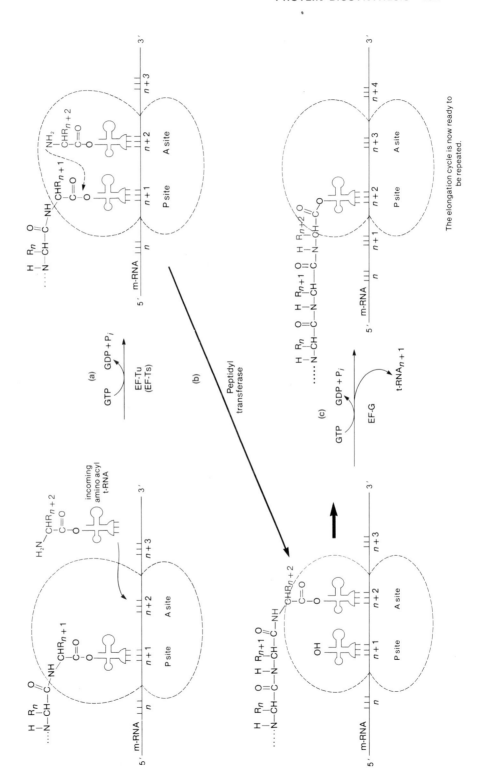

of steps described above and in Figure 13.25 continues until a termination codon is reached.

Termination of the Protein Chain

Certain base triplets in the genetic code do not signify amino acid residues, but rather serve to terminate the growth of the protein chain. Instead of undergoing the peptidyl transferase reaction (Figure 13.25), the peptidyl t-RNA (containing the C-terminal amino acid residue esterified to its t-RNA) hydrolyzes, yielding the free polypeptide chain. The "empty" ribosome then leaves the end of the m-RNA and dissociates into subunits for the formation of another initiation complex. This release process requires certain proteins (*R-factors*) in order to occur.

Figure 13.26

Growth of polypeptide chain in the active polysome (see Figure 13-23). The polypeptide chain spontaneously assumes its tertiary structure as it grows.

Assumption of Native Conformation

We have now seen that proteins are synthesized one residue at a time starting from the N-terminal end. As the protein chain grows, it extends out of the ribosome into the cytoplasm and is free to fold. The tertiary structure of a protein, then, represents the minimum free-energy form of the polypeptide chain, which it spontaneously assumes in this manner. This important concept is illustrated in Figure 13.26.

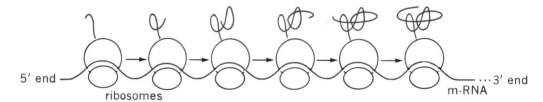

5' end — ribosomes — ···3' end m-RNA

Energy Requirements of Protein Synthesis

The steps of protein synthesis which we have discussed require the energy of at least four high-energy phosphate bonds:

1. 1 GTP for the aminoacyl t-RNA/A site binding process in the elongation cycle.

2. 1 GTP for the mechanical work of translocating the ribosome along the m-RNA strand during the elongation cycle.

3. 2ATP for the formation of aminoacyl t-RNA from t-RNA and amino acid. This reaction produces AMP with pyrophosphate cleavage, and the rephosphorylation of AMP involves the equivalent of forming two high-energy phosphate bonds (see Chapter 7).

These four high-energy phosphate bonds represent about $7.5 \times 4 = 30$ kcal/mole of free energy consumed to produce a peptide bond with a $\Delta G^{0\prime}$ of hydrolysis of about -5 kcal/mole. This means that roughly 25 kcal/mole has been invested to ensure that the formation of each peptide bond is irreversible. This very strong driving force contributes to the high degree of precision in protein biosynthesis (see Problem **9** at the end of this chapter).

Regulation of Protein Synthesis

In Section 3.10 we discussed the various means by which enzyme reactions are controlled. One means, termed *genetic control*, involved the *induction* (turning on) or *repression* (turning off) of the biosynthesis of groups of related enzymes. We discussed how *E. coli* has an inducible enzyme system for the utilization of lactose, a carbon source not normally used by the bacterium (Section 3.10). The synthesis of the key lactose-utilizing enzymes is repressed when lactose is absent. In the presence of lactose, the synthesis of these enzymes is "turned on" or *derepressed*. While the phenomenon of induction and repression of certain groups of enzymes had been known for some time, it was not until 1961 that Jacob and Monod published their *operon model* for gene regulation in procaryotic cells. This model still explains well the observed induction and repression of adaptive enzymes in procaryotic cells. The term *operon* refers to two or more closely spaced genes that are controlled as a unit. In the case of the lactose-utilizing enzymes in *E. coli*, the operon concerned is called the *lac operon*. The action of the lac operon is summarized in Figure 13.27.

Figure 13.27

The action of the lac operon shown schematically.

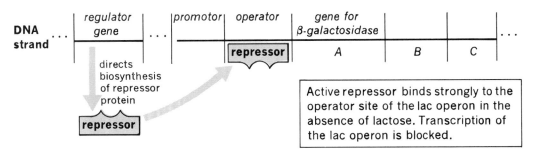

a. *Lactose absent*; lac operon is repressed. There are only 1-2 β-galactosidase molecules per cell.

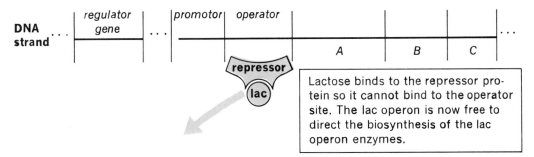

b. *Lactose present*; lac operon is derepressed. There are about 2000 β-galactosidase molecules per cell.

Referring to this figure, we see that when the repressor protein is in an active state, it binds strongly to the operator site of the operon. In this example, the system is said to be inducible because the repressor protein leaves the operator site when lactose (the inducer) binds to a binding site on the repressor molecule. The promoter site of the gene binds RNA polymerase, which then produces m-RNA by transcription from DNA. When the repressor protein is no longer bound to the operator site, synthesis of the lac operon proteins begins. It has recently been shown that a specific cyclic AMP/protein complex must first bind to a specific location on the promoter site before RNA polymerase can bind. This means that there are two levels of control in the expression of the lac operon: derepression by the binding of lactose to the repressor protein, and activation by the binding of the cyclic AMP/protein complex to the promoter site. Cyclic AMP is the same molecule we discussed in Section 12.5 as the "second messenger" in the action of many hormones in higher animals. This connection suggests an evolutionary link between the action of endocrine hormones in higher animals and the regulation of cellular metabolism in procaryotes.

Some enzymes are repressed by the *presence* of a metabolite, in contrast to inducible systems. An example of this is the *his operon*, which contains codes for the enzymes of histidine biosynthesis. Addition of histidine to the growth medium causes the *repression* of the his operon. Thus, interaction of histidyl-t-RNA^{His} with the repressor protein causes it to bind to the operator site. In the case of the his operon, the system is said to be *repressible*. A number of operons in bacteria have been studied; Table 13.5 lists several of these.

Table 13.5   Some Operons in Bacteria.

Operon	Number of proteins	Function
lac	3	Utilization of lactose and related sugars
his	10	Biosynthesis of histidine
leu	4	Synthesis of leucine from ketoisovaleric acid
ile	5	Synthesis of isoleucine from threonine
ara	4	Utilization of the sugar arabinose

Inhibitors of Protein Synthesis

Many antibiotics inhibit protein synthesis at various levels. Some antibiotics specifically inhibit protein synthesis in procaryotes and are therefore useful medicines. Others act on eucaryotic cells and are toxins. Because the action of these substances is highly specific, they have proven useful in understanding the details of ribosomal protein synthesis. Let's discuss some examples.

Protein inhibitors include *ricin*, a toxic plant protein from the castor bean that inactivates the 60 S ribosomal subunits in eucaryotes. *Diphtheria toxin* is an enzyme protein produced by the organism responsible for the deadly childhood disease, diphtheria—a disease now preventable by vaccination. The diphtheria toxin catalyzes the modification and inactivation of a protein which is essential for translocation in mammalian ribosomes. Since one toxin molecule can inactivate many translocation-factor protein molecules, extremely low concentrations (*ca.* $10^{-8}M$) of diphtheria toxin in the cytoplasm can be fatal. The mechanism of this modification is well understood. In eucaryotic cells, an elongation factor called *translocase*, analogous to EF-G in procaryotes, reacts with NAD^+ in the presence of diphtheria toxin to produce a covalently modified and inactive translocase, as shown below

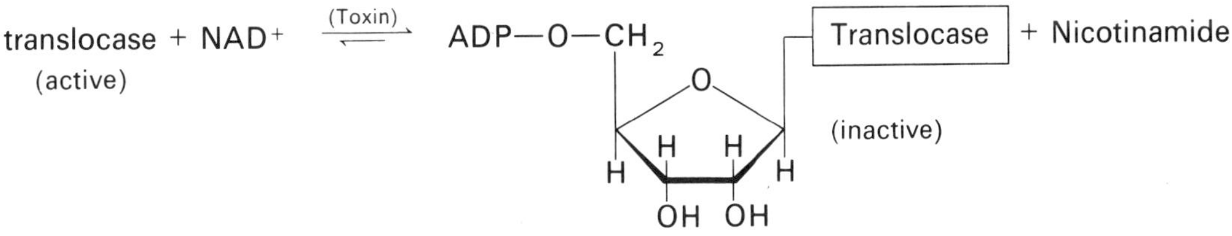

It is noteworthy that NAD^+ is apparently not acting in its usual redox capacity, but rather as a donor of an adenosine diphosphate ribose group, with the release of nicotinamide.

Inhibitors of protein synthesis that are important drugs include the *streptomycin* family of antibiotics. These inhibit protein synthesis in procaryotes by binding to the 30 S ribosomal subunit. *Tetracycline* and its derivatives are widely used broad spectrum antibiotics that inhibit the binding of aminoacyl t-RNA to the A site of the 30 S ribosomal subunit in bacteria. *Chloramphenicol* binds to the 70 S ribosome of procaryotes and inhibits the peptidyl transferase reaction. Because it also inhibits peptidyl transferase in mitochondrial ribosomes in eucaryotes, it is a drug with toxic side effects.

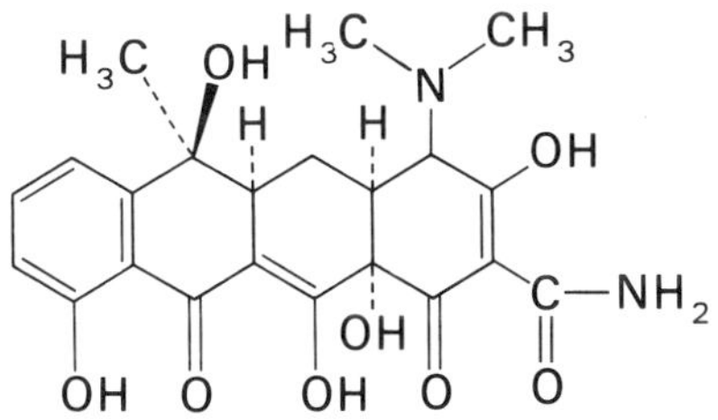

Tetracycline

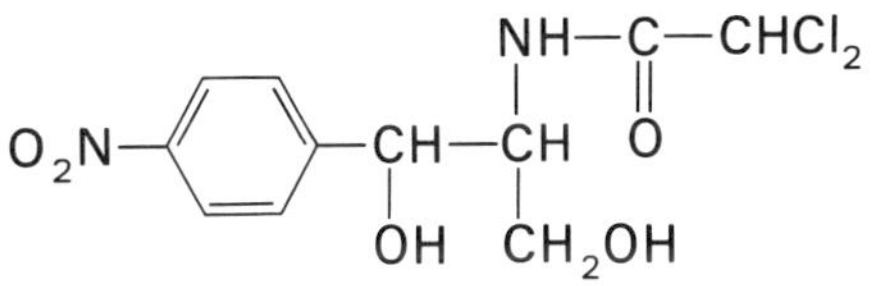

Chloramphenicol

13.8

The Genetic Code

Development of the Genetic Code

In our discussion of protein synthesis we saw that the key factor in generating the correct primary structure for a given protein is the base-pairing between complementary base triplets in m-RNA and aminoacyl t-RNA. We will now describe what the code words of m-RNA are for the various amino acids.

In 1961, M. Nirenberg and his coworkers prepared a number of synthetic m-RNA molecules by the action of *polynucleotide phosphorylase* on various pure ribonucleoside 5′-diphosphates:

$$n \text{ ADP} \; \rightleftharpoons \; \text{poly A} + P_i$$

$$n \text{ CDP} \; \rightleftharpoons \; \text{poly C} + P_i$$

$$n \text{ UDP} \; \rightleftharpoons \; \text{poly U} + P_i$$

$$n \text{ GDP} \; \rightleftharpoons \; \text{poly G} + P_i$$

They found that three of these synthetic RNA molecules could act as templates for protein synthesis in a cell-free system composed of isolated *E. coli* ribosomes, various enzymes, ATP, GTP, t-RNA, and amino acids. When polyuridylic acid (poly U) was used as a template, polyphenylalanine was produced. Therefore, the nucleotide triplet UUU is a codon for phenylalanine. Similarly, they found that AAA coded for lysine and CCC coded for proline. They could not use poly G because extensive intrachain hydrogen bonding prevented it from binding to ribosomes.

Not long after this original work, Nirenberg, Ochoa, and others synthesized various random copolymers of two bases at a time, using polynucleotide phosphorylase to link the mononucleotide units as before. Since the *in vitro* synthesis of RNA by this method produces polynucleotides with random sequences of the two bases, they were able to statistically compute the relative abundance of various possible base triplets based on the relative amounts of the two original ribonucleoside diphosphates. For example, if you were to make poly UG, the various base triplets in the chain, starting from the 5′ end, would include UUU, UUG, UGU, UGG, GUU, GUG, GGU, and GGG. However, these triplets will not be present in equal amounts in poly UG, but rather in amounts based on the ratio of UDP/GDP in the original reaction mixture.

These researchers used such random copolymers as templates for *in vitro* protein synthesis and then compared the amounts of the amino acids present in the polypeptides produced with the relative abundance of the various base triplets in the copolymer template. The most abundant triplet coded for the most abundant amino acid in the protein, and so on. From experiments like these, fifty triplet combinations for the various amino acids were rapidly deduced.

Although these copolymer experiments produced combinations of three bases which code for various amino acids, the exact sequence or "spelling" of these code words was not obtainable from such studies. For example, workers found that different combinations of $2A + 1G$ coded for lysine, glutamic acid, and arginine; this is because there are three ways of arranging this set of bases: AAG, GAA, or AGA. But they were unable to determine which combination coded for which amino acid.

In 1964, Nirenberg, Khorana, Ochoa, and their coworkers in several laboratories devised an experimental way to deduce the "spelling" of the words in the genetic code. They first managed to synthesize pure triribonucleotides with a predetermined base sequence. With these very short templates, they found that the appropriate purified aminoacyl t-RNA formed a complex with the 70 S ribosome. An example of this is:

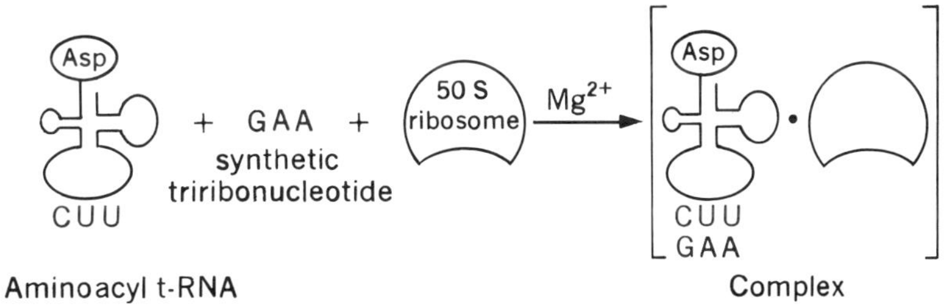

The high specificity of the codon-anticodon interaction pertains, of course, even to such small synthetic m-RNA fragments. This type of binding experiment enabled the deduction of the correct codons for the amino acids. The workers found that 61 of the 64 possible triplet combinations of A, G, C, and U code for specific amino acids, and that many amino acids have more than one codon. The remaining three codons are *termination codons* and cause the release of the completely formed protein molecule from the m-RNA/ribosome complex.

This short discussion cannot do justice to the remarkable biochemical detective work done by many researchers to discover the genetic code. If you wish to learn about these studies in more detail, you should consult the sources listed at the end of this chapter. Many of these were written by people directly involved in the discovery of the genetic code.

The Features of the Genetic Code

Figure 13.28 gives a compilation of the genetic code. You can see that all amino acids except tryptophan and methionine have more than one possible codon. This multiplicity of codons for a given amino acid is termed **degeneracy**. The degenerate nature of the genetic code has vital implications for the process of evolution at the molecular level. Let's list some of the important features of the genetic code:

1. **The code is degenerate.** The degeneracy in the genetic code often involves only the third base in a codon. For example, alanine has four possible codons: GCU, GCC, GCA, and GCG. All four possible alanine codons start with GC. If you refer to Figure 13.28, you will see that this is a general phenomenon.

2. **Nearly all codons have the same general scheme.** The combination of bases in the codon(s) for a given amino acid generally obey the form

$$XY(A/G) \quad \text{or} \quad XY(C/U)$$
$$\text{Purines} \qquad\qquad \text{Pyrimidines}$$

    You should study Figure 13.28 and satisfy yourself that this is indeed the case.

3. **Nonsense triplets.** When 3 of the 64 possible codons were found not to correspond to any amino acids, they were termed "nonsense triplets." We now know that these are actually termination codons which signal the release of the completed protein molecule. Of the three termination codons, the triplet UAA appears to be most commonly used.

4. **The genetic code is nearly universal.** The codons given in Figure 13.28 are applicable to all cells. This is illustrated by the fact that *E. coli* (procaryote) t-RNA will work in mammalian cells. Further, we have seen that eucaryotic DNA fragments can be inserted into procaryotic DNA by the procedure outlined in Figure 13.13. This recombinant DNA can then undergo gene expression within the host procaryote.

    Recently, an interesting exception to the universality of the genetic code has been discovered. Protein synthesis within the mitochondrion is coded for by mitochondrial DNA, a small, circular molecule separate from the cell chromosomes of the nucleus. Experimentation has shown that the mitochondrial genetic code differs in at least two codons from the

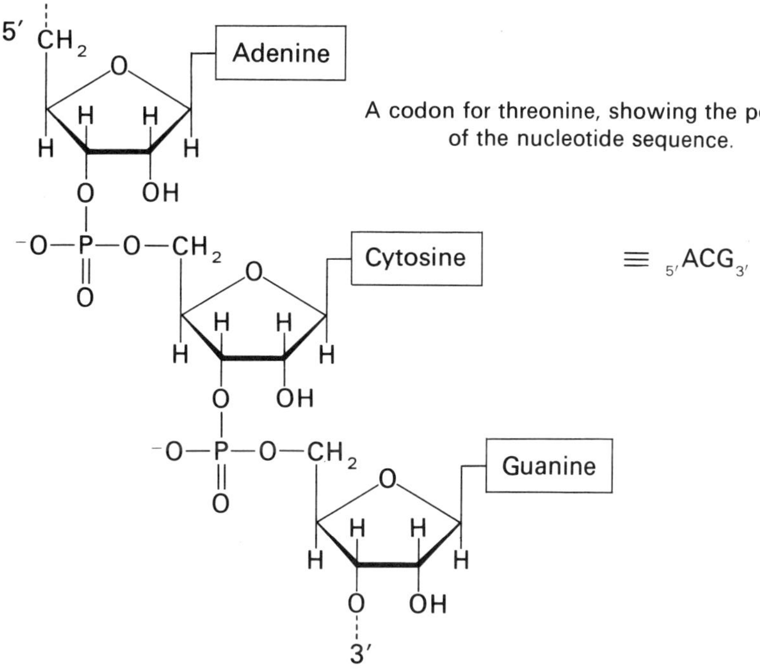

5′ CH₂

Adenine

H H

H H

O OH

⁻O—P—O—CH₂

Cytosine

O

H H

H H

O OH

⁻O—P—O—CH₂

Guanine

O

H H

H H

O OH

3′

A codon for threonine, showing the polarity of the nucleotide sequence.

$$\equiv\ _{5'}ACG_{3'}$$

Figure 13.28

The genetic code. Note that the genetic code is degenerate, meaning that more than one codon can code for a given amino acid. The codons UAA, UAG, and UGA are *termination signals*, denoted in the table as "Stop." The codon AUG codes for either methionine or N-formylmethionine. In the case of N-formylmethionine, the codon AUG serves as a chain-initiating signal. The sequence of the codons runs from 5′ → 3′, as illustrated for the threonine codon ACG.

A Compilation of the Genetic Code

UUU	Phe	UCU	Ser	UAU	Tyr	UGU	Cys
UUC	Phe	UCC	Ser	UAC	Tyr	UGC	Cys
UUA	Leu	UCA	Ser	UAA	Stop	UGA	Stop
UUG	Leu	UCG	Ser	UAG	Stop	UGG	Trp
CUU	Leu	CCU	Pro	CAU	His	CGU	Arg
CUC	Leu	CCC	Pro	CAC	His	CGC	Arg
CUA	Leu	CCA	Pro	CAA	Gln	CGA	Arg
CUG	Leu	CCG	Pro	CAG	Gln	CGG	Arg
AUU	Ile	ACU	Thr	AAU	Asn	AGU	Ser
AUC	Ile	ACC	Thr	AAC	Asn	AGC	Ser
AUA	Ile	ACA	Thr	AAA	Lys	AGA	Arg
AUG	Met	ACG	Thr	AAG	Lys	AGG	Arg
GUU	Val	GCU	Ala	GAU	Asp	GGU	Gly
GUC	Val	GCC	Ala	GAC	Asp	GGC	Gly
GUA	Val	GCA	Ala	GAA	Glu	GGA	Gly
GUG	Val	GCG	Ala	GAG	Glu	GGG	Gly

chromosomal genetic code given in Figure 13.28. In the mitochondrion, the termination codon UGA is translated as tryptophan, and the isoleucine codon AUA is translated as methionine. This finding lends support to the hypothesis that the mitochondrion was once a separate primitive cell that became locked into a symbiotic ("living together") relationship with a larger host cell at some point early in cellular evolution.

These features of the genetic code have a profound effect on the continued evolution of living systems. Let's consider why a degenerate code is a desirable feature.

Desirability of Degeneracy

Suppose that of the 64 possible codons, just one corresponded to each of the 20 amino acids. In other words, what would be the implications of a nondegenerate code?

About 43 codons would have no meaning. Since mutation processes change the base sequence of DNA, and are common in all organisms, many alterations in base sequence would occur, resulting in new sets of triplets having no meaning. This would be disastrous from an evolutionary standpoint because a mutation in a given gene could prevent a protein molecule from being made at all. By using all possible codons, living systems ensure that a mutation in a given gene always results in a new base sequence that will still code for a complete protein. Since the degenerate genetic code can accommodate mutations in DNA without a corresponding complete loss of protein formation, the process of mutation is a powerful factor in the evolution of optimal protein structure over many generations.

The Genetic Code and Evolution

We often think of mutations as being undesirable. In Chapter 2 we saw how considerable changes in protein primary structure could take place without a corresponding loss of function if the changes at amino acid positions were of a conservative nature. This means that if a hydrophobic amino acid residue is replaced by another hydrophobic residue, no drastic overall changes in the protein are likely. Only when a nonconservative amino acid replacement occurs, as in the case of hemoglobin S (sickle-cell anemia), do we see impairment of protein function. Since a given gene codes for a given protein, this means that changes in individual codons are responsible for the genetic variation of protein primary structure.

The most common types of mutations are *point mutations*, where a change at a single nucleotide position in the nucleic acid chain occurs. Point mutations

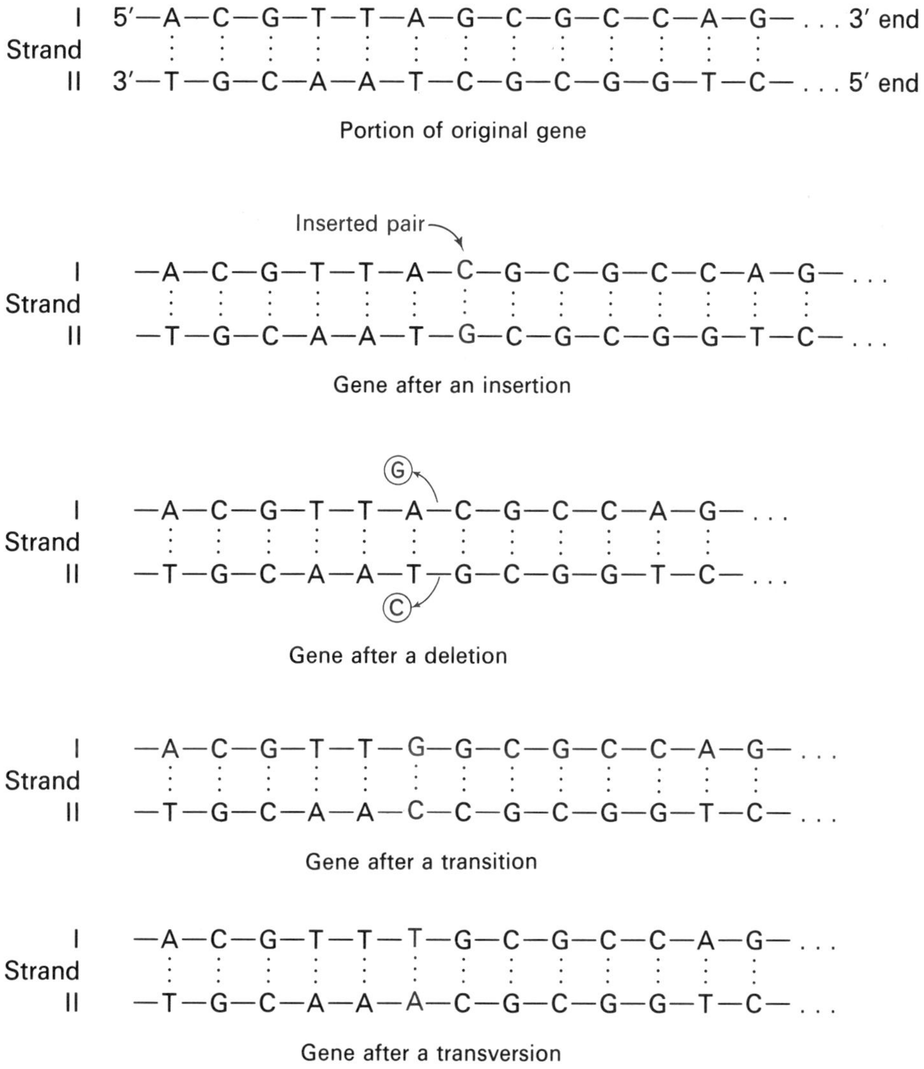

```
 I   5'—A—C—G—T—T—A—G—C—G—C—C—A—G— ... 3' end
         :   :   :   :   :   :   :   :   :   :   :   :   :
Strand
 II  3'—T—G—C—A—A—T—C—G—C—G—G—T—C— ... 5' end
```

Portion of original gene

Inserted pair

```
 I      —A—C—G—T—T—A—C—G—C—G—C—C—A—G— ...
            :   :   :   :   :   :   :   :   :   :   :   :   :   :
Strand
 II     —T—G—C—A—A—T—G—C—G—C—G—G—T—C— ...
```

Gene after an insertion

```
                         G
 I      —A—C—G—T—T—A—C—G—C—C—A—G— ...
            :   :   :   :   :   :   :   :   :   :   :
Strand
 II     —T—G—C—A—A—T—G—C—G—G—T—C— ...
                      C
```

Gene after a deletion

```
 I      —A—C—G—T—T—G—G—C—G—C—C—A—G— ...
            :   :   :   :   :   :   :   :   :   :   :   :   :
Strand
 II     —T—G—C—A—A—C—C—G—C—G—G—T—C— ...
```

Gene after a transition

```
 I      —A—C—G—T—T—T—G—C—G—C—C—A—G— ...
            :   :   :   :   :   :   :   :   :   :   :   :   :
Strand
 II     —T—G—C—A—A—A—C—G—C—G—G—T—C— ...
```

Gene after a transversion

Figure 13.29

Point mutations.
Insertion/deletion
mutations can affect
large portions of a gene
by altering the
registration of the base
sequence. Transitions
and transversions affect
only one codon in a
gene.

fall into two broad classes: nucleotide replacements and insertion/deletion mutations. These are shown schematically in Figure 13.29.

In the case of insertion mutations, an additional nucleotide pair is inserted into the DNA. This increases each of the DNA strands by one unit and alters the normal transcription of the DNA by placing that part of the DNA chain coming after the mutation out of register with respect to the original. The direction of protein synthesis by this altered gene would result in a normal polypeptide chain up to the point mutation and a completely different polypeptide chain following it.

In the case of deletions, a complementary nucleotide base pair is removed from the DNA double helix at a given point. The result of this point mutation is the same as an insertion. Since both deletions and insertions generally result in a major alteration of a large part of a protein molecule, their effect is often lethal; that is, the function of the protein is totally lost.

Replacement point mutations can either be *transitional* mutations or *transversional* mutations, and often occur at the third nucleotide position in a codon. A transitional mutation replaces a purine nucleotide at one DNA position by another purine at the same position, and replaces the complementary pyrimidine nucleotide at the same position on the other strand by another pyrimidine nucleotide. This would correspond to an A—G replacement on one strand and a T—C replacement on the other. Transversions occur when a pyrimidine is replaced by a purine on one strand and a purine by a pyrimidine on the other. In either case, the original triplet sequence of the mutated gene remains unaltered. Only one codon is affected, and the resulting protein would differ at only one amino acid position.

Mutations can be induced by a variety of mutagenic agents, both natural and artificial. Ultraviolet (high-energy) light and radioactivity are known to cause mutations. Many chemical mutagens induce various types of point mutations. You can find further information on mutagenesis in the sources listed at the end of this chapter.

Let's go back to our table of the genetic code in Figure 13.28 and assess the effect of transitions and transversions on some of the codons shown. To begin with, we see that there is considerable variability in the third codon position. This means that a change in the third nucleotide in a degenerate codon can often cause no change in protein structure at all. For example, proline is coded by CCA, CCG, CCC, and CCU. Many hydrophobic amino acids (Chapter 2) are specified by codons beginning with a pyrimidine–purine or purine–pyrimidine sequence. Hence, either a transition or a transversion at the first position can still result in a codon for a hydrophobic amino acid, and thus a conservative replacement. A transition in the second nucleotide position can also result in a codon for a hydrophobic amino acid. However, a transversion in the second position can cause nonconservative amino acid replacement in a protein (see Problem **11** at the end of this chapter).

What all this means is that a gene mutation can often result in a conservative change in protein primary structure and may, in fact, produce a better-functioning protein molecule. In the continuing process of evolution at the molecular level, this aspect of mutations is important because it permits natural selection of the optimal amino acid residue at each position in a protein chain. You can appreciate this by reviewing our discussion of cytochrome c primary structure in Chapter 2 (Section 2.11).

Many genetic mutations have detrimental effects on protein function, sometimes manifested as metabolic abnormalities. In the extreme case, of course, the mutant organism cannot survive. This would be the case for a mutation inactivating an enzyme of a major metabolic process such as glycolysis.

Nonlethal mutations often lead to hereditary disorders of metabolism, producing clearly observable external effects. In the case of bacteria, the artificial creation of mutant strains is an important tool for studying metabolism. For example, many *E. coli* strains exist in which a specific protein has little or no activity. The use of these artificially created metabolic blocks has proven particularly useful for elucidating metabolic pathways. In humans, the genetic impairment of certain enzymes gives rise to a number of inborn errors of metabolism, worth discussing in more detail.

13.9

Inborn Errors of Metabolism

A. E. Garrod recognized in 1908 that certain metabolic disorders were hereditary and involved the body's inability to conduct a particular metabolic reaction. Starting with the four inborn errors of metabolism documented by Garrod, molecular biologists have characterized over 150 hereditary defects in metabolism and identified the particular protein affected. Furthermore, over 1000 hereditary diseases are known for which the metabolic defect has not yet been identified. It is safe to assume that most, if not all, true hereditary disorders must involve some defect in the synthesis of a protein or group of proteins. This is because the direct expression of genetic information is the synthesis of proteins.

Inborn errors of metabolism can be expressed in several ways. Higher organisms contain pairs of chromosomes (diploid). If both chromosomes in a pair have the same genetic defect (homozygous state), no normal protein corresponding to that gene will be produced. If the individual has one normal and one mutant chromosome in a given chromosome pair (heterozygous state), about 50% of the protein produced will be normal (coded by the normal gene) and about 50% will be abnormal protein (coded by the mutant gene). You can see that although genetic defects in higher organisms can result in decreased intracellular levels of a given protein, the levels may still be sufficient to meet the needs of the cell.

Throughout the text we have discussed many examples of inborn errors of metabolism along with the metabolic sequences affected. Table 13.6 lists these examples together with the section where each may be found.

Table 13.6 A Listing of the Inborn Errors of Metabolism Discussed in the Text Together With the Section Where Each May Be Found.

Inborn error of metabolism	Section
Albinism	12.3
Alkaptonuria	11.3
Congenital hyperammonemia	11.3
Diabetes mellitus	2.13, 7.7, 11.2
Fructose intolerance	8.2
Galactosemia	8.2
Glucose 6-phosphate dehydrogenase deficiency	8.3
Glycogen storage diseases	12.5 (Table 12.2)
Von Gierke's disease	
Pompe's disease	
Cori's disease	
Anderson's disease	
McArdle's disease	
Her's disease	
Types VII, VIII and IX	
Gout	11.4 (Figure 11.9)
Hemophilia	7.9
Lactase deficiency	7.7
Maple syrup urine disease	11.3 (Figure 11.8)
Methyl malonic aciduria	11.1
Phenylketonuria	11.3
Refsum's disease	11.1 (Figure 11.3)

Suggestions for Further Reading

Barry, J. M. and Barry, E. M. *Molecular Biology: An Introduction to Chemical Genetics.* Englewood Cliffs, N.J.: Prentice-Hall, 1973. This concise and clearly written paperback gives a good, short treatment of classical genetics and gene mapping.

Freifelder, D., ed. *Recombinant DNA.* San Francisco: W. H. Freeman & Co., 1978. A collection of articles from *Scientific American* dealing with genetic engineering. The background pieces make this collection a good overview of molecular genetics.

Watson, J. D. *Molecular Biology of the Gene.* 3d ed. Menlo Park, Ca.: W. A. Benjamin, 1976. An unusually lucid and complete account of molecular genetics by one of the originators of the double helix model.

White, A.; Handler, P.; Smith, E.; Hill, R. and Lehman, I. *Principles of Biochemistry.* New York: McGraw-Hill, 1978. This standard comprehensive text contains an excellent discussion of hereditary disorders of metabolism in Chapter 27.

Scientific American
Cairns, J. "The Bacterial Chromosome." January 1966.
Clark, B. F. C. and Marcker, K. A. "How Proteins Start." January 1968.

Cohen, S. N. "The Manipulation of Genes." July 1975.
Crick, F. H. C. "The Genetic Code." October 1962.
————. "The Genetic Code III." October 1966.
Gorini, L. "Antibiotics and the Genetic Code." April 1966.
Grobstein, C. "The Recombinant-DNA Debate." July 1977.
Hanawalt, P. C. and Hayes, R. H. "The Repair of DNA." February 1967.
Holley, R. W. "The Nucleotide Sequence of a Nucleic Acid." February 1966.
Kornberg, A. "The Synthesis of DNA." October 1968.
Maniatis, T. and Ptashne, M. "A DNA Operator-Repressor System." January 1976.
Miller, O. L., Jr. "The Visualization of Genes in Action." March 1973.
Nirenberg, M. W. "The Genetic Code II." March 1963.
Nomura, M. "Ribosomes." October 1969.
Ptashne, M. and Gilbert, W. "Genetic Repressors." June 1970.
Rich, Alexander. "Polyribosomes." December 1963.
Stent, G. S. "Prematurity and Uniqueness in Scientific Discovery." December 1972.
Temin, H. M. "RNA-Directed DNA Synthesis." January 1972.
Yanofsky, C. "Gene Structure and Protein Structure." May 1967.

Problems

1. For each of the terms below, provide a short definition or explanation.

Anticodon	Polarity
Chromosome	Polysome
Codon	Recombinant DNA
Degeneracy	Replication
Gene	Ribosome
Histones	Semiconservative
Inborn error of metabolism	Transcription
Mutation	Translation
Operon	Wobble

2. Give the structure of each metabolite listed below.

Adenine	Inosine
Cytosine	Thymine
Guanine	Uracil

3. For each enzyme below, give the reaction catalyzed, showing clearly the functional groups affected in the reaction.

aminoacyl t-RNA synthetase	peptidyl transferase
DNA ligase	polynucleotide phosphorylase
DNA polymerase I	reverse transcriptase
DNA polymerase III	RNA polymerase

4. Is the base composition of DNA consistent with the principle of complementarity? Give the reason for your answer.

5. You have isolated DNA from *Bacillus hypotheticus* and found the percentage of adenine to be 20%. What is the percentage of each of the other DNA bases?

6. Which type of RNA forms a high-energy bond with an amino acid? What is the nature of the amino acid-RNA bond? (Show structures.)

7. In the formation of N-formylmethionyl t-RNA$_f$ from methionyl t-RNA$_f$, what coenzyme would most likely be required?

8. Puromycin is an antibiotic that binds to the 50 S ribosomal subunit. When the growing peptide chain is transferred onto puromycin, peptide synthesis is prematurely terminated. What does puromycin resemble that permits it to act this way?

Puromycin

9. In our discussion of the energetics of protein biosynthesis, the energy required was given in terms of the $\Delta G^{0\prime}$ for ATP hydrolysis. Under cellular conditions, would you expect the actual $\Delta G'$ values to be the same or somewhat different? Give a reason for your answer.

10. In deducing the genetic code, the use of synthetic m-RNA's containing repetitive sequences as a template for protein synthesis was a valuable tool. When poly UUC was used, three different polypeptides resulted, each a polymer of a single amino acid. Name these three products.

11. Using the table in Figure 13.28, answer each of the following:

    (a) List the codons for the hydrophobic (apolar) amino acids. (See Chapter 2.) What generalities can you state concerning any relationships between these codons?
    (b) List the codons for the charged polar amino acids. (See Chapter 2.) Are there any relationships between these codons? If so, what are they?
    (c) Show the result of a transversion at the second base of each of the leucine codons. Identify the amino acids corresponding to the new codons.

12. Given in the table below are just a few of the known hemoglobin mutants (see Chapter 2). For each amino acid replacement listed, state whether the mutation is transitional or transversional.

Type	Chain position	Normal	Mutant
I	16	Lys	Glu
M Boston	58	His	Tyr
S	6	Glu	Val
D Punjab	121	Glu	Gln
K Woolwich	132	Lys	Gln

13. In Figure 13.29 we gave a specific example of each of four types of point mutation, using a portion of a hypothetical gene. Using the table in Figure 13.28, answer the following questions.

(a) Assume that both strand I and strand II of the original gene portion are transcribed and translated. (We know, of course, that only one DNA strand is transcribed in the actual situation.) Give the primary structure of the peptide fragment coded by each strand.

(b) For each of the mutant genes, assuming only strand I is transcribed and translated, give the primary structure of the oligopeptide resulting from the translation of each mutant gene. Compare each to the result in (a).

Active center (active site) Location in the three-dimensional structure of an enzyme where substrate is bound and reacted.

Active transport An energy-requiring transport process that carries solute across a membrane from a region of low concentration to a region of higher concentration.

Aerobic organisms Organisms that utilize oxygen.

Allosteric proteins Proteins whose biological activity changes when certain metabolites (effectors) are bound at binding sites other than the active site. Also called **regulatory** proteins.

Amphibolic A metabolic pathway used for both catabolic and anabolic purposes.

Amphipathic A molecule containing an apolar (hydrophobic) part and a polar (hydrophilic) part.

Anabolism Metabolic pathways of biosynthesis.

Anaerobic Living in the absence of oxygen.

Anaplerosis Reactions that increase the concentrations of the Krebs TCA cycle intermediates.

Angstrom (Å) A unit of length, equal to 10^{-8} cm. This unit is commonly used to designate molecular dimensions.

Antibodies A group of blood plasma proteins that forms the basis for the immune system in vertebrates.

Anticodon The unique base triplet in the t-RNA molecule which base-pairs to a codon triplet of m-RNA.

Antigen A substance, usually a protein, whose injection into the bloodstream of vertebrates stimulates the production of specific antigen-neutralizing antibodies.

Autotrophs Organisms that can produce all necessary biomolecules from simple inorganic precursors.

Backbone The chain of fundamental structural units of a polymer.

Base-pairing Strong hydrogen bonding between a purine and a pyrimidine having the correct mutual molecular geometry.

Calorie An energy unit, equal to the quantity of energy needed to raise the temperature of 1.0 gram of water from 14.5 to 15.5 degrees Centigrade.

Catabolism The metabolic pathways that generate energy in the cell by degrading food molecules.

Cell   The smallest unit of life capable of independent reproduction.

Chromosome   A single intact molecule of double-helical DNA.

Codon   A base triplet on m-RNA coding for an amino acid or a termination signal.

Coenzyme   An organic molecule essential to the catalytic activity of certain enzymes.

Cofactor   A small inorganic molecule or ion required for activity in certain enzymes.

Complementary base sequences   Nucleotide base sequences related by base-pairing relationships.

Conjugated protein   A protein containing a prosthetic group (nonprotein part) in its native form.

Copolymer   A polymer containing more than one kind of recurring structural unit (monomer).

Cytochromes   Heme-containing electron-transferring proteins.

Dalton   A unit of weight equal to the actual weight of one hydrogen atom.

Degeneracy   The case where more than one codon code for the same amino acid.

Dehydrogenase   A redox enzyme that removes the elements of two hydrogen atoms ($2H^+ + 2e^-$) from a substrate molecule.

Denaturation   The disruption of the normal three-dimensional structure of a macromolecule.

Diploid   Cells that contain two chromosomes of each type.

DNA (deoxyribonucleic acid)   A nucleic acid, containing a specific sequence of monodeoxyribonucleotide units, that acts as the carrier of genetic information in all cells.

Electron acceptor   An organic or inorganic molecule that acts as an oxidant.

Electron microscope   An instrument that uses an electron beam instead of visible light for the visualization of minute objects.

Electron transport   The sequential transfer of electrons by electron carriers from a reductant to a terminal electron acceptor.

Endergonic reaction   A reaction that requires energy; $K_{eq} < 1$.

Entropy   A measure of the disorder or randomness of a system.

Enzyme   A protein catalyzing a reaction in a highly specific manner.

Equilibrium   The minimum free-energy state of a system. There is no driving force for change in a system at equilibrium.

Escherichia coli ($E.\,Coli$)   A common aerobic bacteria found in great abundance in the intestinal tracts of animals.

Eucaryote   The class of cells having a well-defined nucleus surrounded by a nuclear membrane and membrane-surrounded subcellular organelles.

Exergonic reaction   An energy-producing reaction; $K_{eq} > 1$.

Facultative anaerobes Cells that can live in the presence or absence of oxygen.

Feedback inhibition Inhibition of the first enzyme in a pathway by the end product of that pathway.

Fermentation The anaerobic catabolism of a fuel molecule such as glucose.

Flavoprotein A protein containing a flavin coenzyme or prosthetic group.

Free energy That part of the total energy of a system available to perform work at constant temperature and pressure. The free-energy change of a reaction is a measure of the chemical potential or driving force behind the reaction.

Gene The segment of a DNA strand that directs the synthesis of a particular RNA molecule.

Gluconeogenesis Biosynthesis of glucose from pyruvate and other precursors.

Glycolysis The fermentation of glucose to yield two lactic acid molecules.

Glycoprotein A conjugated protein containing a carbohydrate prosthetic group.

Heme protein A conjugated protein containing an iron-porphyrin (heme) prosthetic group.

Heterotrophs Cells requiring nutrient biomolecules such as glucose and amino acids for catabolism and anabolism.

High-energy bond In a biochemical context, a bond yielding 5 kcal/mole or more upon hydrolysis.

Hydrogen bond A weak attractive interaction between an electronegative atom (such as O or N) in one group and a hydrogen atom covalently bound to an electronegative atom in another group.

Hydrolysis The cleavage of a bond by the addition of the elements of H_2O.

Hydrophilic "Water loving"; groups that associate with water.

Hydrophobic (apolar) "Water hating"; groups that are insoluble in water.

Inducible enzymes Enzymes whose biosynthesis is promoted by the presence of certain metabolites (inducers).

Informational macromolecules Polymers whose recurring structural units follow a genetically prescribed sequence.

In vitro (Latin, "in glass.") Experiments performed in biochemical systems isolated or extracted from living cells.

In vivo (Latin, "in life.") Experiments on intact living cells, tissues, or organisms.

Isozymes Multiple forms of a given enzyme that catalyze the same chemical reaction but differ in net charge or catalytic activity or both.

K_{eq} The ratio of concentrations of products to concentrations of reactants for a chemical system at equilibrium at a given temperature. This number is the **equilibrium constant**.

Lipoprotein A conjugated protein having a lipid prosthetic group.

Macromolecules Molecules in the molecular weight range of several thousands to many millions.

Messenger RNA The RNA molecule that serves as a working template for protein biosynthesis.

Metabolic pathway A set of consecutive enzyme-catalyzed reactions that converts starting material to end product.

Metabolite A chemical intermediate in a metabolic pathway.

Micron (μ) A unit of length, equal to 10^{-4} cm.

Mole One gram-molecular weight of a substance, equal to the weight in grams of 6.02×10^{23} molecules of the substance.

Mutation An inheritable alteration in a chromosome.

Nanometer (nm) A unit of length, equal to 10^{-7} cm.

Nucleic acids Macromolecules consisting of long chains of mononucleotide units linked by 3′, 5′-phosphodiester links.

Nucleoside The combination of a purine or pyrimidine base with a pentose.

Nucleotide A phosphate ester of a purine- or pyrimidine-pentose derivative.

Oligomeric protein A protein composed of two or more noncovalently linked subunits, or oligomers.

Operon A group of genes whose products are usually metabolically related and whose transcription is regulated as a unit.

Organelles Membrane-surrounded structures having specialized functions in the eucaryotic cell.

Oxidation Loss of electrons from a reductant.

Oxidative phosphorylation The phosphorylation of ADP to ATP coupled to the oxidation/reduction processes of electron transport.

Peptide Two or more amino acids linked by amide (peptide) links.

Polysome A functional complex of messenger RNA and two or more ribosomes.

Primary protein structure The amino acid sequence of a polypeptide.

Procaryotes Unicellular organisms having no well-defined nucleus, no nuclear membrane, no subcellular organelles, and only a single chromosome.

Quaternary protein structure The manner in which the subunits (oligomers) of an oligomeric protein associate to produce the natural three-dimensional structure of the protein.

Replication The synthesis of two daughter DNA molecules having nucleotide sequences identical to those of the parent DNA.

Repressible enzymes Enzymes whose biosynthesis is prevented by the presence of certain metabolites (repressors).

Respiration The aerobic catabolism of foodstuffs with the concurrent consumption of oxygen.

Ribosomes RNA-protein complexes that serve as the sites of protein synthesis.
RNA (ribonucleic acid) Nucleic acids containing specific sequences of mono-ribonucleotide units.

Secondary structure The covalent and hydrogen-bonded structure of a polypeptide.
Somatic cell A cell in a higher organism that is not a germ (sperm or egg) cell.
Spontaneous process A process with $K_{eq} > 1$.
Steady state A nonequilibrium state where the concentration of one or more intermediates remains constant because the rates of formation and breakdown are equal.
Substrate The specific metabolite bound and acted upon by an enzyme active site.

Tertiary protein structure The three-dimensional structure of a folded protein chain in its natural state.
Transcription The biosynthesis of RNA using a DNA template.
Transfer RNA The RNA acting as the amino acid carrier and adaptor in translation.
Translation The biosynthesis of a protein chain using an RNA template.

X-Ray crystallography The use of X-ray diffraction patterns produced from the scattering of X rays by a crystal to determine the three-dimensional locations of the atoms in the crystal molecules.

Zymogen An inactive precursor protein that is covalently modified *in vivo* to yield an active enzyme.

Background Review of Organic Chemistry

An immense variety of organic reaction types and organic compounds participate in chemical processes in living systems. However, several important classes of compounds and reactions do seem to stand out. The basic principles of functional-group organic chemistry are particularly important here. For example, the amino acid alanine is shown in Figure B.1. It contains two functional groups: the amino group and the carboxylic-acid group, each with its own chemical properties. Figure B.1 shows that the amine function of alanine can form an amide through reaction with the carboxylic-acid functional group of acetic acid. Also, the carboxyl function of alanine can react with the hydroxyl group of methanol to form an ester via a separate reaction. Each of these reactions is characteristic of the functional groups involved.

Figure B.1

Illustration of functional-group chemistry, showing alanine, an amino acid.

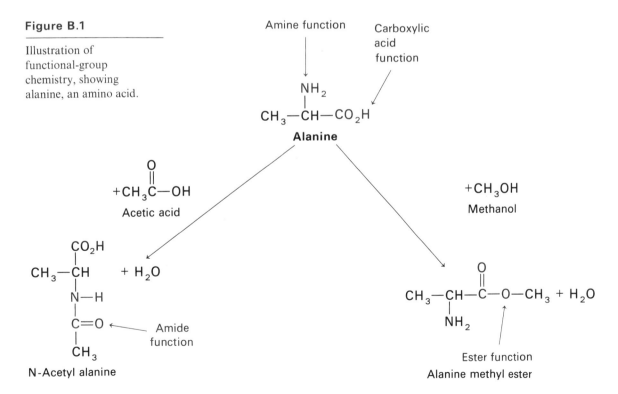

Reactions Involving the Formation of Hydrolyzable Linkages

Esters   Esters are formed from the reaction of an *alcohol* and an *oxy-acid*. The reaction proceeds to completion with the removal of H_2O. Ester formation is summarized by Eq. (1):

$$R-OH + HO-\overset{\displaystyle O}{\overset{\|}{C}}-R' \rightleftharpoons R-O-\overset{\displaystyle O}{\overset{\|}{C}}-R' + H_2O \qquad (1)$$

Alcohol   Carboxylic acid   Ester

The symbols R and R′ in Eq. (1) refer to the rest of the organic molecule, which need not be specified. In accordance with Le Châtelier's Principle, removal of H_2O causes the reaction to proceed to the right; that is, toward ester formation. Similarly, treatment of an ester with excess H_2O, and perhaps a catalyst such as an acid, a base, or an enzyme (see Chapter 3), causes **hydrolysis** of the ester to the parent carboxylic acid and alcohol. Table B.1 shows several representative alcohols and oxy-acids together with the structures of several esters.

Sulfur analogs of alcohols are called **thiols**. Thiols also form esters with oxy-acids. The reaction for the formation of a *thiol-ester* is

$$R-SH + HO-\overset{\displaystyle O}{\overset{\|}{C}}-R' \rightleftharpoons R-S-\overset{\displaystyle O}{\overset{\|}{C}}-R' + H_2O \qquad (2)$$

Thiol   Carboxylic acid   Thiol-ester

Certain thiol-esters are extremely important in metabolism.

Amides   The reaction of an amine with a carboxylic acid function yields an amide. This reaction is analogous to the formation of an ester:

$$R-\overset{\displaystyle H}{\overset{\|}{N}}-H + HO-\overset{\displaystyle O}{\overset{\|}{C}}-R' \rightleftharpoons R-\overset{\displaystyle H}{\overset{\|}{N}}-\overset{\displaystyle O}{\overset{\|}{C}}-R' + H_2O \qquad (3)$$

Amine   Carboxylic acid   Amine

The amide function, $-\overset{\displaystyle O}{\overset{\|}{C}}-\overset{\displaystyle H}{\overset{\|}{N}}-$, is the basic covalent link in the backbone of all proteins. In a protein, an amide link forms between the amino group of one amino acid and the carboxyl function of another (see Figure B.1):

Table B.1   Alcohols, Oxy-acids, and Esters.

Alcohols	Oxy-acids	Esters
CH$_3$CH$_2$OH Ethanol	CH$_3$C(=O)OH Acetic acid	CH$_3$—C(=O)—OCH$_2$CH$_3$ Ethyl acetate
OH OH OH CH$_2$—CH—CH$_2$ Glycerol	CH$_3$—(CH$_2$)$_{14}$—C(=O)OH Palmitic acid	CH$_2$—O—C(=O)—(CH$_2$)$_{14}$CH$_3$ CH—O—C(=O)—(CH$_2$)$_{14}$CH$_3$ CH$_2$—O—C(=O)—(CH$_2$)$_{14}$CH$_3$ Glyceryl tripalmitate
	HO—NO$_2$ Nitric acid	CH$_2$—O—NO$_2$ CH—O—NO$_2$ CH$_2$—O—NO$_2$ Glyceryl trinitrate
	HO—P(=O)(OH)—OH Phosphoric acid	CH$_2$—O—P(=O)(OH)—OH CHOH CH$_2$OH Glycerol 1-phosphoric acid

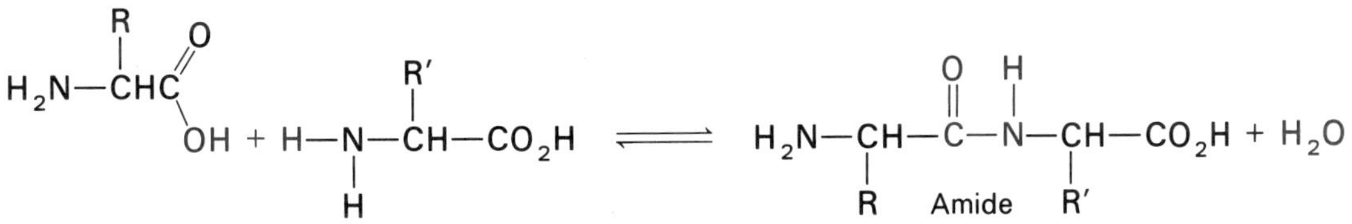

$$H_2N-\overset{R}{\underset{}{CH}}C\overset{O}{\underset{OH}{\diagdown}} + H-\overset{}{\underset{H}{N}}-\overset{R'}{\underset{}{CH}}-CO_2H \rightleftharpoons H_2N-\overset{}{\underset{R}{CH}}-\overset{O}{\underset{}{C}}-\overset{H}{\underset{}{N}}-\overset{}{\underset{R'}{CH}}-CO_2H + H_2O \quad (4)$$

$$\text{Amide}$$

The product still has one free amino group and one free carboxyl group. Thus, another amino acid can react at either end, forming a longer molecule. As this can happen over and over, **polyamides** with many repeat units and high molecular weight are possible.

Amides are subject to hydrolysis under conditions similar to the hydrolysis of esters. When we eat proteins, the amide links in the protein molecules are hydrolyzed during the course of digestion.

Acid anhydrides If we remove the elements of H_2O from two molecules of an oxy-acid, we obtain an acid anhydride. For an organic carboxylic acid, this process is shown in Eq. (5):

$$R-\overset{\overset{\text{O}}{\|}}{C}-OH + HO-\overset{\overset{\text{O}}{\|}}{C}-R \rightleftharpoons R-\overset{\overset{\text{O}}{\|}}{C}-O-\overset{\overset{\text{O}}{\|}}{C}-R + H_2O \quad (5)$$

While anhydrides of organic carboxylic acids are not of great importance in living systems, the anhydrides of the inorganic oxy-acid phosphoric acid are extremely important. Equation (6) shows the formation of pyrophosphoric acid, the anhydride containing two phosphoric-acid units. The anhydride containing three phosphoric-acid units is also extremely important.

$$HO-\overset{\overset{\text{O}}{\|}}{\underset{\underset{\text{OH}}{|}}{P}}-OH + HO-\overset{\overset{\text{O}}{\|}}{\underset{\underset{\text{OH}}{|}}{P}}-OH \rightleftharpoons HO-\overset{\overset{\text{O}}{\|}}{\underset{\underset{\text{OH}}{|}}{P}}-O-\overset{\overset{\text{O}}{\|}}{\underset{\underset{\text{OH}}{|}}{P}}-OH + H_2O \quad (6)$$

Phosphoric acid   Pyrophosphoric acid

Acetals and hemiacetals The chemistry of the aldehyde functional group and of the alcohol functional group is very important in sugars. The reason for this is that sugars can be classified as polyhydroxyaldehydes or polyhydroxyketones. A *hemiacetal* forms from the reversible reaction of an alcohol with the carbonyl group of an aldehyde or ketone:

$$R-\overset{\overset{\text{O}}{\|}}{C}-H + HO-R' \rightleftharpoons R-\overset{\overset{\text{OH}}{|}}{\underset{\underset{\text{OR}'}{|}}{C}}-H \quad (7)$$

Aldehyde   Alcohol   Hemiacetal

Note that one molecule can contain both alcohol and carbonyl functions, in which case a cyclic hemiacetal can be formed:

$$\overset{\displaystyle H\diagdown \diagup\!\!\diagup O}{\underset{\displaystyle CH_2-OH}{C}} \rightleftharpoons \underset{\displaystyle H \qquad H}{\overset{\displaystyle HO\diagdown O\diagup H}{CC}} \quad (8)$$

A bifunctional compound   Cyclic hemiacetal

This chemistry accounts for the cyclic forms of the simple sugars discussed in Chapter 4.

A hemiacetal can react with yet another molecule of an alcohol, producing an *acetal* plus H_2O:

$$\underset{\text{Hemiacetal}}{R-\overset{\overset{\displaystyle OH}{|}}{\underset{\underset{\displaystyle OR'}{|}}{C}}-H} + \underset{\text{Alcohol}}{HO-R''} \rightleftharpoons \underset{\text{Acetal}}{R-\overset{\overset{\displaystyle OR''}{|}}{\underset{\underset{\displaystyle OR'}{|}}{C}}-H} + H_2O \qquad (9)$$

The acetal carbon–carbon bond is hydrolyzable, and serves as the fundamental connecting link between the units of such polymeric molecules as starch, cellulose, and other polysaccharides, discussed in detail in Chapter 4.

Oxidation–Reduction Reactions of Organic Compounds

Oxidation–reduction, or **redox** reactions are responsible for the production of energy in not only the biological world, but in the technological world as well. In an organic chemical sense, we can view a simple oxidation as something akin to burning; that is, the addition of oxygen to carbon, as in the oxidation of methane:

$$CH_4 + 2O_2 = CO_2 + 2H_2O$$

This reaction can also be considered as the removal of four hydrogens from methane: a **dehydrogenation**. In biological oxidations, dehydrogenation of an organic molecule is common.

In an oxidation–reduction reaction, the overall sequence shown in Eq. (10) is always followed:

$$(\text{Oxidant}) + (\text{Reductant}) \rightleftharpoons (\text{Reduced oxidant}) + (\text{Oxidized reductant})$$

$$\begin{array}{l} \text{oxidant} \equiv \text{oxidizing agent: an electron } \textbf{acceptor} \\ \text{reductant} \equiv \text{reducing agent: an electron } \textbf{donor} \end{array} \qquad (10)$$

As Eq. (10) implies, redox reactions involve the *transfer of electrons*. This is easy to see for the reduction of cupric ion by zinc metal (which can also be considered as the oxidation of zinc metal by cupric ion):

$$Zn + Cu^{2+} \rightleftharpoons Cu + Zn^{2+}$$

Zn loses two electrons to Cu^{2+} and is therefore oxidized
Cu^{2+} gains two electrons from Zn and is therefore reduced

While many oxidation reactions involve organic compounds, we can identify several types of oxidation reactions which are particularly common in biochemistry.

Oxidation of alcohols There are three main types of alcohols: primary, secondary, and tertiary (see Figure B.2). Primary and secondary alcohol functional groups are common in nature. The products obtained from the oxidation of primary, secondary, and tertiary alcohol functional groups are summarized in Figure B.2.

In Figure B.2, the dehydrogenations of the primary and secondary alcohols are written with a double arrow. Thus, *reduction* of an aldehyde produces a primary alcohol and *reduction* of a ketone yields a secondary alcohol. To satisfy yourself that oxidation does, in fact, involve a loss of electrons, you should count the electrons in the products and reactants in each of the reactions shown in Figure B.2. This can be done most easily, perhaps, by using *electron-dot notation*:

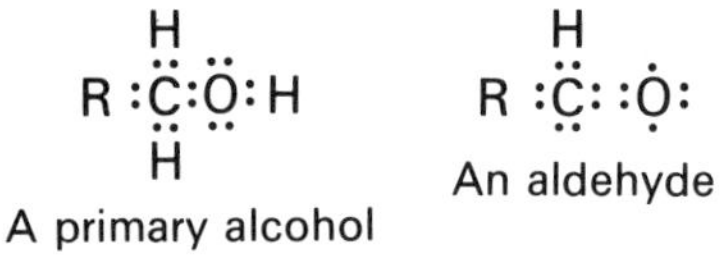

$$
\begin{array}{cc}
\text{H} & \text{H} \\
\text{R} : \ddot{\text{C}} : \ddot{\text{O}} : \text{H} \qquad & \text{R} : \ddot{\text{C}} : : \ddot{\text{O}} : \\
\text{H} & \\
\text{A primary alcohol} & \text{An aldehyde}
\end{array}
$$

Oxidation of aldehydes The aldehyde functional group is associated with many sugars and with a number of intermediates in metabolism. In general, the aldehyde functional group oxidizes smoothly to a carboxylic acid

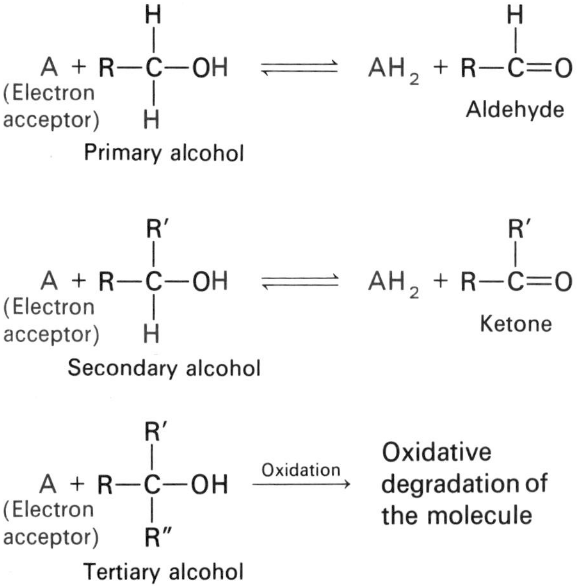

Figure B.2

The oxidation (dehydrogenation) of alcohols. Each alcohol is shown reacting with an electron acceptor, or oxidant, A.

function, as shown in Eq. (11). As before, we refer to the oxidant as A, an electron acceptor. Note that here H_2O can be considered a reactant.

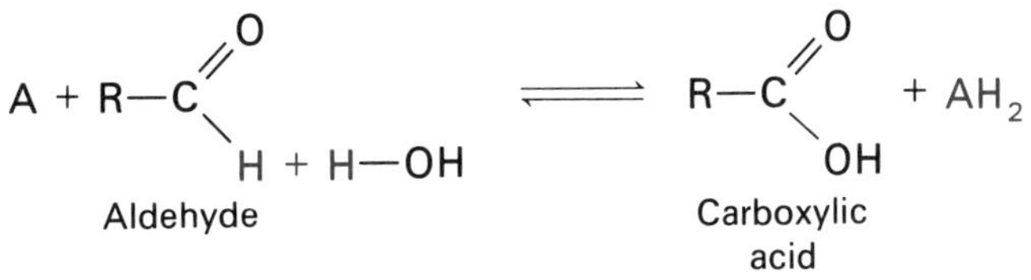

$$A + R-C\underset{H}{\overset{O}{\lessgtr}} + H-OH \quad \rightleftharpoons \quad R-C\underset{OH}{\overset{O}{\lessgtr}} + AH_2 \quad (11)$$

Aldehyde   Carboxylic
acid

Oxidation of alkanes Many biologically important substances have parts of their molecular structure that are hydrocarbon-like in nature: $-CH_2-CH_2-$. The dehydrogenation of an **alkane** to an **alkene** is:

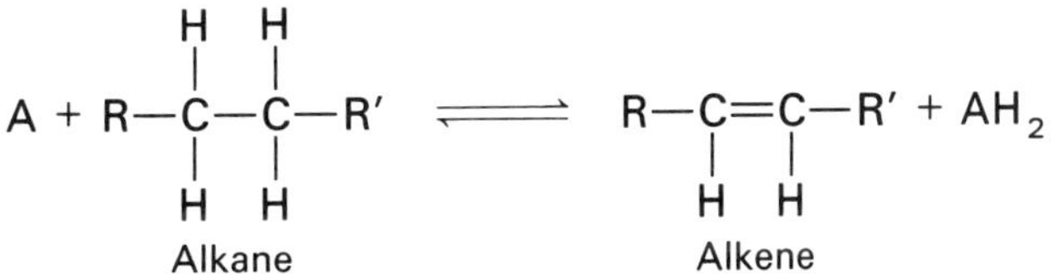

$$A + R-\underset{\underset{H}{|}}{\overset{\overset{H}{|}}{C}}-\underset{\underset{H}{|}}{\overset{\overset{H}{|}}{C}}-R' \quad \rightleftharpoons \quad R-\underset{\underset{H}{|}}{C}=\underset{\underset{H}{|}}{C}-R' + AH_2 \quad (12)$$

Alkane   Alkene

Such a process is an important step in the breakdown of the hydrocarbon parts of fats for energy (see Chapter 10).

It is often the case in the chemistry of a living cell that an intermediate, once formed, undergoes further reaction. For example, an alkene function produced by the dehydrogenation of an alkane can undergo another process characteristic of the chemistry of the alkene functional group. This process is **hydration**. Equation (13) shows a short sequence of chemical reactions in which an alkane converts to an aldehyde:

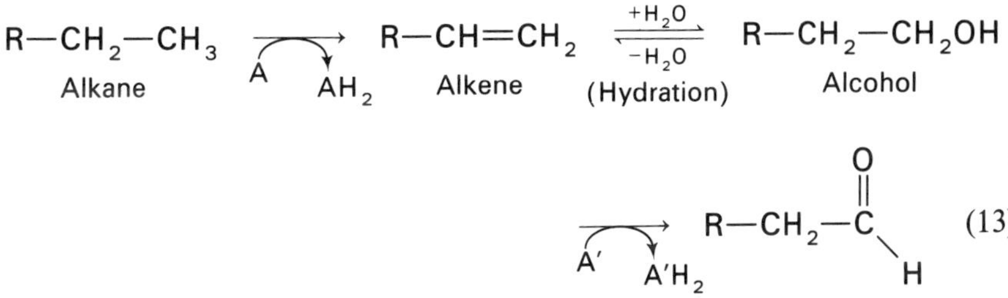

$$R-CH_2-CH_3 \underset{A}{\overset{}{\curvearrowright}} \underset{AH_2}{} R-CH=CH_2 \underset{-H_2O}{\overset{+H_2O}{\rightleftharpoons}} R-CH_2-CH_2OH$$

Alkane   Alkene   (Hydration)   Alcohol

$$\underset{A'}{\overset{}{\curvearrowright}} \underset{A'H_2}{} R-CH_2-C\underset{H}{\overset{O}{\lessgtr}} \quad (13)$$

The overall transformation of the alkane to the aldehyde occurs via three discrete steps. In a cell, each step is catalyzed by a different enzyme, and the overall sequence could represent a **metabolic pathway** for the conversion of the alkane to the aldehyde. The use of curved arrows in Eq. (13) is a shorthand notation used many times in this book. Use of this notation avoids the need to write each step separately. Thus, the first step in Eq. (13) is formally identical with the reaction in Eq. (12).

Oxidation of thiols   The thiol, or sulfhydryl group (—SH) is associated with many biologically important substances. In many cases, the function of a given sulfhydryl-containing substance is based on the redox properties of the —SH group. Oxidation of a thiol yields a disulfide:

$$\text{R—SH} + \text{HS—R} + \text{A} \;\rightleftharpoons\; \text{R—S—S—R} + \text{AH}_2 \qquad (14)$$
$$\text{Thiol} \qquad \text{Thiol} \qquad\qquad\qquad \text{Disulfide}$$

The covalent —S—S— link of the disulfide group is an important factor in the structure of many proteins (see Chapter 2).

This brief review of a few reaction types does not by any means cover all the organic chemistry you should know for studying biochemistry. However, the reactions we have considered are exemplified many times throughout this text, and are thus worthy of particular note.

Answers to Selected Problems

Chapter 2, page 74

3. (a) $[H_3O^+] = 1.85 \times 10^{-5}M$; (b) [acetic acid] $=$ 0.05M, [acetate] $= 0.15M$, $[H_3O^+] = 6.2 \times 10^{-6}M$; (c) $pH_{(a)} = 4.73$, $pH_{(b)} = 5.2$, for adding 0.05 moles of NaOH to one liter of water, pH $=$ 12.7

5. (b) vasopressin

7. (a) 2, 5, 7, 9; (b) 3, 10; (c) 6; (d) 1, 3; (e) 4

8. (a) 4, 21, 25, 29; (b) 2, 23; (c) 1; (d) 7, 15

10. Val-Lys-Gly-Leu-Phe-Leu-Gly

11. Phe-Val-Asp-Glu-His-Leu-Cys-Gly

13. (a) 72%; (b) 18% (c) $9.3 \times 10^{-3}M$; (d) for (a), $[O_2] = 6.7 \times 10^{-3}M$; for (b), $[O_2] = 1.6 \times 10^{-3}M$

Chapter 3, page 138

2. (a) oxidoreductase (oxygenase); (b) transferase (kinase); (c) isomerase (mutase); (d) transferase (transaminase); (e) hydrolase (esterase)

3. See Table 3.8

7. (a) NAD^+, FAD, FMN; (b) FAD, FMN; (c) TPP; (d) Pyridoxal phosphate; (e) NADPH

10. $1/V = 1/V_{max} + (K_M/V_{max})(1/S)$
$V = V_{max} - K_M(V/S)$
$S/V = (K_M/V) + (S/V_{max})$

13. $K_M = 1.5 \times 10^{-3}M$; $V_{max} = 5.2 \times 10^{-4}M$; $K_I = 6.4 \times 10^{-4}M$; competitive

Chapter 4, page 160

2. (a) D-gluconic acid; (b) δ-gluconolactone; (c) α-D-glucose, β-D-glucose, open-chain form; (d) same as (c); (e) β-D-galactose

4. Cyclohexamylose: a cyclic structure of six glucose units joined by α-1 → 4 links.

7. (a) D-glucuronic acid + N-acetyl D-glucosamine; (b) N-acetyl D-glucosamine, N-acetyl muramic acid, L-alanine, D-glutamine, L-lysine, D-alanine, glycine.

Chapter 5, page 181

7. (a) pCp; GpUp; Up; A; (b) pCpGp; UpUpA

8. *Sarcina lutea*

9. length $= 2.1 \times 10^{-1}$ cm $= 2.1$ mm

Chapter 6, page 208

3. C_8: octanoic acid; C_{10}: decanoic acid; C_{12}: dodecanoic acid; C_{14}: tetradecanoic acid; C_{16}: hexadecanoic acid; C_{18}: octadecanoic acid; C_{20}: eicosanoic acid; *unsaturated*: palmitoleic: 9-hexadecenoic acid; oleic: 9-octadecenoic acid; linoleic: 9,12-octadecadienoic acid; linolenic: 9,12,15-octadecatrienoic acid; arachidonic acid; 5,8,11,14-eicosatetrienoic acid

4. Yes

5. (a) glycerol, fatty acid anions, phosphate, choline; (b) myristate anion, oleate anion, glycerol; (c) glucose, palmitate anion, sphingosine

8. (From top to bottom of Figure 6.12): *cis*-7 dodecenyl acetate; *trans, trans*-8,10 dodecadienol; *cis*-9 triacosene

Chapter 7, page 241

3. $G' = -12.8$ kcal/mole

4. (a) $K_{eq} = 6.0 \times 10^{-5}M$; (b) $G' = 1.7$ kcal/mole

5. (a) ATP, ADP, PP_i; (b) phosphocreatine; (c) 1,3-diphosphoglycerate, acetyl phosphate; (d) phosphoenol pyruvate, 1,3-diphosphoglycerate, 3 phosphoglycerate, ATP, ADP, glucose 1-phosphate, fructose 6-phosphate, glucose 6-phosphate, glycerol 1-phosphate

7. greater than

8. 378 calories (kcal); for an "average" 70 kg man, it would be necessary to walk 6.1 miles

Chapter 8, page 268

3. (a) maltose + $H_2O \rightarrow$ 2 D-glucose
   (b) sucrose + $H_2O \rightarrow$ D-fructose + D-glucose
   D-fructose + ATP $\rightarrow$ D-fructose 6-phosphate
   (c) glycerol + ATP $\rightarrow$ glycerol 3-phosphate
   glycerol 3-phosphate + NAD^+
   $\rightarrow$ glyceraldehyde 3-phosphate + NADH + H^+

4. Via pentose phosphate pathway.

6. All reactants are present plus: dihydroxyacetone phosphate, D-glyceraldehyde 3-phosphate

$$
\begin{array}{ccc}
CH_2OPO_3^{2-} & & CH_2OPO_3^{2-} \\
| & & | \\
C{=}O & & C{=}O \\
| & & | \\
HO{-}C{-}H & \text{and} & HO{-}C{-}H \\
| & & | \\
H{-}C{-}OH & & H{-}C{-}OH \\
| & & | \\
CH_3 & & H{-}C{-}OH \\
& & | \\
& & CH_2OH
\end{array}
$$

8. (b) No net ATP produced.
10. No.

Chapter 9, page 297

4. (a) 37 ATP; (b) 12 ATP; (c) 72 ATP; (d) 20 ATP
5. 4 ATP

Chapter 10, page 314

3. (a) H_2O; (b) Pigments of PS I; (c) Pigments of PS I; (d) $NADP^+$; (e) Redox potential should be equal or more positive than that of $NADP^+$.

Chapter 11, page 335

4. 135 ATP
5. 3 moles acetyl CoA, 3 moles propionyl CoA, 1 mole 2-methyl propionyl CoA
8. acetate
9. (i) malate $\rightarrow$ pyruvate; (ii) acetaldehyde $\rightarrow$ pyruvate; (iii) oxaloacetate $\rightarrow$ pyruvate + CO_2; (iv) alanine $\rightarrow$ pyruvate
10. glutamine
12. benzoic acid

Chapter 12, page 375

6. (a) palmitoleic acid; oleic acid
8. (a) blood glucose depressed; blood lactate elevated; (b) yes
10. The cell must employ the glyoxylate cycle.
11. (a) synthetase active; phosphorylase inactive
    (b) synthetase inhibited; phosphorylase active
    (c) synthetase inhibited; phosphorylase active

Chapter 13, page 429

4. yes
5. 20% T, 30% G, 30% C
6. t-RNA; ester link to 3'-OH
7. 5,10-methylene tetrahydrofolate
8. aminoacyl t-RNA
9. different
10. polyphenylalanine; polyleucine; polyserine
11. (a) Except for Trp, middle letter is a pyrimidine
    (b) Middle letter is a purine
    (c) UAA—stop; UAG—stop; CAU—Gln; CAC—His; CAA—Gln; CAG—Gln
12. I: transition; M_{Bos}: transition; S: transversion; D_{Punjab}: transversion; $K_{Woolwich}$: transition
13. Transcription of DNA strand goes from $3' \rightarrow 5'$;
    (a) I: Leu-Ala-Leu-Thr; II: Thr-Leu-Ala-Pro;
    (b) insertion; Leu-Ala-Arg-Asn; deletion; Leu-Ala-*Stop*; Transition; Leu-Ala-Pro-Thr; Transversion; Leu-Ala-Gln-Thr

Index

Key to Numbered Reactions on the
Intermediary Metabolism Chart

Enzyme	Reaction	Coenzyme/Cofactor	Processes in Intermediary Metabolism
Glycogen synthetase	1		
Glycogen phosphorylase	1a	Adenylate, P_i	
Uridine diphosphoglucose pyrophosphorylase	2	UTP	Glycogenesis/ Glycogenolysis (Section 12.5)
Phosphoglucomutase	3	Mg^{2+}	
Galactokinase	4	ATP, Mg^{2+}	
D-Galactose 1-phosphate uridyl transferase	5		
UDP galactose 4-epimerase	5a	NADPH	
Hexokinase	6	ATP, Mg^{2+}	
Glucose 6-phosphatase	7		
Phosphoglucoisomerase	8		
Hexokinase	9	ATP, Mg^{2+}	
Phosphomannose isomerase	10		
Fructose 1,6-diphosphatase	11	Mg^{2+}	
Phosphofructokinase	12	ATP, Mg^{2+}	
Fructokinase	13	ATP, Mg^{2+}	
Aldolase	14		
Liver aldolase	15		Glycolysis: Embden-Meyerhof Glycolytic Pathway (Section 8.2)
Glycerol phosphate dehydrogenase	16	NAD^+	
Triose phosphate isomerase	17		
Glyceraldehyde 3-phosphate dehydrogenase	18	NAD^+	
Phosphoglycerate kinase	19		
Phosphoglyceromutase	20	Mg^{2+}	
Enolase	21		
Pyruvate kinase	22	ADP, Mg^{2+}, K^+	
Lactate dehydrogenase	23	NAD^+	
Alcohol dehydrogenase	24	NAD^+	
Xanthine oxidase	25	FAD, Mo(VI)	
Pyruvate decarboxylase	26	TPP (Thiamine pyrophosphate)	
Pyruvate dehydrogenase	27	NAD^+, TPP, lipoic acid, FAD	
Acetate thiokinase	28		
Acetate thiokinase	29		
Glucose 6-phosphate dehydrogenase	30	$NADP^+$	Pentose Phosphate Pathway (Section 8.3)
Gluconolactonase	31		
6-Phosphogluconate dehydrogenase	32	$NADP^+$	
Citrate synthetase	33		
Aconitase	34	(Fe ion)	
Aconitase	35	(Fe ion)	
Isocitrate dehydrogenase	36	NAD^+, Mg^{2+}	Krebs (TCA) Cycle (Section 9.4)
α-Ketoglutarate dehydrogenase	37	NAD^+, TPP, lipoic acid	
Succinyl thiokinase	38	GDP, Mg^{2+}	
Succinate dehydrogenase	39	FAD, nonheme iron	
Fumarase	40		
Malate dehydrogenase	41	NAD^+	
Malic enzyme	42	$NADP^+$	Glycolytic-Krebs Cycle Interchange Reactions (Anaplerosis) (Section 9.5)
Phosphoenol pyruvate carboxykinase	43	GTP	
Pyruvate carboxylase	44	biotin	

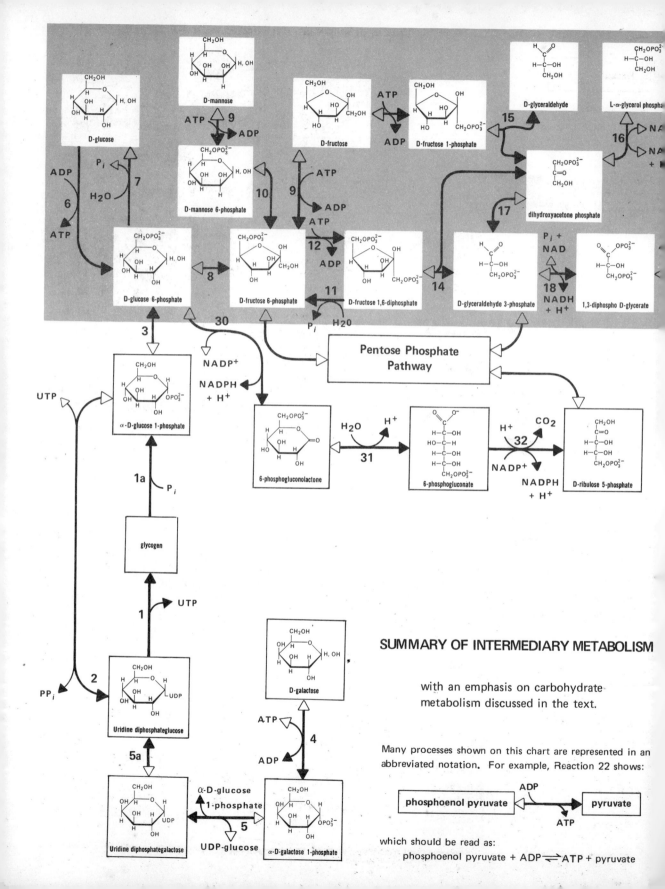

SUMMARY OF INTERMEDIARY METABOLISM

with an emphasis on carbohydrate metabolism discussed in the text.

Many processes shown on this chart are represented in an abbreviated notation. For example, Reaction 22 shows:

phosphoenol pyruvate ⟶ pyruvate

ADP

ATP

which should be read as:

phosphoenol pyruvate + ADP ⇌ ATP + pyruvate